WITHDRAWN

HUMUS CHEMISTRY

HUMUS CHEMISTRY

Genesis, Composition, Reactions

Second Edition

F. J. STEVENSON
Department of Agronomy
University of Illinois

JOHN WILEY & SONS, INC.
New York • Chichester • Brisbane • Toronto • Singapore

This text is printed on acid-free paper.

Copyright © 1994 by John Wiley & Sons, Inc.

All rights reserved. Published simultaneously in Canada.

Reproduction or translation of any part of this work beyond
that permitted by Section 107 or 108 of the 1976 United
States Copyright Act without the permission of the copyright
owner is unlawful. Requests for permission or further
information should be addressed to the Permissions Department,
John Wiley & Sons, Inc., 605 Third Avenue, New York, NY
10158-0012.

This publication is designed to provide accurate and
authoritative information in regard to the subject
matter covered. It is sold with the understanding that
the publisher is not engaged in rendering,
professional services. If legal, accounting, medical, psychological, or
any other expert assistance is required, the services of a competent
professional person should be sought.

Library of Congress Cataloging in Publication Data:
Stevenson, F. J.
　Humus chemistry : genesis, composition, reactions / F.J.
Stevenson. — 2nd ed.
　　p. cm.
　Includes bibliographical references and index.
　ISBN 0-471-59474-1 (cloth : acid-free paper)
　1. Humus.　2. Soil biochemistry.　I. Title.
S592.8.S76　1994
631.4'17—dc20　　　　　　　　　　　　　　　93-46704

Printed in the United States of America

10 9 8 7 6 5 4 3 2 1

*Dedicated to my wife Leda,
with love and appreciation*

PREFACE

Interest in the chemistry and reactions of organic substances in soils, sediments, and natural waters is long-standing and has been fortified in recent years by major advances in methodology and recognition of the important role of organic substances in environmental processes. Since publication of the first edition of *Humus Chemistry*, a considerable body of new information has accumulated and the time has come to provide an update of this material for classroom and general use. A series of reviews and monographs have been published since the first edition (1982), but most of them are addressed to the senior investigator or specialist. To the author's knowledge, there is no comprehensive textbook other than *Humus Chemistry* that is suitable for students or new researchers.

Essentially, the book is designed as a reference text for graduate students and advanced undergraduates taking a one-semester course in soil organic matter chemistry and is based upon the experience of the author in teaching a course on this subject at the University of Illinois. While it is assumed that the student has had prior training in organic and analytical chemistry, background information is provided when possible to make the book more palatable to the nonspecialist. Since emphasis is given to the basic organic chemistry and reactions of naturally occurring organic substances, the book will be of considerable value to scientists in several scientific disciplines, including all aspects of soil science, organic geochemistry, forestry, environmental science, and others. It is hoped that the book will also be of value to advanced researchers, although they may detect omissions and deficiencies arising in part from an attempt to present as broad a coverage as possible in a single volume. That much remains to be done will be evident to all readers of the volume.

The subject matter is conveniently arranged into four sections. The first section (Chapters 1 and 2) covers pools of organic matter in soils, transformations, and methods of extraction and fractionation. The second section (Chapters 3 through 7) deals primarily with the chemistry of known classes of organic compounds in soil, such as saccharides, lipids, and constituents containing nitrogen, phosphorus, and sulfur. The basic organic chemistry of the so-called humic substances (humic acids, fulvic acids, etc.) is presented in Chapters 8 through 15 of the third section. Included are two chapters not found in the first edition, one on NMR spectroscopy and analytical pyrolysis (Chapter 11) and the other on chemical structures (Chapter 12). Portions of this section may be too advanced for the average student but will be of great value to graduate students interested in pursuing a career in the organic chemistry of soils, sediments, or natural waters. A particularly important section of the book (the fourth section, encompassing Chapters 16 through 20) deals with the importance of organic matter associations and interactions, such as combinations with polyvalent cations, clay minerals, and pesticides, the formation of stable aggregates, and the role of organic substances in pedogenesis.

As with most advanced texts, the book contains somewhat more material than can be covered in any one course on soil organic matter or soil chemistry. Arrangement of the subject matter is such that certain sections or chapters can be omitted in formal study. In some respects, the second edition can be looked upon as a companion volume to *Cycles of Soil* (1986, Wiley-Interscience), in which emphasis is given to biochemical transformations.

One of the functions of this book has been to present a critical account of our knowledge of each topic, and, for this reason, rather extensive use has been made of references within the text. Because of space limitations, complete documentation was not possible and selection of references was rather arbitrary; in some cases a full search of the literature for establishing priorities was not possible. The author apologizes for omission of important work.

In an undertaking of this nature errors will undoubtedly creep in and the author may have inadvertently misrepresented some of the research discussed. Suggestions for improvement will be welcome.

Appreciation is expressed to graduate students and staff members at the University of Illinois for encouragement and assistance in preparation of the text. Special thanks are extended to the many individuals who detected errors in the first edition and offered suggestions for this second edition.

F. J. STEVENSON

Urbana, Illinois
March 1994

CONTENTS

1 Organic Matter in Soils: Pools, Distribution, Transformations, and Function   1

Pools of Organic Matter in Soils, 1
Organic Matter (Humus) Content of Soils, 6
Decomposition Processes and Ages of Organic Matter, 11
Function of Organic Matter in Soil, 14
Organic Matter and Sustainable Agriculture 19
Environmental Significance of Humic Substances, 19
Summary, 20
References, 21

2 Extraction, Fractionation, and General Chemical Composition of Soil Organic Matter   24

Brief Historical Review, 24
Associations of Organic Matter in Soil, 34
Extraction Methods, 34
Fractionations Based on Solubility Characteristics, 41
Purification, 45
Classification and Distribution of Dissolved Organic C (DOC), 51
Comparison of Soil Humic Substances with Those of Other Environments, 54
Physical Fractionation of Organic Matter, 54
Summary, 55
References, 56

3 Organic Forms of Soil Nitrogen 59

Fractionation of Soil N, 59
Nitrogen in Humic and Fulvic Acids, 66
Biochemical N Compounds, 71
Biomass N, 89
Natural Variations in N Isotope Abundance, 90
^{15}N–NMR Spectroscopy, 90
Stability of Soil Organic N, 90
Summary, 91
References, 92

4 Native Fixed Ammonium and Chemical Reactions of Organic Matter with Ammonia and Nitrite 96

Levels of Native Fixed NH_4^+ in Soils, 96
The C/N Ratio, 100
Chemical Reactions of NH_3 and NO_2^- with Organic Matter, 102
Summary, 110
References, 111

5 Organic Phosphorus and Sulfur Compounds 113

The C/N/P/S Ratio, 113
Soil Organic P, 115
Soil Organic S, 128
Summary, 137
References, 137

6 Soil Carbohydrates 141

Significance of Soil Carbohydrates, 142
Structure and Classification, 142
State in the Soil, 144
Quantitative Determination of Soil Carbohydrates, 154
Summary, 163
References, 164

7 Soil Lipids 166

Lipid Content of Soil Humus, 166
Function in Soil, 171
Composition, 175
Summary, 185
References, 185

8 Biochemistry of the Formation of Humic Substances — 188

Major Pathways of Humus Synthesis, 188
The Lignin Theory, 191
The Polyphenol Theory, 197
Sugar–Amine Condensation, 206
Scheme for the Formation of Humic Substances, 208
Summary, 209
References, 210

9 Reactive Functional Groups — 212

Elemental Content, 212
Methods of Functional Group Analysis, 214
Distribution of Oxygen-Containing Functional Groups, 226
Diagenetic Transformations, 231
Summary, 234
References, 234

10 Structural Components of Humic and Fulvic Acids as Revealed by Degradation Methods — 236

Experimental Approaches, 236
Hydrolysis Methods, 238
Reductive Cleavage, 240
Oxidative Methods, 248
Miscellaneous Chemical Approaches, 254
Thermal Methods, 255
Summary, 256
References, 257

11 Characterization of Soil Organic Matter by NMR Spectroscopy and Analytical Pyrolysis — 259

Theory of Nuclear Magnetic Resonance Spectroscopy (NMR), 259
Techniques for ^{13}C–NMR Spectroscopy, 263
^{13}C–NMR Spectroscopic Analysis of Humic and Fulvic Acids, 265
^{13}C–NMR Spectroscopic Analysis of Soil Organic Matter *In Situ*, 272
^{1}H–NMR Spectroscopy, 274
^{15}N–NMR Spectroscopy, 276
^{31}P–NMR Spectroscopy, 277
Analytical Pyrolysis, 277
Summary, 281
References, 281

12 Structural Basis of Humic Substances 285

General Considerations, 286
Type Structures for Humic and Fulvic Acids, 287
Unique Feature of Humic Substances in Seawater, 294
A Simplified Structural Representation of Soil Humic Substances, 294
Summary, 301
References, 301

13 Spectroscopic Approaches 303

Ultraviolet (UV) and Visible Regions, 304
Infrared (IR) Spectroscopy, 307
Electron Spin Resonance (ESR) Spectroscopy, 317
Fluorescence Spectroscopy, 322
Summary, 322
References, 323

14 Colloidal Properties of Humic Substances 325

The Colloidal State, 325
Molecular Weights and Particle Sizes, 326
Morphological Features by Electron Microscopy, 339
Structural Evaluation by X-Ray Diffraction, 341
Small-Angle X-Ray Scattering, 342
Size Evaluation by Gel Filtration, 343
Summary, 347
References, 348

15 Electrochemical and Ion-Exchange Properties of Humic Substances 350

Acidic Nature of Humic and Fulvic Acids, 351
Selectivity of Humic Substances for Exchange Cations, 367
Contribution of Organic Matter to the Cation-Exchange Capacity (CEC) of the Soil, 367
Coagulation of Humic Substances by Polyelectrolytes, 372
Electrophoresis, 373
Oxidation–Reduction Potential, 374
Summary, 375
References, 375

16 Organic Matter Reactions Involving Metal Ions in Soil 378

Properties of Metal–Organic Matter Complexes 379
Significance of Chelation Reactions in Soil, 381
Forms of Transition Metal Ions in Soil, 384

Biochemical Compounds as Chelating Agents, 388
Trace Metal Interactions with Humic Substances, 394
Summary, 401
References, 401

17 Stability Constants of Metal Complexes with Humic Substances   405

General Considerations, 405
Modeling Approaches, 408
Summary, 427
References, 427

18 Clay-Organic Complexes and Formation of Stable Aggregates   429

Nature of Clay Colloids, 429
Adsorption of Defined Organic Compounds by Clay, 431
Naturally Occurring Clay-Organic Complexes, 439
Role of Organic Matter in Forming Stable Aggregates in Soil, 445
Soil Wettability, 449
Summary, 450
References, 450

19 Organic Matter Reactions Involving Pesticides in Soil   453

Humus Chemistry in Relation to Pesticide Behavior, 455
Adsorption Mechanisms, 462
Relative Affinities of Pesticides for Soil Organic Matter, 469
Summary, 469
References, 469

20 Role of Organic Matter in Pedogenic Processes   472

Weathering of Rocks and Minerals, 473
Neogenesis of Minerals, 477
Translocation of Mineral Matter and Horizon Differentiation, 478
Organic Matter and Soil Classification, 483
Paleohumus, 485
Effects of Acidic Inputs (Acid Rain) on Pedogenic Processes, 485
Summary, 486
References, 486

Index   **489**

HUMUS CHEMISTRY

1

ORGANIC MATTER IN SOILS: POOLS, DISTRIBUTION, TRANSFORMATIONS, AND FUNCTION

The decomposition of plant and animal remains in soil constitutes a basic biological process in that carbon (C) is recirculated to the atmosphere as carbon dioxide (CO_2) and associated elements (nitrogen, N; phosphorus, P; sulfur, S; and various micronutrients) appear in forms required by higher plants. In the process, some of the C is assimilated into microbial tissues (i.e., the soil biomass); part is converted into stable humus. Some of the native humus is mineralized concurrently; consequently, total organic matter content is maintained at some steady-state level characteristic of the soil and management system.

This introductory chapter places into perspective those aspects of C transformations that relate to humus synthesis and composition. Emphasis is given to pools of organic matter, the organic matter content of the soil, decomposition processes, mean-residence times (MRT), and the role of organic matter in soil and the environment. Several books and reviews provide excellent material for background reading.[1-8]

POOLS OF ORGANIC MATTER IN SOILS

As used herein, the term "soil organic matter" refers to the whole of the organic material in soils, including litter, light fraction, microbial biomass, water-soluble organics, and stabilized organic matter (humus). The concept of pools of organic matter that differ in their susceptibilities to microbial decomposition has provided a basis for understanding the dynamic nature of soil organic matter and how the availability of nutrients (i.e., N, P, S, trace elements) is influenced by management practices and changes in the soil environment. Ideally, the pools should have a real existence and be measurable so that fluxes of nutrients through the pools can be determined.

From an agronomic point of view, organic matter has often been partitioned into two major pools: "active" (labile) and "stable." Included in the "active" pool are comminutive plant litter, the light fraction (discussed below), the biomass, and nonhumic substances not bound to mineral constituents. Interest in the "active" fraction is that it serves as a ready source of nutrients (N, P, and S) for plant growth; the "stable" fraction (i.e., passive humus) functions as a "reservoir" of plant nutrients and is important from the long-term balance of the soil.

As will be noted later, the "active" fraction is of particular important in maintaining soil productivity under sustainable farming, where maximum use is made of organic residues under conditions of minimum tillage and reduced fertilizer inputs. The size of the "active" pool is related to residue inputs and is influenced by decomposition patterns as affected by climate and those factors that affect microbial activity.

Litter

Litter is defined as the macroorganic matter that lies on the soil surface. This pool is particularly important to the cycling of nutrients in forest soils and natural grasslands. In cultivated soils, crop residues are normally incorporated into the soil well in advance of planting and the litter becomes part of the light fraction (described below). With greater acceptance of cultural practices designed to reduce erosion (e.g., minimal tillage or no-till), greater amounts of crop residues will be left on the soil surface and for a longer time, ultimately to be mixed with the top soil through the activities of macrofaunal organisms or through leaching of partial decomposition products.

Herbaceous litter is normally determined by collecting the macroorganic matter from measured plot areas of the ecosystem under study. Litter quantities are expressed on an oven-dry basis in terms of grams per square meter per year ($g/m^2/yr.$) or metric tons per hectare per year ($t/ha/yr.$). Guidelines for quantification of litter inputs are given by Anderson and Ingram.[9]

Light Fraction

The "light" fraction consists largely of plant residues in varying stages of decomposition and that exists within the soil proper. The chemical composition of the "light" fraction, as determined by ^{13}C–NMR spectroscopy (see Chapter 11), has been found to be comparable to that of plant litter.[10] The "light" fraction has been used as an indicator of changes in labile organic matter as affected by tillage, crop rotation, and environmental factors affecting microbial activity.[10-14] Due to its dynamic nature, the "light" faction has a rapid turnover rate in soil and thus serves as a source of nutrients for plant growth.

Liquids with densities of from 1.6 to 2.0 are commonly used to separate the light fraction. The liquid can be either inorganic (i.e., NaI, $ZnBr_2$, CsCl) or organic (i.e., tetrabromomethane, tetrabromoethane, or dibromochlorpropane). Organic liquids are toxic to microorganisms and can cause problems when isolated material is to be used in biochemical studies; some inorganic liquids are also toxic.

The size of the "light" fraction is considerably less than for stable humus but can constitute as much as 30 percent of the organic matter in some soils.[2] In a recent study, Janzen et al.[13] found that the "light" fraction accounted for from 2.0 to 17.5 percent of the organic matter in the soils of three long-term rotations in Saskatchewan, Canada. Differences in the "light" fraction among sites and treatments were attributed to variable residue inputs and substrate decomposition.

As one might expect, the amount of "light" fraction material in any given soil will vary with the season, and will be greatest immediately following incorporation of crop residues through tillage. Factors affecting the "light" fraction include the quantity of plant litter produced and extent of decomposition as affected by the soil variables of pH, temperature, texture, and moisture.

Microbial Biomass

A vast array of microorganisms live in the soil. Numbers of bacteria are particularly high, often reaching one billion per gram and higher. Actinomycetes are the next numerous, usually of the order of several hundred million per gram. Numbers of fungi vary widely but the normal population is 10–20 million/g. Algal numbers are somewhat lower, with between 10,000 and 3 million/g being typical. Protozoa are found in numbers up to one million/g; nematodes number 50 or more per gram.

The microbial biomass plays a dual role in the soil, first, as an agent for the decomposition of plant residues, with concurrent release of nutrients, and second, as a labile pool of nutrients.

The amounts of C and N in the microbial biomass for cultivated soils from three locations are given in Table 1.1. Biomass values were lower in the tropical soil even though C inputs were much higher. This result was attributed to an accelerated decay rate for the organic matter in the tropical soil, as indicated by a shorter microbial turnover rate (see the last column of the table). The turnover rate for the biomass is considerably higher than for stable humus.

Figure 1.1 shows the relationship between microbial C and total organic C for a number of soils from temperate and tropical regions. Ratios for biomass C/soil organic C (lines marked A and B on the diagram) show that, for most soils, from 1 to 3 percent of the soil organic C occurred as biomass C.

The importance of the microbial biomass in nutrient cycling has been well-established and many methods exist for the determination of biomass C and the nutrients contained therein (see Chapters 3 and 5). Methods for biomass C include respiration measurements for CO_2 production, O_2 consumption, and

TABLE 1.1 Amounts and Turnover Rates of C and N in the Microbial Biomass for Cultivated Soils for Three Locations

Soil and Location	Microbial C	Microbial N	C Inputs	Microbial Turnover Time[a]
	kg/ha	kg/ha	mg/ha/yr	yr
Temperate				
England	570	95	1.2	25
Canada	1600	300	1.6	6.8
Tropical				
Brazil	460	84	13.0	0.24

[a]Calculated from simulation modeling.
Source: Duxbury et al.,[15] as adapted from Paul and Voroney.[16]

estimates for a given enzyme or biochemical compound. Biochemical methods for the determination of the biomass are discussed by Tunlid and White.[17]

Most current approaches for determining the microbial biomass are based on the chloroform ($CHCl_3$) fumigation technique initially developed by Jenkinson and Powlson.[18] The basis for this method is that fumigation of the soil

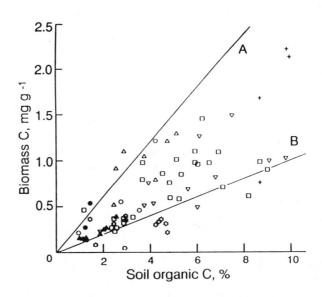

Fig. 1.1 Relationship between microbial biomass C and total soil organic C for a range of temperate and tropical soils. Included are four New Zealand soils (△, ▽, □, +) and one each from England (○), Scotland (△), Nigeria (●), and India (▲). Values outside the two demarkation lines are for soils in which pH, moisture, and other soil factors affected the results. From Theng et al.[7]

leads to an increase in respiration rate, with release of CO_2 and nutrients from dead microbial cells. The assumption is made that the extra CO_2 and nutrients released by incubation of the fumigated soil are derived from cells of microorganisms killed by fumigation and that are decomposed by recolonizing microbes. Biomass C is calculated from the relationship

$$\text{Biomass C} = F_c/k_c$$

where F_c in the difference between the amount of CO_2–C released by incubation of the fumigated and unfumigated soil and k_c is the fraction of the biomass C mineralized to CO_2 over the incubated period (usually 10 to 14 days).

The value chosen for k_c is of some importance and is not precisely the same for all soils. However, k_c values obtained for pure cultures of bacteria and fungi span a rather narrow range, with most values being within the range of 0.43 to 0.50.[19] For biomass C calculations, a k_c value of 0.45 is commonly used.[18]

Assumptions underlying the $CHCl_3$ fumigation–incubation method are:

1. All microorganisms are killed during fumigation
2. Death of organisms in the control soil is negligible as compared to the fumigated soil
3. The C in dead organisms is more rapidly mineralized than in living organisms
4. Microbial decomposition of native soil organic matter occurs at the same rate in fumigated and unfumigated soil

The last assumption is not valid for all soils, notably those receiving large inputs of fresh substrates.[18,20] An evaluation of methods for measuring the microbial biomass in soils following incorporation of plant residues has been given by Ocio and Brookes.[21]

In addition to the above, an extraction method has been described for the direct determination of biomass C.[22,23] The basis for this method is that the amounts of organic C in extracts of $CHCl_3$ fumigated soils are reasonably well-correlated with the corresponding amounts released to CO_2 or soluble N forms by incubation. The reagent used for extraction is $0.5M$ K_2SO_4. An advantage of the extraction method is that there is no need for complete removal of fumigant or for prolonged incubation of the soil under carefully controlled conditions.

Conversion factors for estimating biomass C from extracted organic C are highly variable; for mineral soils, a provisional value of 0.33 has been recommended.

Water Soluble Organics

Considerable attention has been given in recent years to organics contained in the soil solution, both from the standpoint of plant nutrition (source of nutrients) and the environment (alleviation of metal ion toxicities through chelation, acidifying effects on natural waters, and carriers of xenobiotics).

Recovery of the aqueous phase can be achieved by applying suction to a porous collector inserted into the soil. The sampler (i.e., ceramic, alumdum, or fritted glass plate) is installed to the desired depth, a vacuum is applied, and pore water is drawn into the sample chamber thorough the porous section.

Methods have also been described for recovery of the soil solution from field-moist soil under laboratory conditions. They include column miscible displacement, centrifugation with and without an immiscible liquid, and use of ceramic or plastic filters. The reader is referred to Elkhatib et al.[24] and Wolt and Graveel[25] for a review and evaluation of methods.

Soil Enzymes

All biochemical action is dependent upon, or related to, the presence of enzymes. Because of the complex and variable substrates that serve as energy sources for microorganisms, soils contain a wide array of enzymes.[2,26] Each soil may have its own characteristic pattern of specific enzymes.

It should be noted that enzymes in soil are not determined by direct analysis but indirectly through their ability to transform a given substrate into a known product(s). For example, urease activity is estimated from the reaction: urea $\rightarrow 2NH_3 + CO_2$.

Like microbial numbers, enzymatic activity is not static but fluctuates with biotic and abiotic conditions. Activities would be expected to be particularly high in productive soils rich in organic matter. Such factors as moisture, temperature, aeration, soil structure, organic matter content, seasonal changes, and soil treatment all have an influence on the presence and abundance of enzymes. A marked change occurs in the kinds and amounts of enzymes when virgin lands are first placed under cultivation.

Stable Humus

In most agricultural soils, the bulk of the organic matter occurs as stable humus. Both chemical and physical fractionation procedures have been used in attempts to separate the various components of humus and to ascertain their location within the soil matrix. These methods are described in Chapter 2.

ORGANIC MATTER (HUMUS) CONTENT OF SOILS

Soils vary greatly in organic matter content. A typical prairie grassland (e.g., Mollisol) may contain 5 to 6 percent organic matter in the top 15 cm; a sandy soil less than 1 percent. Poorly drained soils (Aquepts) often have organic matter contents approaching 10 percent. Tropical soils (Oxisols) are known for their low contents of organic matter.

For all practical purposes, the organic matter content of the soil parallels the N content. The C/N ratio of organic matter generally falls within the range of 10 to 12, although higher values are not unusual. Because of the ease with which total soil N can be determined, this is the parameter most often used as an index of organic matter content; i.e., organic matter = C × 1.727 ≈ N × 17.27.

The quantities of C in various reservoirs at the earth's surface, and to a depth of 16 km, are recorded in Table 1.2. It can be seen that the mass of soil C exceeds the amount found in all other "surface reservoirs" combined. However, the reservoir of soil C (30 to 50 × 10^{14} kg) is small when compared to the total amount contained in sediments (200,000 × 10^{14} kg). For supplementary information regarding global aspects of the C cycle, the works of Bolin[27] and Garrels et al.[28] are recommended.

Natural processes leading to the development of soils having variable organic matter contents are related to the so-called factors of soil formation.

$$O.M. = f_{(time, climate, vegetation, parent material, topography, ...)}$$

where f stands for "depends" or "function of" and the dots indicate that other factors may be involved.

In the classical studies of Jenny[29-32] each soil forming factor was treated as an independent variable.

$$O.M. = (climate)_{time, vegetation, parent material, topography, ...}$$

Evaluation of any given factor requires that all other factors remain constant, which seldom if ever occurs under natural soil conditions. Despite this short-

TABLE 1.2 Amount of C in Various Reservoirs[a]

Reservoir	Amount of C, 10^{14} kg
At Earth's surface	
Atmospheric CO_2	7
Biomass	4.8
Fresh water	2.5
Marine, above thermocline	5–8
Soil organic matter	30–50
At depths to 16 km	
Marine organic detritus	30
Coal and petroluem	100
Deep sea solute carbon	345
Sediments	200,000

[a] From Bolin.[27]

coming, studies of the type pioneered by Jenny have contributed substantially to our understanding of factors influencing the organic matter content of soils. The order of importance of the soil-forming factors in determining the organic matter (and N) content of loamy soils within the United States is climate > vegetation > topography = parent material > age.

The Time Factor

Organic matter does not accumulate indefinitely in well-drained soils, and, with time, an equilibrium level is attained that is governed by the other soil-forming factors. The numerous combinations under which the factors operate account for the great variability in the organic matter content of soils, even in a very localized area.

Information on the rate of organic matter accumulation during soil formation has come from studies of time sequences, or chronosequences, on mud flows, spoil banks, sand dunes, road cuts, and the moraines of receding glaciers. This work, reviewed by Stevenson,[5] shows that the rate of organic matter accumulation is rapid during the first few years, diminishes slowly, and reaches equilibrium in periods of time that vary from as few as 110 years for fine-textured parent material to as many as 1500 years for sandy areas.

Although several reasons have been given for the establishment of equilibrium levels or organic matter in soil, none has proven entirely satisfactory. Included with the explanations are:

1. Humic substances (e.g., humic acids) are produced that resist attack by microorganisms
2. Humus is protected from decay through its interaction with mineral matter (e.g., polyvalent cations and clay)
3. A limitation of one or more essential nutrients (N, P, S) places a ceiling on the quantity of stable humus that can be synthesized

Influence of Climate

Climate is the most important single factor that determines the array of plant species at any given location, the quantity of plant material produced, and the intensity of microbial activity in the soil; consequently, this factor plays a prominent role in determining organic matter levels. Considering climate in its entirety, a humid climate leads to forest associations and the development of Spodosols and Alfisols; a semiarid climate leads to grassland associations and the development of Mollisols. Grassland soils exceed all other well-aerated soils in humus content; desert, semidesert, and certain tropical soils have the lowest.

Soils formed under restricted drainage (Histosols and Inceptisols) do not follow a climatic pattern. In these soils, oxygen deficiency during prolonged periods prevents complete microbial decay of plant remains over a wide temperature range.

Extensive studies were made by Jenny and his coworkers[29-32] on the effect of climate on soil N (and organic matter) levels. For soils along a north to south transect of the semihumid region of central United States, the N (and organic matter) content of the soil decreased 2 to 3 times for each rise of 10°C in mean annual temperature. Whereas this relationship cannot be extrapolated directly to other areas, it is well known that soils of the warmer climatic zones generally have very low organic matter contents.

The effect of increasing rainfall (moisture component of climate) on soil organic matter content is to promote greater plant growth, and, consequently, the production of larger quantities of raw material for humus synthesis. The quantity of plant tissue produced, and subsequently returned to the soil, can vary from a trace in arid and arctic regions to several tons per hectare in warm climates where plant growth occurs throughout the year. Both roots and tops serve as energy sources for humus synthesis.

Vegetation

It is a well-known fact that, other factors being constant, grassland soils (e.g., Mollisols) have substantially higher organic matter contents than forest soils (e.g., Alfisols). Reasons given for this include:

1. Larger quantities of raw material for humus synthesis are produced under grass
2. Humus synthesis occurs in the rhizosphere, which is more extensive under grass than under forest vegetation. A closely related theory is that nitrification is inhibited in grassland soils, thereby leading to the preservation of N (and C)
3. The harsh climatic conditions under which grassland soils are formed (cold winters; hot, dry summers) lead to organic matter preservation

A combination of several factors is probably involved, with the third item listed being of some importance.

In the case of forest soils, differences in the profile distribution of organic matter occur by virtue of the manner in which the leaf litter becomes mixed with the mineral layer. It soils formed under deciduous forests on sites that are well drained and well supplied with calcium (Alfisols), litter is incorporated into the mineral layer through the activities of earthworms and faunal organisms. For these soils, mineral particles in the top 10 to 15 cm of soil become coated with humus. On the other hand, on sites low in available calcium (Spodosols), the leaf litter does not become mixed with the mineral layer but forms a mat on the soil surface. An organic-rich layer of acid (mor) humus accumulates at the soil surface and only the topmost portion of the mineral layer becomes stained with humus.

Parent Material

Parent material is effective mainly through its influence on texture. It has been well established that, for any given climate zone, and provided vegetation and topography are constant, organic matter content depends upon textural properties. The fixation of humic substances in the form of organo–mineral complexes serves to preserve organic matter. Thus, heavy-textured soils have higher organic matter contents than loamy soils, which in turn have higher contents than sandy soils.

Topography

Topography, or relief, affects organic matter content through its influence on rainfall runoff/retention. Local variations in topography, such as knolls, slopes, and depressions, have a profound effect on moisture retention in the upper portion of the soil, and thus organic matter content. Soils occurring in depressions, particularly those with restricted drainage, usually have high organic matter contents because the anaerobic conditions that prevail during wet periods of the year lead to the preservation of organic matter.

Effect of Cropping

Marked changes are brought about in the organic matter content of the soil through the activities of man. Usually, but not always, organic matter levels decline when soils are first placed under cultivation. However, destruction of organic matter is far from complete and new equilibrium levels are attained characteristic of the cropping system employed. For productive agricultural soils of the north central region of the United States, about 25 percent of the organic matter (N) was lost during the first 20 years of cultivation, 10 percent the second 20 years, and 7 percent during the third 20 years.[29,32] History has shown that when nutrients (notably N) are not replaced by organic or fertilizer inputs, the soil no longer remains sustainable and crop yields decline.

The decline in organic matter content through cropping cannot be attributed entirely to a reduction in the quantity of plant residues available for humus synthesis. Improved aeration, resulting from cultivation, may lead to an increase in microbial activity and loss of organic matter through mineralization. Since substantial amounts of fresh soil are subjected to cycles of wetting and drying during repeated and intensive cultivation, losses of organic matter can be appreciable. A major effect of cultivation in stimulating microbial activity may be the exposure of organic matter previously inaccessible to microbial attack (such as in micropores).

Increases in plant yield brought about by improved varieties, more widespread (and prudent) use of fertilizers, and adoption of better management practices have a positive effect on equilibrium levels of organic matter through return of larger quantities of plant residues. However, under intensive farming conditions, the increases have been slight. For most agricultural soils, organic matter can only be maintained at high levels by inclusion of a sod crop in the cropping sequence, by frequent addition of large quantities of organic residues (e.g., animal manure), or by minimizing tillage operations, and maximizing the amount of plant material returned to the soil.

DECOMPOSITION PROCESSES AND AGES OF ORGANIC MATTER

The Overall Decay Process

Several stages can be delineated in the decay of organic residues in soil. Earthworms and other soil animals play an important role in reducing the amount of plant residue material contained in the litter layer and the light fraction; further transformations are carried out by microorganisms. The initial phase of microbial attack is characterized by rapid loss of readily decomposable organic substances. Depending on the nature of the soil microflora and quantity of synthesized microbial cells, the amount of substrate C utilized for cell synthesis will vary from 10 to 70 percent. By-products include CO_2, NH_3, H_2S, organic acids, and other incompletely oxidized substances. In subsequent phases, organic intermediates and newly formed biomass tissues are attacked by a wide variety of microorganisms, with production of new biomass and further C loss as CO_2. The final stage of decomposition is characterized by gradual decomposition of the more resistant plant parts, such as lignin, for which actinomycetes and fungi play a major role.

The special conditions that exist in poorly drained soils and wet sediments alter considerably the activities of macro- and microfaunal organisms. Wet sediments not only support a different flora of microorganisms but decomposition of organic matter proceeds at a greatly reduced rate; furthermore, the end products of metabolism are different.

Organic compound peculiar to wet sediments are given in Table 1.3. They include synthesis of a variety of carcinogenic compounds, such as methylmercury, dimethylarsine, dimethylselenide, and nitrosamines of various types.

Fate of Carbon in Crop Residues: Regeneration of Humus

By using ^{14}C-labeled substrates, it has been possible to follow the decomposition of plant residues under conditions existing in the field and to identify the residue C as it becomes incorporated into stable humus. This work has shown that, for soils of the temperate zone, approximately one-third of the C in plant residues remains behind in the soil after the first growing season, mostly as labile and stable components of humus. Results obtained for C retention after the first year are summarized in Table 1.4.

Results of ^{14}C studies have shown that, with time, the residual C becomes increasingly resistant to decomposition. For example, in the field experiments carried out by Jenkinson,[33,34] about one-half of the residual C was still present

TABLE 1.3 Organic Compounds Peculiar to Wet Sediments

Class	Comments
Fermentation products	Incomplete oxidation leads to production of CH_4, organic acids, amines, mercaptans, aldehydes, and ketones.
Modified or partially modified remains of plants	In addition to slightly altered lignins, carotenoids, sterols, and porphyrins of chlorophyll origin are preserved.
Synthetic organic chemicals	Many man-made chemicals (e.g., DDT) decompose slowly if at all under anaerobic conditions.
Carcinogenic compounds	Synthesis of methylmercury, dimethylarsine, dimethylselenide, and nitrosamines of various types.

in the soil after a four-year period, indicating further stabilization. As noted below, the average age of the residual C is of the order of five to 25 years and gradually approaches that of native humus.

Lignins and related aromatic substances are much more resistant to decomposition than aliphatic substances, such as cellulose, but stabilization of nonaromatic C also occurs through their transformations to aromatic constituents by soil microbes, as noted in Chapter 8, on the biochemistry of the formation of humic substances.

TABLE 1.4 Carbon Retained from ^{14}C-Labeled Plant Material Applied to Field Soils

Location	Type	Carbon Retained (%)	Reference
England	Ryegrass tops and roots	Approximately 33% first year irrespective of soil type or plant material	Jenkinson[33,34]
West Germany	Wheat straw and chaff	31% after first year for fallow and cropped soil	Führ and Sauerbeck[35]
Austria	Maize (corn)	33 to 47% after first year, depending on time applied	Oberländer and Roth[36]
Canada	Wheat straw	35 to 45% after first growing season	Shields and Paul[37]
Colorado, U.S.	Blue grama 1. Herbage	43 to 46% after 412 days	Nyhan[38]
	2. Roots	63 to 74% after 412 days	Nyhan[38]

^{14}C Dating of Soils

Organic materials formed from atmospheric CO_2 acquire a small amount of cosmic-ray-produced ^{14}C during their synthesis. When photosynthesis ceases, slow radioactive decay of the ^{14}C begins. This gradual loss of radioactivity provides a method for estimating the age (or average age) of soil organic matter. The ^{14}C-dating method is suitable for accurate age estimates up to 50,000 years, but extension to 70,000 years is possible.

The fundamental assumption underlying the ^{14}C-dating method is that production of ^{14}C by cosmic radiation has been constant for at least 70,000 years and that C exchange between the reservoirs (atmosphere, biosphere, and oceans) has followed the same rate. These conditions appear to have been fulfilled, although deviations are known to have occurred. Since about 1870, the distribution of ^{14}C has been complicated by consumption of fossil fuel, which has diluted the ^{14}C in the atmosphere by release of ^{12}C as CO_2. More recently, the explosion of thermonuclear bombs has added large amounts of ^{14}C to the atmosphere; in 1964, the ^{14}C level of the atmosphere was twice the natural level. At present, the ^{14}C content of the atmosphere is still much higher than the previous normal level. Enrichment of the atmosphere with "bomb" radiocarbon has been used as the basis for determining rates of movement and turnover of organic C in soils.[39]

Mean Residence Time (MRT) of Modern Humus Absolute ages for organic matter in soils cannot be determined by ^{14}C-dating procedures because of continued decomposition of old humus and resynthesis of new humus by microorganisms. The term "mean residence time" (MRT) has been used to express the results of ^{14}C measurements for the average age of modern humus.[40-42] The resistance of humus to biological decomposition has long been known but not until the advent of ^{14}C dating was it possible to express average age on a quantitative basis. A major factor contributing to stabilization is the formation of clay–humus and metal–humus complexes.

Typical MRTs for organic C in the surface layer of soils from western Canada and the United States are recorded in Table 1.5. While considerable variation in mean ages have been reported (250 to 1900 years), the findings attest to the high resistance of humus to microbial attack. In general, the humus of Mollisols appears to be more stable than that for other mineral soils. As one might expect, the MRT of organic matter increases with soil depth. Results of MRT measurements show that much of the humus in present-day soils of North America was derived from the vegetation (grasses, forests) that existed on them long before the land was colonized by Europeans.

On the basis of ^{14}C measurements, Hsieh[43] determined mean ages for the "active" and stable pools of organic matter in the Morrow Plots at the University of Illinois and the Sanborn Field at the University of Missouri. Mean ages for the stable pool were 2973 and 853 years, respectively.

A key feature of many models used to simulate the dynamics of soil organic matter and the cycling of N, P, and S in the soil–plant system is the division

TABLE 1.5 Mean Residence Time of Organic Matter in Some Typical Soils

Sample Description and Location	Mean Residence Time, Years	Reference
Mollisol		
a. Saskatchewan	1000	Paul et al.[42]
Saskatchewan	870 ± 50	Campbell et al.[40]
Saskatchewan	545	Martel and Paul[41]
b. North Dakota		
Virgin soil	1175 ± 100	Paul et al.[42]
Clean cultivated	1900 ± 120	Paul et al.[42]
Manured soil	880 ± 74	Paul et al.[42]
Bridgeport Loam, Wyoming		
a. Surface sod layer	3280	Paul et al.[42]
b. Continuous wheat plot	1815	Paul et al.[42]
Alfisol, Saskatchewan	250 ± 60	Paul et al.[42]

of soil organic matter into three major fractions, namely, 1) an active soil fraction with a 0.14-year turnover time under ideal conditions, 2) a protected fraction with a 5-year turnover time, and 3) the stable fraction with a long turnover time of 150 years under ideal conditions.[44]

Absolute Ages of Buried Soils The humus of buried soils offer the possibility for establishing the chronology of climatic changes during Wisconsinian time through dating of organic matter by the ^{14}C method. However, this approach is usually attempted only as a last resort. One major limitation is that the amount of humus found in buried soils is normally extremely low. Also, contamination by recent organic matter is nearly always a problem, such as by downward movement of soluble organics in percolating water and penetration of the buried soil by plant roots.

FUNCTION OF ORGANIC MATTER IN SOIL

This brief section places into proper perspective the significance of organic matter in soil. Interest in this subject is long-standing and the pertinent aspects have been reviewed elsewhere.[1-4] The properties of soil humus and associated effects on soil are given in Table 1.6.

Organic matter contributes to plant growth through its effect on the physical, chemical, and biological properties of the soil. It has a nutritional function in that it serves as a source of N, P, and S for plant growth, a biological function in that it profoundly affects the activities of microflora and microfaunal organisms, and a physical function in that it promotes good soil structure, thereby improving tilth, aeration, and retention of moisture.

The dark brown to black color of soils is due to their stable humus content.

FUNCTION OF ORGANIC MATTER IN SOIL   15

TABLE 1.6 General Properties of Humus and Associated Effects in the Soil

Property	Remarks	Effect on soil
Color	The typical dark color of many soils is caused by organic matter	May facilitate warming
Water retention	Organic matter can hold up to 20 times its weight in water	Helps prevent drying and shrinking. Improves moisture-retaining properties of sandy soils
Combination with clay minerals	Cements soil particles into structural units called aggregates	Permits exchange of gases Stablizes structure Increases permeability
Chelation	Forms stable complexes with Cu^{2+}, Mn^{2+}, Zn^{2+}, and other polyvalent cations	Enhances availability of micronutrients to higher plants
Solubility in water	Insolubility of organic matter is due to its association with clay. Also, salts of divalent and trivalent cations with organic matter are insoluble.	Little organic matter is lost by leaching
Buffer action	Exhibits buffering in slightly acid, neutral, and alkaline ranges	Helps to maintain a uniform reaction in the soil
Cation exchange	Total acidities of isolated fractions of humus range from 300 to 1400 cmoles/kg	Increases cation exchange exchange capacity (CEC) of the soil. From 20 to 70% of the CEC of many soils (e.g., Mollisols) is caused by organic matter
Mineralization	Decomposition of organic matter yields CO_2, NH_4^+, NO_3^-, PO_4^{3-}, and SO_4^{2-}	Source of nutrients for plant growth
Combines with xenobiotics	Affects bioactivity, persistence, and biodegradability of pesticides	Modifies application rate of pesticides for effective control

Highly productive soils often have a characteristic rich odor that can be attributed to organic constituents.

Many of the benefits attributed to organic matter have been well documented, but it should be noted that the soil is a multi-component system of interacting materials. Accordingly, soil properties represent the net effect of the various interactions and not all benefits cannot be ascribed solely to the organic component.

The functions of humus are elaborated briefly below and in greater detail in

16 ORGANIC MATTER IN SOILS

subsequent chapters. The importance of any given factor will vary from one soil to another and will depend upon such environmental conditions as soil type, drainage characteristics, and cropping history.

Availability of Nutrients for Plant Growth

Organic matter has a profound effect on the availability of nutrients for plant growth. In addition to serving as a source of N, P, and S through its mineralization by soil microorganisms, organic matter influences the supply of nutrients from other sources. For example, organic matter is required as an energy source for N-fixing bacteria; accordingly, the amount of molecular N_2 fixed by free-living fixers will be influenced by the quantity of available energy in the form of carbohydrates.

Humus substances have the following indirect effects on the mineral nutrition of crop plants:

1. Through incorporation of N and S into stable structures of humic substances during mineralization/immobilization
2. Chemical transformations of inorganic N forms, namely, stabilization of N through chemical fixation of NH_3 and conversion of $NO_2^- $-N to N_2 and N_2O through nitrosation (see Chapter 4)
4. Solubilization of phosphates by complexation of Ca in calcareous soils and Fe and Al in acid soils
5. Alleviation of metal ion toxicities, including Al^{3+} in acid soils

A factor that needs to be taken into consideration in evaluating humus as a source of nutrients is the cropping history. When soils are first placed under cultivation, the humus content generally declines over a period of 10 to 30 years until a new equilibrium level is attained. Since the C/N/P/S ratio is rather constant (Chapter 5) any nutrients liberated by microbial activity must be compensated for by incorporation of equal amounts into newly formed humus.

The availability of phosphate in soil is often limited by fixation reactions, which convert the monophosphate ion to various insoluble forms. Insoluble Ca-phosphates predominate in calcareous soils while insoluble Fe- and Al-phosphates are formed in acidic soils. Adsorption by clay minerals can affect phosphate availability under neutral or slightly acid conditions. As noted later in Chapter 16, the availability of soil phosphate in enhanced by additions of organic matter, presumably due to chelation of polyvalent cations by organic acids and other decay products. The availability of many micronutrient cations is also affected by fixation and precipitation reactions, typical examples being the low availability of Fe and Mn in calcareous soils. Naturally occurring chelating agents may influence the availability of micronutrients in much the same way as for inorganic phosphate.

Effect on Soil Physical Condition

Humus has a profound effect on the structure of many soils (Chapter 18). The deterioration of structure that accompanies intensive tillage is usually less severe in soils adequately supplied with humus. When humus is lost, soils tend to become hard, compact, and cloddy. Seedbed preparation and tillage operations are usually easier to carry out and are more effective when humus levels are adequate.

Aeration, water-holding capacity, and permeability are all favorably affected by humus. The frequent addition of easily decomposable organic residues leads to the synthesis of complex organic compounds (e.g., polysaccharides) that bind soil particles into structural units called aggregates. These aggregates help to maintain a loose, open, granular condition. Water is then better able to infiltrate and percolate downward through the soil. The roots of plants need a continual supply of O_2 in order to respire and grow. Large pores permit better exchange of gases between soil and atmosphere.

Soil organic matter can absorb and hold a substantial amount of water. However, the fact that organic matter increases the water-holding capacity of the soil does not necessarily mean that more water will be available to the plant, for the reason that organic matter may increase the permanent wilting percentage. Organic matter may increase the amount of plant-available water in sandy soils.

Soil Warming

Organic matter imparts a dark color to the soil that can enhance soil warming and thereby promote early plant growth.

Soil Erosion

Humus usually increases the ability of the soil to resist erosion. First, it enables the soil to hold more water. Even more important is its effect in promoting soil granulation and maintaining large pores through which water can enter and percolate downward. In a granular soil, individual particles are not easily carried along by moving water.

Source of Energy for Soil Organisms

Organic matter, particularly litter and the "light" fraction, serves as a source of energy for both macro- and microfaunal organisms. Numbers of bacteria, actinomycetes, and fungi in the soil are related in a general way to humus content. Earthworms and other faunal organisms are strongly affected by the quantity of plant residue material returned to the soil.

The role played by the soil fauna has not been completely elaborated but the functions they perform are multiple and varied. For example, earthworms may be important agents in producing good soil structure. They construct extensive channels that serve not only to loosen the soil but to improve aeration and drainage. Earthworms flourish only in soils that are well provided with organic matter.

Growth of Higher Plants

Organic substances in soil can have a direct physiological effect on plant growth (Chapter 7). Some compounds, such as certain phenolic acids, have phytotoxic properties; others, such as the auxins, enhance plant growth. Under certain conditions, substances toxic to plants can arise either directly or indirectly during the decomposition of plant residues in soil.

A major unanswered question at the present time is whether organic substances have a positive beneficial effect on plant growth, and hence, yields, in normally productive agricultural soils. From time to time, claims are made to the effect that crops grown on soils well supplied with organic matter are more vigorous, more healthy, and more nutritious than crops grown on soils treated with equivalent amounts of inorganic nutrients in the form of fertilizers. However, as Whitehead[45] pointed out, many of these claims are based on empirical observations and have not been substantiated. A similar conclusion has been reached regarding the value of commercial humates (lignites or their humic acids) when used as soil amendments.[46] Numerous studies (see Chapter 3 of MacCarthy et al.,[3]) have shown that humic acids can stimulate plant growth, but most of this work has been carried out in nutrient solutions or in a sand culture. Most mineral soils have the ability to inactivate humic acid (see Chapter 18).

Buffering and Exchange Capacity

From 20 to 70 percent of the exchange capacity of many soils (e.g., Mollisols) is due to colloidal humic substances (Chapter 15). Total acidities of isolated fractions of humus range from 300 to 1400 cmols/kg. As far as buffer action is concerned, humus exhibits buffering over a wide pH range.

Adsorption of Pesticides and Other Organic Chemicals

The immense importance of organic matter in influencing the behavior of organic pesticides in soil, including effectiveness against the target species, phytotoxicity to subsequent crops, leachability, volatility, and biodegradability, has been well documented (see Chapter 19). Organic matter content has been found to be the soil factor most directly related to the sorption of most herbicides.

Incidence of Plant Pathogens

It is well known that many of the factors influencing the incidence of pathogenic organisms in soil are directly or indirectly influenced by organic matter. For example, a plentiful supply of organic matter may favor the growth of saprophytic organisms relative to parasitic ones and thereby reduce populations of the latter. Biologically active compounds in soil, such as antibiotics and certain phenolic acids, may enhance the ability of certain plants to resist attack by pathogens.

ORGANIC MATTER AND SUSTAINABLE AGRICULTURE

It has been said that proper management of organic matter is the heart of sustainable agriculture.[47] By maximizing the amount of organic residues returned to the soil, and by minimizing tillage operations, benefits accrue due to reduced erosion, better soil tilth, preservation of stable humus, and improved nutrient cycling through conservation of nutrients that would otherwise be lost through leaching. The "active" fraction (discussed earlier) plays a particularly important role in maintaining the productivity of soils under sustainable agriculture. A stimulating discussion of the role of organic matter in sustainable farming has been given by Weil.[47]

ENVIRONMENTAL SIGNIFICANCE OF HUMIC SUBSTANCES

Humic acids and related substances are among the most widely distributed organic materials in the earth. They are found not only in soils but in natural waters, sewage, compost heaps, marine and lake sediments, peat bogs, carbonaceous shales, lignites, brown coals, and miscellaneous other deposits. The amount of C in the earth as humic acids (60×10^{11} t) exceeds that which occurs in living organisms (7×10^{11} t).[48]

Humic substances are important in geochemistry and the environment for the following reasons:

1. They may be involved in the transportation and subsequent concentration of mineral substances, such as bog ores and nodules of marine strata. Also, they may be responsible for the enrichment of uranium and other metals in various bioliths, including coal
2. Humic substances serve as carriers of organic xenobiotics (as well as trace elements) in natural waters. Humic substances per se are not believed to be physiologically harmful, but they are aesthetically unacceptable because they impart a reddish-black color to potable waters and recreational lakes. Humic substsances play a role in reducing the toxic-

ities of certain heavy metals (e.g., Cu^{2+} and Al^{3+}) to aquatic organisms, including fish
3. Humic substances act as oxidizers or reducing agents, depending on environmental conditions. They may affect photochemical processes in natural waters, including photoalteration of xenobiotics. Humic substances have been shown to reduce Hg(II) to volatile Hg° under natural pH conditions, thereby providing a potential pathway for the mobilization of Hg in the environment
4. The sorption capacity of the soil for a variety of organic and inorganic gases is strongly influenced by humus. The ability of the soil to function as a "sink" for N and S oxides in the atmosphere may be due in part to reactions involving organic colloids
5. Humic-like materials in waste waters treated by biological secondary treatment processes create problems of considerable importance in many water works. For example, they can react with chlorine (added during chlorination) to produce the carcinogen chloroform and other undesirable halogenated organics. A major problem in the treatment of waste waters for reuse involves removal of humic substances and other organics; techniques such as coagulation and carbon adsorption have been proposed for this purpose[49]

Add Humic substances have a number of industrial applications, including use as well-drilling fluids. They are under investigation as potential therapeutics.

As noted in later chapters, humic substances occurring in soils have many properties in common with those found in other natural systems. Accordingly, the investigations of soil scientists are of interest to scientists in several other scientific disciplines, including organic geochemistry, limnology, and sanitary engineering.

SUMMARY

Organic matter performs many useful functions in soil and it is responsible for many of the reactions that take place in sediments and natural waters.

Pools of soil organic matter that can currently be measured, albeit with difficulty, include litter, "light" fraction, microbial biomass, water soluble organics, and stable humus. A major goal in the study of soil organic matter is to relate information regarding the size (and composition) of the various pools to soil productivity. In particular, there is a critical need for accurate methods for determining the size of the so-called "active" fraction (biomass plus "light" fraction) and the rates at which the nutrients contained therein are released to available mineral forms.

Soils vary widely in their organic matter contents. In undisturbed (uncultivated) soils, the amount present is governed by the soil-forming factors of age (time), parent material, topography, vegetation, and climate. The order of importance of the soil-forming factors in determining the organic matter contents of loamy soils within the United States is believed to be: climate > vegetation > topography = parent material > age.

Organic matter is usually lost when soils are first placed under cultivation, and a new equilibrium is reached that is characteristic of cultural practices and soil type. For most soils, organic matter can only be maintained at high levels by inclusion of a sod crop in the rotation, by no-till or minimal tillage practices, or by frequent addition of large quantities of organic residues (e.g., animal manure).

Considerable information has been obtained on decomposition processes in soil using ^{14}C-labeled plant residues. For soils of the temperate zone, approximately one-third of the C remains behind in the soil after the first growing season, mostly as labile and stable components of humus. The MRT of the residual C is initially short but approaches that of the native humus C within a relatively short time. The MRT of stable humus varies from several hundred to somewhat over 1000 years.

REFERENCES

1. Y. Chen and Y. Avnimelech, Eds., *The Role of Organic Matter in Modern Agriculture*, Martinus Nijhoff, Dordrecht, 1986.
2. D. C. Coleman, J. Malcolm Oades, and G. Uehara, Eds., *Dynamics of Soil Organic Matter in Tropical Ecosystems*, University of Hawaii Press, Honolulu, 1989.
3. P. MacCarthy, C. E. Clapp, R. L. Malcolm and P. R. Bloom, Eds., *Humic Substances in Soil and Crop Sciences: Selected Readings*, Soil Science Society of America, Madison, 1990.
4. F. J. Stevenson, *Cycles of Soil: C, N, P, S, Micronutrients*, Wiley, New York, 1986.
5. F. J. Stevenson, "Origin and Distribution of Nitrogen in Soil," in F. J. Stevenson, Ed., *Nitrogen in Agricultural Soils*, American Society of Agronomy, Madison, 1982, pp. 1–42.
6. R. L. Tate, Jr., *Soil Organic Matter: Biological and Ecological Effects*, Wiley, New York, 1987.
7. B. K. G. Theng, K. R. Tate, and P. Sollins, "Constituents of Organic Matter in Temperate and Tropical Soils, in D. C. Coleman, J. M. Oades, and G. Uehara, Eds., *Dynamics of Soil Organic Matter in Tropical Ecosystems*, University of Hawaii Press, Honolulu, 1989, pp. 5–32.
8. W. S. Wilson, Eds., *Advances in Soil Organic Matter Research: The Impact on Agriculture and the Environment*, Redwood Press Ltd., Wiltshire, England, 1991.
9. J. M. Anderson and J. S. I. Ingram, Eds., *Tropical Soil Biology and Fertility Handbook of Methods*, Commonwealth Agricultural Bureau, Wallingford, England, 1989.
10. J. O. Skjemstad, R. C. Dalal, and P. F. Barron, *Soil Sci. Soc. Amer. J.*, **50**, 354 (1986).

11 R. C. Dalal and R. J. Mayer, *Aust. J. Soil Res.*, **24**, 293 (1986).
12 H. H. Janzen, *Can. J. Soil Sci.*, **67**, 845 (1987).
13 H. H. Janzen, C. A. Campbell, S. A. Brandt, G. P. Lafond, and L. Townley-Smith, *Soil Sci. Soc. Amer. J.* **56**, 1799 (1992).
14 G. Spycher, P. Sollins, and S. Rose, *Soil Sci.*, **135**, 79 (1983).
15 J. M. Duxbury, M. S. Smith, and J. W. Doran, "Soil Organic Matter as a Source and Sink of Plant Nutrients," in D. C. Coleman, J. M. Oades, and G. Uehara, Eds., *Dynamics of Soil Organic Matter in Tropical Ecosystems*, University of Hawaii Press, Honolulu, 1989, pp. 33–67.
16 E. A. Paul and R. P. Voroney, "Field Interpretation of Microbial Biomass Activity Measurements," in M. J. Klug and C. A. Reddy, Eds., *Current Perspectives in Microbiology Ecology*, American Society for Microbiology, Washington DC, 1983, pp. 509–514.
17 A. Tunlid and D. C. White, "Biochemical Analyses of Biomass, Community Structure, Nutritional Status, and Metabolic Activity of Microbial Communities in Soil," in G. Stotzky and J.-M. Bollag, Eds., *Soil Biochemistry*, Vol. 7, Marcel Dekker, New York, 1992, pp. 229–262.
18 D. S. Jenkinson and D. S. Powlson, *Soil Biol. Biochem.* **8**, 209 (1976).
19 J. P. E. Anderson and K. H. Domsch, *Soil Sci.*, **130**, 211 (1980).
20 R. Martens, *Soil Biol. Biochem.*, **17**, 57 (1985).
21 J. A. Ocio and P. C. Brookes, *Soil Biol. Biochem.*, **22**, 685 (1990).
22 G. P. Sparling and A. W. West, *Soil Biol. Biochem.*, **20**, 337 (1988).
23 K. R. Tate, D. J. Ross, and C. W. Feltham, *Soil Biol. Biochem.*, **20**, 319 (1988).
24 E. A. Elkhatib, O. L. Bennett, V. C. Baligar, and R. J. Wright, *Soil Sci. Soc. Amer. J.*, **50**, 297 (1986).
25 J. Wolt and J. G. Graveel, *Soil Sci. Soc. Amer. J.*, **50**, 602 (1986).
26 S. Kiss, M. Drägan-Bularda, and D. Radulescu, *Adv. Agron.*, **27**, 25 (1975).
27 B. Bolin, *Science*, **196**, 613 (1977).
28 R. M. Garrels, F. T. Mackenzie, and C. Hunt, *Chemical Cycles and the Global Environment*, Kauffman, Los Altos, California, 1975.
29 H. Jenny, *Missouri Agr. Exp. Sta. Res. Bull.*, **152**, 1 (1930).
30 H. Jenny, *Soil Sci.*, **31**, 247 (1931).
31 H. Jenny, *Soil Sci.*, **69**, 63 (1950).
32 H. Jenny, *Missouri Agr. Exp. Sta. Res. Bull.*, **765**, 1 (1960).
33 D. S. Jenkinson, *J. Soil Sci.*, **17**, 280 (1966).
34 D. S. Jenkinson, *J. Soil Sci.*, **16**, 104 (1965).
35 F. Führ and D. Sauerbeck, "Decomposition of Wheat Straw in the Field as Influenced by Cropping and Rotation," in *Isotopes and Radiation in Soil Organic Matter Studies*, International Atomic Energy Agency, 1968, pp. 241–250.
36 H. E. Oberländer and K. Roth, "Transformations of Carbon-14 Labeled Plant Materials in a Grassland Soil under Field Conditions," in *Isotopes and Radiation in Soil Organic Matter Studies*, International Atomic Energy Agency, 1968, pp. 351–361.
37 J. A. Shields and E. A. Paul, *Can. J. Soil Sci.*, **58**, 297 (1973).

38 J. W. Nyhan, *Soil Sci. Soc. Amer. Proc.*, **39,** 643 (1975).
39 B. J. O'Brien and J. D. Stout, *Soil Biol. Biochem.*, **10,** 309 (1978).
40 C. A. Campbell, E. A. Paul, D. A. Rennie, and K. J. McCallum, *Soil Sci.*, **104,** 81, 217 (1967).
41 Y. A. Martel and E. A. Paul, *Soil Sci. Soc. Amer. Proc.*, **38,** 501 (1974).
42 E. A. Paul, C. A. Campbell, C. A. Rennie, and K. J. McCallum, *Trans. 8th Intern. Congr. Soil Sci.*, **3,** 201 (1964).
43 Y.-P. Hsieh, *Soil Sci. Soc. Amer. J.*, **56,** 460 (1992).
44 W. J. Parton, R. L. Sanford, P. A. Sanchez, and J. W. B. Stewart, "Modeling of Soil Organic Matter in Tropical Soils," in D. C. Coleman, J. M. Oades, and G. Uehara, Eds., *Dynamics of Soil Organic Matter in Tropical Ecosystems*, University of Hawaii Press, Honolulu, 1989, pp. 153-171.
45 D. C. Whitehead, *Soils Fert.*, **26,** 217 (1963).
46 F. J. Stevenson, *Crops and Soils Magazine*, **31,** 14 (1979).
47 R. R. Weil, "Inside the Heart of Sustainable Farming," in *The New Farm*, Jan. 1992, pp. 43-48.
48 A. Szalay, *Geochim. Cosmochim. Acta*, **22,** 1605 (1964).
49 J. R. Stukenberg, *J. Water Pollution Control Federation*, **47,** 338 (1975).

2

EXTRACTION, FRACTIONATION, AND GENERAL CHEMICAL COMPOSITION OF SOIL ORGANIC MATTER

The chemical and colloidal properties of soil organic matter can only be studied in the free state, that is, when freed of inorganic soil components. Thus, the first task of the researcher is to separate organic matter from the inorganic matrix of sand, silt, and clay. Alkali, usually 0.1 to $0.5N$ NaOH, has been a popular extractant of soil organic matter, but this reagent may alter the organic matter through hydrolysis and autoxidation. In recent years, milder but less efficient extractants have been used with variable success. A second task of the researcher is to reduce the heterogeneity of the extracted material so that analytical techniques can be applied.

Methods for the extraction and fractionation of soil organic matter have evolved from the research and thinking of many scientists. A brief historical account of this work is presented as a prelude to modern studies on humus chemistry.

BRIEF HISTORICAL REVIEW

Initial Experimental Period, 1786–1900

The first attempt to isolate humic substances from soil appears to have been made by Achard,[1] who extracted peat with alkali and obtained a dark, amorphous precipitate upon acidification. This alkali-soluble, acid-insoluble material later came to be known as humic acid. Achard observed that larger amounts of humic material could be extracted from the lower, more humified layers of the peat than from the upper, less decomposed layers.

De Saussure[2] is usually credited with introducing the term "humus" (Latin

equivalent of soil) to describe the dark colored organic materi
noted that humus was richer in C and poorer in H and O than th(
from which it was derived.

In 1822, Döbereiner[3] designated the dark colored component of soil organic matter as "humussäure," or "humus acid." This term came to be used synonymously with "humic acid," although Waksman[4] concluded that "humus acid" was, for the most part, an inclusive term for all "humic acids" whereas the latter was applied to the precipitate obtained from the alkali extract by acidification. This definition of humic acid has persisted through the present time.

The first comprehensive study of the origin and chemical nature of humic substances was carried out by Sprengel.[5,6] Many of the procedures he developed for the preparation of humic acids became generally adopted, such as pretreatment of the soil with dilute mineral acids prior to extraction with alkali. Sprengel concluded that, for soils rich in bases, humic acid was in a bound form, and consequently, the soil had a neutral reaction (contained "mild humus"). This soil was regarded as highly fertile. On the other hand, for soils poor in bases, the humic acid was believed to be in the free form, with the result that the soil was acid and unproductive (contained "acid humus"). Thus, Sprengel felt that the different forms of humus were of considerable importance in soil fertility. A major contribution of Sprengel to humus chemistry was his extensive studies on the acidic nature of humic acids.

Research on the chemical properties of humic substances was extended by the Swedish investigator Berzelius,[7] whose main contribution was the isolation of two light-yellow colored humic substances from mineral waters and a slimy mud rich in iron oxides. They were obtained from the mud by extraction with alkali, which was then treated with acetic acid and copper acetate. A brown precipitate was obtained called "copper apocrenate." When the acetic acid solution was neutralized, another precipitate was obtained, called "copper crenate." The free acids, "apocrenic" and "crenic" acids, were then brought into solution by decomposition of the Cu complexes with alkali. These newly described humic substances were examined in considerable detail, including isolation, elementary composition, and properties of their metal complexes (Al, Fe, Cu, Pb, Mn, etc.). The chemical formulas were $C_{24}H_{12}O_{16}$ and $C_{24}H_6O_{12}$, respectively. Berzelius established the great mobilities of crenic and apocrenic acids as compared to humic acids. A popular view of the time was that the various preparations represented distinct chemical compounds.

The concept that isolated components of humus (humic acid, crenic acid, apocrenic acid) were chemically individual compounds was challenged by the Russian investigator German (cited by Kononova[8]), who carried out more extensive fractionations and isolated a variety of substances that differed from one another in elementary composition. German was of the opinion that N was a constituent component of humic substances.

The investigations of Berzelius were developed further by his contemporaries and former students, especially Mulder,[9] who classified humic substances on the basis of solubility and color into the following groups: 1) ulmin and humin—insoluble in alkali; 2) ulmic acid (brown) and humic acid (black)—soluble in alkali; and 3) crenic and apocrenic acids—soluble in water. A number of new preparations were described for which various names were given (glucic acid, apoglucic acid, chlor–humic acid, etc). These terms were later abandoned. Like his contemporaries, Mulder was of the opinion that the different humic fractions were chemically individual compounds and that they did not contain N. His concept of soil humus was undoubtedly influenced by his extensive studies on the brown and black substances obtained by the action of acids on carbohydrates, which he believed to be identical to natural humic substances.

The views of Mulder were strongly criticized by Hermann,[10] who felt that all plants adsorbed N from the air and that even artificially prepared humic substances could adsorb N. Hermann concluded that when humic acids were oxidized to crenic and apocrenic acids, still more N was adsorbed.

The second half of the nineteenth century was characterized by the proliferation of classification schemes for the isolation of new products from decomposing plant residues, from soil, and from artificial mixtures generated in the laboratory. These additional studies have been admirably reviewed by Kononova[8] and Waksman.[4] As Kononova[8] pointed out, each investigator repeated the mistakes of his predecessor by regarding the substance isolated as a chemical individual compound of specific composition. However, towards the end of the century, more and more investigators came to doubt the validity of chemical formulas attached to the various preparation. Also, the concept that synthetic laboratory products were similar to those occurring naturally came under considerable criticism. Of the many names proposed during this period, only the hymatomelanic acid of Hoppe-Seyler[11] has survived. This substance was obtained from humic acid by extraction with alcohol.

By the end of the century, it had been firmly established that humus was a complex mixture of organic substances that were mostly colloidal in nature and which had weakly acidic properties. Information had also been obtained on their interactions with other soil components.

Early Twentieth Century, 1900–1940

During the period 1900–1940, renewed efforts were made to classify humic substances and to determine their chemical nature and structure. Among the more important contributions are those of Oden,[12,13] who classified humic substances into the following groups: humus coal, humic acid, hymatomelanic acid, and fulvic acid. Humus coal corresponds to the humin and ulmin of Berzelius and Mulder; this fraction is insoluble in water, alcohol, and alkali. The term humic acid was reserved for the dark-brown to black material that is soluble in alkali but insoluble in acid, and which has a C content of about 58 percent. Hymatomelanic acid, earlier named by Hoppe-Seyler, is the alcohol-soluble component of humic acid and was believed by Oden to be formed when humic acid is decomposed during extraction with alkali. This material is lighter in color than humic acid (chocolate brown) and has a higher C content (about 62 percent). The yellow to yellow–brown fulvic acids, which were recovered from marsh waters by ultrafiltration, were recognized as being similar to the crenic and apocrenic acids of Berzelius. Oden believed that humus coal and fulvic acids were group designations while humic acid and hymatomelanic acids were chemically individual compounds. Oden was also of the opinion that N was a contaminant of humic acid rather than part of its structure.

Although workers of earlier periods recognized the occurrence of specific organic compounds in soil,[7,9] attention had been focused almost exclusively on the so-called humic substances. Starting in 1908, Schreiner and Shorey[14] initiated a series of studies on identifiable organic chemicals in soil. In the following three decades, they established the existence in soil of over 40 compounds belonging to the well-known classes of organic chemistry, including organic acids, hydrocarbons, fats, sterols, aldehydes, carbohydrates, and specific N-containing substances. Their studies on the toxic effects of organic compounds on plant growth attracted wide attention during this and subsequent periods.

A detailed study of the nature and structure of humic substances was carried out by Shmook,[15] who in his final work in 1930, reviewed the more important aspects of humus chemistry and its formation by soil microorganisms. He considered humic acids to be the most characteristic constituents of humus and regarded them not as specific compounds but as mixtures of closely related substances having similar structural features. By esterification of humic acids with alcohol in the presence of dry HCl, Shmook demonstrated the occurrence of esters, thereby indicating the presence of COOH groups. Nitrogen was regarded as a structural component, and, in this respect, his views differed from those of Oden.[12,13] The source of N was believed to be the protein of microorganisms. Shmook concluded that humic acids contained two major components: an organic N-containing compound (protein) and the aromatic ring, the latter being derived from lignin.

The origin of humus was also a popular subject during this period. Two main theories prevailed, one being that they were derived from the lignified tissues of plant residues and the second that cellulose or simple sugars were sources. With regard to the latter, Maillard's[16] work deserves special consideration. He was of the opinion that humic substances were products synthesized from simple compounds produced during decomposition of plant remains. Extensive studies were carried out on the production of dark-colored, humic-like substances (melanins) from mixtures of reducing sugars and amino acids. The melanins were believed to result from the interaction of the $C=O$ group of the sugar with NH_2 and COOH groups of the amino acid, with the elimination of water from the former and CO_2 from the latter (through decarboxylation). The "humic acid" complexes formed in this way contained from 4 to 6 percent N. According to Maillard, the humic substances found in nature are the result of purely chemical reactions in which microorganisms do not play a direct role except to produce sugars from carbohydrates and amino acids from proteins.

The reaction between reducing sugars and amino acids or amines to form brown nitrogenous polymers has since been shown to be of great significance in the commercial dehydration of food products and is commonly referred to as the "Maillard reaction."

Maillard's concept that humic substances represented products formed from simple compounds is in essential agreement with the modern view of humus formation, as described later. However, Trusov's concept of humification (reviewed by Kononova[8]) comes even closer to present-day concepts. He postulated the following sequence in the humification process: 1) hydrolytic decomposition of plant remains with the synthesis of simple substances of an aromatic nature, 2) oxidation of the latter by microbial enzymes to form hydroxyquinones, and 3) condensation of the quinones into dark-colored products (humic substances). Williams,[17] a well-known Russian investigator, also postulated the existence of two stages in the humification process, the first being the decomposition of the original plant residues to simpler compounds, and the second being the synthesis of substances of a more complex nature. In contrast to the views of Maillard, both processes were believed to result from the enzymatic activity of microorganisms. Williams was critical of drastic methods of isolating humic substances and conducted investigations on the humus in lysimeter waters.

The view that lignin was the precursor of humic acids was further advanced by Fuch's,[18,19] Springer,[20] Hobson and Page,[21] Waksman,[4] and others. The theory became generally adopted and had a dominating influence on humus chemistry for several decades. Waksman[4] was of the opinion that proteins, derived from microorganisms, combined with modified lignins through the Schiff reaction (see Chapter 8) to form a "ligno–protein" complex that was recalcitrant to microbial attack.

Waksman[4] regarded humus as a mixture of plant-derived materials, including fats, waxes, resins, hemicellulose, cellulose, and "ligno–protein." His method on "proximate analysis" was widely applied to mineral soils, to peats, to compost mixtures, and to the organic layers of forest soils. The sequential treatments used, together with the usual range of values obtained for mineral soils, are given in Table 2.1. Waksman's method was generally unsuited for characterizing humus because, when tested on genetically different soils, only small differences were observed. It should be noted that, for most soils, three-fourths of the organic matter was accounted for in the vague group designated as "protein plus lignin–humus." The practice of estimating protein from the N analysis was of dubious validity.

Concurrent investigations into the chemistry of brown coals and their humic acids, such as those carried out by the German scientist Fuchs,[18,19] had a great influence on the study of soil humus. Kononova[8] questioned the value of this work as it applied to soil humus on the basis that the natural conditions of peat and coal do not correspond with those found in aerobic soils. Nevertheless, many of the techniques that were applied to coal and peat humic acids had general applicability to the study of soil humic substances.

TABLE 2.1 Waksman's Proximate Method of Analysis of Soil Organic Matter[a]

Fraction	Treatment	% of Organic Matter
Fats, waxes, oils	Ether extraction	0.5–4.7
Resins	Alcohol extraction	0.3–3.0
Hemicellulose	Hydrolysis (2% HCl)	5–12
Cellulose	Hydrolysis (80% H_2SO_4)	3–5
Protein plus "lignin-humus"	Analysis of final residues for C and N	
a. Protein ($N \times 6.25$)		30–35
b. "Lignin-humus"		30–50

[a] After Waksman.[4]

Finally, toward the end of the pre-modern era, additional fractions were added to the list of substances that could be separated from humus. On the basis of optical properties and behavior towards electrolytes, Springer[20] subdivided humic acids into Braunhuminsäure (brown humic acid) and Grauhuminsäure (gray humic acid). This was accomplished by redissolving the original humic acid in alkali and adding an electrolyte (KCl) to a final concentration of $0.1N$. The gray humic acids were easily coagulated, had relatively high C contents, and exhibited a low degree of dispersion. This group was typical of the humic acids of Altoll (Chernozem) and Rendoll (Rendzina) soils. The brown humic acids, which were characteristic of peats, Alfisols and brown coals, remained in solution. This group had lower C contents than the gray humic acids and showed a higher degree of dispersion.

At an even earlier date, Waksman[4] isolated a separate component from the fulvic acid fraction. When the filtrate from the separation of humic acid was adjusted to pH 4.8 another precipitate was formed. This was designated as the β-fraction of humus, or the "neutralization fraction" of Hobson and Page.[21] This material was rich in Al and was considered to be an "Al-humate." No mention is made of the β-fraction in the book by Kononova.[8]

Gradually, many investigators came to realize that the various preparations obtained on the basis of solubility characteristics were not chemically individual compounds but group designations. Also, the concept arose that a significant portion of soil humus does not exist as dark-colored, amorphous colloids but as compounds belonging to the well-known classes of organic chemistry, such as the carbohydrates.

Despite the problems involved, it is advantageous for the soil chemist to work with fractionated rather than with crude humus extracts. The three principal fractions are humic acid, fulvic acid, and humin but mention is frequently made to hymatomelanic acid, brown humic acid, and gray humic acid. Table 2.2 summarizes the derivation of these names and their relation to terms proposed by early investigators into humus chemistry.

TABLE 2.2 Names Suggested for the Various Humus Fractions on the Basis of Solubility Characteristics

Preparation	Solubility Characteristics	Sprengel[6]	Berzelius[7]	Mulder[9]	Hoppe-Seyler[11]	Oden[13]	Springer[20]
Humin	Insoluble in alkali	Humus coal	Humin	Humin, ulmin	Humin	Humus coal	—
Humic acid	Soluble in alkali, precipitated by acid	Humus acid	Humic acid	Humic acid	Humic acid	Humus acid	Humic acid
a. Brown humic acid	Not coagulated from alkali solution in the presence of electrolyte	—	—	—	—	—	Braun-huminsäure
b. Gray humic acid	Coagulated in the presence of electrolyte	—	—	—	—	—	Grau-huminsäure
Hymatomelanic acid	Soluble in alkali, precipitated by acid, soluble in alcohol	—	—	—	Hymatomelanic acid	Hymatomelanic acid	—
Fulvic acid	Soluble in alkali, not precipitated by acid	—	Crenic acid, Apocrenic acid	Crenic acid, Apocrenic acid	—	—	—

From time to time, questions arose as to the validity of classification schemes for identifying components of soil organic matter. Page[22] suggested that the term "humus" be discontinued altogether because of the widely different meanings attached to the word. The term "humic matter" was proposed to describe the dark-colored, high-molecular-weight organic colloids; "nonhumic matter" was suggested for the colorless organic substances resulting from the biological decomposition of plant and animal residues, such as waxes and cellulose. The fulvic acids were included with the "nonhumic" matter. The two terms suggested by Page[22] are somewhat analogous to the present-day use of humic and nonhumic substances. Fulvic acids, as usually defined, consist of mixtures of both types.

Waksman,[4] on the other hand, recommended abandonment of all terms except "humus," as shown by the following quotation:

> One may, therefore, feel justified in abandoning without reservation the whole nomenclature of "humic acids," beginning with the "humins," and "ulmins," through the whole series of "humus," "hymatomelanic," "crenic," "apocrenic," and numerous other acids, and ending with the "fulvic acid" and "humal acids," the last additions to the list. These labels designate, not definite chemical compounds but merely certain preparations which have been obtained by specific procedures. The only name warranting preservation is Humus, because of its historical importance; it should be used to designate the organic matter of the soil as a whole.

Despite Waksman's recommendation, terms such as humic acid, humin, fulvic acids, and others have survived and will undoubtedly continue to be used in the future. The majority of studies conducted on humus involve preliminary separations on the basis of solubility characteristics, and abandonment of these terms would cause even greater confusion than their continued use. For example, reference to the acid insoluble material as humic acid is considerably less cumbersome than repeated reference to the "alkali-soluble, acid-insoluble fraction."

With regard to other terms, the designation of humin as a separate fraction may be inappropriate, as this material may consist in part of portions of the other fractions that are so intimately associated with mineral matter that they cannot be extracted with alkali. Also, it is not known whether hymatomelanic acid is a distinct chemical entity. This material may be an artifact produced from humic acid during fractionation. The simple process of redissolving the alcohol insoluble material in alkali followed by reprecipitation with acid results in a further increase in alcohol-soluble material.

The fulvic acid "fraction" has a straw yellow color at low pH values and turns to wine red at high pH's, passing through an orange color at a pH near 3.0. There is little doubt that biochemical constituents are also present, such as polysaccharides. The term fulvic acid should be reserved as a generic (class) name for the pigmented components of the acid-soluble fraction.

Modern-Day Concepts

Humus or soil organic matter is now known to include a broad spectrum of organic constituents, many of which have their counterparts in biological tissues. Thus, two major types of compounds can be distinguished:

1. Nonhumic substances, consisting of compounds belonging to the well-known classes of organic chemistry. Included with this group are such constituents as amino acids (Chapter 3), carbohydrates (Chapter 6), and lipids (Chapter 7)
2. Humic substances, a series of high-molecular-weight, yellow to black substances formed by secondary synthesis reactions (Chapter 8). They can be generally characterized as being rich in oxygen-containing functional groups, notably COOH but also phenolic and/or enolic OH, alcoholic OH, and C=O of quinones

Compounds in the second group are distinctive to the soil or sediment environment—they have characteristics that are dissimilar to polymers from other natural sources. The two groups are not easily separated, because some nonhumic substances, such as carbohydrates, may be bound covalently to the humic matter. Humus probably contains most, if not all, of the biochemical compounds synthesized by living organisms.

Some Definitions

As noted earlier, the term humus is generally used synonymously with soil organic matter and refers to those organic substances that do not occur in the form of plant residues or their partial decay products (i.e., "light" fraction, see Chapter 1). The "light" fraction is sometimes included with the definition of "soil organic matter," in which case the term "humus" has a restricted meaning and refers to humic substances plus resynthesis products of microorganisms that have become stabilized, and are thus an integral part of the soil. Intermediate between the two groups is the organic material present in living microorganisms (the microbial biomass).

In this discourse, the terms humus and soil organic matter will be used as recommended by Waksman,[4] namely, synonymously and to exclude the remains of plant residues and their partial decomposition products. Other definitions are listed in Table 2.3 but it should be noted that strict adherence will not always be possible. This applies in particular to the term fulvic acid, which has often been used to designate the alkali-soluble, acid-soluble fraction, rather than the colored material contained therein (i.e., "generic" fulvic acid).

A variety of terms have been used from time to time to describe specific types of humus in sedimentary environments.[23] The brown or gray, pulpy, coprogenic substance formed from the dead remains of microscopic plants and animals in the top muds of eutrophic lakes and marshes is called "copropel."

TABLE 2.3 Definitions

Term	Definition
Litter	Macroorganic matter (e.g., plant residues) that lies on the soil surface
Light fraction	Undecayed plant and animal tissues and their partial decomposition products that occur within the soil proper and that can be recovered by flotation with a liquid of high density (see Chapter 1)
Soil biomass	Organic matter present as live microbial tissue
Humus	Total of the organic compounds in soil exclusive of undecayed plant and animal tissues, their "partial decomposition" products, and the soil biomass
Soil organic matter	Same as humus
Humic substances	A series of relatively high-molecular-weight, yellow to black colored substances formed by secondary synthesis reactions. The term is used as a generic name to describe the colored material or its fractions obtained on the basis of solubility characteristics. These materials are distinctive to the soil (or sediment) environment in that they are dissimilar to the biopolymers of microorganisms and higher plant (including lignin)
Nonhumic substances	Compounds belonging to known classes of biochemistry, such as amino acids, carbohydrates, fats, waxes, resins, organic acids, etc. Humus probably contains most, if not all, of the biochemical compounds synthesized by living organisms
Humin	The alkali insoluble fraction of soil organic matter or humus
Humic acid	The dark-colored organic material that can be extracted from soil by dilute alkali and other reagents and that is insoluble in dilute acid
Hymatomelanic acid	Alcohol-soluble portion of humic acid
Fulvic acid fraction	Fraction of soil organic matter that is soluble in both alkali and acid
Generic fulvic acid	Pigmented material in the fulvic acid fraction

A black mass of humus occurring in deeper hypolimnetic areas of lakes and bays is termed "sapropel." A pondweed type of sapropel believed to originate from cellulose-rich plants is called "förna." A deposit in dystrophic lakes consisting of an allochthonous precipitate of humic acids and detritus is known as "dy." Marine slime resulting from settled planktonic detritus has been referred to as "pelogoea," and an amorphous, gummy concentration of humic substances beneath or within certain peat bogs is "dopplerite."

"Mor" is a type of forest humus in which there is practically no mixing of surface organic matter with the mineral portion of the soil. When the leaf litter becomes mixed with mineral matter so that decomposition occurs in the superficial layer of the soil instead of on the surface the material is referred to as "mull."

ASSOCIATIONS OF ORGANIC MATTER IN SOIL

The bulk of the organic matter in soils occurs in water insoluble forms. Methods for extraction must take into account the various ways in which organic matter is retained, as follows:

1. As insoluble macromolecular complexes.
2. As macromolecular complexes bound together by di- and trivalent cations, such as Ca^{2+}, Fe^{3+}, and Al^{3+}
3. In combination with clay minerals, such as through bridging by polyvalent cations (clay–metal–humus), H-bonding, and in other ways as discussed in Chapter 18
4. As organic substances held within the interlayers of expanding-type clay minerals. This organic matter is not solubilized by conventional extraction, but is released by destruction of clay with HF

Item 1 is particularly important in peat and other organic-rich layers, where clay and metal complexes are present in very low amounts in relation to the humus component. A typical example of organic matter bound by polyvalent cations (item 2) is the illuvial (B) horizon of Spodosols (see Chapter 20). This horizon is a rich source of fulvic acid, which can be readily separated from the sesquioxides to which it is bound by mild extractants. Allophanic materials, which have the general structure $xSiO_2 \cdot AlO_3 \cdot yH_2O$, are strong adsorbents of humic substances, which accounts for the exceptionally high levels of organic matter in soils derived from volcanic ash. In agricultural soils of the temperate zone, most of the organic matter is intimately bound to clay minerals (item 3).

As noted below, the usual procedure for recovering material bound to polyvalent cations is by extraction with a chelating agent; more drastic extractants are required for extracting clay-bound organic matter.

EXTRACTION METHODS

A résumé of methods used to recover organic substances from soil is given in Table 2.4. A variety of techniques have been employed, depending upon the nature of the material to be examined. Thus, nonpolar compounds (fats, waxes, resins, etc.) can be extracted with such organic solvents as hexane, ether, carbon tetrachloride, alcohol–benzene mixtures, and others. Hydrolysis procedures have been used for isolating individual monomers, such as amino acids

TABLE 2.4 Reagents Used for Extraction of Organic Constituents from Soil

Type of Material	Extractant	Organic Matter Extracted (%)
Humic substances[a]	Strong bases	
	NaOH	To 80%
	Na_2CO_3	To 30%
	Neutral salts	
	$Na_4P_2O_7$, NaF, organic	To 30%
	acid salts	To 30%
	Organic chelates	
	Acetylacetone	To 30%
	Cupferron	
	8-hydroxyquinoline	
	Formic acid (HCOOH)	To 55%
	Acetone–H_2O–HCl solvent	To 20%
Hydrolyzable compounds		
1. Amino acids, amino sugars	Hot 6N HCl	25–45%
2. Sugars	Hot 1N H_2SO_4	5–25%
Polysaccharides	NaOH, HCOOH, hot water	<5%
Clay-bound biochemicals	HF	5–50%
"Free" biochemicals (amino acids, sugars)	H_2O, 80% alcohol, ammonium acetate	1%
Fats, waxes, resins	Usual "fat" solvents	2–6%

[a]Considerably higher amounts of organic matter can be extracted from Spodosol B horizons with most reagents.

and sugars. The approaches used for isolating humic substances will be emphasized herein, although nonhumic substances of various types are invariably extracted as well.

The ideal extraction method is one which meets the following objectives:

1. The method leads to the isolation of unaltered material
2. The extracted humic substances are free of inorganic contaminants, such as clay and polyvalent cations
3. Extraction is complete, or nearly so, thereby insuring representation of fractions from the entire molecular-weight range
4. The method is universally applicable to all soils

It is safe to say that these objectives have yet to be realized. In addition, there is the troublesome problem of removing coadsorbed organic impurities (carbohydrates, proteinaceous compounds) from the "true" humic substances. The subject of extraction and fractionation of soil organic matter has been the topic of several reviews.[24-26]

Alkali Extraction

Sodium hydroxide and Na_2CO_3 solutions of 0.1 to $0.5N$ concentration in water and a soil to extractant ratio of from 1:2 to 1:5 (g/mL) have been widely used for recovering organic matter. Repeated extraction is required to obtain maximum recovery. The solubility of humic substances in alkali is believed to be caused by disruption of bonds holding organic material to inorganic soil components and conversion of acidic components to their soluble salt forms: salts of monovalent cations are soluble while those of di- and trivalent cations are insoluble. Leaching the soil with dilute HCl, which removes Ca and other polyvalent cations, increases the efficiency of extraction of organic matter with alkaline reagents. However, a certain amount of organic matter, normally less than 5 percent of the total for surface soils but somewhat more for certain subsoils, is removed in the process, part of which may be in the form of low-molecular-weight fulvic acids. If desired, the acid extract can be saved and used to neutralize the alkali-soluble material during recovery of humic acids. As a general rule, extraction of soil with 0.1 or $0.5N$ NaOH leads to recovery of approximately two-thirds of the soil organic matter.

Undesirable features of alkali extraction are as follows:

1. Alkali solutions dissolve silica from the mineral matter and this silica contaminates the organic fractions separated from the extract
2. Alkali solutions dissolve protoplasmic and structural components from fresh organic tissues and these becomes mixed with the humified organic matter
3. Under alkaline conditions, autoxidation of some organic constituents occurs in contact with air, both during extraction and when the extracts are allowed to stand
4. Other chemical changes can occur in alkaline solution, including condensation between amino acids and the C=O group of reducing sugars or quinones to form humic-type compounds through the browning (Maillard) reaction

The more alkaline the solution and the longer the extraction period the greater will be the chemical changes.[27] As can be seen from Table 2.5, little O_2 is consumed by extraction under neutral conditions (pH 7.0) but consumption becomes appreciable at pHs above 8.0. The amount of organic matter extracted from soil with caustic alkali increases with the concentration of NaOH and time of extraction, a result that may be due to slow depolymerization of high-molecular-weight complexes. Swift and Posner[28] found that the breakdown of humic acids under alkaline conditions was greatly enhanced in the presence of O_2; other changes produced by O_2 included increases in the cation-exchange capacity and in the oxidation state of the humic acid. To minimize chemical changes due to autoxidation, all steps should be carried out in the presence of an inert gas (e.g., N_2).

TABLE 2.5 Uptake of O_2 During Extraction of Soil Organic Matter as Influenced by pH of the Extracting Solution[a,b]

Reagent	O_2 Uptake, mm^3/0.2 g	
	Soil A	Soil B
0.5M NaOH	896	712
0.5M Na$_2$CO$_3$, pH 10.5	56	71
0.2M Na-citrate, pH 7.0	39	58
0.1M Na$_4$P$_2$O$_7$		
pH 7.0	7	37
pH 8.0	12	—
pH 9.0	31	52

[a]From Bremner.[27]
[b]The reaction time was 7 hours.

Mild Extractants

Several milder and more selective extractants have been recommended in recent years as alternatives for the classical extraction with strong alkali.[24-26] Included are salts of complexing agents, such as sodium pyrophosphate (Na$_4$P$_2$O$_7$), organic complexing agents in aqueous media (e.g., acetylacetone), dilute acid mixtures containing HF, and organic solvents of various types. Whereas less alteration of organic matter may result, these extractants are much less effective than alkali hydroxides in removing organic matter, the main exception being the illuvial (B) horizon of the Spodosol. As was the case with alkali extraction, recovery of organic matter can often be increased by pretreating the soil with mineral acids to remove carbonates (HCl) or silicates (HCl-HF mixtures).

For certain investigations, a mild extractant is definitely preferred; for others, a more complete extraction with caustic alkali may be desired even though this will be achieved at the expense of some changes in the organic matter. Many investigators are now using a sequence of extractants in which part of the organic matter is recovered by a mild reagent (such as neutral Na$_4$P$_2$O$_7$) prior to alkali extraction.

Sodium Pyrophosphate (Na$_4$P$_2$O$_7$) and Other Neutral Salts In many soils, Ca and other polyvalent cations (e.g., Fe and Al) are responsible for maintaining organic matter in a flocculated and insoluble condition. Accordingly, reagents that inactivate these cations by forming insoluble precipitates, or soluble coordination complexes, lead to solubilization of the organic matter in a medium containing Na$^+$, K$^+$, or NH$_4^+$ ions. Such reagents as ammonium oxalate, Na$_4$P$_2$O$_7$, and salts of weak organic acids have been used for this purpose. The Na-salt of the chelating agent EDTA would appear to be a good extractant of organic matter but little work has been done on this reagent because neither the C or N content of the soil extracts can be used to measure the quantity of organic matter extracted.

Of the various reagents, aqueous $Na_4P_2O_7$ at 0.1 to 0.15M concentration has been the most widely used. As noted earlier, the amount of organic matter recovered (<30 percent) is considerably less than with caustic alkali but less alteration occurs. To minimize chemical modification of the humic material, extraction should be carried out at pH 7.0. Acid pyrophosphate has been used as an extractant for fulvic acids.[29]

Reactions leading to the extraction of organic matter by $Na_4P_2O_7$ was postulated by Aleksandrova[30] to be as follows:

$$R(COO)_4Ca_2 + Na_4P_2O_7 \rightarrow R(COONa)_4 + Ca_3P_2O_7\downarrow \quad [1]$$

$$2[RCOOX(OH)_2](COO)_2Ca + Na_4P_2O_7 \rightarrow 2[RCOOX(OH)_2](COONa)_2 + Ca_2P_2O_7\downarrow \quad [2]$$

where X is a trivalent cation. Humic acids recovered from soil by extraction with $Na_4P_2O_7$ usually contain Fe and Al as contaminants.

Anhydrous Formic Acid Extensive research on the extraction of soil organic matter with formic acid has been conducted by Tinsley and his students. This work, summarized by Tinsley and Walker[31] shows that under certain circumstances up to 55 percent of the organic matter in mineral soils and as much as 80 percent of that in composts can be extracted with formic acid containing LiF, LiBr, or HBF_4 (fluoroboric acid) to disrupt H-bonds or complex metal ions. The latter reagent would also be expected to release organic matter bound to mineral particles. Some typical results are given in Fig. 2.1.

Advantages of anhydrous formic acid for extraction of organic matter is that it is a polar compound that exhibits neither oxidizing nor hydrolytic properties. Furthermore, formic acid is a good solvent for a wide variety of compounds, including polysaccharides. Large quantities of Ca, Fe, Al, and other inorganic components are dissolved from the soil along with organic matter and thus far it has not been possible to remove the inorganic material completely, even by repeated precipitation with diisopropyl ether containing acetyl chloride to keep the metal ions in solution. The work of Jones and Parsons[32] indicates that formic acid is most efficient with soils where much of the organic matter is only partially humified.

Organic Chelating Agents Organic compounds such as acetylacetone, cupferron, and 8-hydroxyquinoline, which form chelate complexes with polyvalent metal ions, have been used for extracting illuvial organic matter from Spodosols. The organic matter in the B horizon of these soils occurs as complexes with Fe and Al and the complexing of these metals by chelating agents releases

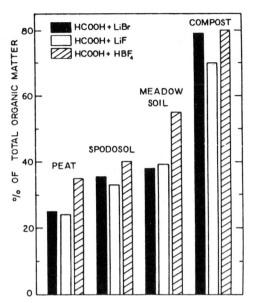

Fig. 2.1 Percentage extraction of organic matter from four soils with formic acid. Adapted from Tinsley and Walker[31]

the organic matter to soluble forms. Organic chelating agents are rather ineffective for extracting organic matter from other soil types and the humic material thus isolated still contains residual metal cations. A chelating resin has also been used as an extractant of soil organic matter.

Dilute Mineral Acids Except for acid mixtures containing HF, very little organic matter can be extracted from soil with dilute mineral acids. Hydrofluoric acid probably releases organic matter from inorganic combinations through dissolution of hydrated silicate minerals and the formation of complexes with Fe and Al.

Acid mixtures containing HF have been used for the release of clay-fixed NH_4^+ from soils. In the process, some organic N is also solubilized, as shown in Fig. 2.2. The percentage of the soil organic N extracted with HF increases with depth and can account for over 50 percent of the organic N in some subsoils.[33]

Other Reagents and Approaches Porter[34] found that organic colloids could be extracted quickly and easily from soils using aqueous HCl mixtures of such solvents as acetone, dioxane, cyclopentanone, dimethylformamide, and others. The HCl was believed to facilitate the separation of organic colloids from mineral portions of the soil by breaking polyvalent salt bridges. After breaking

40 EXTRACTION, FRACTIONATION, GENERAL CHEMICAL COMPOSITION

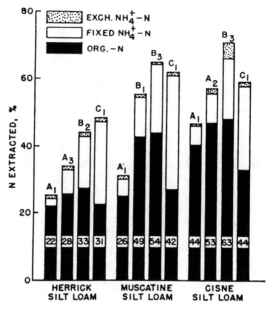

Fig. 2.2 Organic N and NH_4^+ removed from three soil types by extraction with $2.5N$ HF:$0.1N$ HCl. Values in the solid portion of the bars represent the percentage recovery of organic N. From Stevenson et al.[33]

of ionic bonds, the organic solvents were believed to compete with soil organic colloids for sorption sites. Up to 23 percent of the total C could be extracted from soils with an acetone–H_2O–HCl solvent mixture. The acetone was easily removed at low temperatures and by dialysis. However, the extracted organic matter still appears to contain appreciable amounts of contaminating mineral matter (unpublished observations).

Acidified dimethyl sulfoxide has been shown to be effective in the isolation of humic substances from soil, as well as components from the humin fraction.[24] Another promising approach is by superficial fluid extraction with an organic solvent;[35,36] only nonionic molecules are recovered and it is necessary to hydrogen saturate the soil in order to improve extraction. Schnitzer and Preston[36] found that superficial gas extraction of an Aquoll soil with ethanol and ethanol/H_2O mixtures (250°C at a pressure of 14.0 MPa for two hours) led to removal of as much as 50 wt/wt % of organic matter in the initial soil sample. The technique of supercritical fluid extraction with aqueous organic solvents has been used for recovery of organic matter from peat.[37]

Hayes[24] has provided a comprehensive review of the properties of reagents that are good solvents for soil organic matter, including dimethyl sulfoxide and other organic extractants. A major disadvantage of S- or N-containing solvents is the difficulty of completely removing the solvent from the final product.

FRACTIONATIONS BASED ON SOLUBILITY CHARACTERISTICS

Fractionation is a necessary step in the characterization of soil organic matter. The primary objective of fractionation is to facilitate the application of analytical techniques by reducing heterogeneity of the isolated material.

As noted earlier, the organic matter extracted from soils or sediments with such reagents as $0.5N$ NaOH and $0.1M$ $Na_4P_2O_7$ is usually fractionated on the basis of solubility characteristics. The fractions commonly obtained include "humic acid," "fulvic acid" and "humin;" other fractions include "hymatomelanic acid" and the "gray and brown humic acids" (see Table 2.2 for definitions).

A complete fractionation scheme is given in Fig. 2.3. Included in the scheme is the recovery of "generic" fulvic acid from the fulvic acid "fraction" by sorption/desorption from an XAD-8 resin (discussed later). Instructions for extraction and recovery of humic acids are as follows.[26]

Place a 40-g sample of acid-washed ($0.1N$ HCl) soil (20 g for peat and organic rich samples) into an 8-oz polyethylene centrifuge bottle, add 200 ml. of $0.5N$ NaOH solution, and stopper the bottle tightly with a rubber stopper. Shake the mixture for 12 hours on a mechanical shaker, wash down the sides of the bottle with distilled water, and centrifuge the mixture. Decant off the dark-colored supernatant liquor, filter it through glass wool to remove suspended plant material, and adjust the pH of the solution to about 1.0 with concentrated HCl. All

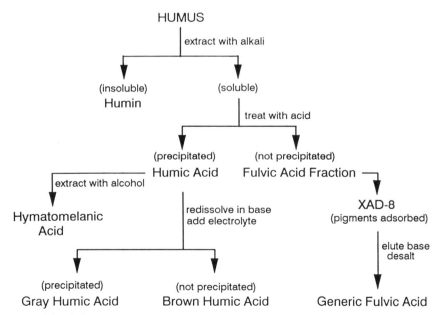

Fig. 2.3 Scheme for the fractionation of soil organic matter (humus)

steps should be carried out under a blanket of N_2 gas so as to minimize chemical changes in the extracted organic matter.

Add an additional 200 ml of $0.5N$ NaOH solution to the soil, shake the mixture for 1 hour, and repeat the centrifuging and decanting procedures. Disperse the residue in 200 ml of distilled water, centrifuge the mixture, and add the supernatant liquor to the previous extracts. Adjust the pH of the resulting solution to 1.0 with concentrated HCl, and allow the humic acid to settle. If a larger quantity of soluble organic matter is required, repeat the extraction procedure with a fresh sample of soil.

Siphon off the excess supernatant liquor (fulvic acid) from the acidified extract, transfer the remainder of the suspension to an 8-oz polyethylene bottle, and centrifuge off the humic acid. Redissolve the humic acid in $0.5N$ NaOH solution, reprecipitate it by adjusting the pH of the solution to 1.0 with concentrated HCl, and then centrifuge out the humic acid. In each case, add the supernatant liquor to the original acid filtrate. Repeat the purification procedure a second time, and then wash the humic acid precipitate several times with small quantities of H_2O to remove residual HCl. Dry the humic acid (preferably by freeze-drying), and grind it to a brown powder.

An alternate procedure for recovery of humic acids is to subject the wet precipitate to freezing and thawing to exude water (and residual HCl), which can then be removed by centrifugation. A technique preferred by the author is to dialyze a suspension of the humic acid against deionized H_2O until free of chlorides, followed by freeze-drying. Air-dried preparations of humic substances are brittle and difficult to redissolve; those that have been freeze-dried occur as soft, pliable powders that dissolve readily in neutral or slightly alkaline aqueous solvents. The humic acids should be stored in amber colored bottles to minimize changes induced by photochemical reactions.

Crude humic acids, as obtained by the above procedure, invariably contain inorganic matter that must be removed prior to chemical characterization. Procedures for reducing the ash content, and for removing coadsorbed nonhumic substances, are discussed in a later section.

Several factors influence the separations obtained on the basis of solubility characteristics. Sequi et al.[38] found that metals contained in soil organic matter extracts affected the amount of organic matter precipitated by acid and thereby the humic acid/fulvic acid ratio. The concentration of NaOH used for extraction may also affect this ratio.[39] The effect of increasing concentration of NaOH (and $Na_4P_2O_7$) may be due to enhanced disruption of intermolecular forces binding "fulvic acids" to "humic acids," such as by H-bonding or possibly ester-type linkages.

Kumada[40] has devised a classification scheme for humic acids based on spectral characteristics (see Chapter 13). Four types were recognized: A, B, P, and R_p. A-type humic acids have high aromatic C contents and are associated with grassland soils. P-type humic acids were associated with forest soils; R_p-type represented an early stage of humification. Aromaticity followed the order: $A > B \sim P > R_p$.

In other work, Kumada and Sato[41] recovered a "green humic acid" from an alkaline extract of a Spodosol. Chromatography of the extract on a cellulose column gave a brown band that descended rapidly through the column and a green band that remained in the upper part of the column. The fraction producing the green band was termed "green humic acid." A fungal origin of this material has been suggested.[42]

High molecular weight biopolymers and generic fulvic acids in the acid filtrate from removal of humic acids can be recovered by dialysis and subsequent chromatography on an XAD-8 resin column, as described later. Generic fulvic acids can also be recovered by sorption–desorption from activated charcoal, although extensive elution with base is required for recovery.[43]

Fulvic acids, if not handled properly, can acquire properties characteristic of humic acids, such as by prolonged standing at elevated pHs or during concentration at moderate high temperatures. Acid-induced polymerization can also occur during storage in acidic solutions.[44]

The Humin Fraction

As commonly defined, the humin fraction is that portion of soil humus that remains behind after extraction of the soil with dilute alkali. The humin fraction may consist of one or more of the following:

1. Humic acids so intimately bound to mineral matter (e.g., clay) that the two cannot be separated
2. Highly condensed (humified) humic matter with a high C content (>60 percent) and thereby insoluble in alkali
3. Fungal melanins
4. Paraffinic substances. Recent work using ^{13}C–NMR (see Chapter 11) indicates that a significant amount of the alkali insoluble organic matter consists of paraffinic substances, some of which cannot be easily extracted with solvents commonly used for recovery of lipids from soil (Chapter 7)

To concentrate humin by removal of mineral matter (clay), the residue from alkali extraction is often treated repeatedly with an HF–HCl acid mixture to remove the mineral component.

Soil humin has been divided into two fractions based on separations obtained by ultrasonic disruption of soil microaggregates.[45] The humin isolated by the ultrasonic treatment (suspension following centrifugation) was referred to as "inherited humin" and was believed to consist of altered lignin-like polymers and/or microscopic (subcellular) particles of plant origin retained within microaggregates.

In a recent study, Rice and MacCarthy[46] subdivided humin into four separate fractions by extraction with an organic solvent (methylisobutylketone) that forms an immiscible phase with water. The four fractions were:

1. A solvent-extractable lipid fraction referred to as bitumen
2. A humic acid-like fraction referred to as bound humic acid
3. An unextractable lipid fraction referred to as bound lipids
4. A mineral component referred to as the insoluble residue

Two of the four fractions (1 and 3) are constituents of lipids, which are discussed in Chapter 7. Preextraction of the soil with an organic solvent to remove lipids would be expected to reduce the amounts of lipid material in the humin fraction.

Humic Substances as a System of Polymers

A particularly interesting concept of the nature of humic substances is that the various fractions obtained on the basis of solubility characteristics are part of a heterogeneous mixture of molecules, which, in any given soil, range in molecular weight from as low as several hundred to perhaps over 300,000 and that exhibit a continuum of any given chemical property.[47,48] Some postulated relationships are depicted in Fig. 2.4, where it can be seen that C and O contents, acidity, and degree of polymerization all change systematically with increasing molecular weight. The low-molecular-weight fulvic acids have higher O contents but lower C contents than the high-molecular-weight humic acids. Also, the total acidities of fulvic acids (usual range 900 to 1400 cmole/kg) are considerably higher than for humic acids (usual range 500 to 870 cmole/kg).

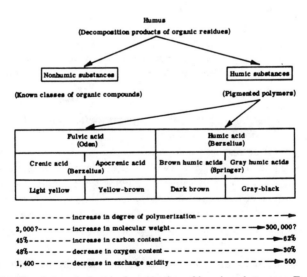

Fig. 2.4 Classification and chemical properties of humic substances. From Stevenson and Butler[48] as adapted from Scheffer and Ulrich.[47]

As will be shown later, both COOH and acidic OH groups (presumed to be phenolic OH) contribute to the acidic nature of humic substances, with COOH being the most important. In as much as fulvic acids are less susceptible to precipitation with acids and polyvalent cations than humic acids, they are the constituents usually responsible for the brownish-yellow color of many natural waters.

Evidence in support of the view that the humic substances of any given soil are part of a system of polymers and that those of soils in general conform to this pattern has been given by Kononova.[8] They include results of comparative studies showing consistent differences in elemental composition (C, H, O), optical properties, behavior towards polyelectrolytes, exchange acidities, electrophoretic properties, molecular weight characteristics, and others. These numerous studies show that humic and fulvic acids are not only extremely heterogeneous but that each fraction contains intermediate forms that have characteristics overlapping the other.

As will be shown in later chapters, differences between humic and fulvic acids cannot be explained entirely by differences in molecular weight and content of acidic functional groups. Notwithstanding, the concept of humic substances as a system of polymers has merit and is useful in showing the heterogeneity and range in chemical and physical properties of humic substances.

The Humic Acid/Fulvic Acid Ratio

The percentage of the humus that occurs in the various humic fractions varies considerably from one soil type to another. Results recorded by Kononova[8] show that the humus of forest soils (Alfisols, Spodosols, and Ultisols) generally have lower humic acid/fulvic acid ratios than the humus of peat and grassland soils (Mollisols). Another interesting point is that chemical differences also exist. The humic acids of Spodosols and Alfisols are less aromatic in nature, and more closely resemble fulvic acids, than do the humic acids of Mollisols. Also, the humic acids of forest soils have lower C but higher H contents than those of grassland soils, and optical density in the visible region is lower. As noted earlier, humic acids of forest soils are mostly of the brown humic-acid-type; those of grassland soils are of the gray humic-acid-type.

The forms of humus in the soils of four great soil groups are illustrated in Fig. 2.5, where FA refers to the fulvic acid "fraction." Humus of the Mollisol is shown to be somewhat richer in humic acids than the other soils; humus of Spodosol B horizons is rich in components associated with the fulvic acid "fraction." A unique feature of the Vertisol is its high content of alkali-insoluble organic matter (i.e., humin).

PURIFICATION

Organic matter, as normally recovered from soil, contains considerable quantities of inorganic constituents (salts, Si, sesquioxides, and clay) that must be

46 EXTRACTION, FRACTIONATION, GENERAL CHEMICAL COMPOSITION

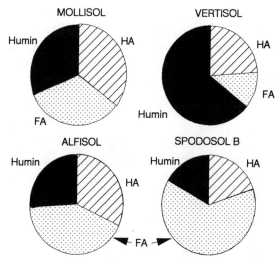

Fig. 2.5 Distribution of humus forms in the soils of four great soil groups. Values for FA are for the fulvic acid "fraction"

eliminated before characterization studies can be initiated. For humic substances, there is the additional problem of removing organic impurities, such as carbohydrates and proteins.

Humic Acids

Removal of Inorganic Contaminants The humic acid fraction, extracted from soil with alkali or neutral salts and precipitated under acidic conditions, nearly always contain a considerable amount of inorganic material. Crude humic acids from grassland soils appear to have especially high ash content, often exceeding 30 percent. Considerable reduction in ash can be achieved by repeated precipitation with mineral acids and by passage through ion-exchange resins. An effective way for removing clay is by high-speed centrifugation after redissolving the humic acid in dilute alkali and adjusting the pH to 7.0.

Success in obtaining humic acids with low ash content has been attained by treatment with acid solutions containing HF. A typical procedure is shown in Fig. 2.6. The crude humic acid material is first dissolved in base (pH 7), the ionic strength is adjusted to 0.1 with respect to KCl, and the sample is centrifuged to remove clay. The solution is then acidified to precipitate humic acids and the sample is treated several times with a dilute acid solution consisting of $0.3N$ HF : $0.1N$ HCl. The sample is then dialyzed extensively against distilled water and freeze dried. To ensure removal of metal ions, the sample can be dialyzed in the presence of a cation-exchange resin in the H^+-form, which is contained in a second dialysis bag.[49]

The effectiveness of HF in reducing the ash content of humic acids can be

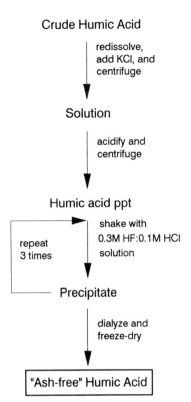

Fig. 2.6 Scheme for the removal of inorganic impurities from humic acid

explained by its ability to dissolve hydrated clay minerals and to form complexes with di- and trivalent cations. As illustrated earlier in reaction [2], part of the humic material extracted from soil with $Na_4P_2O_7$ (and other extractants) may contain metal ions in coordination linkages. Reduction in ash is usually accompanied by significant loss of humic matter,[38] possibly by liberation of low-molecular-weight fulvic acids bound to the unpurified humic acid through metal linkages.

Removal of Organic Impurities Separation of coprecipitated or coadsorbed nonhumic substances (e.g., proteins and carbohydrates) from crude humic acid preparations is more easily accomplished on paper than in the laboratory. Because these constituents are not easily removed by mild separation methods, it is sometimes thought that they are covalently bound to and an integral part of humic acids.

As implied above, attempts to separate organic impurities from humic acids have not been entirely successful. Methods for removing carbohydrate and protein constituents include hydrolysis with mineral acids, gel filtration, and phenol extraction. In the latter case, treatment of humic acid with aqueous phenol has led to the separation of a protein-rich component.[50,51]

Lipids (hydrocarbons, fats, waxes, resins, etc.) can be partially removed with such solvents as ether or alcohol–benzene. Efficiency of extraction can be improved by supercritical gas extraction, such as with n-pentane.[52,53] Gases in the critical state have physical characteristics intermediate between those of the corresponding gases and liquids, thereby giving the solvent exceptional ability to penetrate complex organic matrices to solubilize specific compounds otherwise inaccessible or only slightly soluble in the solvent used.

Boiling humic (and fulvic) acids with water has been reported to extract polysaccharides, polypeptides, and small amounts of phenolic acids and aldehydes.[54] Adsorbed materials can also be removed by acid hydrolysis, such as with 6N HCl. Among objections to the latter procedure is that large weight losses and chemical changes can occur. As indicated by the data reported in Table 2.6, hydrolysis of humic acid with 6N HCl resulted in a weight loss of over 40 percent. However, the treatment had little effect on C and H content, content of oxygen-containing functional groups, or the E_4/E_6 ratio. The latter is a ratio between absorbance at 465 and 665 nm and is regarded as an index of molecular condensation (see Chapter 13). Drastic changes were observed when the fulvic acid was hydrolyzed with 6N HCl. The C content was increased (from 49.5 to 53.9 percent) while total acidity decreased (from 1240 to 890 cmole/kg), apparently due to decarboxylation. Schnitzer[55] concluded that hydrolysis procedures were satisfactory for the purification of humic acids but not fulvic acids.

Fractionation Crude humic acids, as recovered by precipitation under acidic conditions, represent a mixture of molecules covering a wide range in molecular weights. The subfractions known as hymatomelanic acid (alcohol extraction)

TABLE 2.6 Effects of 6N HCl Hydrolysis on Analytical Characteristics of Humic and Fulvic Acids[a]

Characteristics	Humic Acid		Fulvic Acid	
	Untreated	After Hydrolysis	Untreated	After Hydrolysis
C (%)	57.2	58.1	49.5	53.9
H (%)	4.4	3.9	4.5	4.1
N (%)	2.4	0.7	0.8	0.1
Total acidity (meq/g)	8.1	7.0	12.4	8.9
Carboxyl (COOH) (meq/g)	4.8	4.3	9.1	4.9
Phenolic OH (meq/g)	3.3	2.7	3.3	4.0
Ketonic C=O (meq/g)	2.2	2.9	—	—
Quinoid C=O (meq/g)	2.1	2.2	—	—
E_4/E_6 ratio	4.3	4.0	7.1	4.3
Ash (%)	7.9	0.3	2.0	0.0

[a]From Schnitzer.[55]

and brown- and gray humic acids (partial precipitation with electrolyte under alkaline conditions) represent early attempts to obtain more homogeneous preparations.

In recent years, additional separation methods have been applied, including fractionation using a nearly linear pH gradient,[56] hydrophobic interaction chromatography,[57] and stepwise precipitation with $(NH_4)_2SO_4$,[58] dimethylformamide in water,[59] and alcohol in basic medium.[60] Size separation by gel filtration has also been applied and is discussed in Chapter 14.

Fulvic Acids

In most characterization studies of soil organic matter, the fulvic acid fraction is discarded; only the material precipitated with acid (humic acids) is saved for examination. The major reason for this is that components of the fulvic acid fractions cannot easily be recovered from the acid filtrate following removal of humic acids. The filtrate not only contains appreciable amounts of mineral matter dissolved during extraction, but of salts formed by neutralization of the base used for extraction (NaOH + HCL → NaCl). Salts can be removed by dialysis but with considerable loss of low-molecular-weight organic constituents, including generic fulvic acids.

True ("generic") fulvic acids have less carbohydrate and peptide constituents than the so-called fulvic acid "fraction," although "generic" fulvic acid may contain saccharides intimately bound to the macromolecule and sometimes classed as "pseudopolysaccharides." Unbound polysaccharides have been recovered from the fulvic acid "fraction" by removal of humic substances through adsorption on a Polyclar AT (polyvinylpyrrolidone) column, followed by gel filtration of the filtrate (see Chapter 6).

In early work, generic fulvic acids were recovered from the acidified solution, concentrated, and partially purified by selective adsorption–desorption on activated charcoal.[43] A small amount of dark-colored organic matter was recovered from the charcoal by elution with 90 percent acetone; the greater part required elution with strong base. A limitation of this approach is that a considerable volume of base was required for elution and that substantial amounts of humic material were irreversibly adsorbed by the charcoal and thus not recovered. Watanabe and Kuwatsuka[61] separated the fulvic acid fraction into soluble and insoluble humic and nonhumic substances by extraction with ethanol.

A more complex fractionation scheme using activated charcoal has been devised by Dragunov and Murzakov.[62] In all, 13 separate components were recovered from the fulvic acid fraction by successive elution of the charcoal with 2 percent NH_3, ethanol, ethanol–benzene, acetone, water, and acetone–water followed by rechromatography of some of the extracts, also on activated charcoal. It is unknown as to whether the 13 components represent distinct and meaningful components of fulvic acid.

A relatively new approach for recovery of generic fulvic acids is through adsorption–desorption from Amberlite XAD-8, a macroreticular, nonionic acrylic ester polymer with excellent sorption properties for humic substances. At low pH, weak acid polyelectrolytes such as humic and fulvic acids are protonated and are thereby adsorbed on the resin; at high pH, acidic groups (i.e., COOH) are ionized and desorption occurs.

In using the method, the acid filtrate remaining after removal of humic acids is passed through a column of XAD-8 resin (see Fig. 2.3). The resin binds the H^+-form of the fulvate macromolecules; inorganic salts, other macromolecular organic substances (e.g., polysaccharides), and low molecular weight biochemical compounds pass through. The adsorbed humic material is then recovered by elution with base (or other suitable solvent), desalted using a cation-exchange resin, and freeze-dried.[63]

The XAD-8 procedure yields fulvic acids of low ash contents and that are essentially free of nonhumic substances. Accordingly, the method permits a distinction to be made between true ("generic") fulvic acids and the fulvic acid "fraction." The method has also been used for the fractionation of humic acids.[64-65] In the study of MacCarthy et al.,[64] a peat humic acid was separated into two fractions by pH-gradient elution from the XAD-8 resin. Fraction 1, which was eluted in the pH range 4-6, had a much higher COOH content than fraction 2, which was eluted in the pH range 8-11.

A scheme found by the author for obtaining more homogeneous preparations from the fulvic acid "fraction" is shown in Fig. 2.7. The acid filtrate following removal of humic acids is first dialyzed extensively against distilled water. The nondialyzable part (material retained inside the dialysis bag) consists of a mixture of high-molecular-weight fulvic acids, polysaccharides, and possibly proteinaceous constituents; the dialysate contains lower-molecular-weight fulvic acids, as well as amino acids, simple sugars, and miscellaneous other organic compounds. Both preparations are passed through a column of XAD-8. For the dialysate, essentially all of the dark-colored organic matter has been found to be retained by the resin; polysaccharides, hydrophilic biochemical compounds and inorganic constituents pass through and are removed in the effluent. On the other hand, retention of colored organic matter in the colloidal fraction is not quantitative (for results with stream and groundwaters, see next section).

Recovery of "generic" fulvic acids from the resin is achieved by elution with base (e.g., 0.1N NaOH), from which the inorganic cation (Na^+) can be removed using a cation-exchange resin in the H^+-form. An alternate procedure found useful by the author is to elute the XAD-8 resin first with 90% acetone and then an acetone-water-HCl solvent (95-4-1, v/v/v). Hydrophobic biochemicals (e.g., hydrocarbons, long-chain fatty acids), if present, would be partially removed and recovered in the 90 percent acetone effluent. An advantage of the acetone-water-HCl solvent over alkali for recovery of sorbed humic material is that less volume of solution is required for elution. Also, there is no need to remove the inorganic cation (i.e., Na^+) using a cation exchange resin. The effluent is diluted with water to reduce the concentration of HCl, acetone is removed by volatilization, and the sample is freeze-dried. Further size frac-

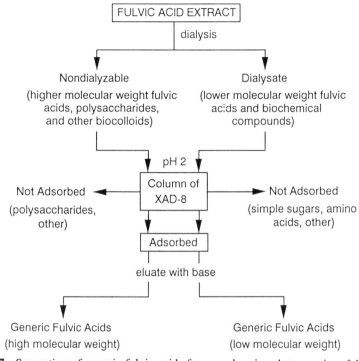

Fig. 2.7 Separation of generic fulvic acids from nonhumic substances in a fulvic acid extract of soils

tionations may be possible by use of molecular sieves or membranes containing different pore sizes, as described later in Chapter 4.

An extraction scheme specific for recovery of humic and fulvic acids from surface and groundwaters is outlined in Table 2.7.[66] The method is straightforward and simple and is suitable for recovery of humic substances from leachate waters and the soil solution. A review of the use of XAD resins to concentrate organic compounds in natural waters has been given by Daignault et al.[67]

CLASSIFICATION AND DISTRIBUTION OF DISSOLVED ORGANIC C (DOC)

A general scheme for the classification and distribution of dissolved organic C (DOC) in streams and groundwaters is outlined in Fig. 2.8.[66,68] By this scheme, DOC is classified into hydrophobic and hydrophilic acid, base and neutral fractions. "Generic" fulvic acids occur in the fraction defined as "weak hydrophilic acids" and are recovered from the XAD-8 resin by elution with dilute base. Hydrophobic neutrals, consisting of such compounds as hydrocarbons,

TABLE 2.7 Extraction Scheme Using XAD-8 to Concentrate and Recover Humic and Fulvic Acids from Natural Waters. From Aiken[66]

1. Filter sample through a 0.45 μm silver-membrane filter to remove particulate matter and lower pH to 2.0 with HCl.
2. Pass the acidified sample through a column of XAD-8: humic substances adsorb to the resin.
3. Elute XAD-8 resin in reverse direction with 0.1N NaOH; acidify immediately to avoid oxidation of humic substances.
4. Reconcentrate on a smaller XAD-8 column until the DOC concentration is greater than 500 mg C/L.
5. Adjust pH to 1.0 with HCl to precipitate humic acid and remove by centrifugation. Rinse with distilled H_2O until $AgNO_3$ test shows no Cl^- in the wash water. Dissolve humic acid in 0.1N NaOH and hydrogen saturate by passage through the H^+ form of a cation-exchange resin.
6. Reapply the fulvic acid fraction at pH 2 to the XAD-8 column. Rinse the column with 1-void volume of distilled water to remove inorganic salts and HCl and recover fulvic acids by back-elution with 0.1N NaOH.
7. Remove Na^+ and convert fulvic acids to their H^+-forms by passing the eluate through the H^+-form of a cation-exchange resin.
8. Freeze-dry the humic acid and fulvic acid samples.

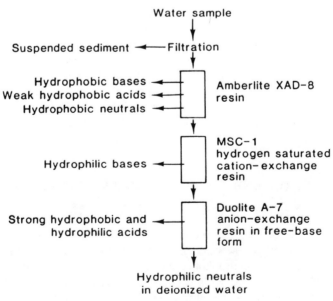

Fig. 2.8 Fractionation scheme for organic C (DOC) in natural waters. From Aiken[73]

long-chain fatty acids, and esters are recovered by elution with an organic solvent. Polysaccharides are found in the fraction designated as "hydrophilic" neutrals. The material adsorbed to the MSC-1 resin, defined as hydrophobic bases, is believed to consist in part of amphoteric proteinaceous constituents.

The various fractions are "operational definitions" in that they describe the sorption characteristics of the various components as affected by pH and the nature of the resins to which they adsorb. Among other studies, the method has been applied for the fractionation of DOC in leachate waters of forested ecosystems.[69-72]

As noted in the previous section, XAD-8 is a macroreticular, nonionic resin with excellent binding properties for humic substances under acidic conditions. The MSC-1 cation-exchange resin is a strong acid, sulfonated, polystyrene macroporous resin with high sorption capacity for organic cations. The Duolite A-7 resin is a macroporous weak-base resin with a high sorption capacity for anionic organic constituents. Inorganic anions are also retained and coeluted with organic anions in the column eluate.

A critical evaluation of the fractionation method has been given by Aiken.[73] In using the method, the following points must be kept in mind:

1. Fractionation of DOC in the sample varies depending on the volume of solution analyzed and the concentration of organic matter in the sample. The method has been defined for samples in the concentration range 2–10 mg C/L
2. Separation of the different fractions is not sharp, resulting in an overlap between fractions. Certain hydrophobic neutrals (i.e., long-chain fatty acids) are not neutral in the classic sense but are operationally defined as hydrophobic neutrals because of their interaction with the XAD-8 resin. Sorption of these compounds is not pH dependent and they are thus recovered along with the weak hydrophobic acids (i.e., generic fulvic acids) during their elution with base
3. Fulvic acid-like substances of low aromatic content, often accounting for 20 percent or more of the DOC, are not retained on the XAD-8 resin but are recovered in the hydrophilic acid fraction. Additional separations will need to be applied for the recovery of "true" fulvic acids from other components in this fraction
4. When the concentration of weak hydrophobic acids (i.e., fulvic acids) in the eluate from the XAD-8 resin is high, the acids can interact with and solubilize some of the hydrophobic neutrals, thereby incorporating them into the final product
5. Contamination of samples can arise due to impurities derived from the resin (e.g., residual monomers, artifacts of the polymerization pathway, chemical preservatives). The problem of bleed contamination is particularly serious when the concentration of DOC in the experimental sample is low

COMPARISON OF SOIL HUMIC SUBSTANCES WITH THOSE OF OTHER ENVIRONMENTS

Humic substances have traditionally been characterized by elemental composition, functional group analysis, acidic properties, molecular weight, spectral characteristics, and in other ways as noted in subsequent chapters. These same characterizations have been applied to humic substances in sediments and stream water and have shown similarities and variations between and among humic substances from the different sources. In comparison to stream water and groundwater, soil humic substances are higher in C content, of higher molecular weight (i.e., higher content of humic acids), of greater ^{14}C age, of greater percentage in aromatic C, of more intense color per C atom and of higher polysaccharide content.[74]

Humic substances from all environments have the following in common:[74]

1. An extreme complexity of molecular structures
2. An abundance and wide array of acidic functional groups
3. An ability to form complexes with metal ions and to bind xenobiotics
4. An ability to fluoresce
5. A refractory nature to microbial decay

PHYSICAL FRACTIONATION OF ORGANIC MATTER

Interest in the "physical fractionation" of organic matter arises from the observation that nutrient turnover depends not only on the kind (and amount) of organic matter in soil but on its "location" within the soil. Also, location of organic matter has implications to the physical properties of the soil. A review of physical fractionation methods has been given by Christensen[75] and Stevenson and Elliott.[76]

Physical fractionation procedures have been used in the following types of studies:

1. To recover the "light fraction," consisting largely of undecomposed plant residues and their partial decomposition products (see section on POOLS in Chapter 1)
2. To determine the types of organic matter involved in the formation of water-stable aggregates[77]
3. To establish the nature and biological significance of organic matter in organo–mineral complexes[78-83]

A typical fractionation procedure is shown in Fig. 2.9. The light fraction is separated using a liquid of high density (see Chapter 1), following which further separations are made by sedimentation. Common fractions are sand, coarse

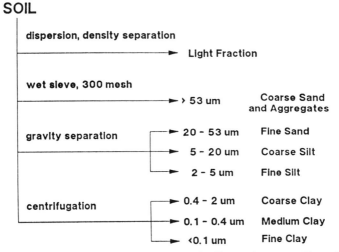

Fig. 2.9 Scheme for the physical fractionation of soil organic matter. From Stevenson and Elliot[75]

silt, fine silt, coarse clay, and fine clay, although more complete separations are often made. The various fractions have been subjected to chemical,[78-80] microbiological,[81] and microscopic examination.[82]

There is a large body of information supporting the quantitative importance and resistant nature of organic matter associated with the fine silt/coarse clay fractions. Organic matter of the fine clay, as obtained by sonication/sedimentation, has been shown to have a faster turnover rate than for the silt and coarse clay fraction during decades of cultivation.[82]

Results of incubation studies on mineralization rates of organic N, P, and S have shown that differences exist in the recalcitrance of organic matter in the various size fractions. Apparently, C and N of the silt fraction is more recalcitrant than for the clay fractions.[83]

SUMMARY

Improved methods have been developed in recent years for the extraction, fractionation, and purification of humic substances, but much more needs to be done. Both inorganic and organic impurities are invariably present and these must be removed by procedures that do not cause alteration in the extracted organic matter. An ideal method for extraction of soil organic matter has yet to be devised.

Humic substances represent a complex mixture of molecules having various sizes and shapes. The ultimate aim of research in this area is to obtain homogeneous fractions that will be more suitable than unfractionated materials for characterization studies, such as functional group analysis, molecular weight determination, infrared spectroscopy, electron spin resonance (ESR) spectrometry, nuclear magnetic resonance (NMR) spectrometry, thermal analysis, and mass spectrometric analysis of products obtained by degradative procedures. Results obtained by application of these methods are described in subsequent chapters.

REFERENCES

1. F. K. Achard, *Crell's Chem. Ann.*, **2**, 391 (1786).
2. T. de Saussure, *Recherches Chemiques sur la Végetation*, Paris, 1804.
3. J. W. Döbereiner, *Phytochemie,* **1822,** 64 (1822).
4. S. A. Waksman, *Humus*, Williams and Wilkins, Baltimore, 1936.
5. C. Sprengel, *Kastner's Arch. Ges. Naturlehre*, **8,** 145 (1826).
6. C. Sprengel, *Die Bodenkunde oder die Lehre vom Boden*, Müller, Leipzig, 1837.
7. J. J. Berzelius, *Lehrbuch der Chemie*, Wöhler, Dresden and Leipzig, 1839.
8. M. M. Kononova, *Soil Organic Matter*, Pergamon Press, Oxford, 1966.
9. G. J. Mulder, *Die Chemie der Ackerkrume*, Muller, Berlin, 1862.
10. R. Hermann, *J. Prakt. Chem.* **34,** 156 (1845).
11. F. Hoppe-Seyler, *Z. Physiol. Chem.*, **13,** 66 (1889).
12. S. Oden, *Kolloid Z.*, **14,** 123 (1914).
13. S. Oden, *Kolloidchem. Beih.*, **11,** 75 (1919).
14. O. Schreiner and E. C. Shorey, *U.S. Dept. Agr. Bureau Soils Bull.*, **53, 70, 74, 77, 80, 83, 87, 89, 90, 108** (1909-1914).
15. A. Shmook, *Pedology,* **25,** 5 (1930).
16. L. C. Maillard, *Ann. Chim. Phys.*, **5,** 258 (1916).
17. V. R. Williams, *Pochvovedenie, Vol. 1*, 1914.
18. W. Fuchs, *Kolloid Ztschr.*, **52,** 248; **53,** 124 (1930).
19. W. Fuchs, *Die Chemie der Kohle*, Springer, Berlin, 1931.
20. U. Springer, *Bodenk. PflErnähr.*, **6,** 312 (1938).
21. R. P. Hobson and H. J. Page, *J. Agr. Sci.*, **22,** 297, 497, 516 (1932).
22. H. J. Page, *J. Agr. Sci.*, **20,** 455 (1930).
23. F. M. Swain, "Geochemistry of Humus," in I. A. Breger, Ed., *Organic Geochemistry*, Pergamon Press, New York, 1963, pp. 81-147.
24. M. H. B. Hayes, "Extraction of Humic Substances from Soil," in G. R. Aiken, D. M. McKnight, R. L. Wershaw, and P. MacCarthy, Eds., *Humic Substances in Soil, Sediment, and Water*, Wiley, New York, 1985, pp. 329-362.
25. J. W. Parson, "Isolation of Humic Substances from Soils and Sediments," in F. H. Frimmel and R. F. Christman, Eds., *Humic Substances and Their Role in the Environment*, Wiley, New York, 1988, pp. 3-14.
26. F. J. Stevenson, "Gross Chemical Fractionation of Organic Matter," in C. A. Black et al., Eds., *Methods of Soil Analysis*, American Society of Agronomy, Madison, 1965, pp. 1409-1421.

27 J. M. Brenner, *J. Soil Sci.*, **1**, 198 (1950).
28 R. S. Swift and A. M. Posner, *J. Soil Sci.*, **23**, 381 (1972).
29 J. E. Gregor and H. K. J. Powell, *J. Soil Sci.*, **37**, 577 (1986).
30 L. N. Alexsandrova, *Soviet Soil Sci.*, **1960**, 190 (1960).
31 J. Tinsley and C. H. Walker, *Trans. 8th Intern. Congr. Soil Sci.*, **2**, 149 (1964).
32 M. J. Jones and J. W. Parsons, *J. Soil Sci.*, **23**, 119 (1972).
33 F. J. Stevenson, G. Kidder, and S. N. Tilo, *Soil Sci. Soc. Amer. Proc.*, **31**, 71 (1967).
34 L. K. Porter, *J. Agr. Food Chem.*, **15**, 807 (1967).
35 E. M. Thurman, "Isolation of Soil and Aquatic Humic Substances Group Report," in F. H. Frimmel and R. F. Christman, Eds., *Humic Substances and Their Role in the Environment*, Wiley, New York, 1988, pp. 31-46.
36 M. Schnitzer and C. M. Preston, *Soil Sci. Soc. Amer. J.*, **51**, 639-646 (1987).
37 W. P. Scarrah and K. L. Mykleburst, *Fuel*, **55**, 274 (1986).
38 P. Sequi, G. Guidi, and G. Petruzzelli, *Geoderma*, **13**, 153 (1975).
39 G. J. Gascho and F. J. Stevenson, *Soil Sci. Soc. Proc.*, **32**, 117 (1968).
40 K. Kumuda, *Chemistry of Soil Organic Matter* (English translation), Japan Scientific Societies Press, Elsevier, Tokyo, 1987.
41 K. Kumuda and O. Sato, *Soil Sci. Plant Nutr.*, **8**, 31 (1962).
42 K. Kumada and H. M. Hurst, *Nature*, **214**, 631 (1967).
43 W. G. C. Forsyth, *Biochem J.*, **41**, 176 (1947).
44 K. M. Goh and M. R. Reid, *J. Soil Sci.*, **26**, 207 (1975).
45 G. Almendros and F. J. Gonzalez-Vile, *Soil Biol. Biochem.*, **19**, 513 (1987).
46 J. A. Rice and P. MacCarthy, *Sci. Total Environ.*, **81/82**, 61 (1989).
47 F. Scheffer and B. Ulrich, *Humus and Humusdüngung*, Ferdinand Enke, Stuttgart, 1960.
48 F. J. Stevenson and J. H. A. Butler, "Chemistry of Humic Acids and Related Substances," in G. Eglinton and M. T. J. Murphy, Eds., *Organic Geochemistry*, Springer-Verlag, New York, 1969, pp. 534-557.
49 S. U. Khan, *Soil Sci.*, **114**, 73 (1971).
50 V. O. Biederbeck and E. A. Paul, *Soil Sci.*, **115**, 357 (1973).
51 P. Simonart, L. Batistic, and J. Mayoudon, *Plant Soil*, **27**, 153 (1967).
52 M. Schnitzer, "Selected Methods for the Characterization of Soil Humic Substances," in P. MacCarthy, C. E. Clapp, R. L. Malcolm, and P. R. Bloom, Eds., *Humic Substances in Soil and Crop Sciences: Selected Readings*, Soil Science Society of America, Madison, 1990, pp. 65-89.
53 M. Schnitzer, C. A. Hindle, and M. Meglic, *Soil Sci. Soc. Amer. J.*, **50**, 913 (1986).
54 R. D. Haworth, *Soil Sci.*, **111**, 71 (1971).
55 M. Schnitzer, "Chemical, Spectroscopic, and Thermal Methods for the Classification and Characterization of "Humic Substances," in *Proc. Intern. Meeting Humic Substances*, Wageningen, 1972, pp. 193-310.
56 M. A. Curtis, A. F. Witt, S. B. Schram, and L. B. Rogers, *Anal. Chem.*, **53**, 1195 (1981).

57 R. Blondeau and E. Kalinowski, *J. Chromatogr.*, **351**, 585 (1986).
58 B. K. G. Theng, J. H. R. Wake, and A. M. Posner, *Plant Soil*, **29**, 305 (1968).
59 A. Otsuki and T. Hanya, *Nature*, **212**, 1462 (1966).
60 K. Kyuma, *Soil Sci. Plant Nutr.*, **10**, 33 (1964).
61 A. Watanabe and S. Kuwatsuka, *Soil Sci. Plant Nutr.*, **38**, 391, 401 (1992).
62 S. S. Dragunov and B. G. Murzakov, *Soviet Soil Sci.*, **1970**, 220 (1970).
63 G. R. Aiken, E. M. Thurman, R. L. Malcolm, and A. F. Walton, *Analy. Chem.*, **51**, 799 (1979).
64 P. MacCarthy, M. J. Peterson, R. L. Malcolm, and E. M. Thurman, *Anal. Chem.*, **51**, 2041 (1979).
65 K. L. Cheng, *Microchim. Acta*, **1977-II**, 398 (1977).
66 G. R. Aiken, "Isolation and Concentration Techniques for Aquatic Humic Substances," in G. R. Aiken, D. M. McKnight, R. L. Wershaw, and P. MacCarthy, Eds., *Humic Substances in Soil, Sediment, and Water*, Wiley, New York, 1985, pp. 363–385.
67 S. A. Daignault, D. K. Noot, D. T. Williams, and P. M. Huck, *Wat. Res.*, **22**, 803 (1988).
68 J. A. Leenheer, *Environ. Sci. Technol.*, **15**, 578 (1981).
69 C. S. Cronan and G. R. Aiken, *Geochim. Cosmochim. Acta*, **49**, 1697 (1985).
70 G. F. Vance and M. B. David, *Soil Sci. Soc. Amer. J.*, **53**, 1242 (1989).
71 M. B. David, G. F. Vance, J. M. Rissing, and F. J. Stevenson, *J. Environ. Qual.*, **18**, 212 (1989).
72 A. A. Pohlman and J. G. McColl, *Soil Sci. Soc. Amer. J.*, **52**, 265 (1988).
73 G. R. Aiken, "A Critical Evaluation of the Use of Macroporous Resins for the Isolation of Aquatic Humic Substances," in F. H. Frimmel and R. F. Christman, Eds., *Humic Substances and Their Role in the Environment*, Wiley, New York, 1988, pp. 15–28.
74 R. L. Malcolm, "Variations Between Humic Substances Isolated from Soils, Stream Waters, and Groundwaters as Revealed by ^{13}C-NMR Spectroscopy," in G. R. Aiken, D. M. McKnight, R. L. Wershaw, and P. MacCarthy, Eds., *Humic Substances in Soil, Sediment, and Water*, Wiley, New York, 1985, pp. 13–35.
75 B. T. Christensen, *Adv. Soil. Sci.*, **20**, 1 (1992).
76 F. J. Stevenson and E. T. Elliott, "Methodologies for Assessing the Quantity and Quality of Soil Organic Matter," in D. C. Coleman, J. M. Oades, and G. Uehara, Eds., *Dynamics of Soil Organic Matter in Tropical Ecosystems*, University of Hawaii Press, Honolulu, 1989, pp. 173–199.
77 L. W. Turchenek and J. M. Oades, *Geoderma*, **21**, 311 (1979).
78 D. W. Anderson, S. Saggar, J. R. Bettany, and J. W. B. Stewart, *Soil Sci. Soc. Amer. J.*, **45**, 767 (1981).
79 B. T. Christensen and L. H. Sørensen, *J. Soil Sci.*, **36**, 219 (1985).
80 J. M. Oades, A. M. Vassallo, A. G. Waters, and M. A. Wilson, *Aust. J. Soil Res.*, **25**, 71 (1987).
81 M. Ahmed and J. M. Oades, *Soil Biol. Biochem.*, **16**, 465 (1984).
82 H. Tiessen and J. W. B. Stewart, *Soil Sci. Soc. Amer. J.*, **47**, 509 (1983).
83 B. T. Christensen, *Soil Biol. Biochem.*, **19**, 429 (1987).

3

ORGANIC FORMS OF SOIL NITROGEN

Somewhat over 90 percent of the nitrogen (N) in the surface layer of most soils occurs in organic forms, with most of the remainder being present as clay-fixed ammonium (NH_4^+). The surface layer of most cultivated soils contains between 0.06 and 0.3 percent N. Peat soils have a relatively high N content (to 3.5 percent).

The importance of organic N from the standpoint of soil fertility has long been recognized, and our knowledge concerning the nature and chemical composition of organic N is extensive. Nevertheless, approximately one-half of the soil organic N has not been adequately characterized and little is known of the chemical bonds linking nitrogenous constituents to other soil components.

Consideration is given in this chapter to the kinds and amounts of organic N compounds in soil. Chemical reactions of organic matter with ammonia (NH_3) and nitrite (NO_2^-) are covered in Chapter 4. The reader is referred to several recent reviews[1-4] for background information.

FRACTIONATION OF SOIL N

Most studies on the forms of organic N in soils have been based on the use of hot mineral acids (or bases) to liberate nitrogenous constituents from clay and organic colloids. In a typical procedure, the soil is heated with 3N or 6N HCl for from 12 to 24 h., after which the N is separated into the fractions shown in Table 3.1. Methods for determining the various forms of N in soil hydrolysates have been described by Bremner[5] and Stevenson.[6]

The N that is not solubilized by acid hydrolysis is usually referred to as "acid insoluble-N;" that recovered by distillation with MgO is NH_3-N. The

Table 3.1 Fractionation of Soil N Based on Acid Hydrolysis

Form of N	Definition and method	% of Soil N
Acid insoluble-N	Nitrogen remaining in soil residue following acid hydrolysis. Usually obtained by difference (total soil N-hydrolyzable N)	(Usual range) 20–35
NH_3—N	Ammonia recovered from hydrolysate by steam distillation with MgO	20–35
Amino acid-N	Usually determined by the ninhydrin-CO_2 or ninhydrin-NH_3 methods. Recent workers have favored the latter	30–45
Amino sugar-N	Steam distillation with phosphate-borate buffer at pH 11.2 and correction for NH_3—N. Colorimetric methods can also be used. Sometimes referred to as hexosamine-N	5–10
Hydrolyzable unknown-N (HUN fraction)	Hydrolyzable N not accounted for as NH_3, amino acids, or amino sugars. Part of this N occurs as non-α-amino-N of arginine, tryptophan, lysine, and proline	10–20

soluble N not accounted for as NH_3 or in known compounds is the "hydrolyzable unknown (HUN) fraction."

The main identifiable organic N compounds in soil hydrolysates are the amino acids and amino sugars. Soils also contain nucleic acids and other nitrogenous biochemicals but specialized techniques are required for their separation and identification.

The procedure for hydrolyzing the soil has not been standardized and many variations in hydrolytic conditions have been employed. The variables include: 1) type and concentration of acid; 2) time and temperature of hydrolysis; 3) ratio of acid to soil; and 4) pretreatment. A two-stage acid hydrolysis procedure has been used by Russian soil scientists,[7] thereby enabling them to subdivide soil organic N into 3 main fractions; easily hydrolyzable N, difficulty hydrolyzable N, and nonhydrolyzable N. The easily hydrolyzable N was regarded as being the fraction most readily available for plant growth.

An unusually large amount of the soil N, usually of the order of 20 to 35 percent, is recovered in acid-insoluble forms. At one time it was thought that this fraction was an artifact resulting from the condensation of amino acids with reducing sugars during hydrolysis, but it is now believed that part of this N occurs as a structural component of humic substances. The percentage of the soil N accounted for in acid-insoluble forms can be reduced through pretreatment of the soil with HF prior to hydrolysis.[8] Also, some of the insoluble N can be extracted with dilute base and subsequently solubilized by acid hydrolysis, thereby reducing the acid-insoluble fraction to about 15 percent of the total soil N.[9]

Another unique feature of N fractionation schemes is that a large proportion of the soil N, usual range of 20 to 25 percent for surface soils, is recovered as NH_3. Some of the NH_3 is derived from indigenous clay-fixed NH_4^+ (see Chapter 4), part comes from amino sugars and the amino acid amides, asparagine and glutamine. It is also known that NH_3 can arise from the breakdown of certain amino acids during hydrolysis. Tryptophan is lost completely; others, such as serine and threonine, are destroyed to a lesser extent.

An accounting of all potential sources of NH_3 in soil hydrolysates shows that the origin of approximately one-half of the NH_3, equivalent to 10 to 12 percent of the total organic N, is still obscure. Some NH_3 may be derived from complexes formed by fixation reactions of the types described in Chapter 4.

By chromatographic assay of soil hydrolysates for aspartic acid and glutamic acid, and by assuming that the ratio of NH_3-N to (aspartic acid + glutamic acid)-N is the same for the soil "protein" as for pure proteins (about 0.5), it has been possible to estimate the amount of the NH_3-N that is derived from the amide N of asparagine and glutamine. From the work of Aldag,[10] it would appear that as much as two-thirds of the NH_3-N may occur in this form.

The percentage of the soil N in hydrolyzable unknown compounds (HUN fraction) can be appreciable, often > 20 percent of the total N. The possibility that arginine, tryptophan, lysine, and proline are present in sufficient quantities to account for most of the unknown N is remote. The nonamino N in these amino acids is not included with the amino acid-N values as determined by the ninhydrin–NH_3 or ninhydrin–CO_2 methods (discussed later). Goh and Edmeades[11] concluded that from one-fourth to one-half of the hydrolyzable unknown N in the soils they examined occurred as non-α-amino acid-N. In other work, Schnitzer et al.[12] subjected the acid hydrolysates of soil humic and fulvic acids to gel filtration and obtained several fractions in which > 90 percent of the N occurred in unknown forms.

Distribution of N Forms in Mineral Soils

Relatively little is known of the factors affecting the distribution of the forms of organic N in soils. Wide variations in N distribution patterns have been observed but no consistent trend has emerged relating composition to soil properties.

As this work is discussed, it should be noted that several problems are encountered in evaluating published data on N distribution patterns in soils. They include inaccuracies in analytical methods, especially for amino acid-N. Discrepancies also arise because part of the NH_3 liberated by hydrolysis results from partial destruction of amino sugars and this has not always been taken into account in estimating NH_3-N and amino sugar-N. Still another factor is that a significant amount of the N in many soils occurs as clay-fixed NH_4^+. In some cases, results obtained by different investigators cannot be compared directly due to variations in hydrolysis conditions.

Typical data showing the distribution of the forms of N in soils from different sections of the world are recorded in Table 3.2.[13–22] In general, differences in

Table 3.2 Distribution of the Forms of N in Representative Surface Soils of the World

Location[a]	Form of N (%)					Reference
	Acid Insoluble	NH$_3$	Amino Acid	Amino Sugar	HUN	
Africa						
Sierra Leone (8)	19.8–24.5	9.3–20.3	22.8–33.3	4.1–13.8	17.4–40.2	Amara and Stevenson[b]
Tanzania (4)	14.4–30.7	10.8–21.4	18.6–31.2	5.2–11.5	24.1–36.0	Singh et al.[22]
Argentina						
Misc. group (4)	22.2–32.7	15.1–21.3	13.3–20.2	2.0–10.9	22.2–32.7	Rosell et al.[21]
Canada						
Alberta (14)	18.4–26.4	13.7–21.2	26.8–35.8	4.8–11.9	12.9–25.0	Khan and Sowden[18]
Quebec (20)	16.2–33.6	15.7–28.2	23.4–37.6	4.8–9.2	6.9–22.5	Kadirgamathryah and MacKenzie[16]
Japan						
Misc. group (6)	11.4–43.6	15.3–24.2	16.9–43.3	1.2–8.8	19.4–26.6	Kyuma et al.[19] Hayashi and Harada[15]
United Kingdom						
Misc. group (20)	14.0–34.0	16.0–37.0	20.0–41.0	4.0–12.0	13.0–34.0	Greenfield[13]
United States						
Iowa (20)	18.4–36.7	18.6–29.0	17.8–34.3	3.3–7.1	17.9–28.9	Keeney and Bremner[17]
Nebraska (8)	18.0–22.0	18.0–28.0	31.0–46.0	5.0–9.0	6.0–17.0	Meints and Peterson[20]
West Indies						
Volcanic (4)	11.5–41.3	11.6–17.4	25.4–45.7	0.8–3.0	4.2–35.0	Dalal[14]
Nonvolcanic (3)	6.9–22.7	21.5–32.1	20.4–49.8	3.6–7.9	6.5–20.7	Dalal[14]

[a]Numbers in parentheses indicate number of soils.
[b]Unpublished observations.

N composition within a similar group of soils are fully as great as for soils having contrasting chemical and physical properties. For the soils listed, amino acids accounted for from 13.3 to 49.8 percent of the N; amino sugars accounted for from 0.8 to 13.8 percent. The percentage of the N in acid-insoluble forms ranged from 6.9 to 43.6; the range for hydrolyzable unknown forms (HUN) was from 4.2 to 40.2 percent.

Slight variations in N distribution exist within different size fractions of the soil[21] and changes are brought about through incubation[23] or when organic matter becomes decomposable through the effect of drying the soil.[24] The percentage of the N in acid-insoluble forms has been found to increase with an increase in "degree of humification,"[25] as well as in "degree of decomposition."[26]

Data obtained by Sowden et al.[27] for the distribution of the forms of N in soils from widely different climatic zones (Table 3.3) indicate that higher percentages of the N in soils from warmer climates occur as amino acid-N and as amino sugar-N; lower percentages occur as hydrolyzable NH_3.

Effect of Cultivation It is a well-known fact that the N content of most soils declines when land is subjected to intense cultivation (Chapter 1). This loss of N is not spread uniformly over all N fractions, although it should be pointed out that neither long-term cropping nor the addition of organic amendments to the soil greatly affects the relative distribution of the forms of N.[17,20,28,29]

Data showing the effect of cultivation and cropping systems on the distribution of the forms of N in soil are recorded in Table 3.4. Cultivation generally leads to a slight increase in the proportion of the N as hydrolyzable NH_3 but this effect is due in part to an increase in the percentage of the soil N as clay-fixed NH_4^+. The proportion of the soil N as amino acid-N generally decreases with cultivation whereas the percentage as amino sugar-N changes very little or increases.

The observation that cultivation has little effect on N composition emphasizes that all N forms are biodegradable and that fractionation of soil N by acid hydrolysis will be of little practical value as a means of testing soils for available N or for predicting crop yields.[16,17,23]

Depth Distribution Patterns Several studies have dealt with changes in the distribution of the forms of N with depth in the soil profile.[20,30] Much of this information cannot be interpreted directly in terms of organic N because a significant amount of the subsoil N may occur as clay-fixed NH_4^+.

Sowden's[30] results for amino acid-N in the A and B horizons of five Canadian soils (Table 3.5) suggest that the "protein" content of the organic matter may be higher in surface soil than in subsoil. The values shown for the Mollisol and Alfisol are undoubtedly affected by clay fixed NH_4^+, but Spodosols generally contain little N in this form. Recalculation of date published by Nømmik[31] for a Spodosol profile from Sweden gives the following values for percentage

Table 3.3 Nitrogen Distribution in Soils from Widely Different Climatic Zones[a]

Climatic Zone[b]	Total Soil N (%)	Form of N (%)			
		Acid Insoluble	NH_3	Amino Acid	Amino Sugar	HUN
Arctic (6)	0.02–0.16	13.9 ± 6.6	32.0 ± 8.0	33.1 ± 9.3	4.5 ± 1.7	16.5
Cool temperate (82)	0.02–1.06	13.5 ± 6.4	27.5 ± 12.9	35.9 ± 11.5	5.3 ± 2.1	17.8
Subtropical (6)	0.03–0.30	15.8 ± 4.9	18.0 ± 4.0	41.7 ± 6.8	7.4 ± 2.1	17.1
Tropical (10)	0.24–1.61	11.1 ± 3.8	24.0 ± 4.5	40.7 ± 8.0	6.7 ± 1.2	17.6

[a]Adapted from Sowden et al.[27]
[b]Numbers in parentheses indicate number of soils.

Table 3.4 Effect of Cultivation and Cropping System on the Distribution of the Forms of N in Soil

Location[a]	Form of N (%)					Reference
	Acid Insoluble	NH$_3$	Amino Acid	Amino Sugar	HUN	
Alberta, Canada (Brenton plots)[b]						
Rotation of grains and legumes (6)	21.1	15.1	30.9	10.4	22.6	Khan[29]
Wheat-fallow sequence (6)	25.0	17.8	28.8	9.3	19.2	Khan[29]
Germany[c]						
Grass sod	16.4	27.6	27.6	4.2	24.2	Fleige and Beaumer[28]
Arable (tilled)	16.2	32.1	22.1	4.2	25.4	Fleige and Beaumer[28]
Illinois, USA (Morrow plots)[d]						
Grass border and COCL rotation (2)	20.3	16.6	42.0	10.5	10.7	Stevenson[f]
Continuous corn and CO rotation (2)	20.2	16.7	35.0	14.4	13.9	Stevenson[f]
Iowa, USA						
Virgin (10)	25.4	22.2	26.5	4.9	21.0	Keeney and Bremner[17]
Cultivated (10)	24.0	24.7	23.4	5.4	22.5	Keener and Bremner[17]
Nebraska, USA[e]						
Virgin (4)	20.8	19.8	44.3	7.3	7.8	Meints and Peterson[20]
Cultivated (4)	19.3	24.5	35.8	7.0	13.4	Meints and Peterson[20]

[a]Numbers in parentheses indicate number of soils.
[b]Treatments for each sequence include a control, manure plot, NPKS plot, NS plot, lime plot, and a P plot.
[c]Average for 0–5, 5–10, and 10–15 cm depths.
[d]COCl = corn, oats, clover rotation with lime and P additions. CO = corn oats rotation.
[e]Soils of the Ustoll suborder.
[f]Unpublished observations.

Table 3.5 Nitrogen Accounted For As Amino Acids in the A and B Horizons of Five Canadian Soils[a]

Soil Type	Total N as Amino Acids (%)	
	A Horizon	B Horizon
Black Solonetz (Mollisol)	31.5	19.5
Gray Wooded (Alfisol)	17.1	14.0
Podzol (Spodosol)	22.5	19.9
Podzol (Spodosol)	30.1	22.8
Podzol (Spodosol)	41.8	24.9

[a]Adapted from Sowden.[30]

organic N as amino acids: A_0 horizon, 50.8 percent: A_1 horizon, 39.5 percent; A_2 horizon, 30.2 percent; B horizon, 27.1 percent.

An accurate accounting of organic forms of N in the lower soil horizons will require improvements in methods for determining clay-fixed NH_4^+, as well as for specific organic N forms when present in low amounts (e.g., amino acid-N and amino sugar-N). Conventional hydrolysis procedures may lead to incomplete recoveries of amino acids from clay-rich subsurface soils.[32]

Distribution of N in Histosols

Environmental conditions in aquatic sediments are particularly suitable for the preservation of microbially produced substances. Thus, while the same forms of N occurs in organic soils (Histosols) as for mineral soils, the N distribution is usually different.[33,34] In general, a higher percentage of the N in Histosols occur in the form of amino acids, a result that can be explained by less extensive microbial turnover of organic N in aquatic sediments. An exception to this rule occurs for well-humified peats, where amino acid-N values are relatively low and a high percentage of the N occurs in the acid-insoluble fraction.

Data obtained by Sowden et al.[34] show that the percentage of the total N as amino acid-N is generally higher for mesic peats than for humic peats (Table 3.6). The latter type of peat is in an advanced stage of decomposition and more of the N has been incorporated into unknown humus forms. When Histosols are incubated in the laboratory, the percentage of the N as amino acid-N decreases while the percentage as acid insoluble-N increases, as also shown in Table 3.6.

NITROGEN IN HUMIC AND FULVIC ACIDS

Humic and fulvic acids (see Chapter 2) contain those some forms of N that are obtained when soils are subjected to acid hydrolysis.[18,21,35,36] However, N

Table 3.6 Distribution of the Forms of N in Histosols

Distribution[a]	Form of N, Percentage of Total N					Reference
	Acid Insoluble	NH_3	Amino Acid	Amino Sugar	HUN	
Canada						
Humic peat profile	20.3–29.5	11.1–14.7	37.7–43.3	8.7–11.0	6.5–19.4	Sowden et al.[34]
Mesic peat profile	21.2–24.5	8.6–12.1	46.3–56.3	4.3– 5.8	6.8–17.7	Sowden et al.[34]
Wisconsin						
Histic materials as collected (9)	4.6–37.0	9.3–23.6	15.9–58.8	3.5–11.5	13.7–30.6	Isirimah and Keeney[33]
Following incubation (9)	21.6–45.4	13.0–17.1	24.3–39.2	3.9– 4.8	11.6–17.2	Isirimah and Keeney[33]

[a]Numbers in parentheses indicate number of soils analyzed.

distribution patterns vary somewhat. For example, as compared to unfractionated soil, lower percentages of the N in humic acids occur as NH_3–N and in hydrolyzable unknown compounds (HUN fraction); higher percentages occur as amino acid-N and as acid insoluble-N. Results obtained by Rosell et al.[21] for the distribution of N in some Argentine soils and their humic acids are shown in Fig. 3.1.

A rather high percentage of the N in the fulvic acid fraction of soil organic matter occurs as amino acids,[37,38] and sometime as NH_3–N. By contrast, a rather low percentage occurs as acid insoluble-N. The fulvic acid fraction contains a broad spectrum of organic compounds, including those belonging to the well-known classes of organic compounds (see Chapter 2). Thus, differences in N distribution patterns observed for "fulvic acids" may partially reflect variations in the amounts of "true fulvic acids" that are present relative to nonhumic substances.

Data given in Table 3.7 show that humic acids extracted from soils with $0.5N$ NaOH have higher N contents than those extracted with $0.1M$ $Na_4P_2O_7$. Also, a higher percentage of the N in the alkali-extracted humic acids occurs in the form of amino acids; a lower percentage occurs in acid-insoluble forms.

As much as one-half of the total N in humic substances can be accounted for as amino acid-N.[35-41] Additional quantities of amino acids can be recovered from humic acids by subjecting the acid-hydrolyzed residues to a second hydrolysis with $2.5N$ NaOH.[9,42,43] Sodium amalgam reduction of the alkali-treated residue leads to a further release of amino acids.[42] Acid hydrolysis would be expected to remove amino acids bound by peptide bonds, **I**, as well as those linked to quinone rings, **II**. On the other hand, amino acids bonded directly to phenolic rings, **III**, would not be released without subsequent alkaline hydrol-

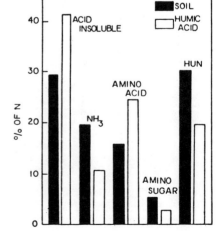

Fig. 3.1 Comparison of the distribution of the forms of nitrogen by acid hydrolysis of soils and their humic acids. HUN = hydrolyzable unknown. Adapted from Rosell et al.[21]

Table 3.7 Distribution of the Forms of N in Humic Acids[a]

Extractant[b]	N (%)	Acid Insoluble	NH_3	Amino Acids	Amino Sugars	HUN
0.1M $Na_4P_2O_7$ (6)	1.79–2.63	41.3–59.0	8.8–12.8	19.5–34.5	2.6–5.0	5.2–10.6
0.5N NaOH (5)	2.31–3.74	32.6–43.7	8.4–13.7	31.2–44.7	3.4–8.1	4.7– 8.6
0.5N NaOH (4)	2.11–2.69	35.9–50.8	8.2–14.0	22.1–26.5	1.8–3.9	16.2–21.8

[a]Adapted from Bremner[35] and Rosell et al.[21]
[b]Numbers in parentheses indicate number of samples.

ysis. The effect of alkaline hydrolysis was believed to be due to oxidation of the phenol to the quinone form, with release of the amino acid.

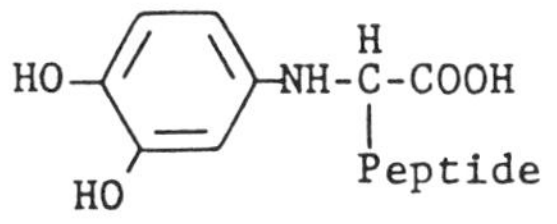

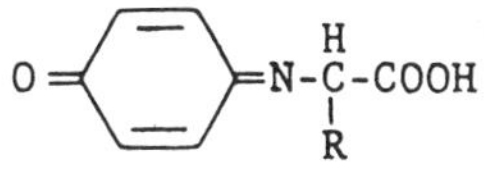

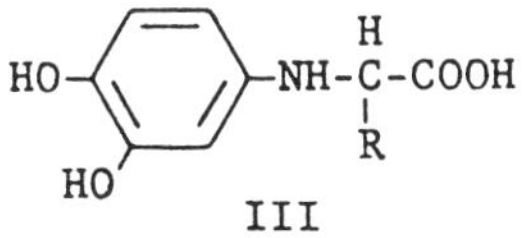

Other possible types of quinoid-amino acid structures in humic and fulvic acids are shown below, where **R** = amino acid or peptide.

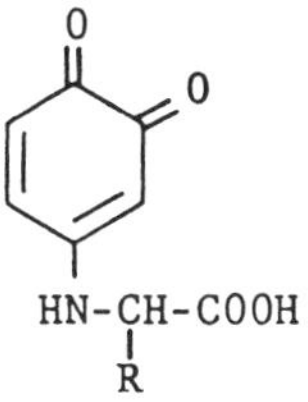

For peptides, the amino acid directly attached to the aromatic ring may be less available to plants and microorganisms than amino acids of the peripheral peptide chain, as depicted in Fig. 3-2.

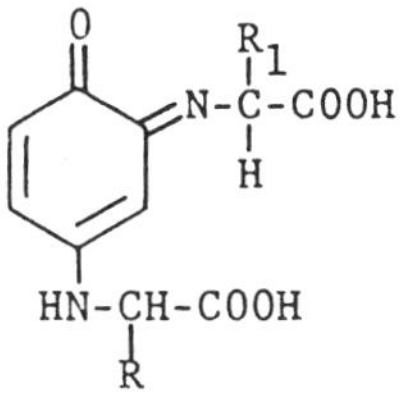

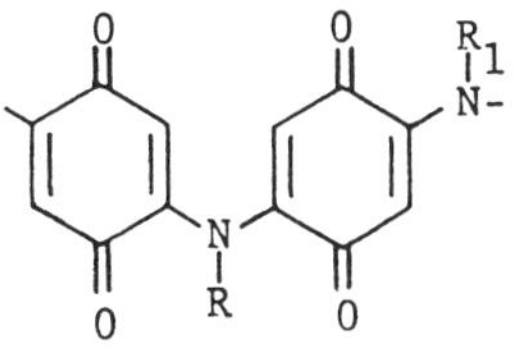

An interesting result was recently obtained by Aldag,[10] who observed an

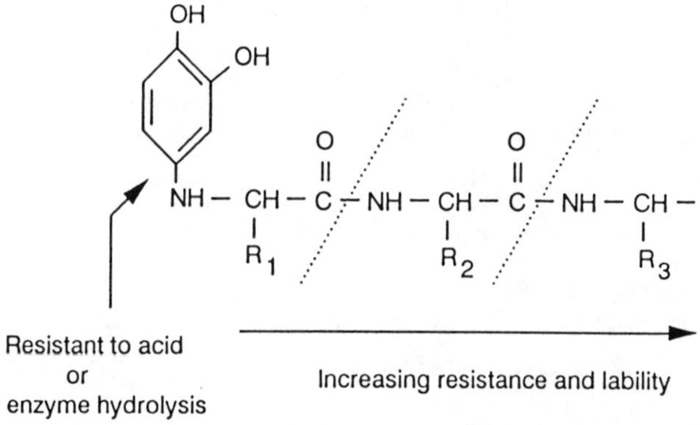

Fig. 3.2 Biological availability of amino acid-N in peptide–phenolic complexes.

increase in amino acid-N at the expense of the HUN fraction when an acid hydrolysate of humic acid (6N HCl for 24 h.) was subjected to a second hydrolysis with 6N HCl containing 3 percent H_2O_2. The increased release of amino acids was attributed to the presence of phenolic–amino acid addition products in the hydrolysate (e.g., see structure **III**).

Part of the N associated with humic acids may exist as peptides or proteins linked to the central core by H-bonding. Protein-rich fractions have been obtained from humic acids by extraction with phenol[44] and formic acid.[45]

Peptide-like substances have been determined in humic acids by chromatographic assay of proteinaceous constituents liberated by cold hydrolysis with concentrated HCl.[42] Infrared spectra of some, but not all, humic acids show absorption bands typical of the peptide linkage (see Chapter 13). As was the case for soil hydrolysates, some of the N accounted for in hydrolyzable unknown compounds (HUN fraction) may occur as the non-α-amino acid-N of such amino acids as arginine, tryptophan, lysine, and proline.

In concluding this section, it can be said that significant amounts of the N associated with humic and fulvic acids cannot be accounted for in known compounds. This N may occur in the following types of linkages:

1. As a free amino ($-NH_2$) group
2. As an open chain ($-NH-$, $=N-$) group
3. As part of a heterocyclic ring, such as an $-NH-$ of indole and pyrrole or the $-N=$ of pyridine. The possible occurrence of heterocyclic N in humic acids has been indicated by the work of Müller-Wegener.[46]
4. As a bridge constituent linking quinone groups together
5. As an amino acid attached to aromatic rings in such a manner that the intact molecule is not released by acid hydrolysis

BIOCHEMICAL N COMPOUNDS

Amino Acids

The amino acids associated with proteins are α-amino acids; that is, they have an NH_2 group attached to the same C that holds the COOH group. A list of the more common amino acids, together with their structures, is given in Table 3.8. As will be shown later, soils also contain nonprotein amino acids.

The occurrence of amino acids in soil has been known since the turn of the century, when Suzuki[47] reported the presence of aspartic acid, alanine, aminovaleric acid, and proline in an acid hydrolysate of humic acid. By 1917, several other amino acids had been isolated, including glutamic acid, valine, leucine, isoleucine, tyrosine, histidine, and arginine. In more recent times, numerous other amino acids have since been identified, many of which are not normal constituents of proteins.

Extraction and Quantitative Determination Amino acids are normally extracted from soil by treatment with hot mineral acids, usually $6N$ HCl under reflux for from 16 to 24 h. Hydrolysis with base is required if tryptophan is to be determined. Conditions for maximum recovery of amino acids have been studied in considerable detail.[8,48–50]

The efficiency of conventional acid hydrolysis for recovering amino acids from soils containing fine-grained mineral matter is suspect. It was pointed out earlier that an HF pretreatment may be required for quantitative release of amino acids that are bound to clay minerals and that amino acids chemically attached to humic acids may not be released by acid hydrolysis.

Although several methods have been used for estimating amino acid-N in hydrolysates of soil or soil organic matter preparations, most investigators favor methods based on the ninhydrin reaction.

$$\text{ninhydrin} + R\text{-}CH(NH_2)\text{-}COOH \xrightarrow{pH < 2.5} \text{product} + R\text{-}CHO + CO_2 + NH_3$$

The classical ninhydrin–CO_2 method is highly specific because it requires a free NH_2 group adjacent to a COOH group. However, aspartic acid yields two moles of CO_2; proline and hydroxyproline give low values. Bremner[5] applied the ninhydrin–NH_3 method to determine amino acid-N in soil hydrolysates. Careful control of pH is required, otherwise NH_3 forms a colored product with ninhydrin as described below.

Methods based on colorimetric analysis of the blue-colored product produced when the ninhydrin reaction is carried out at pH 5.0 have also been used.[32,51] This approach is especially useful for subsurface soils and sediments where amino acid levels are low.

Table 3.8 Chemical Structures of Some Protein Amino Acids

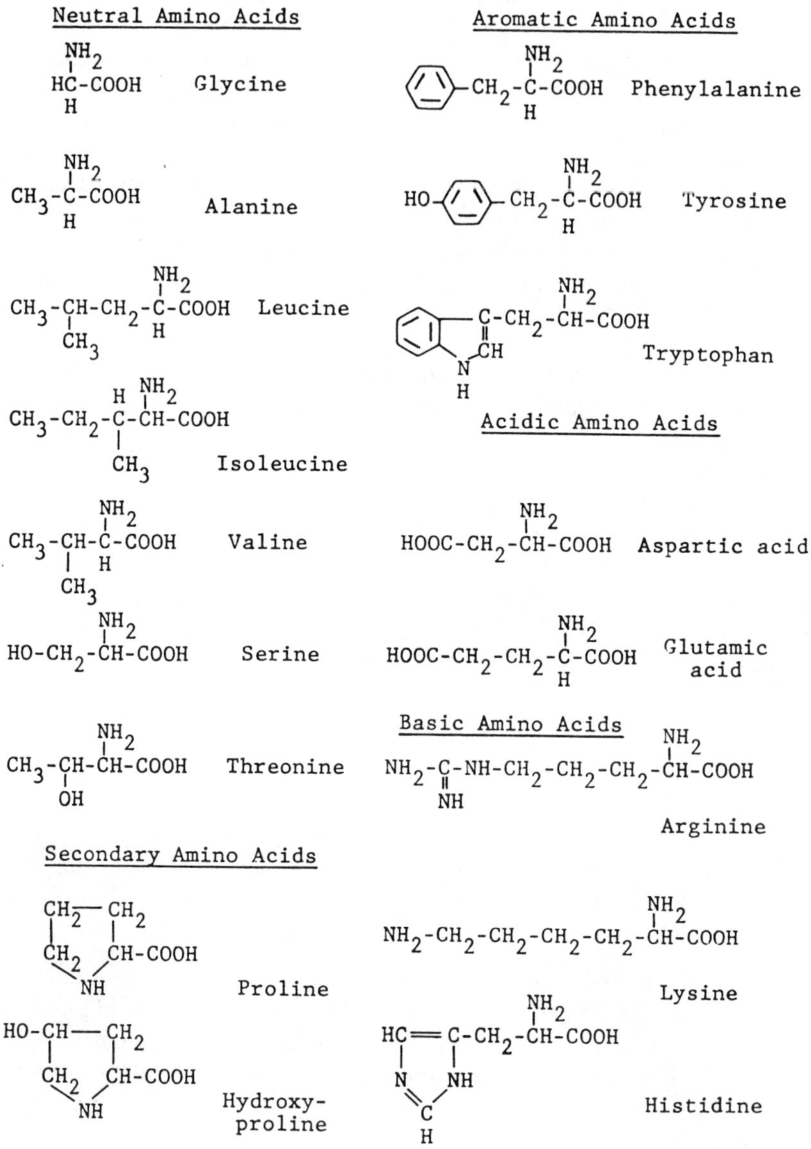

Neutral Amino Acids

$$\underset{H}{\overset{NH_2}{HC-COOH}} \quad \text{Glycine}$$

$$\underset{H}{\overset{NH_2}{CH_3-C-COOH}} \quad \text{Alanine}$$

$$CH_3-\underset{CH_3}{CH}-CH_2-\underset{H}{\overset{NH_2}{C}}-COOH \quad \text{Leucine}$$

$$CH_3-CH_2-\underset{CH_3}{\overset{H\;NH_2}{C}}-CH-COOH \quad \text{Isoleucine}$$

$$CH_3-\underset{\underset{CH_3}{|}H}{\overset{NH_2}{CH-C}}-COOH \quad \text{Valine}$$

$$HO-CH_2-\overset{NH_2}{CH}-COOH \quad \text{Serine}$$

$$CH_3-\underset{OH}{CH}-\overset{NH_2}{CH}-COOH \quad \text{Threonine}$$

Secondary Amino Acids

Proline

Hydroxy-proline

Aromatic Amino Acids

Phenylalanine — $\text{C}_6\text{H}_5-CH_2-\underset{H}{\overset{NH_2}{C}}-COOH$

Tyrosine — $HO-\text{C}_6\text{H}_4-CH_2-\underset{H}{\overset{NH_2}{C}}-COOH$

Tryptophan (indole ring)–$C-CH_2-\overset{NH_2}{CH}-COOH$

Acidic Amino Acids

$$HOOC-CH_2-\overset{NH_2}{CH}-COOH \quad \text{Aspartic acid}$$

$$HOOC-CH_2-CH_2-\underset{H}{\overset{NH_2}{C}}-COOH \quad \text{Glutamic acid}$$

Basic Amino Acids

$$NH_2-\underset{NH}{\overset{\|}{C}}-NH-CH_2-CH_2-CH_2-\overset{NH_2}{CH}-COOH \quad \text{Arginine}$$

$$NH_2-CH_2-CH_2-CH_2-CH_2-\overset{NH_2}{CH}-COOH \quad \text{Lysine}$$

Histidine (imidazole ring)–$CH_2-\overset{NH_2}{CH}-COOH$

[Chemical reaction scheme showing phthalic acid derivative + NH₃ + hydroxy compound → (−3H₂O, pH 5) → blue-colored product with C=N−CH linkage]

blue-colored product

A less popular procedure is the nitrous acid method, which is based on the ability of aliphatic amines to react with nitrous acid to form N_2.

$$R-NH_2 + HNO_2 \rightarrow R-OH + H_2O + N_2$$

The method is subject to interference by many compounds, such as phenols, tannins, lignins, alcohols, and keto acids.

Results obtained for amino acid-N based on the analysis of soil hydrolysates must be regarded as minimal. In addition to incomplete extraction and destruction of amino acids, part of the N of glutamine and asparagine is liberated as NH_3. Furthermore, some amino acids contain N other than in an NH_2 group adjacent to a COOH group, typical examples being arginine, histidine, and lysine. As noted earlier, the percentage of the soil N as amino acids varies considerably from one soil to another and within different horizons of the soil profile.

Identification of Amino Acids A variety of chromatographic techniques have been used for isolating amino acids from soil hydrolysates, and an impressive number of compounds have been detected. The first modern study was conducted by Bremner,[52] who detected the following compounds by paper partition chromatography: glycine, alanine, valine, leucine, isoleucine, serine, threonine, aspartic acid, glutamic acid, phenylalanine, arginine, histidine, lysine, proline, hydroxyproline, α-amino-*n*-butyric acid, α,ε-diaminopimelic acid, β-alanine, and γ-amino butyric acid. Later, Stevenson[53,54] applied the more precise technique of ion-exchange chromatography. The identifications made by Bremner were confirmed, and, in addition, ornithine, cysteine, methionine sulfone, and methionine sulfoxide were identified. A typical elution diagram for the separation of amino acids from soil by ion-exchange chromatography is shown in Fig. 3.3.

Methods are now available for the ultrarapid analysis of amino acid by high-pressure liquid chromatography; they provide complete separation of ninhydrin-positive compounds in biological samples containing a multiplicity of constituents. Gas chromatographic methods are also available and may have advantages for certain applications.[8,55]

A rather large number of unidentified ninhydrin-reacting substances have been observed in soil hydrolysates. In the study conducted by Stevenson,[53] 33 amino compounds were isolated, of which 29 were identified. Young and

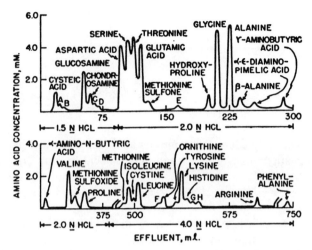

Fig. 3.3 Separation of amino compounds from an acid hydrolysate of soil using ion-exchange chromatography. From Stevenson[54]

Mortensen[56] reported 57 ninhydrin-reacting substances in acid hydrolysates of some Ohio soils, only 24 of which were identified. Many of the unidentified compounds occur in minute amounts and can only be found using special isolation techniques.

It is of interest that many of the amino acids found in soils, specifically α,ϵ-diaminopimelic acid **(VII)**, ornithine **(VIII)**, β-alanine **(IX)**, α-amino-n-butyric acid **(X)**, and γ-amino-butyric acid **(XI)**, are not normal constituents of proteins. Several of these compounds may represent waste products of microbial metabolism; others occur in a variety of natural products synthesized by microorganisms. Ornithine and β-alanine, for example, are constituents of certain antibiotics; the latter is also a component of pantothenic acid, an important vitamin. The occurrence of α,ϵ-diaminopimelic acid is of interest because this amino acid appears to be confined to bacteria, where it occurs as part of the cell wall. It should be noted that well over 100 amino acids and amino acid derivatives have been identified as constituents of living organisms.

$$\text{HOOC-CH(NH}_2\text{)-CH}_2\text{-CH}_2\text{-CH}_2\text{-CH(NH}_2\text{)-COOH} \qquad \text{NH}_2\text{-CH}_2\text{-CH}_2\text{-CH}_2\text{-CH(NH}_2\text{)-COOH}$$

α,ϵ-Diaminopimelic acid $\qquad\qquad$ Ornithine

VII $\qquad\qquad$ VIII

$$\text{NH}_2\text{-CH}_2\text{-CH}_2\text{-COOH} \qquad \text{CH}_3\text{-CH}_2\text{-CH(NH}_2\text{)-COOH} \qquad \text{NH}_2\text{-CH}_2\text{-CH}_2\text{-CH}_2\text{-COOH}$$

β-Alanine $\qquad$ α-Amino-n-butyric acid $\qquad$ γ-Aminobutyric acid

IX $\qquad\qquad$ X $\qquad\qquad$ XI

Distribution Patterns in Soil Because of the advanced state of the art of chromatography, no great ingenuity is required to demonstrate the occurrence of amino acids in soil or soil organic matter preparations. Of greater significance than the detection of these compounds is a knowledge of precise distribution patterns. Unfortunately, data on this subject are difficult to evaluate because of analytical errors associated with incomplete extraction, losses of specific amino acids during hydrolysis and desalting, possible improper identification, and others.

Sowden[49] found that most methods used for desalting soil hydrolysates led to selective losses of aspartic acid, glutamic acid, and some of the basic amino acids. The conventional approach has been to pass the hydrolysate through a cation-exchange resin in the H^+-form, followed by elution of adsorbed amino acids with dilute NH_4OH. Some loss of amino acids can occur by this procedure.[49]

A novel desalting procedure has been proposed by Cheng et al.,[8] and is based on the fact that the Fe^{3-}, Al^{3-}, and Si^{4-} form polyanion complexes with the F^- ion and are retained as FeF_6^{3-}, AlF_6^{3-}, and SiF_6^{2-} when the hydrolysate is passed through an anion exchange resin in the F^--form.

The extreme variability that has been reported for the amino acid composition of soils is indicated in Table 3.9 where a tabulation is given of some of the early work on the relative distribution of amino acid-N as acidic, neutral and basic compounds in the soils of different climatic zones.[56-60] It can be seen that over one-third of the amino acid-N in some soils has been reported in the form of basic amino acids (lysine, histidine, arginine, and ornithine); in others, less than one-tenth has been reported in these compounds. Equally divergent results have been recorded for individual amino acids in each group. Aspartic acid has been reported to constitute practically all of the acidic amino acids in some soils while in other glutamic acid seems to predominate.

A comparison of early data for amino acid distribution patterns in temperate zone soils,[56,58,59] with those for tropical and subtropical soils[57,60] indicates that the percentage of the soil amino acid-N as basic amino acids is higher in the tropical and subtropical soils. The explanation given for this result was that

Table 3.9 Percentage of α-Amino Acid N as Acidic, Neutral, and Basic Compounds for Soils of Different Climatic Zones[a]

Zone[b]	Distribution of α-Amino Acid-N (%)		
	Acid Compounds	Neutral Compounds	Basic Compounds
Temperate (7)	12.6–25.0	66.6–76.2	8.4– 9.8
Semitropical (5)	0.8–10.9	61.2–70.8	10.3–35.5
Tropical (5)	1.2– 6.4	65.0–85.6	8.0–29.1

[a] Adapted from Almeida et al.,[57] Stevenson,[59] Wang et al.,[60] and Young and Mortensen.[56]
[b] Numbers in parentheses indicate the number of soils.

greater microbial activity in the warmer soils results in more extensive turnover of proteinaceous material with selective preservation of basic compounds due to interaction with other soil components.

Later results by Sowden and his associates failed to confirm the above-mentioned findings.[26,27,61-63] Their data, summarized by Sowden et al.[27] indicate that it is the acidic amino acids (particularly aspartic) rather than the basic amino acids that predominate in tropical soils. Reasons for the discordant results are unknown but may be due to differences in the nature of the soils examined. Another possibility is that selective losses of amino acids have occurred during extraction and desalting, particularly in some of the earlier investigations.

Several studies have been conducted to determine the effects of cultivation practices on the amino acid composition of the soil. Stevenson[59] examined some soils of the Morrow Plots at the University of Illinois and concluded that long-term cultivation (plots established in 1901) without organic matter additions led to an increase in the relative proportion of the amino acid material in the form of basic amino acids. A similar result was obtained by Yamoshita and Akiya[64] for some Japanese soils. On the other hand, Young and Mortensen[56] failed to detect quantitative differences in amino acid composition of some rotation plots in Ohio. These plots had been in operations for only about three decades, as compared to nearly seven decades for the Morrow Plots; consequently, differences due to cropping would not be as great. Khan[29] found identical amino acid distribution patterns for the soils of two cropping systems on a "Gray Wooded" soil in Alberta, Canada (plots continuous for 39 y. at time of sampling).

Factors Affecting the Distribution of Amino Acids A variety of complex factors may affect the distribution of amino acids in soils, including synthesis and destruction by the indigenous biota, adsorption by clay minerals, and reactions with quinones and reducing sugars.

The predominant amino acids in soils appear to be those present in the cell walls of microorganisms.[15,65] Glycine, alanine, aspartic acid, and glutamic acid are often the dominant amino acids in bacterial cells, along with ornithine, lysine, and diaminopimelic acid. Thus, it appears that much of the amino acid material that accumulates in soil is derived from peptides, mucopeptides, and teichoic acids of microbial cells.

A priori reactions between amino acids and reducing sugars or quinones would be expected to play a significant role in amino acid composition. For the reactions involved, the reader is referred to Chapter 8. Basic amino acids, for example, react with reducing sugars and quinones at considerably higher rates than neutral and acidic amino acids. Thus, if these reactions occurred in soils, the more basic compounds, such as lysine, would be affected to a greater extent than other amino acids. Another factor to consider is that the accessory amino group of basic amino acids, when present in peptides, is capable of combining with carbonyl-containing substances whereas the amino group of neutral and acidic amino acids is inaccessible because of participation in peptide linkages. Mention should also be made of the likelihood that the amino acids in soil are linked in some manner to polymeric constituents of humus.

Adsorption reactions may also be a factor affecting amino acid distribution patterns. Carter and Mitterer[66] found that the amino acid composition of carbonaceous sediments was characterized by elevated levels of acidic amino acids (i.e., aspartic and glutamic acids). This was attributed to the formation of insoluble complexes or salts between these amino acids and Ca^{2+}. Greater attention needs to be given to the effect of adsorbing surfaces on the amino acid composition of soils.

Amino Acids in Humic and Fulvic Acids Humic and fulvic acids contain the same amino acids that are found in soils but not necessarily in the same proportions. Differences also exist in the amino acid distribution patterns of humic and fulvic acids. In comparison to fulvic acids, humic acids contain relatively higher amounts of basic amino acids but relatively lower amounts of acidic amino acids.[63] This effect can be seen in Table 3.10, where a compilation is given of values recorded in the literature for acidic, neutral, and basic amino acids in humic and fulvic acids.[21,61,63,66,67]

The percentage of the amino acid-N in humic acids as basic compounds may increase with an increase in "degree of humification" but information on this point is lacking. Changes in amino acid composition may also occur when soils are subjected to long-term cultivation, as noted earlier for the distribution of amino acids in the soil proper.

Table 3.10 Percent Molar Distribution of Amino Acids in Hydrolysates of the Humic Acid and Fulvic Acid Fractions of Organic Matter

Source or Location[a]	Group			Reference
	Acidic	Neutral	Basic	
Humic Acids				
Argentine soils (4)	26.9–27.5	56.6–62.0	11.1–16.4	Rosell et al.[21]
Canadian soils (1)	18.3–19.8	71.2–74.0	7.0– 9.0	Khan and Sowden[61]
Tropical soils (3)	23.6–25.2	65.1–67.8	8.6– 9.7	Sowden et al.[63]
Carbonate mud (4)	27.5–33.5	59.2–67.3	5.2– 9.2	Carter and Mitterer[66]
Fulvic Acids				
Italian soils (5)	29.6–37.7	56.8–67.7	2.7– 7.5	Guidi et al.[67]
Tropical soils (2)	43.8–45.6	51.8–51.9	2.7– 4.5	Sowden et al.[63]
Carbonate mud (4)	34.3–43.5	47.9–58.8	3.5– 8.6	Carter and Mitterer[66]

[a]Numbers in parentheses indicate number of samples.

Stereochemistry of Amino Acids The percentage [(D-amino acid × 100)/ (D-amino acid + L-amino acid)] for D-alanine, D-aspartic acid, and D-glutamic acid in soils has been found to be significantly greater than for other protein amino acids.[55] This is to be expected because these D-amino acids are known to be significant components of bacterial cell walls and metabolites. As noted earlier, the predominate amino acids in soils appear to be those associated with cell walls of microorganisms.

Amino acid racemization–epimerization reactions have been used as a geochronological tool for dating fossil bones, shells, deep-sea foraminerferal deposits, and marine sediments. This work has been reviewed by Dungworth[68] and Schroeder and Bada.[69]

Free Amino Acids Practically all of the proteinaceous material in soil is intimately combined with clay minerals or humus colloids. From the standpoint of N availability to microorganisms and higher plants, free amino acids may be of greater importance. The occurrence of small amounts of amino acids in the soil solution is expected because these compounds are formed during the conversion of protein N to NH_3 by heterotrophic organisms.

$$\text{Proteins} \rightarrow \text{peptides} \rightarrow \text{amino acids} \rightarrow NH_3$$

Free amino acids may modify the biological properties of the soil through their effect on specific types or groups of living organisms. Some loss of soil N is possible through leaching of amino acids but the extent of such losses, if any, is unknown.

The term "free amino acids" has been used to refer to those amino acids in soil that do not exist in peptide linkages. In this sense, the term is a misnomer, because soils may contain amino acids that are sorbed on mineral surfaces and that are not removed by the solvents commonly used to extract free amino acids.

In addition to serving as a source of N, amino acids may play a role in rock weathering and in pedogenic processes through their ability to form chelate complexes with metal ions (see Chapter 16).

Solvents used as extractants for free amino acids include distilled water, 80 percent ethyl alcohol, and dilute aqueous solutions of $Ba(OH)_2$ and ammonium acetate (NH_4OAc). Barium hydroxide and NH_4OH appear to be the most efficient, since they remove large quantities of amino acids absorbed on clay particles. Water is rather inefficient in removing amino acids. Paul and Schmidt[70] found that $Ba(OH)_2$ and NH_4OAc extracted from 5 to 25 times more amino acids from soil than 80 percent ethyl alcohol. A disadvantage of $Ba(OH)_2$ as an extractant is that hydrolysis of organic N compounds may occur during extraction. Accordingly, NH_4OAc would appear to be the preferred solvent for recovering "free amino acids" from soil. Extraction of soil with water in the presence of CCl_4 increases the quantity of amino acids extracted by 25 to 100 times, presumably due to the effect of CCl_4 on release of amino acids from microbial cells.

The difficulty of obtaining quantitative release of free amino acid is emphasized by Paul and Schmidt's[70] findings that only 31 to 83 percent of the added amino acids could be removed from soil with NH_4OAc; recovery of basic amino acid was somewhat less than that for neutral and acidic compounds. Similar results were obtained by Gilbert and Altman[71] using 20 percent ethyl alcohol as extractant. The possibility that amino acids may be held on interlamellar surfaces of expanding-lattice clays further complicates the problem of complete extraction.[8,32] Under natural soil conditions, some free amino acids may be held in small voids or micropores and thereby be inaccessible to microorganisms.

Amino acids, being readily decomposed by microorganisms, have only an ephemeral existence in soil. Thus, the amounts present in the soil solution at any one time represent a balance between synthesis and destruction by microorganisms; levels will be highest when microbial activity is intense.

The free amino acid content of the soil is strongly influenced by weather conditions, moisture status of the soil, type of plant and stage of growth, additions of organic residues and cultural conditions. Levels seldom exceed 2 μg/g, or 4.5 kg/ha plow depth, but they may be sevenfold higher in rhizosphere soil. Ivarson and Sowden[72] showed that freezing brought about a ten- to fourteen-fold increase in free amino acid content. Marked increases have also been observed by prolonged storage of soil at low temperature,[73] as well as by treatment of soil with fungicides.[74] On the assumption that NH_4OAc-extractable amino acids exist largely in water-soluble forms, their concentration in the soil solution (20 percent moisture level) would be of the order of 10^{-5} to $10^{-6}M$.

Healthy plant roots excrete amino acids into the soil. The kinds and amounts of amino acids exuded from roots varies with plant type and maturity, and, for any given plant, upon environmental conditions affecting growth, such as temperature, light intensity, moisture status of the soil, and availability of nutrient elements.[75]

Most estimates for amino acids exuded by roots have been made with culture solutions under laboratory conditions, and, for this reason, the amounts exuded under field conditions cannot be accurately estimated. As indicated earlier, the free amino acid content of rhizosphere soil is normally many times higher than for nonrhizosphere soil. The release of amino acids from roots may be an important factor affecting microbial activity, and, in some cases, plant growth. These aspects of amino acids in soil, discussed in detail by Rovira and Mcdougall,[75] are beyond the scope of this chapter.

Summary of the State of Amino Acids in Soil The deliberations of the previous sections indicate that amino acids exist in soils in several different forms, including the following:

1. As free amino acids
    a. In the soil solution
    b. In soil micropores

2. As amino acids, peptides or proteins bound to clay minerals
   a. On external surfaces
   b. On internal surfaces
3. As amino acids, peptides or proteins bound to humic colloids
   a. H-bonding and van der Waal's forces
   b. In covalent linkage as quinoid–amino acid complexes, as exemplified by structures **I** to **VI**
4. As mucoproteins (amino acids combined with N-acetylhexosamines, uronic acids, and other sugars)
5. As muramic acid-containing mucopeptide (**XII**) derived from bacterial cell walls

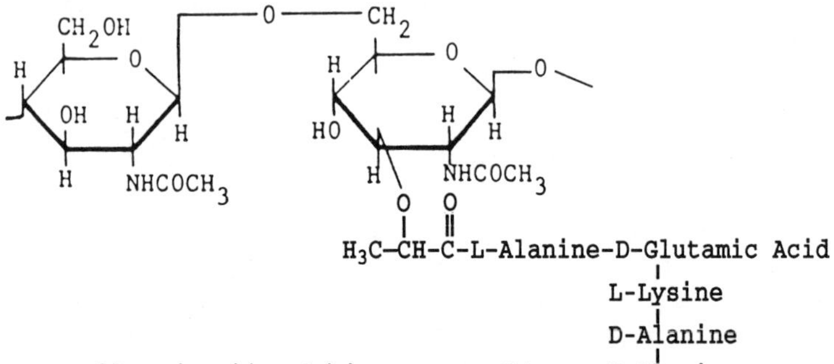

Muramic acid-containing mucopeptide
XII

6. As teichoic acids (linear polymers of polyol, ribitol, or glycerophosphate containing ester-linked alanine)

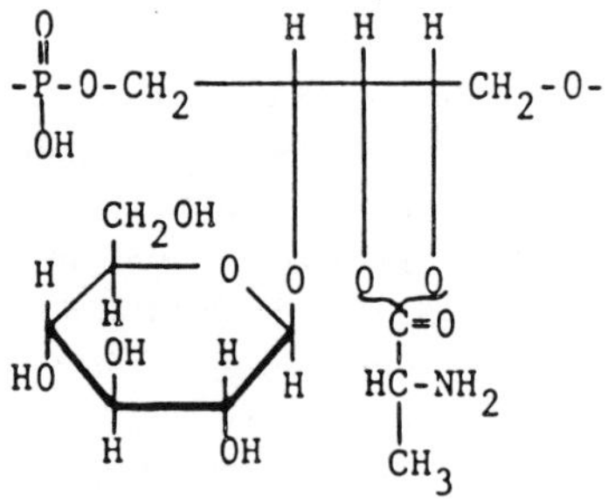

Teichoic acid
XIII

Considerable controversy exists as to whether proteins as such occur in significant amounts in soil organic matter. The well-known ligno–protein theory of soil humus has yet to be confirmed, and as mentioned earlier, many investigators believe that the theory in its original form is obsolete. Failure to account for a significant amount of the soil amino acids as protein or as lignin–protein complexes has led to the conclusion that the amino acids are directly bound to complex polymers formed from phenols or quinones.

Amino Sugars

Amino sugars occur as structural components of a broad group of substances, the mucopolysaccharides, and they have been found in combination with mucopeptides and mucoproteins, as well as with small molecules, such as antibiotics. Some of the amino sugar material in soil may exist in the form of an alkali-insoluble polysaccharide referred to as chitin. This substance, which is a polymer of *N*-acetylglucosamine, comprises the cell walls, structural membranes, and skeletal component of fungal mycelia, where it plays a structural role analogous to the cellulose of higher plants. The amino sugars in soil have often been referred to as "chitin," but this practice cannot be justified because it is known that soils contain amino sugars other than glucosamine; furthermore, the capsular material encasing the bodies of many bacteria consists of complex polysaccharides bearing amino sugars but that cannot be classified as chitin. It is generally assumed that the amino sugars in soil are of microbial origin. Lower members of the animal world, including insects, contain chitin in structural tissue but it is unknown as to whether any of this material persists in soil.

Amino sugars may play the dual role in soils of serving as a source of N for plant growth and promoting good soil structure. The ability of microbial polysaccharides to bind soil particles into aggregates of high stability is well-known, and special aggregating properties have sometimes been attributed to nitrogenous polysaccharides (see Chapter 18).

From 5 to 10 percent of the N in the surface layer of most soils can be accounted for as N-containing carbohydrates, or amino sugars (see section on Distribution of N Forms in Mineral Soils). Data of Gallali et al.[76] suggest that a relationship may exist between the percentage of the soil N as amino sugars and Ca^{2+} content, with the higher percentages being typical of Ca^{2+}-saturated soils. On the other hand, mean values for a broad selection of soils from widely different climatic zones indicate higher percentages for subtropical and tropical soils, which are usually acidic.[27] These relationships are shown in Table 3.11. The observed results may be due to extensive decomposition and turnover of organic matter in soils of the warmer climates, with selective preservation of amino sugars. The situation is analogous to the observed increase in amino sugar–N (relative to total organic N) when soils are subjected to intensive cultivation (see Table 3.4)

Early studies indicated that the proportion of the soil N as amino sugars may increase with increasing depth in the soil profile, reaching a maximum in the B horizon.[77,78] These observations require confirmation since other investigations have failed to show such an effect.[22,26,76]

82 ORGANIC FORMS OF SOIL NITROGEN

Table 3.11 Distribution of Amino Sugars (Glucosamine and Galactosamine) in Soils from Widely Different Climatic Zones (Means and Standard Deviations)[a]

Climatic Zone[b]	% of Soil N			Glucosamine
	Amino Sugar-N	Glucosamine	Galactosamine	Galactosamine
Arctic (6)	4.5 ± 1.7	2.8 ± 0.4	1.7 ± 0.5	1.7
Cool temperate (82)	5.3 ± 2.1	3.4 ± 0.9	1.9 ± 0.7	1.8
Subtropical (6)	7.4 ± 2.1	5.0 ± 1.5	2.4 ± 0.7	2.1
Tropical (10)	6.7 ± 2.1	3.9 ± 0.8	2.8 ± 0.5	1.4

[a] Adapted from Sowden et al.[27]
[b] Numbers in parentheses indicate number of soils.

Extraction and Quantitative Determination Amino sugars are normally recovered from soil by hydrolysis with HCl (usually $3N$ or $6N$ solution) for 6 to 9 h. Some degradation occurs during hydrolysis and a correction factor must be applied to account for these losses. The liberated amino sugars are then analyzed by the standard colorimetric method of Elson and Morgan or by alkaline distillation. Both methods give similar results.[6]

With the Elson–Morgan colorimetric method, the amino sugar is first heated with an alkaline solution of acetylacetone, and then an acid, alcoholic solution of p–dimethylaminobenzaldehyde (Ehrlich's reagent) is added. A chromogen is formed by the first reaction, following which the addition of Ehrlich's reagent produces a red solution. The overall reaction is shown in Fig. 3.4.

Many substances, including Fe and mixtures of amino acids and simple sugars, produce colors that interfere with the determination. However, these interfering substances are removed by treatment of the soil hydrolysate on ion exchange resin columns.[6]

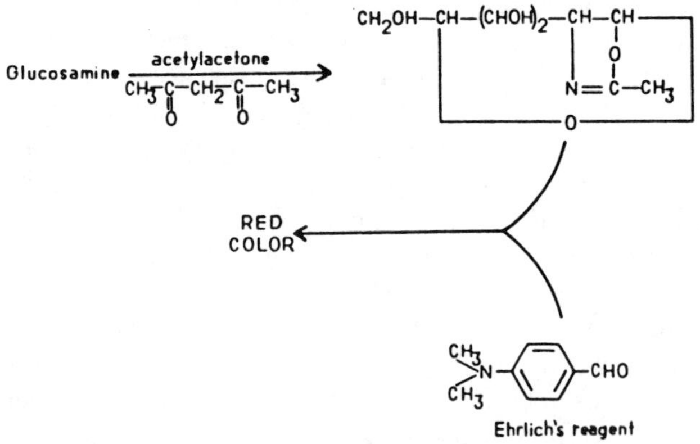

Fig. 3.4 Determination of amino sugars by the Elson–Morgan colorimetric method

The alkaline distillation method is based on the observation that amino sugars are readily deaminated by heating with alkali. In Bremner's[5] distillation procedure, amino sugar-N plus NH_3-N is determined by steam distillation with borate buffer of pH 8.7. Preformed NH_3 is then estimated by steam distillation with MgO, a procedure that recovers the NH_3 without deaminating amino sugars. The difference in values obtained by the two distillation procedures is amino sugar-N.

Isolation of Amino Sugars Individual amino sugars have been isolated from soil hydrolysates by paper partition chromatography,[52] ion-exchange chromatography,[79,80] and gas-liquid chromatography.[81,82] These studies indicate that most of the amino sugar material occurs as *D*-glucosamine and *D*-galactosamine, with the former occurring in greatest amounts. Structures of some of the more common amino sugars are as follows:

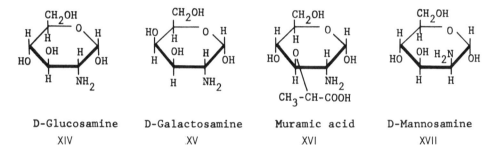

D-Glucosamine   D-Galactosamine   Muramic acid   D-Mannosamine
XIV   XV   XVI   XVII

Sowden's[77] findings indicate that the percentage of the amino sugar-N as glucosamine is rather high in acidic soils, an effect which can be explained by enhanced growth of fungi in acidic environments. Unlike bacteria, fungi synthesize glucosamine only.

Variations in the glucosamine/galactosamine ratios for the soils of several suborders are shown in Fig. 3.5. The highest ratios are found in those soils that are either highly acidic (Spodosol) or that have natural pH values in the neutral or slightly alkaline ranges (e.g., Albolls and Ustolls). As suggested above, the high ratio for the highly acidic Spodosol can be accounted for by a high population of fungi; high ratios for the near-neutral or alkaline soils may be due to high populations of actinomycete. Numbers of actinomycete are known to be especially high in soils that are near-neutral or slightly alkaline in reaction. Stevenson and Braids[79] observed increases in the glucosamine/galactosamine ratios with increasing depth in some soil profiles.

Muramic acid has also been observed in soil.[80,83] Miller and Casida[83] found that muramic acid levels were 10 to 1000 times the amounts that could be accounted for by microbial tissue as estimated by the plate-counting technique, but roughly comparable to the amounts expected on the basis of the microscopic counting method.

By using an improved ion-exchange chromatographic procedure, a large number of compounds have been isolated from soil that gives color reactions

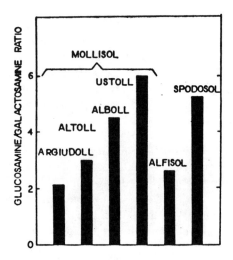

Fig. 3.5 Variations in the glucosamine/galactosamine ratio in the soils of several soil orders

of amino sugars.[80] A typical elution curve showing the separations obtained is given in Fig. 3.6. In addition to D-glucosamine and D-galactosamine, D-mannosamine and muramic acid were positively identified. Unknown 6 was tentatively identified as D-fucosamine. The presence of a wide variety of amino sugars in soil is expected because over 25 compounds of this class are known to exist in products synthesized by microorganisms.[84]

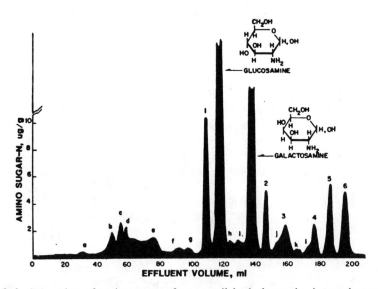

Fig. 3.6 Separation of amino sugars from a soil hydrolysate by ion-exchange chromatography. Note the occurrence of many compounds other than glucosamine and galactosamine, six of which occur in rather high amounts. From Stevenson[80]

Other Biochemical Compounds

A wide array of naturally occurring nitrogenous compounds other than amino acids and amino sugars have been found in soil, but in very low amounts. They include the nucleic acids and their derivatives, chlorophyll and chlorophyll-degradation products, phospholipids, amines, and vitamins.

Nucleic Acids and Derivatives Nucleic acids, which occur in the cells of all living organisms, consist of individual mononucleotide units (base–sugar–phosphate) joined by a phosphoric acid ester linkage through the sugar. Two types are known, ribonucleic acid (RNA) and deoxyribonucleic acid (DNA). The two types are identified by the nature of the pentose sugar (ribose or deoxyribose, respectively). Both contain the purine bases adenine (**XVIII**) and guanine (**XIX**) and the pyrimidine base cytosine (**XX**). In addition, RNA contains the pyrimidine uracil (**XXI**); DNA also contains thymine (**XXII**). Plant DNA contains 5-methyl cytosine (**XXIII**).

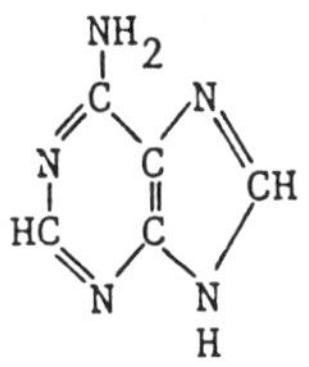

Adenine
XVIII

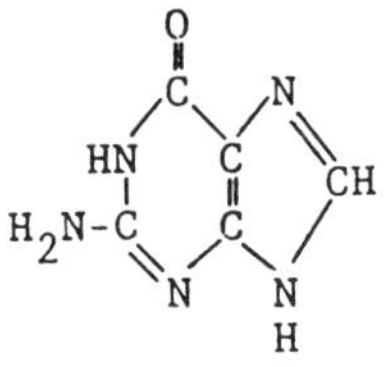

Guanine
XIX

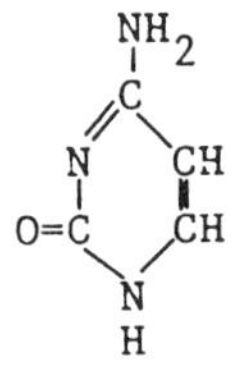

Cytosine
XX

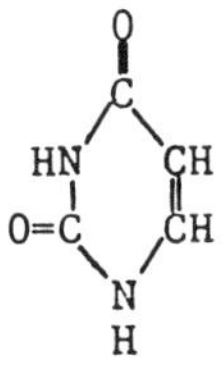

Uracil
XXI

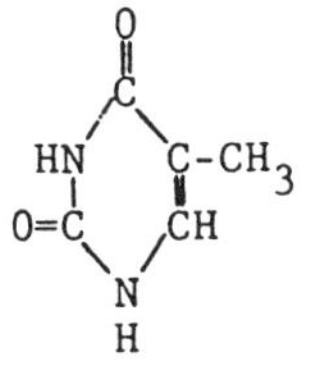

Thymine
XXII

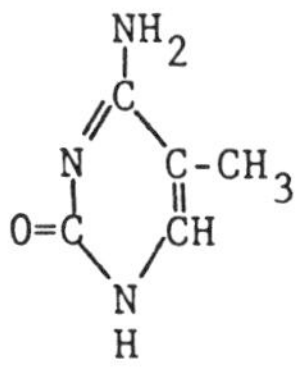

5-Methylcytosine
XXIII

The application of chromatographic methods has led to the identification of all of the above-mentioned bases in hydrolysates of soil organic matter preparation.[85,86] This work indicates that the bases occur as polynucleotides derived primarily from bacterial DNA. The occurrence of 5-methyl cytosine in some, but not all, soils suggests a plant source of some of the DNA.

The N in purine and pyrimidine bases is usually considered to account for less than 1 percent of the total soil N.[1-4] Somewhat higher percentages were found by Cortez and Schnitzer.[86] Their data (Table 3.12) indicated that up to 7.4 percent of the soil N and up to 18.6 percent of the N in fulvic acid occurred as purine and pyrimidine bases. On this basis, significant amounts of the N in the HUN fraction (see Table 1) may occur in nucleic acid bases.

Table 3.12 Percentage of the Total Soil N and of the N in the Humic Acid, Fulvic Acid, and Humin Fractions That Occurred as Purine and Pyrimidine Bases[a]

Soil[b,c]	% of N			
	Soil	Humic Acid	Fulvic Acid	Humin
Chernozemic (5)	0.9–2.3	1.7– 2.6	1.2– 6.9	0.4–10.4
Podzolic (5)	2.0–7.4	1.3–10.7	2.9–13.3	1.6– 6.2
Organic (3)	0.2–3.1	0.7– 2.6	0.8–18.6	0.2– 2.8
Regosolic (1)	4.4	2.1	2.2	7.7
Gleysolic (1)	0.7	0.4	0.8	0.8

[a] Adapted from Cortez and Schnitzer.[86]
[b] Equivalents in the Comprehensive Soil Classification System for the first four soil groups are Mollisol, Spodosol, Histosol, and Entisol, respectively.
[c] Numbers in parentheses indicate number of soils.

Chlorophyll and Chlorophyll-Degradation Products The green color of the landscape largely disappears in the autumn in temperate climate zones, and, as Hoyt[87] pointed out, plants in all climates usually lose their green color before their leaves fall on the soil surface. Accordingly, chlorophyll in its original state, is not an obvious residue constituent in most soils. Degradation of chlorophyll is apparently initiated by tissue enzymes and is very rapid.[87] Nevertheless, significant quantities of chlorophyll or its derivatives are added to the soil each year in plant remains, as well as in animal feces.[87,88] The amount of this material that persists depends upon a variety of soil conditions including moisture content and pH.

Chlorophyll and its derivatives in soil are estimated on the basis of the intensity of absorption peaks near 665 nm in 90 percent aqueous acetone extracts. A similar procedure has been used for the determination of these compounds in lake and marine sediments. The geochemical significance of chlorophyll has been discussed by Orr et al.[89]

A deficiency of O_2 inhibits complete destruction of chlorophyll; thus, poorly drained soils contain larger quantities of chlorophyll-type compounds than well-drained soils. Cornforth[90] found that the chlorophyll content of some soils of India was inversely related to acidity, and Gorham's[91] data showed that the amounts contained in the acidic mor humus layers of forest soils was considerably higher than in the more neutral mull humus layers. The higher chlorophyll content of acidic soils may be due, in part, to substitution of Fe or Mn for Mg in the molecule.[90] Hoyt[88] found that soil under grass grazed for 100 years contained higher amounts of chlorophyll than soil under grass cut for hay, a result that was attributed to chlorophyll added in the feces.

Phospholipids Small amounts of N are extracted from soil with the so-called "fat solvents," and it has been established that this N occurs in the form of glycerophosphatides. Hance and Anderson[92] found glycerophosphate, choline, and ethanolamine in approximate molar ratios of 1:1:0.2, respectively, which indicates that phosphatidyl choline (lecithin, **XXIV**) is the most abundant soil phospholipid, followed by phosphatidyl ethanolamine (**XXV**). Microorganisms contain variable amounts of phosphatidyl choline and phosphatidyl ethanolamine, the latter comprising over one-third of the lipid material in some bacteria.

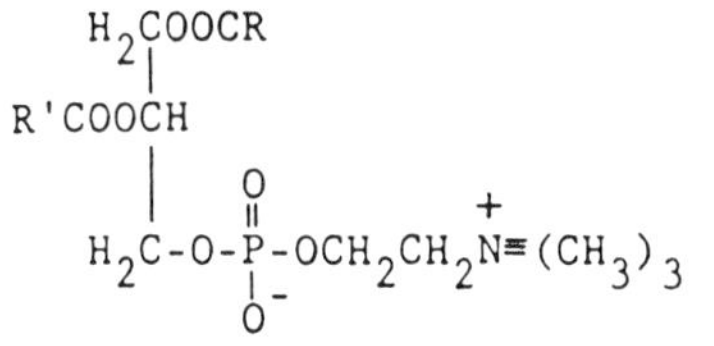

$$\begin{array}{c} H_2COOCR \\ | \\ R'COOCH \\ | \quad O \\ | \quad \| \\ H_2C-O-P-OCH_2CH_2\overset{+}{N}\equiv(CH_3)_3 \\ | \\ O^- \end{array}$$

L-α-Lecithin
XXIV

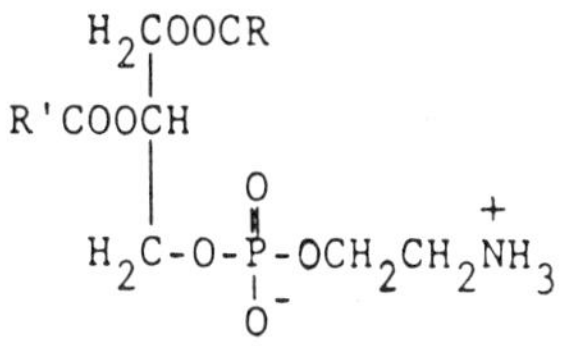

$$\begin{array}{c} H_2COOCR \\ | \\ R'COOCH \\ | \quad O \\ | \quad \| \\ H_2C-O-P-OCH_2CH_2\overset{+}{N}H_3 \\ | \\ O^- \end{array}$$

Phosphatidyl ethanolamine
XXV

Amines, Vitamins, and Other Compounds A wide variety of amines and other organic N compounds have been detected in trace amounts in soil or soil extracts, including choline, $CH_2N(CH_3)_3$, ethanolamine, $CH_2OH-CH_2NH_2$, trimethylamine, $(CH_3)_3N$, urea $CO(NH_2)_2$, histamine (**XXVI**), creatine (**XXVII**), allantoin (**XXXVIII**), cyanuric acid (**XXIX**), and α-picoline-γ-carboxylic acid (**XXX**).

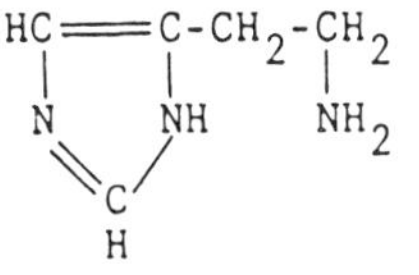

Histamine
XXVI

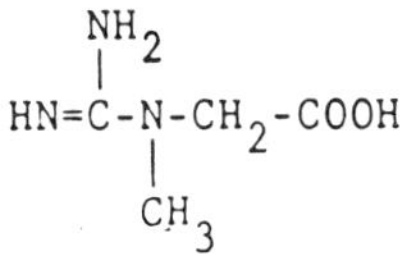

Creatine
XXVII

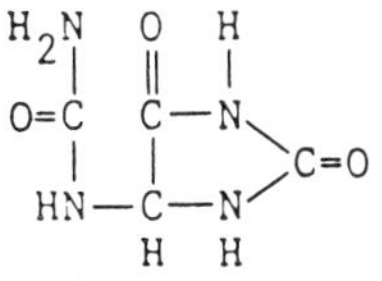

Allantoin
XXVIII

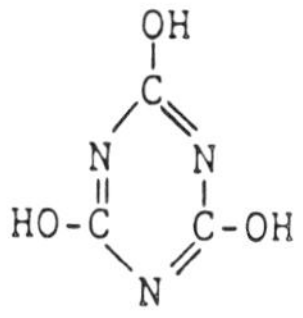

Cyanuric acid
XXIX

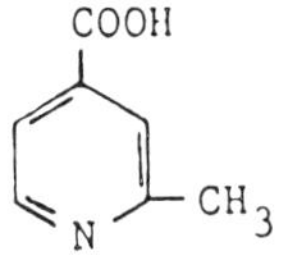

α-Picoline-γ-carboxylic acid
XXX

Anaerobic or water-logged conditions are particularly suitable for the formation and preservation of amines in soil. The amines identified by Fugii et al.[93] in plant residues and sand incubated under waterlogged conditions were putrescine (**XXXI**), cadaverine (**XXXII**), methylamine (**XXXIII**), ethylamine

(XXXIV), *n*-propylamine **(XXXV)**, and isobutylamine **(XXXVI)**. Temporary accumulations were also noted by aerobic incubations.

$NH_2-(CH_2)_4-NH_2$
Putrecine
XXXI

$NH_2-(CH_2)_5-NH_2$
Cadaverine
XXXII

CH_3-NH_2
Methylamine
XXXIII

$CH_3-CH_2-NH_2$
Ethylamine
XXXIV

$CH_3-CH_2-CH_2-NH_2$
n-Propylamine
XXXV

$CH_3-CH(CH_3)-CH_2-NH_2$
Isobutylamine
XXXVI

Secondary amines, both aliphatic and aromatic, react with nitrous acid to form N–nitrosamines (compounds containing the N–N=O group). They represent a group of substances that are carcinogenic and mutagenic at low concentrations. Thus, a potential hazard to the health of man and animals would exist if nitrosamines were formed in soil from pesticide degradation products or from precursors present in manures and sewage sludge. A health hazard would become a reality, however, only if the nitrosamines thus formed were leached into water supplies or taken up by plants used as food by livestock or humans. Trace quantities of nitrosamines have been detected in soils amended with known amines (dimethylamine, trimethylamine) and NO_2^- or NO_3^-, but, for most part, this work has been done under ideal conditions for nitrosamine formation in the laboratory, such as high additions of reactants.[94,95] Evidence is lacking that the synthesis of nitrosamines in field soil represents a threat to the environment.

A host of water-soluble, nitrogen-containing B vitamins have also been reported in soils, including biotin **(XXXVII)**, thiamine **(XXXVIII)**, nicotinic acid **(XXXIX)**, pantothenic acid **(XL)**, and cobamide coenzyme (vitamin B_{12}). These constituents are of special importance because they may act as growth factors for numerous organisms. Levels of B vitamins in soil are directly related to those factors influencing microbial activity.

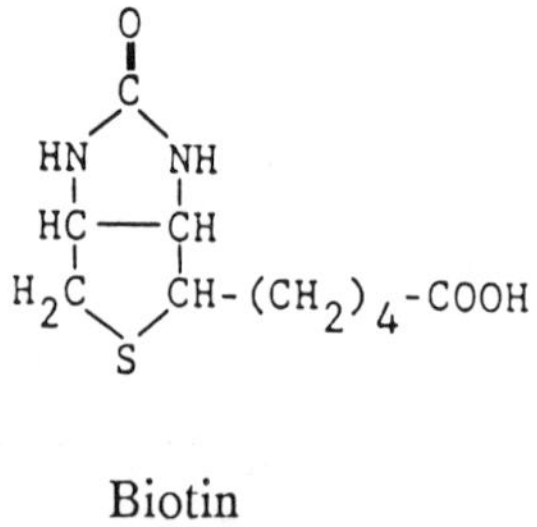

Biotin
XXXVII

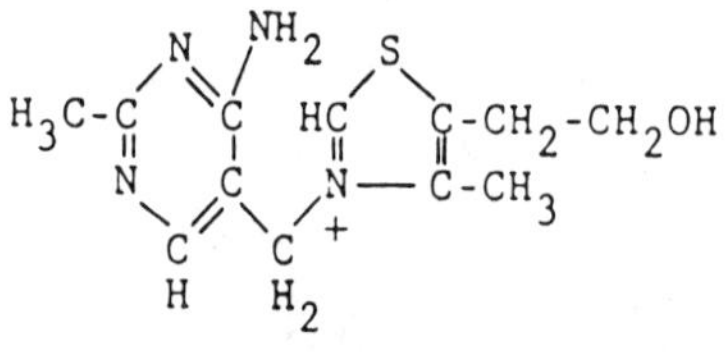

Thiamine
XXXVIII

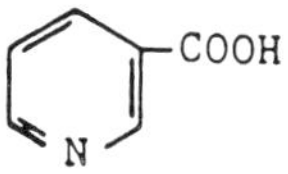

Nicotinic acid
XXXIX

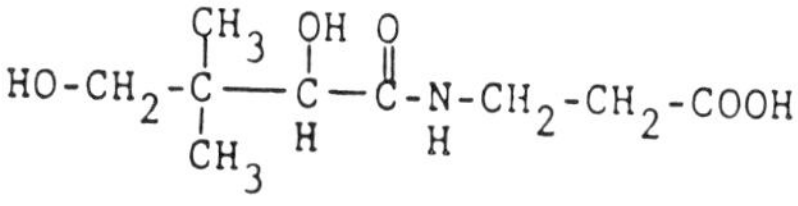

Pantothenic acid
XL

Pesticide and Pesticide Degradation Products Substantial evidence exists to indicate that many pesticides or their partial degradation products can form stable linkages with components of soil organic matter and that such binding increases their persistence in soil. Many of these pesticides contain N as part of their structures, such as the *s*-triazines, phenylcarbamates, substituted ureas, amides, and quaternary ammonium derivatives. The latter, which contain two positive charges and are strongly bound to soil colloids, may persist in soils for years. The adsorption and retention of pesticides and their partial decomposition products is discussed in some detail in Chapters 18 and 19.

BIOMASS N

Biomass N has been determined using incubation and extraction procedures similar to those used for biomass C (see Chapter 1). In the incubation method for biomass N, measurements are made for the amount of N (mineral + organic) released following incubation of $CHCl_3$-fumigated soil, the extractant being 0.5M K_2SO_4.[96] Biomass N is obtained from the relationship:

$$\text{Biomass N} = F_n/k_n$$

where F_n is the difference between the amount of N released by incubation of fumigated and unfumigated soil and k_n is the fraction of the biomass N that is released to soluble forms during incubation. Estimates for k_n have been somewhat variable but a value for k_n of 0.68 has been recommended.

A problem with the fumigation–incubation method is that denitrification and/or immobilization may alter the amount of N accounted for in soluble forms. This problem is eliminated using a method in which a direct extraction is made of the fumigated soil, also with 0.5M K_2SO_4.[97] An additional advantage of this approach is that there is no need for complete removal of fumigant or for prolonged incubation of the soil under carefully controlled conditions. Joergensen and Brookes[98] have described a colorimetric method for biomass N based on the extra amount of ninhydrin-reactive N released during fumigation (e.g., α-amino acids and NH_4^+-N).

As a general rule, from to 1 to 6 percent of the soil organic N residues in the microbial biomass at any one time.

NATURAL VARIATIONS IN N ISOTOPE ABUNDANCE

Slight variations occur in the N isotope composition of natural substances, including soil, peat, coal, petroleum, rocks, minerals, and the proteins of plants and animals. These variations result from isotopic effects during biochemical and chemical transformations, such as nitrification and denitrification. The overall effect of these isotope effects in soil is a slight increase in the average ^{15}N content of soil N and its fractions, as compared to atmospheric N_2.

Natural variations in N isotope abundance are usually expressed in terms of the per mil excess ^{15}N, or delta ^{15}N ($\delta^{15}N$). The equation is:

$$\delta^{15}N = \frac{\text{atom \% }^{15}N \text{ in sample} - \text{atom \% }^{15}N \text{ in standard}}{\text{atom \% }^{15}N \text{ in standard}} \times 1000.$$

Thus, a $\delta^{15}N$ values of +10 indicates that the experimental sample is enriched by 1 percent compared with the atom % ^{15}N of the standard, usually atmospheric N_2. A negative value indicates that the sample is depleted in ^{15}N relative to the standard.

Numerous studies have shown that the $\delta^{15}N$ value of the soil N generally falls within the range of +5 to +12, although higher and lower values are by no means rare. For any given soil, variations exist in the $\delta^{15}N$ value of the various N fractions.[99] Kanazawa and Yoneyama[100] found that the ^{15}N abundance of soil amino acids was higher than for the total soil N.

Natural variations in ^{15}N content may ultimately prove useful in evaluating N cycle processes in soil, such as biological N_2 fixation and denitrification. The approach has been used in attempts to estimate the relative contribution of soil and fertilizer N to the NO_3^- in surface waters. A serious problem in using the technique to evaluate the contribution of fertilizer N to NO_3^- levels in drainage waters is that the ^{15}N content of mineralized N may not be the same as the soil humus N.

^{15}N–NMR SPECTROSCOPY

The technique of ^{15}N–NMR spectroscopy has the potential for the characterization of organic N compounds in soil. This subject is discussed in Chapter 11.

STABILITY OF SOIL ORGANIC N

The high resistance of organic N complexes in soil to microbial attack is of considerably significance to the N balance of the soil and several theories have been given to account for this phenomenon. Explanations often given to explain the stability of organic N include the following:

1. Proteinaceous constituents (e.g., amino acids, peptides, proteins) are stabilized through their reaction with other organic constituents, such as lignins, tannins, quinones, and reducing sugars. Some of the reactions believed to be involved with quinones and reducing sugars have been outlined in Chapter 8. For lignins, the main reaction is believed to be one involving NH_2 groups of the protein and $C=O$ groups of lignin. The high stability of proteins, peptides, and amino acids when linked to aromatic rings has been demonstrated by Verma et al.[101]

2. Biologically resistant complexes are formed in soil by chemical reactions involving NH_3 or NO_2^- with lignins or humic substances. The complexes thus formed have been shown to be highly resistant to mineralization by soil microorganisms. Reactions involved in the chemical fixation of NH_3 and NO_2^- by soil organic matter are discussed in Chapter 4.

3. Adsorption of organic N compounds by clay minerals protects the molecule from decomposition. It is well known that the N content of fine-textured soils is higher than coarse-textured soils, and that clays, particularly montmorillinitic types, reduce the rate at which proteins and other nitrogenous compounds are decomposed by microorganisms or by proteinase enzymes. Treatment of mineral soils with HF to decompose clay minerals results in the solubilization of considerably quantities of organic N (see Chapter 2), which indicates that some of the organic N may be entrapped within the lattice structures of clay minerals. The reader is referred to Chapter 18 for additional information on the nature of clay–humus complexes.

4. Complexes formed between organic N compounds and polyvalent cations, such as Fe, are biologically stable. One explanation given for the high stability of organic matter in allophanic soils is that reactive groups of humic substances combine with Al in such way that the surface of the humic molecule no longer provides a suitable fit for enzymes capable of attacking them.[102] Because N is an integral part of the humic molecule, the availability of N in allophanic soils would be restricted by the same mechanism.

5. Some of the organic N occurs in small pores or voids and is physically inaccessible to microorganisms. A typical soil bacterium would be about 0.5 μm in diameter and 1 μm or so in length; actinomycete, fungi, and soil faunal organisms are even larger. Enzymes have the potential for penetrating very small pores but their movement may be restricted through adsorption.

SUMMARY

Despite the excellent progress that has been made in the past two decades toward the characterization of soil N complexes, relatively little is known regarding the main structural components as they exist in the natural soil. Also, very little is known as to the manner by which nitrogenous substances are bound to organic and inorganic soil colloids. The problems are exceedingly complex, because a variety of chemical and physical processes are involved.

A significant fraction of the organic N in soil can be recovered as simple biochemical compounds, such as amino acids, by hydrolysis procedures and extensive use has been made of chromatographic techniques in attempts to characterize these compounds. Unfortunately, most of the work cannot be evaluated because of errors associated with incomplete extraction, improper sample preparation, and lack of precision during chromatography. The state of the art has reached the stage where accurate quantitative values are of greater importance than the mere detection of compounds. A wide variety of new separation and analytical techniques are available that should facilitate research in this area.

It is noteworthy that over one-half of the organic N in most soils cannot be accounted for in known compounds. The chemistry of this N will be understood only when more is known about the structures of humic and fulvic acids. Establishment of the composition of these constituents constitutes one of the most challenging problems facing the modern soil scientist. Other major problems remaining to be solved include: 1) nature of clay mineral–organic N complexes; 2) role of nitrogenous constituents in the binding of metal ions; 3) resistance of soil–N complexes to attack by microorganisms; 4) availability of organic N to higher plants; and 5) role of organic N compounds in geochemical and pedogenic processes.

REFERENCES

1. H. A. Anderson, W. Bick, A. Hepburn, and M. Stewart, "Nitrogen in Humic Substances," in M. H. B. Hayes, P. MacCarthy, R. L. Malcolm, and R. S. Swift, Eds., *Humic Substances II: In Search of Structure*, Wiley, New York, 1989, pp. 223–253.
2. M. Schnitzer, "Nature of Nitrogen in Humic Substances," in G. R. Aiken, D. M. McKnight, R. L. Wershaw, and P. MacCarthy, Eds., *Humic Substances in Soil, Sediment, and Water*, Wiley, New York, 1985, pp. 303–325.
3. F. J. Stevenson, "Organic Forms of Soil Nitrogen," in F. J. Stevenson, Ed., *Nitrogen in Agricultural Soils*, American Society of Agronomy, Madison, Wisconsin, 1982, pp. 67–122.
4. F. J. Stevenson and X.-T. He, "Nitrogen in Humic Substances as Related to Soil Fertility" in P. MacCarthy, C. E. Clapp, R. L. Malcolm, and P. R. Bloom, Eds., *Humic Substances in Soil and Crop Sciences: Select Readings*. American Society of Agronomy, Madison, Wisconsin, 1990, pp. 91–109.
5. J. M. Bremner, "Organic Forms of Nitrogen," in C. A. Black et al., Eds., *Methods of Soil Analysis*, American Society of Agronomy, Madison, Wisconsin, 1965, pp. 1148–1178.
6. F. J. Stevenson, "Nitrogen–Organic Forms," in A. L. Page et al., Eds., *Methods*

of Soil Analysis: Part 2, 2nd ed., American Society of Agronomy, Madison, Wisconsin, 1982, pp. 6251–6411.
7 R. F. Bunyakina, *Soviet Soil Sci.*, **8,** 438, 1976.
8 C.-N. Cheng, R. C. Shufeldt, and F. J. Stevenson, *Soil Biol. Biochem.*, **7,** 143 (1975).
9 S. M. Griffith, F. J. Sowden, and M. Schnitzer, *Soil Biol. Biochem.*, **8,** 529 (1976).
10 R. W. Aldag, "Relations Between Pseudo-amide Nitrogen and Humic Acid Nitrogen Released Under Different Hydrolytic Conditions," in *Soil Organic Matter Studies*, International Atomic Energy Agency, Vienna, 1977, pp. 293–299.
11 K. M. Goh and D. C. Edmeades, *Soil Biol. Biochem.*, **11,** 127 (1979).
12 M. Schnitzer, P. R. Marshall, and D. A. Hindle, *Can. J. Soil Sci.*, **63,** 625 (1983).
13 L. G. Greenfield, *Plant Soil*, **36,** 191 (1972).
14 R. C. Dalal, *Soil Sci.*, **125,** 178 (1978).
15 R. Hayashi and T. Harada, *Soil Sci. Plant Nutr.*, **15,** 226 (1969).
16 S. Kadirgamatharyah and A. F. Mackenzie, *Plant Soil*, **33,** 120 (1970).
17 D. R. Keeney and J. M. Bremner, *Soil Sci. Soc. Amer. Proc.* **28,** 653 (1964).
18 S. U. Khan and F. J. Sowden, *Can. J. Soil Sci.*, **52,** 116 (1972).
19 K. Kyuma, A. Hussain, and K. Kawaguchi, *Soil Sci. Plant Nutr.*, **15,** 149 (1969).
20 V. W. Meints and G. A. Peterson, *Soil Sci.*, **124,** 334 (1977).
21 R. A. Rosell, J. C. Salfeld, and H. Söchtig, *Agrochimica*, **22,** 98 (1978).
22 B. R. Singh, A. P. Uriyo, and B. J. Lontu, *Soil Biol. Biochem.*, **10,** 105 (1978).
23 D. R. Keeney and J. M. Bremner, *Soil Sci. Soc. Amer. Proc.*, **30,** 714 (1966).
24 Z. Ahmad, Y. Yahiro, H. Kai, and T. Harada, *Soil Sci. Plant Nutr.*, **19,** 287 (1973).
25 K. Yonebayashi, K. Kyuma, and K. Kawaguchi, *Soil Sci. Plant Nutr.*, **20,** 421, 423 (1973).
26 F. J. Sowden, *Can. J. Soil Sci.*, **57,** 445 (1977).
27 F. J. Sowden, Y. Chen, and M. Schnitzer, *Geochim. Cosmochim. Acta*, **41,** 1524 (1977).
28 H. Fleige and K. Baeumer, *Agro-Ecosystems*, **1,** 19 (1974).
29 S. U. Khan, *J. Soil Sci*, **51,** 431 (1971).
30 F. J. Sowden, *Soil Sci*, **82,** 491 (1956).
31 H. Nõmmik, *J. Soil Sci*, **18,** 301 (1967).
32 F. J. Stevenson and C-N. Cheng, *Geochim. Cosmochim. Acta*, **34,** 77 (1970).
33 N. O. Isirimah and D. R. Keeney, *Soil Sci.*, **115,** 123 (1973).
34 F. J. Sowden, H. Morita, and M. Levesque, *Can. J. Soil Sci.*, **58,** 237 (1978)
35 J. M. Bremner, *J. Agr. Sci.*, **46,** 247 (1955).
36 K. Tsutsuki and S. Kuwatsuka, *Soil Sci. Plant Nutr.*, **24,** 29, 561 (1978).
37 H. Otsuka, *Soil Sci. Plant Nutr.*, **21,** 420 (1975).
38 P. Sequi, G. Guidi, and G. Petruzzelli, *Can. J. Soil Sci.*, **55,** 439 (1975).
39 A. A. Batsula and N. K. Krupskiy, *Soviet Soil Sci.*, **4,** 456 (1974).
40 F. J. Stevenson, *Soil Sci. Soc. Amer. Proc.*, **24,** 472 (1966).

41 R. Kickuth and F. Scheffer, *Agrochimica,* **20,** 373 (1976).
42 T. J. Piper and A. M. Posner, *Soil Sci.,* **106,** 188 (1968).
43 T. J. Piper and A. M. Posner, *Plant Soil,* **36,** 595 (1972).
44 V. O. Biederbeck and E. A. Paul, *Soil Sci.,* **115,** 357 (1973).
45 P. Simonart, L. Batistic, and J. Mayaudon, *Plant Soil,* **27,** 153 (1967).
46 U. Müller-Wegener, *Sci. Total Environ.,* **62,** 297 (1987).
47 S. Suzuki, *Bull. Coll. Agr. Tokyo,* **7, 95, 419, 513** (1906–1908).
48 J. M. Bremner, *J. Agr. Sci.,* **39,** 183 (1949).
49 F. J. Sowden, *Soil Sci.,* **107,** 264 (1969).
50 C.-N. Cheng, *Soil Biol. Biochem.,* **7,** 319 (1975).
51 F. J. Stevenson and C.-N. Cheng, *Geochim. Cosmochim. Acta,* **36,** 653 (1972).
52 J. M. Bremner, *Biochem. J.,* **47,** 538 (1950).
53 F. J. Stevenson, *Soil Sci. Soc. Amer. Proc.,* **18,** 373 (1954).
54 F. J. Stevenson, *Soil Sci. Soc. Amer. Proc.,* **20,** 201 (1956).
55 G. E. Pollack, C.-N. Cheng, and S. E. Cronin, *Anal. Chem.,* **49,** 2 (1977).
56 J. L. Young and J. L. Mortensen, *Ohio Agr. Exp. Sta. Res. Circular,* **61,** 1 (1958).
57 L. A. V. Almeida, R. P. Ricardo, and M. B. M. C. Rouy, *Agrochimica,* **13,** 358 (1969).
58 F. J. Sowden, *Can. J. Soil Sci.,* **50,** 227 (1970).
59 F. J. Stevenson, *Soil Sci. Soc. Amer. Proc.,* **20,** 204 (1956).
60 T. S. C. Wang, T-K. Yang, and S-Y. Cheng, *Soil Sci.,* **103,** 67 (1967).
61 S. U. Khan and F. J. Sowden, *Can. J. Soil Sci.,* **51,** 185 (1971).
62 Y. Chen, F. J. Sowden, and M. Schnitzer, *Agrochimica,* **21,** 7 (1977).
63 F. J. Sowden, S. M. Griffith, and M. Schnitzer, *Soil Biol. Biochem.,* **8,** 35 (1976).
64 T. Yamashita and T. Akiya, *J. Soil Sci. Manure, Japan,* **34,** 255 (1963).
65 H. Kai, Z. Ahmad, and T. Harada, *Soil Sci. Plant Nutr.,* **19,** 275 (1973).
66 P. W. Carter and R. M. Mitterer, *Geochim. Cosmochim. Acta,* **42,** 1231 (1978).
67 G. Guidi, G. Petruzzelli, and P. Sequi, *Can. J. Soil Sci.,* **56,** 159 (1976).
68 G. Dungworth, *Chem. Geol.,* **17,** 135 (1976).
69 R. A. Schroeder and J. L. Bada, *Earth Sci. Rev.,* **12,** 347 (1976).
70 E. A. Paul and E. L. Schmidt, *Soil Sci. Soc. Amer. Proc.,* **24,** 195 (1960).
71 R. G. Gilbert and J. Altman, *Plant Soil,* **24,** 229 (1966).
72 K. C. Ivarson and F. J. Sowden, *Can. J. Soil Sci.,* **50,** 191 (1970).
73 K. C. Ivarson and F. J. Sowden, *Can. J. Soil Sci.,* **46,** 115 (1966).
74 M. Wainwright and G. J. F. Pugh, *Soil Biol. Biochem.,* **7,** 1 (1979).
75 A. D. Rovira and B. M. McDougall, "Microbiological and Biochemical Aspects of the Rhizosphere," in A. D. McLaren and G. H. Petersen, Eds., *Soil Biochemistry,* Marcel Dekker, New York, 1967, pp. 417–463.
76 T. Gallali, A. Gluckert, and F. Jacquin, *Bull. de l'Ecole Nationale Superieure d'Agronomie et des Industries Alimentaires,* **17,** 53 (1975).
77 F. J. Sowden, *Soil Sci.,* **88,** 138 (1959).

78. F. J. Stevenson, *Soil Sci.*, **83**, 113; **84**, 99 (1957).
79. F. J. Stevenson and O. C. Braids, *Soil Sci. Soc. Amer. Proc.*, **32**, 598 (1968).
80. F. J. Stevenson, *Soil Sci. Soc. Amer. J.*, **47**, 61 (1983).
81. L. Benzing-Purdie, *Soil Sci. Soc. Amer. J.*, **45**, 66 (1981).
82. L. Benzing-Purdie, *Soil Sci. Soc. Amer. J.*, **48**, 219 (1984).
83. W. N. Miller and L. E. Casida, *Can. J. Microbiol.*, **16**, 299 (1970).
84. N. Sharon, "Distribution of Amino Sugars in Microorganisms, Plants, and Invertebrates," in R. W. Jeanloz and E. A. Balazs, Eds., *The Amino Sugars, Vol. IIA*, Academic Press, New York, 1965, pp. 1–45.
85. G. Anderson, "Nucleic Acids, Derivatives, and Organic Phosphates," in A. D. McLaren and G. H. Petersen, Eds., *Soil Biochemistry*, Marcel Dekker, New York, 1967, pp. 67–90.
86. J. Cortez and M. Schnitzer, *Can. J. Soil Sci.*, **59**, 277 (1979).
87. P. B. Hoyt, *Soil Sci.*, **111**, 49 (1971).
88. P. B. Hoyt, *Plant Soil*, **25**, 167, 313 (1966).
89. W. L. Orr, K. O. Emery, and J. R. Grady, *Bull. Amer. Assoc. Petrol. Geol.*, **42**, 925 (1958).
90. I. S. Cornforth, *Experimental Agric.*, **4**, 193 (1968).
91. E. Gorham, *Soil Sci.*, **87**, 258 (1959).
92. R. J. Hance and G. Anderson, *Soil Sci.*, **96**, 157 (1963).
93. K. Fujii, M. Kobayashi, and E. Takahashi, *Soil Sci. Plant Nutr.*, **20**, 101 (1974).
94. A. L. Mills and M. Alexander, *J. Environ. Qual.*, **5**, 437 (1976).
95. S. K. Pancholy, *Soil Biol. Biochem.*, **10**, 27 (1978).
96. S. M. Shen, G. Pruden, and D. S. Jenkinson, *Soil Biol. Biochem.*, **16**, 437 (1984).
97. P. C. Brookes, A. Landman, G. Pruden, and D. S. Jenkinson, *Soil Biol. Biochem.*, **17**, 837 (1985).
98. R. G. Joergensen and P. C. Brookes, *Soil Biol. Biochem.*, **22**, 1023 (1990).
99. H. H. Cheng, J. M. Bremner, and A. P. Edwards, *Science*, **146**, 1574 (1964).
100. S. Kanazawa and T. Yoneyama, *Soil Sci. Plant Nutr.*, **24**, 153 (1978).
101. L. Verma, J. P. Martin, and K. Haider, *Soil Sci. Soc. Amer. Proc.*, **39**, 279 (1975).
102. F. E. Broadbent, R. H. Jackman, and J. McNicoll, *Soil Sci.*, **98**, 118 (1964).

4

NATIVE FIXED AMMONIUM AND CHEMICAL REACTIONS OF ORGANIC MATTER WITH AMMONIA AND NITRITE

At one time it was thought that essentially all of the N in soils occurred in organically bound forms but it is now known that virtually all soils that contain fine-grained mineral matter (clay) will also contain ammonium (NH_4^+) that is held within the lattice structures of clay minerals. This clay-fixed NH_4^+ is not readily leached, and it is generally unavailable to plants and microorganisms.

Still another facet of the soil N picture is that of chemical reactions of organic matter with ammonia (NH_3) and nitrite (NO_2^-), and this subject is also covered herein.

LEVELS OF NATIVE FIXED NH_4^+ IN SOILS

Rodrigues[1] has been credited as being the first to report the occurrence of fixed NH_4^+ in soils. He concluded that some tropical soils of the Caribbean region contained between 282 to 1920 µg/g of fixed NH_4^+-N. Although these values have since been shown to be high because of inadequacies in methods available at that time, his theory that many soils contain appreciable amounts of fixed NH_4^+ is essentially correct.[2-9]

The retention of NH_4^+ in a form that is not exchangeable with K^+ (also a fixable cation) results from the substitution of NH_4^+ for such interlayer cations as Na^+, Ca^{2+}, and Mg^{2+} within the expandable lattice of clay minerals. According to one popular theory, the reason NH_4^+ is fixed is that the ion fits snugly into hexagonal holes formed by oxygen atoms on exposed surfaces between the sheets of 2:1 lattice-type clay minerals. This causes the lattice layers to contract and be bound together, thereby preventing hydration and expansion. Other common cations, notably Na^+ and Ca^{2+}, are not fixed because they are either too large to fit into the voids or they are so small that they cannot bind the layers tightly together; consequently, these cations move more freely in and out of the clay sheets. Ammonium and the K^+ ion have nearly identical ionic radii and are fixed by the same mechanism.

Data given in Table 4.1 show that a wide range of values have been recorded for the amounts of fixed NH_4^+ in soils.[9] The surface layer of the soil will often contain 100 μg/g of fixed NH_4^+-N, equivalent to 224 kg/ha (200 lb./acre) per furrow slice of soil. Clay and clay loam soils generally contain more fixed NH_4^+ than silt loams, which in turn contain larger amounts than sandy soils. Spodosols contain rather low amounts of fixed NH_4^+, which can be attributed to their very low clay contents. Organic soils (i.e., Histosols) contain very little clay-fixed NH_4^+.

Table 4.1 Some Typical Values for Fixed NH_4^+ in Soils[a]

Location	Range, μg/g	Comments
Australia	41–1,076	From 221 to 1,076 μg/g (5–90% of N) in profiles developed on Permian phyllite and from 41 to 315 μg/g (5–82% of N) in soils formed on other parent materials
Canada		
Saskatchewan	158–330	From 7.7 to 13.3% of N in surface soil and up to 58.6% in subsoil; cultivation did not affect fixed NH_4^+ content
Alberta	110–370	From 7 to 14% of N in a wide variety of surface soils; percentage increased with depth
England	52–252	From 4 to 8% of N in surface soils and from 19 to 45% in subsoils
Nigeria	32–220	From 2 to 6% of N in surface layers and from 45 to 63% in surface layers
Russia	14–490	From 2 to 7% of the N in surface soil but the percentage increases with depth
Sweden	10–17	Values are for a Spodosol profile low in clay
Taiwan	140–170	From 10.6 to 32.6% of the N in surface layer of nine soils
United States		
North Central	7–270	A wide range has been recorded, the lowest being in Spodosols and the highest in silt loams and soils rich in illite; from 4 to 8% of the N in the surface layer with the proportion increasing with depth
Pacific Northwest	17–138	From 1.1 to 6.2% of N in surface soils with the proportion increasing with depth in some soils but not others
Hawaii	0–585	Volcanic ash soils contained less (4–178 μg/g) than soils from basalt (up to 585 μg/g)

[a] For references see Young and Aldag.[9]

The highest value thus far reported for fixed NH_4^+ in the surface layer of the soil appears to be that of Dalal,[3] who recorded a value of 1300 $\mu g/g$ in a Trinidad soil formed from micaceous schist and phyllite. Martin et al.[4] recorded high values for fixed NH_4^+ in some subtropical soils derived from phyllite in Australia; the content of fixed NH_4^+-N in one soil ranged from 415 $\mu g/g$ in the surface layer to over 1000 $\mu g/g$ at a depth of 120 cm (4 ft.).

The amount of N in the soil as fixed NH_4^+ on an acre-profile basis can be appreciable, as shown by Fig. 4.1. The soil volume occupied by plant roots may contain over 1700 kg N/ha (1520 lb./acre) as fixed NH_4^+. The profiles analyzed by Young[5] generally contained a total of from 470 to 2385 kg of fixed NH_4^+-N/ha (420 to 2130 lb./acre) to depths of from 0.76 to 1.52 m; those examined by Hinman[6] ranged from 2960 to 5200 kg to a 1.22 m depth. Moore and Ayeke[7] recorded a range of from 1345 to 2350 kg of fixed NH_4^+-N/ha to a 1.52 m depth in some Nigerian soils while Mogilevkina[8] observed a range of from 526 to 1608 kg/ha (470 to 1440 lb./acre) for the top 100 cm (3.3 ft.) of some Russian soils.

Results obtained for the vertical distribution of fixed NH_4^+ in representative forest and prairie grassland soils of the United States are depicted in Fig. 4.2. In contrast to the sharp decline in total N with depth, the soils' content of fixed

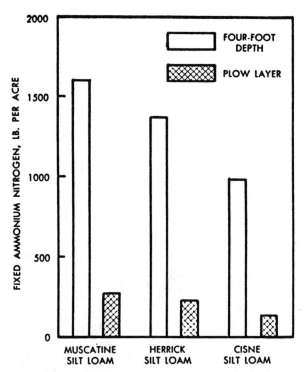

Fig. 4.1 Context of fixed NH_4^+-N in the plow layer and to a depth of 1.22 m (4 ft.) in three agricultural soils of Illinois

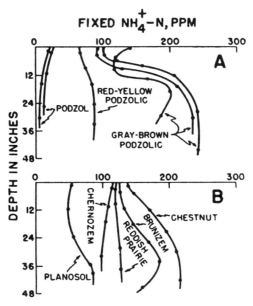

Fig. 4.2 Vertical distribution of fixed NH_4^+ in soils representative of several Great Soil Groups in central United States. The top part (A) includes two Spodosols (Podzol), an Ultisol (Red-Yellow Podzolic), and three Alfisols (Gray-Brown Podzolic). All five soils in the bottom part (B) are Mollisols. From Stevenson and Dhariwal[10]

NH_4^+ either changes very little or increases. Accordingly, the proportion of the soil N as fixed NH_4^+ progressively increases with increasing depth, as shown in Table 4.2. In general, less than 10 percent of the N in the plow layer of the soil occurs as fixed NH_4^+, although higher percentages are not uncommon. In some subsurface soils, over 50 percent of the N may occur as fixed NH_4^+.

Table 4.2 Distribution of Fixed NH_4^+ in Representative Forest and Prairie Grassland Soils of the United States[a]

Horizon	Mollisols (5)[b]		Alfisols (5)		Others[c]	
	µg/g	% of N	µg/g	% of N[d]	µg/g	% of N
A_1	96.4–128.8	4.3–5.6	112.6–154.6	5.6	57.4–140.6	3.5–7.9
A_2	105.6–131.2	8.2–9.3	109.2–168.0	9.4	50.4–154.0	7.8–10.8
B_2	99.4–155.4	11.5–18.4	142.8–210.0	20.1	54.6–182.0	12.7–17.6
B_3	103.6–173.6	21.4–27.7	184.8–224.0	26.5	84.0–210.0	13.1–17.0
C	98.0–187.6	28.8–44.4	102.2–224.0	34.9	89.6–210.0	27.2–36.1

[a] Adapted from Stevenson and Dhariwal[10]
[b] Numerals refer to number of profiles examined.
[c] Includes an Ulltisol and two Mollisols with argillic horizons.
[d] For one profile only.

Such factors as drainage, type of vegetative cover, and extent of leaching of the profile by percolating water seems to have little effect on the soils' content of fixed NH_4^+; the amounts are more closely related to the kinds and amounts of clay minerals that are present (note low values for the Spodosols in Fig. 4.2). Regarding clay mineral type, fixed NH_4^+ content follows the order: vermiculite > illite > montmorillonite > kaolinite. The high levels of fixed NH_4^+ in the lower horizons of some soils of the north central United States can be explained by the fact that the parent material from which these soils were formed contained large amounts of illite.[10] As noted earlier, high values for fixed NH_4^+ have been observed in soils derived from mica-rich parent material. A relationship has been noted between K^+ content and fixed NH_4^+ in some but not all soils.

Changes in fixed NH_4^+ content have been observed through long-term cropping but they have been slight and variable.[10,11] In contrast, organic matter is lost at a rapid rate when soils are first placed under cultivation. Accordingly, the percentage of the soil N as fixed NH_4^+ is lower in virgin soils than in cropped soils. The continuous application of ammoniacal fertilizers sometimes leads to slight increases in fixed NH_4^+ content.

THE C/N RATIO

A distinguishing characteristic of soils, sediments, and sedimentary rocks is the constancy of the C/N ratio.[5,12] For surface soils, and for the top layer of lake and marine sediments, the ratio generally falls within well-defined limits, usually from about 10 to 12. In most soils, the C/N ratio decreases with increasing depth, often attaining values less than 5.0.

Native humus would be expected to have a lower C/N ratio than most plant residues for the following reasons. The decay of organic residues by soil organisms leads to incorporation of part of the C into microbial tissue, with the remainder being liberated as CO_2. As a general rule, about one-third of the applied C in fresh residues will remain in the soil after the first few months of decomposition. The decay process is accompanied by conversion of organic form of N to NH_3 and soil microorganisms utilize part of this N for synthesis of new cells. The N content of microbial tissue varies widely but will be of the order of 6 to 13 percent for bacteria and 3 to 6 percent for fungi. Thus, the gradual transformation of plant raw material into stable organic matter (humus) leads to the establishment of a reasonably consistent relationship between C and N. Other factors that may be involved in narrowing of the C/N ratio include chemical fixation of NH_3 or amines by lignin-like substances.

The C/N ratio of virgin soils formed under grass vegetation is normally lower than for soils formed under forest vegetation, and, for the latter, the C/N ratio of the humus layers is usually higher than for the mineral soil proper. Also, the C/N ratio of a well-decomposed muck soil is lower than for a fibrous peat. As a general rule, it can be said that conditions that encourage decomposition of organic residues result in narrowing of the C/N ratio. The ratio nearly always narrows sharply with depth in the profile; for certain subsurface soils, C/N ratios lower than 5 are not uncommon.

Traditionally, the C/N ratio has been assumed to be characteristic of the indigenous organic matter. This concept is invalid, for the reason that a significant amount of the N in many soils occurs as fixed NH_4^+ (see previous section). The unusually low C/N ratios reported for many subsoils is undoubtedly due to NH_4^+ held by clay minerals. The C/N ratio of surface soils is only slightly affected by fixed NH_4^+ because of the presence of relatively large amounts of organic matter in this zone.

The relationship between clay fixed NH_4^+ and the C/N ratio of several soils is given in Table 4.3. An interesting point brought out by the data is that the C/organic N ratios of many of the soils also narrowed with increasing depth.

Table 4.3 Influence of Fixed NH_4^+ on the C/N Ratio of the Soil

Depth (cm)	Total N (%)	Fixed NH_4^+ (μg/g)	C/N	C/Organic N
Elliott Silt loam[a]				
0–25	0.290	125	10.4	10.9
25–36	0.140	130	9.4	10.5
36–61	0.095	155	6.9	8.6
61–74	0.081	174	6.2	7.9
107–122	0.036	147	3.9	6.6
Blount Silt loam[a]				
0–10	0.205	115	11.1	11.8
10–18	0.116	109	9.8	10.9
18–36	0.071	143	6.6	8.4
64–76	0.053	185	5.5	8.5
Cisne Silt loam[a]				
8–15	0.163	57	8.7	9.1
30–38	0.065	50	7.5	8.3
56–64	0.031	43	6.8	8.4
84–91	0.064	84	5.9	6.9
114–122	0.033	90	5.8	7.9
Walla Walla[a]				
0–18	0.096	48	11.3	12.0
18–33	0.056	48	10.2	11.3
33–58	0.039	59	7.9	9.6
58–114	0.027	58	5.6	7.3
114–145	0.019	55	5.3	7.8
Steiver[b]				
0–13	0.147	111	12.3	13.5
13–25	0.123	112	11.5	12.9
25–48	0.031	111	10.5	12.4
48–81	0.047	112	8.7	11.9
81+	0.029	121	7.2	13.7

[a] From Stevenson.[12]
[b] From Young.[5]

A divergent result was obtained by Hinman[6] for some Canadian soils, where marked increases were noted in the C/organic N ratio with depth (from 9.4 to 12.4 in the surface layers to from 15.5 to 31.5 at the lower depths).

A word of caution must be given to interpreting results for C/organic N ratios. A standard method has not yet been established for determining clay-fixed NH_4^+ and the methods currently in use give widely variable results.[13] A low estimate for fixed NH_4^+ would give abnormally low C/organic N ratios whereas a high estimate would give abnormally high ratios.

CHEMICAL REACTIONS OF NH_3 AND NO_2^- WITH ORGANIC MATTER

The fate of mineral forms of N in soil is determined to some extent by non-biological reactions involving NH_3 and NO_2^-. Although both types of reactions can proceed over to a wide pH range, chemical fixation of NH_3 is favored by a high pH (>7.0). In contrast, NO_2^- interactions occur most readily under highly acidic conditions (pHs of 5.0 to 5.5 or below).

Chemical Fixation of NH_3

The chemical reaction of NH_3 with soil organic matter is frequently referred to as "NH_3 fixation," and this convention will be adopted herein. The term should not be confused with retention or fixation of the NH_4^+ ion by clay minerals. For complete coverage of NH_3 fixation by organic matter, the reader is referred to reviews by Broadbent and Stevenson[14] and Nömmik and Vahtras.[15]

The ability of lignins and soil organic matter to react chemically with NH_3 has been known for more than 50 years, and numerous patents have been issued over this period for the conversion of peat, sawdust, lignaceous residues (corn cobs, etc.), and coal products into nitrogenous fertilizers by treatment with NH_3. Fixation is associated with oxidation (uptake of O_2), and is favored by an alkaline reaction. Thus, the application of alkaline fertilizers to soil, such as aqueous or anhydrous NH_3, may result in considerable fixation. Injection of anhydrous NH_3, for example, results in a pronounced increase in soil pH, with the highest pH being along the injection line with a gradient extending outward from that line. Similarly, the highest concentration of NH_3 will be found in the injection zone. These conditions are highly favorable for NH_3 fixation by organic matter.

As expected, NH_3 fixation by soil organic matter increases not only with an increase in pH, but with an increase in the amounts of NH_3 applied. Nömmik and Nilsson[16] treated a series of mineral and organic soil types with aqueous NH_3 and noted a close correlation between NH_3 fixation and organic matter content.

While greatest attention has been given to fixation of fertilizer NH_3, the possibility that the reaction occurs under natural soil conditions should not be

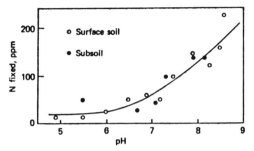

Fig. 4.3 Effect of pH on NH_3 fixation by a muck (Histosol) soil. From Broadbent et al.[18]

overlooked. As mentioned in Chapter 3, part of the unaccounted-for N in soils may exist in complexes formed by the reaction of NH_3 with lignin or lignin-like substances. In early work, Mattson and Koutler-Andersson[17] suggested that the higher N contents and lower C/N ratios of grassland soils, as compared to forest soils, was due to the fact that the higher base status (higher pH) of the former was more favorable for NH_3 fixation and preservation of N.

Mechanisms of Fixation Little is known concerning the NH_3 fixation reaction, although several plausible mechanisms have been proposed. These are based on the observation that fixation proceeds most favorable at high pH values (Fig. 4.3)[18] and is accompanied by the uptake of O_2 (Fig. 4.4). As can be seen from Fig. 4.5, the higher the C content of the soil the greater is the amount of NH_3 that can be fixed. According to Burge and Broadbent,[19] for organic

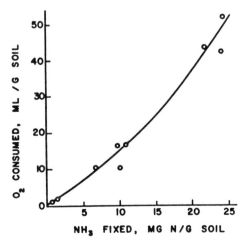

Fig. 4.4 Relationship between NH_3 fixation and O_2 consumption. From Nömmik and Nilsson[16]

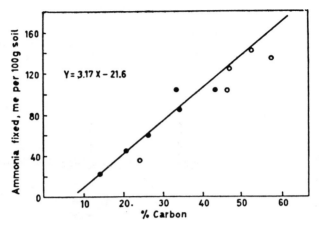

Fig. 4.5 Relationship between carbon content and NH_3 fixation in several organic soils. Adapted from Burge and Broadbent[19]

soils, one molecule of NH_3 can be fixed for every 29 C atoms. Lesser amounts would be expected to be fixed in mineral soils due to association of organic matter with clay and metal ions.

Fixation of NH_3 by soil organic matter has often been attributed to phenolic constituents associated with humic and fulvic acids. The initial step is believed to involve O_2 consumption by the phenol to form a quinone, which subsequently reacts with NH_3 to form complex addition products in which part of the N occurs in heterocyclic linkages. The reaction can be illustrated with catechol (**I**), which under alkaline conditions and in the presence of O_2 is converted first to the o-quinone (**II**) and then benzentriol (**III**) through hydration. Further oxidation was postulated by Flaig[20] to produce a mixture of o-hydroquinone (**IV**) and p-hydroxy-o-quinone (**V**), both of which are capable of reacting with NH_3.

The incorporation of NH_3 into p-hydroxy-o-quinone (**V**) was postulated by Flaig[20] to produce structure of the types represented by **VI** and **VII**.

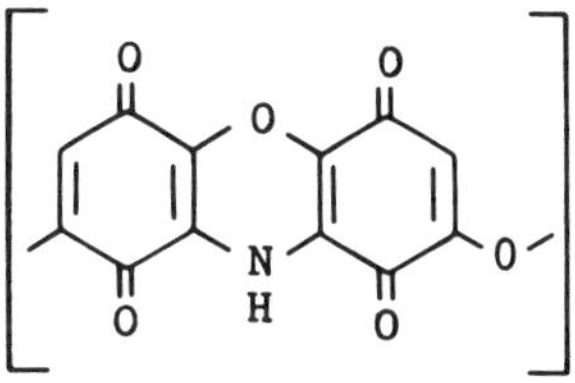

[Structures VI and VII shown]

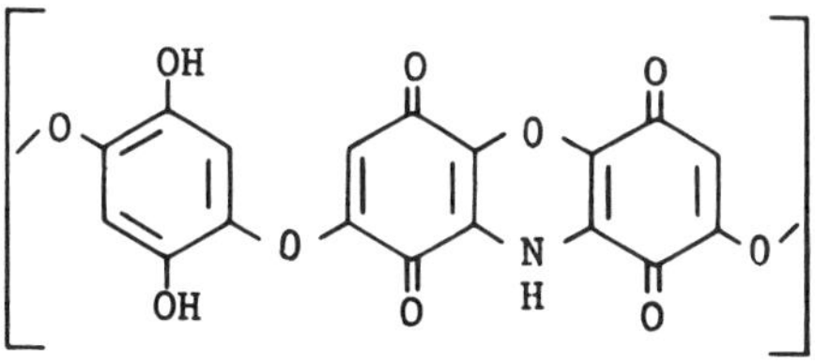

Figure 4.6 illustrates a series of reactions for NH_3 fixation as proposed by Lindbeck and Young.[21] This scheme is based on investigations of *p*-quinone-aqueous NH_3 systems, using polarographic techniques. An interesting feature of the model is that a relatively small amount of quinone (or phenol) will fix a considerable amount of NH_3.

Ammonia is also known to react with reducing sugars under alkaline conditions to form brown-colored nitrogenous polymers. A wide variety of ketones, aldehydes, and other carbonyl-containing compounds also react chemically with NH_3 under alkaline conditions. In practice, fixation by soil organic matter may involve the participation of a wide variety of compounds, including some of a refractory nature (e.g., lignin).

Availability of Chemically Fixed NH_3 to Plants An important question with respect to NH_3 fixation by organic matter is whether the fixed N is available to plants. Limited research indicates that availability is relatively low. In a greenhouse experiment using ^{15}N labeled NH_3, Burge and Broadbent[19] observed that over 95 percent of the fixed NH_3 was no more available than the indigenous soil N. He et al.,[22] also using ^{15}N labeled NH_3, found that organic-matter-fixed NH_3 was initially more labile than the native soil N but became less labile with time.

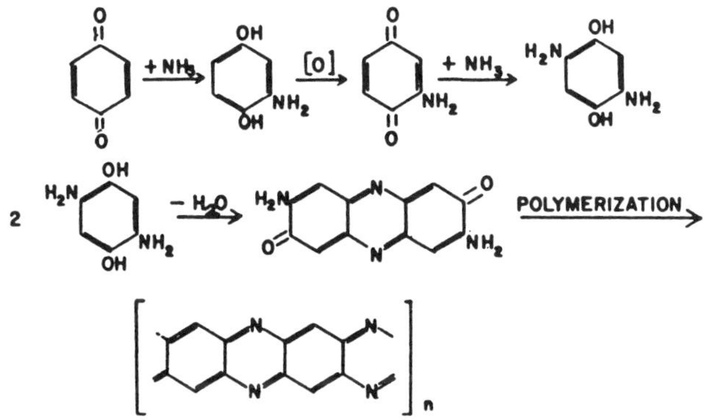

Fig. 4.6 Mechanism of NH_3 fixation by quinones as postulated by Lindbeck and Young[21]

Noticeable solubilization of organic matter has been observed following application of anhydrous (or aqueous) NH_3 to soil,[23-25] a result that may enhance the availability of the native soil N to plants and microorganisms.

Nitrite Reactions

Components of soil organic matter have been shown to react chemically with NO_2^- to form stable organic N complexes and N gases, as depicted in Fig. 4.7.[26] A condition for the reaction to occur in the field is a temporary accumulation of NO_2^- due to inhibition of nitrification at the NO_2^- stage. Under normal soil conditions, the oxidation of NH_3 to NO_2^- by *Nitrobacter* proceeds at a faster rate than the conversion of NO_2^- to NO_3^- by *Nitrosomonas*; consequently, NO_2^- is seldom present in detectable amounts. However, high levels are sometimes found when anhydrous NH_3 or NH_3-type fertilizers are applied to the soil at high application rates,[27-31] and attributed to inhibition of nitrification at the NO_2^- stage, presumably due to NH_3 toxicity to *Nitrobacter* (see Fig. 4.7). The buildup of NO_2^- must be regarded as undesirable because of phytotoxicity[30] and because NO_2^- is relatively unstable and can undergo a series of reactions leading to the formation of N gases. The term "chemodentrification" has frequently been used to designate N loss by this mechanism.

A factor of some importance in determining the activity of NO_2^- is the change in pH accompanying nitrification. Nitrite is particularly reactive at low pH's, a condition that may be attained in localized zones in soil following application of NH_3 or NH_4^+-type fertilizers. Nitrification of applied anhydrous NH_3, for example, starts in peripheral zones of moderately high NH_3 concentrations and proceeds inward toward the center of the retention zone. As a result of the conversion of NH_3 to NO_2^- and NO_3^-, the pH of the soil is lowered; in peripheral zones, values as low as 4.2 to 4.8 have been observed. The pH of the

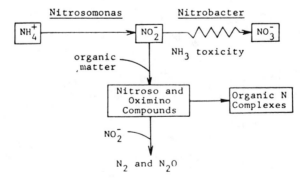

Fig. 4.7 Possible role of organic matter in promoting the decomposition of NO_2^-. From Stevenson[26]

CHEMICAL REACTIONS OF NH_3 AND NO_2^- WITH ORGANIC MATTER   107

soil immediately outside the retention zone may also be lowered because of migration of NO_2^- and NO_3^-. In these regions of low pH, further oxidation of NO_3^- may be hindered due to sensitivity of *Nitrobacter* to high concentrations of H^+. A similar sequence may occur at the soil–particle interface of an individual urea or $(NH_4)_2SO_4$ granule.

With Amino Acids   Claims have been made from time to time that HNO_2 (or NO_2^-) reacts with amino acids and other reduced forms of soil N, such as NH_3, to form N gases. The reaction with an amino acid is

$$RNH_2 + HNO_2 \rightarrow ROH + H_2O + N_2 \qquad [1]$$

The reaction of HNO_2 with amino acids is often referred to as the Van Slyke reaction. Rather low pH values are required, and since unbound amino acids are normally present in soil in only trace quantities (see Chapter 3), serious losses of N by this mechanism would appear unlikely under most soil conditions.

With Humic Substances   Components of the organic matter other than amino compounds (i.e., humic and fulvic acids, lignins) also have the ability to react chemically with NO_2^-,[33-39] with formation of N gases. Evidence for a NO_2^--organic matter interaction has been obtained in studies where ^{15}N-labeled NO_2^- has been applied to soil,[33,38] in which case part of the NO_2^- was fixed by the organic matter and part converted to N gases. Führ and Bremner[33] found that the quantity of NO_2^--N converted to organic forms increased with increasing NO_2^- concentration and decreasing pH, and was related to the C content of the soil.

The gases that have been obtained through the reaction of NO_2^- with lignin, some soil humus preparations, and some model aromatic compounds in buffer solutions at pH 6 are shown in Fig. 4.8. The main N gas identified in these studies was nitric oxide (NO); other gases included N_2, nitrous oxide (N_2O), and CO_2. Still another gas (methyl nitrite, CH_3ONO) has been observed in the gases produced from lignins under highly acidic conditions[39] but there is little evidence that this gas is formed under conditions existing in the natural soil. Smith and Chalk,[38] however, were unable to account for all of the $^{15}NO_2^-$-N added to soil, either as fixed N or in known N gases. They postulated that part or all of the deficit was due to evolution of CH_3ONO from the soil.

Mechanisms leading to the evolution of N gases by nitrosation of lignins and humic substances are not fully understood but the initial reaction probably involves the formation of nitroso and oximino derivatives as shown below, where $O=N-X$ is the nitrosation species (see Austin[40] for a discussion of mechanisms involved in nitrosation).

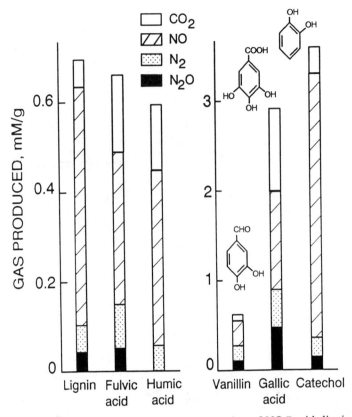

Fig. 4.8 Composition of the gases obtained by reaction of NO_2^- with lignins, humic substances, and some aromatic compounds in buffer solutions of pH 6. Adapted from Stevenson et al.[34]

The nitroso and oximino derivatives are believed to react with excess NO_2^- to give N gases, such as shown below for the formation of N_2 from the nitroso derivative.

The following scheme has been postulated for the formation of N_2O by reaction of HNO_2 with oximes.

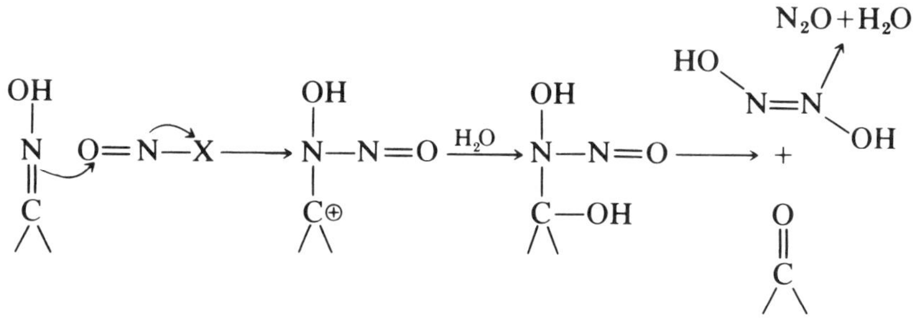

[Reaction scheme showing oxime + HNO₂ intermediates leading to N₂O + H₂O]

Nitric oxide can be formed by reaction of HNO_2 with enolic compounds, as shown below:

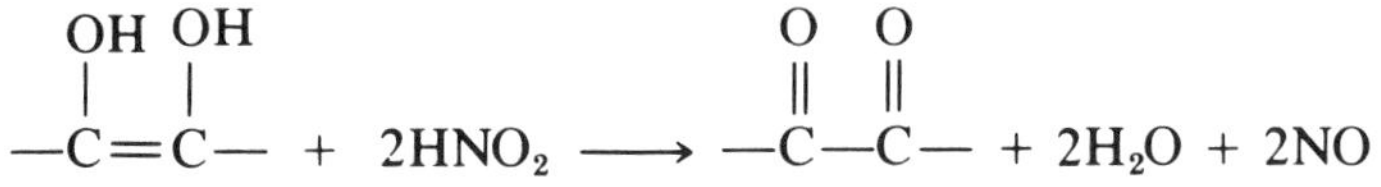

$$-\underset{\underset{OH}{|}}{C}=\underset{\underset{OH}{|}}{C}- + 2HNO_2 \longrightarrow -\underset{\underset{O}{\|}}{C}-\underset{\underset{O}{\|}}{C}- + 2H_2O + 2NO$$

The presence of CO_2 in the gaseous products obtained by nitrosation of lignins and humic substances can be accounted for by oxidation of aromatic rings, as well as through loss of a COOH group through decarboxylation, as illustrated by the following reaction.[40]

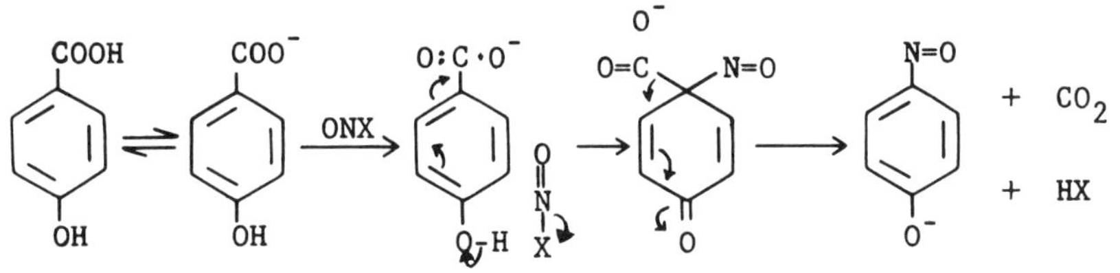

[Reaction scheme showing decarboxylation of hydroxybenzoic acid with ONX to yield CO₂ + HX]

The sequence leading to the formation of CH_3ONO by reaction of HNO_2 with substances containing an OCH_3 group (gas produced only under highly acidic conditions) is as follows.

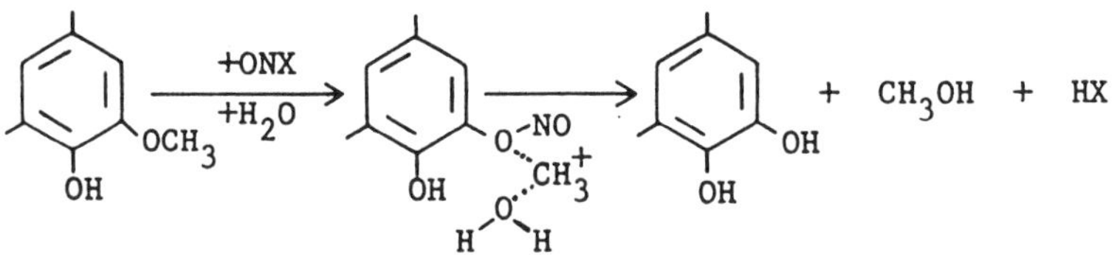

[Reaction scheme showing nitrosation of methoxyphenol yielding CH₃OH + HX]

$$CH_3OH + ONX \longrightarrow CH_3ONO + H_2O$$

Formation of Nitrosamines The reaction of NO_2^- with amines to form N-nitroso compounds is of interest because these substances have been demonstrated to be carcinogenic, mutagenic, and acutely toxic at very low concentrations.

Primary aliphatic amines react with NO_2^- to form diazonium salts, which are unstable and break down to yield N_2 and a complicated mixture of organic compounds:

$$R-NH_2 + NaNO_2 + HX \longrightarrow [RN_2^+] \xrightarrow{H_2O} N_2 + \text{alcohols and alkenes}$$

Primary   Nitrite   Acid
aliphatic
amine

In contrast, secondary aliphatic and aromatic amines react with NO_2^- to form nitrosamines, according to the following reaction:

$$\underset{R}{\overset{R}{>}}N-H + NaNO_2 + HX \longrightarrow \underset{R}{\overset{R}{>}}N-N=O + NaX + H_2O$$

Secondary   Nitrite   Acid   Nitrosamine
amine

In as much as NO_2^- is produced in soil during nitrification, and since amines of various types can be formed during decay of organic residues and metabolism of certain pesticides, the question has arisen as to whether nitrosamines can be generated in soil or assimilated by plants. Traces of nitrosamines have been detected in soil artificially amended with amines and pesticides, but this work was done under ideal conditions in the laboratory (see Chapter 3). Any nitrosamines produced under natural soil conditions would be expected to have a transitory existence.

SUMMARY

Most fine-textured soils contain appreciable amounts of inorganic N in the form of clay-fixed NH_4^+. The soils' content of fixed NH_4^+ is related to the kinds and amounts of clay mineral that are present, with the highest contents being typical of soils rich in micaceous clays. The low C/N ratios frequently observed for subsurface soils (<10) are due, at least in part, to clay-fixed NH_4^+.

Biochemical processes such as ammonification, nitrification, denitrification, and assimilation are responsible for many of the transformations that occur within the soil. Notwithstanding, chemical reactions between inorganic N (NH_3 and NO_2^-) and organic matter also play a role. Ammonia fixation by organic matter is particularly important when anhydrous NH_3 is applied to soils rich in organic matter, an effect that has not been fully investigated under field conditions.

REFERENCES

1. G. Rodrigues, *J. Soil Sci.*, **5**, 264 (1954).
2. J. M. Bremner, *J. Agr. Sci.*, **52**, 147 (1959).
3. R. C. Dalal, *Soil Sci.*, **124**, 323 (1977).
4. A. E. Martin, R. J. Gilkes, and J. O. Skjemstad, *Aust. J. Soil Res.*, **8**, 71 (1970).
5. J. L. Young, *Soil Sci.*, **93**, 397 (1962).
6. W. C. Hinman, *Can. J. Soil Sci.*, **44**, 151 (1964).
7. A. W. Moore and C. A. Ayeke, *Soil Sci.*, **99**, 335 (1965).
8. I. A. Mogilevkina, *Agrokhimiya.*, **7**, 26 (1965).
9. J. L. Young and R. W. Aldag, "Inorganic Forms of Nitrogen in Soil," in F. J. Stevenson, Ed., *Nitrogen in Agricultural Soils*, American Society of Agronomy, Madison, 1982, pp. 43–66.
10. F. J. Stevenson and A. P. S. Dhariwal, *Soil Sci. Soc. Amer. Proc.*, **23**, 121 (1959).
11. D. R. Keeney and J. M. Bremner, *Soil Sci. Soc. Amer. Proc.*, **28**, 653 (1964).
12. F. J. Stevenson, *Soil Sci.*, **88**, 201 (1959).
13. J. M. Bremner, D. W. Nelson, and J. A. Silva, *Soil Sci. Soc. Amer. Proc.*, **31**, 466 (1967).
14. F. E. Broadbent and F. J. Stevenson, "Organic Matter Interactions," in M. H. McVickar et al., Eds., *Agricultural Anhydrous Ammonia*, American Society of Agronomy, Madison, 1966, pp. 169–187.
15. H. Nömmik and K. Vahtras, "Retention and Fixation of Ammonium in Soils," in F. J. Stevenson, Ed., *Nitrogen in Agricultural Soils*, American Society of Agronomy, Madison, 1982, pp. 198–258.
16. H. Nömmik and K. O. Nilsson, *Acta. Agr. Scand.*, **13**, 205 (1963).
17. S. Mattson and E. Koutler-Andersson, *Lantbruks. Hogskol. Ann.*, **11**, 107; **12**, 70 (1943–1944).
18. F. E. Broadbent, W. D. Burge, and T. Nakashima, *Trans. 7th Intern. Congr. Soil Sci.*, **2**, 509 (1960).
19. W. D. Burge and F. E. Broadbent, *Soil Sci. Soc. Amer. Proc.*, **25**, 199 (1961).
20. W. Flaig, *Z. Pflanzennähr. Dung. Bodenk.*, **51**, 193 (1950).
21. M. R. Lindbeck and J. L. Young, *Anal. Chem. Acta*, **32**, 73 (1965).
22. X.-T. He, R. L. Mulvaney, and F. J. Stevenson, *Biol. Fertil. Soil*, **11**, 145 (1991).
23. R. J. Norman, L. T. Kurtz, and F. J. Stevenson, *Soil Sci. Soc. Amer. J.*, **51**, 235, 809 (1987).
24. D. J. Tomasiewicz and J. L. Henry, *Can. J. Soil Sci.*, **65**, 737 (1995).
25. R. G. Meyers and S. J. Thien, *Soil Sci. Soc. Amer. J.*, **52**, 516 (1988).
26. F. J. Stevenson, *Cycles of Soils: Carbon, Nitrogen, Phosphorus, Sulfur, Micronutrients*, Wiley, New York, 1986.
27. H. D. Chapman and G. F. Liebig, *Soil Sci. Soc. Amer. Proc.*, **16**, 276 (1952).
28. R. D. Hauck and H. F. Stephenson, *J. Agr. Food Chem.*, **13**, 486 (1965).
29. D. F. Bezdicek, J. M. MacGregor, and W. P. Martin, *Soil Sci. Soc. Amer. Proc.*, **35**, 397 (1971).
30. M. N. Court, R. C. Stephen, and J. S. Waid, *Nature*, **194**, 1263 (1962).

31. F. E. Clark, *Trans. Inter. Soil Sci. Conf.*, *New Zealand*, 1962, pp. 173–176.
32. H. Vine, *Plant Soil,* **17,** 109 (1962).
33. F. Führ and J. M. Bremner, *Atompraxis,* **10,** 109 (1964).
34. F. J. Stevenson, R. M. Harrison, R. Wetselaar, and R. A. Leeper, *Soil Sci. Soc. Amer. Proc.,* **34,** 430 (1970).
35. D. W. Nelson and J. M. Bremner, *Soil Biol. Biochem.,* **1,** 229 (1969).
36. D. W. Nelson and J. M. Bremner, *Soil Biol. Biochem.,* **2,** 203 (1970).
37. J. M. Bremner and D. W. Nelson, *Trans. 9th Intern. Congr. Soil Sci.,* **2,** 495 (1968).
38. C. J. Smith and P. M. Chalk, *Soil Sci. Soc. Amer. J.,* **44,** 277, 288 (1980).
39. F. J. Stevenson and R. J. Swaby, *Soil Sci. Soc. Amer. Proc.,* **34,** 773 (1964).
40. A. T. Austin, *Sci. Prog.,* **XLIX,** 619 (1961).

5

ORGANIC PHOSPHORUS AND SULFUR COMPOUNDS

Phosphorus (P) and sulfur (S) rank in importance with nitrogen (N) and potassium (K) as major plant nutrients. Practically all of the S in most soils, and up to three-fourths of the total P, occurs in organically bound forms.

Detailed reviews are available on the various aspects of soil organic P[1-4] and organic S.[5-8]

THE C/N/P/S RATIO

A definite relationship has been observed between organic C, total N, organic P, and total S in soils.[9-12] While considerable variation is found in the C/N/P/S ratio for individual soils, the mean for soils from different regions of the world is remarkably similar (Table 5.1). As an average, the proportion of C/N/P/S in soils is approximately 140:10:1.3:1.3. The close relationship between C, N, and total S has been taken as evidence that most of the S in soils of humid regions of the earth occurs in organic combinations.

The P content of soil organic matter varies from as little as 1.0 percent to well over 3.0 percent, which is reflected by the variable C/organic P ratios that have been reported for soils. Carbon/organic-P ratios ranging from 46 to 648 were obtained by John et al.[13] for 38 British Columbia soils. His ratios are more variable than those reported for Finnish soils by Kaila,[14] who observed C/organic-P ratios of from 61 to 276 for cultivated mineral soils, 141 to 526 for cultivated humus soils, and 67 to 311 for virgin soils. Relatively low C/organic-P ratios have been reported for some British and New Zealand soils, with mean C/P ratios of about 60.[10,11]

One explanation for the wide C/P ratios, as compared to those for C/N and

Table 5.1 Organic C, Total N, Organic P, and Organic S Relationships in Soil

Location	Number of Soils	C/N/P/S	Reference
Iowa	6	110:10:1.4:1.3	Neptune et al.[9]
Brazil	6	194:10:1.2:1.6	Neptune et al.[9]
Scotland[a]			
Calcareous	10	113:10:1.3:1.3	Williams et al.[11]
Noncalcareous	40	147:10:2.5:1.4	Williams et al.[11]
New Zealand[b]	22	140:10:2.1:2.1	Walker and Adams[10]
India	9	144:10:1.9:1.8	Somani and Saxena[12]

[a]Values for S given as total S.
[b]Values for subsurface layers (35–53 cm) were 105:10:3.5:1.1

C/S, is that less of the P occurs as structural components of humic and fulvic acids. Goh and Williams[15] found that C/P ratios were lower in the low-molecular-weight components of soil organic matter than in the high-molecular-weight components. In other work, Elliott[16] found that the C/P ratios narrowed with decreasing aggregate size, as determined by wet sieving.

Despite the variable C/organic P ratios, the organic P content of the soil follows rather closely that for organic C, as illustrated for some tropical soils is Fig. 5.1.[17] From the slope of the regression line, the overall C/organic P

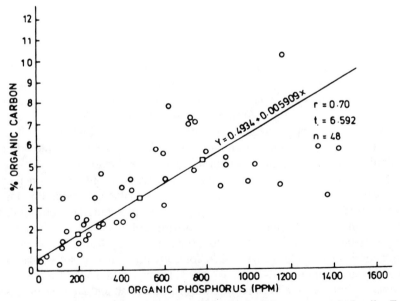

Fig. 5.1 Relationship between organic P and organic C for some tropical soils. From Uriyo and Kesseba,[17] reproduced by permission of Elsevier Scientific Publishing Co.

ratio can be seen to be of the order of 50, which falls within the lower range of the ratios mentioned above.

Factors influencing the proportion of P in soil organic matter include parent material, climate, drainage, cultivation, pH, and depth of soil. The effect of each is not known with certainty and contradictory results are the rule rather than the exception. Some workers have concluded that a low P content of organic matter is a characteristic of P deficient soils, but this hypothesis has not been confirmed. The P content of organic matter appears to be higher in fine-textured soils than in coarse-textured ones.

Ratios recorded for C and total S are somewhat less variable than those observed for C and organic P, the usual range being 60 to 120.[7,9,18] Differences in the C/S ratio between soil groups has been attributed to such factors as parent material and type of vegetative cover. Bettany et al.[18] found that the C/S ratios of some grassland soils of Saskatchewan, Canada, were lower than for some comparable forest soils of the region; mean C/S ratios for the two groups were 58 and 130, respectively.

SOIL ORGANIC P

Phosphorus compounds in soil can be placed into the following three classes: 1) organic compounds of the soil humus, 2) inorganic compounds in which the P is combined with Ca, Mg, Fe, Al, and clay minerals, and 3) organic and inorganic P compounds associated with the cells of microorganisms, the soil biomass.

The total P content of soils varies considerably, depending on the nature of the parent material, degree of weathering, and extent to which P has been lost through leaching and erosion. For surface soils of the United States, the range varies from < 100 μg/g (dry weight) for sandy soils of the Atlantic and Gulf coastal plains to over 1000 μg/g for a large area of the Northwest. Most values, however, fall within the range of 200 to 900 μg/g. A soil containing 500 μg/g P will contain about 1120 kg P/ha (1000 lb./acre) to plow depth.

From 15 to 80 percent of the total P in soils occurs in organic forms, the exact amount being dependent upon the nature of the soil and its composition. The higher percentages are typical of Histosols and uncultivated forest soils, although much of the P in tropical soils may occur in organic forms. Some values for total organic P and the percentage of the total P in this form are recorded in Table 5.2.[4,17]

The organic P content of the soil decreases with depth in much the same way as organic C. Depth distribution patterns for organic P and organic C in two Mollisol soils from Iowa are shown in Fig. 5.2[19]

As shown in Fig. 5.3, four general types of compounds make up the bulk of the organic P in plants, namely, phytin, sugar phosphates, phospholipids, and nucleic acids. Except possibly for inositol phosphates (phytin), these compounds are subject to rapid decomposition in soil and new products are syn-

Table 5.2 Some Values for Organic P in Soil[a]

Location	Organic P μg/g	Percentage of Total P
Australia	40–900	—
Canada	80–710	9–54
Denmark	354	61
England	200–920	22–74
New Zealand	120–1360	30–77
Nigeria	160–1160	—
Scotland	200–920	22–74
Tanzania	5–1200	27–90
United States	4–100	3–52

[a]Adapted from the review of Halstead and McKercher,[3] where specific references can be obtained. The range shown for the U.S. soils must be regarded as a minimum. Data for the Tanzania soils are from Uriyo and Kesseba.[17]

thesized by microorganisms. Much of the P in plants is present as nucleic acids; in soil, very little of the organic P occurs in this form. In contrast, rather high amounts of the organic P in some soils occur as inositol phosphates, and accounted for their tendency to form insoluble complexes with soil components, as discussed later.

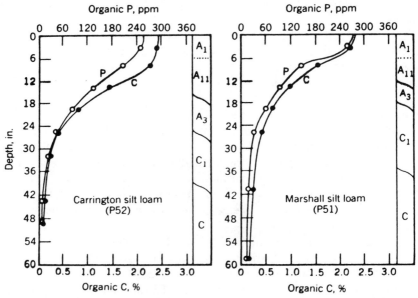

Fig. 5.2 Distribution of organic P and C in two Mollisol soils from Iowa. Adapted from Pearson and Simonson[19]

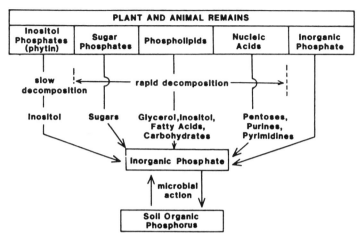

Fig. 5.3 Phosphorus transformations in soil. Adapted from an unpublished diagram by W. Flaig (personal communication)

Determination of Organic P

Methods used for the determination of organic P in soil are of two main types, as follows:

1. *Extraction methods:* In this approach, organic and mineral forms of P are recovered from the soil by extraction with acid and base. Organic P is converted to orthophosphate and the content of organic P is determined from the increase in inorganic phosphate as compared to a dilute acid extract of the original soil.

   Organic P = total P in alkaline extract − inorganic P in acid extract

2. *Ignition methods:* Organic P is converted to inorganic P by ignition of the soil at elevated temperatures and is calculated as the difference between inorganic P in acid extracts of ignited and nonignited soil.

   Organic P = inorganic P of ignited soil − inorganic P of untreated soil

Advantages and disadvantages can be pointed out for both approaches. In extraction methods, pretreatment of the soil with a mineral acid (HCl is usually used) is essential for removing polyvalent cations which render the P compounds insoluble but this treatment may cause some hydrolysis of organic P compounds. Other extractants (8-hydroxyquinoline and EDTA) have been used in attempts to eliminate this problem. The alkali used to dissolve organic matter (dilute NaOH) also causes some hydrolysis of organic P.

Of the two approaches, the ignition method is the most simple and is often the one used for survey types of investigations. Limitations of the method include alteration in the solubility of the native inorganic P and hydrolysis of organic phosphate during acid extraction of the nonignited soil. Studies on the determination of organic P in soils include those of Dick and Tabatabai[20] and Stewart and Oades.[21]

Because organic P is determined by the difference between total and inorganic P, accuracy is poor when the soil is low in organic P, such as in the subsoil.

Specific Organic P Compounds

Less than 50 percent of the organic P in most soils can be accounted for in known compounds. Principal forms include inositol phosphates, nucleic acids, and phospholipids; small quantities of P may be present as phosphoproteins and metabolic phosphates. Approximate recoveries of organic P in these forms are as follows:

Inositol phosphates	2–50 percent
Phospholipids	1–5 percent
Nucleic acids	0.2–2.5 percent
Phosphoproteins	trace
Metabolic phosphates	trace

Data for the forms of organic P in the soils from widely geographical areas are recorded in Table 5.3. Results for the Bangladesh soils are unique in that all values were obtained on the same soil. For these soils, recovery of organic P in known compounds appears to be the highest yet reported (for 3 of 10 soils, 61 to 85 percent of the organic P was accounted for).[22]

Table 5.3 Distribution of the Forms of Organic P in Soils[a]

	Percentage of Organic P		
Source	Inositol Phosphates	Nucleic Acids	Phospholipids
Australia	0.4–38	—	—
Bangladesh[b]	9–83	0.2–2.3	0.5–7.0
Britain	24–58	0.6–2.4	0.6–0.9
Canada	11–23	—	0.9–2.2
New Zealand	5–43	—	0.7–3.1
Nigeria	23–30	—	—
United States	3–52	0.2–1.8	—

[a]From Dalal,[3] where specific references can be obtained.
[b]Results for the soils of Bangladesh are from Islam and Ahmed.[22]

From the above, it can be seen that recovery of soil organic P in the three principle forms follows the order: inositol phosphates >> phospholipids > nucleic acids. The same types of organic P compounds are found in plants, but the order is reversed. The somewhat higher abundance of inositol phosphates in soils, as contrasted to plants, may be due to their tendency to form insoluble complexes with polyvalent cations, such as Fe and Al in acid soils and Ca in calcareous soils.

Inability to account for the bulk of the organic P in most soils may be due to:[23]

1. Occurrence of polymeric phosphate-containing compounds, such as teichoic acids from bacterial cell walls and phosphorylated polysaccharides. The former consists of a polyol, ribitol, or glycerol phosphate in which adjacent polyol residues are linked by a phosphodiester bond. A typical repeating unit of a cell wall teichoic acid is given as structure **III** of Chapter 3.
2. Occurrence of stable complexes of inorganic phosphate with soil organic matter, which is recovered only after destruction of organic matter and is measured along with the organic P.
3. Occurrence of P in chemical association with humic and fulvic acids, as discussed later.

Inositol Phosphates Inositol phosphates are esters of hexahydrocyclohexane, commonly referred to as inositol **(I)**. A variety of esters are possible, one of the most common being the hexaphosphate. In plants, mono-, di-, and triphosphates are sometimes found in rather large quantities. The hexaphosphate ester, or phytic acid **(II)** occurs in cereal grains as the mixed Ca- and Mg-salt called phytin. For many years, the inositol hexaphosphate in soil was thought to be derived from the phytin of higher plants but recent work has shown that microorganisms constitute a major source.

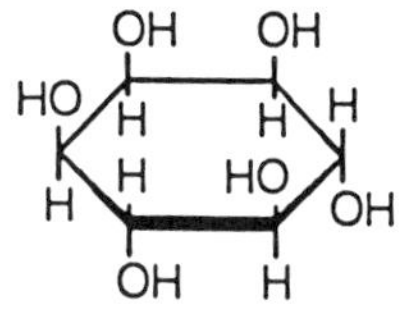

myo-Inositol

I

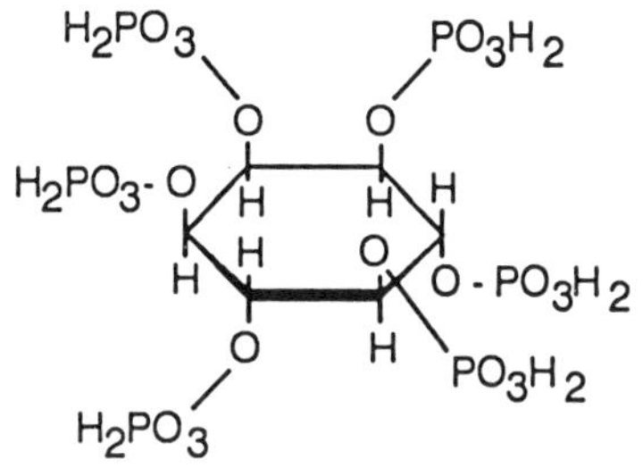

Phytic acid

II

Inositol phosphates, although produced by living organisms in smaller amounts than nucleic acids, become stabilized through formation of insoluble complexes with metal ions and other organic substances, with the result that they gradually accumulate until they form a larger proportion of the soil organic P. Much of the P in farmyard manures may be in the form of inositol phosphates.[24]

A unique feature of inositol phosphates, particularly the hexaphosphate esters, is their high stabilities in acids and bases. Advantage is taken of this property for their extraction and preparation from soil. The steps involved have usually included the following:

1. Extraction with hot 0.5N NaOH after pretreatment of the soil with 5 percent HCl
2. Oxidation of the extracted organic matter with alkaline hypobromite
3. Precipitation of inositol phosphates as their Fe (or Ca) salts
4. Alkaline hypobromite oxidation of the precipitate material
5. A second Fe (or Ca) precipitation of the inositol phosphates
6. Decomposition of the second Fe precipitate with NaOH solution and removal of the $Fe(OH)_3$

The P recovered in the final solution is sometimes considered to be in the form of inositol phosphates, but other P compounds are usually present (e.g., inorganic phosphate). Caldwell and Black[24] pointed out that "if the product is pure, the recovery is not quantitative, and if the recovery is quantitative the product is not pure."

Chromatographic procedures have now replaced chemical fractionation methods for determining inositol phosphates in soil.[24-32] Phosphates function as acids with pK values of 1 and 6 for the first and second dissociations and are subsequently readily adsorbed by anion-exchange resins. Elution with HCl of increasing concentration results in separations of the type shown in Fig. 5.4.

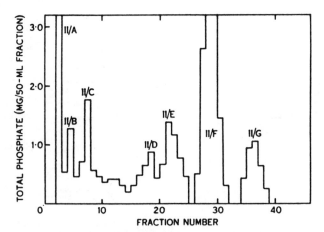

Fig. 5.4 Elution of inositol phosphates by anion exchange chromatography. II/A = orthophosphate; II/C = inositol tetraphosphate; II/D and II/E = inositol pentaphosphate; II/F and II/G = inositol hexaphosphates. From Cosgrove[27]

Table 5.4 Range of Inositol Mono-, Di- and Tri-, Tetra-, Penta-, and Hexaphosphates in 10 Soils of Bangladesh[a]

Form	Percentage of Soil Organic P	Percentage of Inositol Phosphates
Monophosphate	1.3– 6.7	1.8–10.9
Di- and Triphosphate	4.0–11.0	6.5–15.8
Tetraphosphate	5.0–18.1	8.1–29.5
Pentaphosphate	9.4–17.3	15.4–27.3
Hexaphosphate	11.1–49.2	18.2–65.6

[a]From Mandle and Islam.[26]

For chromatographic analysis by anion-exchange chromatography, the inositol phosphates are extracted from soils with acid and alkali and precipitated as insoluble Fe-salts from acid medium or Ba-salts from alkaline medium. The latter gives better recoveries of the lower phosphate esters but more extraneous material is present that interferes later with the chromatographic separations. Subjection of the initial extracts to hypobromite oxidation improves the resolution of inositol phosphates but losses may occur.

By anion-exchange chromatography, Smith and Clark[25] demonstrated that only a part of the inositol phosphates consisted of the hexaphosphate; lower phosphate esters were also present. As can be seen from the data presented in Table 5.4, the percentage of inositol phosphates as the mono-, di-, tri-, tetra-, penta-, and hexaphosphate forms varies widely in soils.[26] Absolute amounts of inositol phosphate P in soils can range from a trace to as much as 460 μg/g, as shown in Table 5.5[4]

Caldwell and Black[24] analyzed 49 Iowa surface soils by anion exchange chromatography and found that inositol hexaphosphate, the most abundant ester, accounted for from 3 to 52 percent of the organic P (see range listed for the United States in Table 5.5). However, only 4 of the 49 soils had values exceeding 30 percent and the average for all soils was only 17 percent. The percentage was higher for soils developed under forest vegetation than under grass vegetation.

Table 5.5 Some Values for Inositol P in Soils[a]

	Inositol P	
Location	μg/g	Percentage of Organic P
Australia	1–356	0.5–38
Canada	10–150	10–30
Denmark	163	46
England	56–460	24–58
New Zealand	22–340	5–26
Nigeria	25–145	16–25
Scotland	54–460	24–58
United States	4–100	3–52

[a]From the summary of Halstead and McKercher.[4]

From the structural formula of inositol, it can be demonstrated that nine positional stereoisomers are possible, depending on the arrangements of H and OH groups. They include seven optically inactive forms and one pair of optically active isomers. The best known form is *myo*-inositol (shown earlier, **I**), which is widely found in nature. This is the isomer from which phytin is constituted. Other naturally occurring isomers of more limited biological distribution are *D-chiro*- and *L-chiro*-inositol (**III** and **IV**), *scyllo*-inositol (**V**), and *neo*-inositol (**VI**).

D-chiro-Inositol	L-chiro-Inositol	scyllo-Inositol	neo-Inositol
III	IV	V	VI

The advent of chromatographic techniques has not only led to more precise values for inositol phosphates in soil but a number of unusual isomers have been isolated. Smith and Clark[25] first demonstrated the presence of an unknown inositol phosphate in soil, subsequently shown by Caldwell and Black[24] to be an inositol hexaphosphate other than *myo*-inositol (**I**). The unknown isomer accounted for an average of 46 percent of the inositol hexaphosphate material and was found to be synthesized by a variety of soil microorganisms.

The unknown isomer was subsequently shown by Cosgrove[27] to be the hexaphosphate of *scyllo*-inositol (**V**). Cosgrove[28,29] also demonstrated the occurrence in soil of the hexaphosphate of *D*- and *L-chiro*-inositol (**III** and **IV**) and of *neo*-inositol (**VI**). Pentaphosphates of *myo*-, *chiro*-, and *scyllo*-inositol were also found. The results of these and other studies[30,31] show that considerable differences exist in the isomeric composition of inositol phosphates in soil and that some, if not all, of the soil inositol phosphates are synthesized *in situ* by microorganisms.[32] It should be noted that some of the stereoisomers found in the soil (e.g., *scyllo*-inositol) are found in microorganisms, but not in higher plants.

Nucleic Acids and Derivatives Nucleic acids are found in all living cells, and they are synthesized by soil microorganisms during the decomposition of plant and animal residues. Two types are known, ribonucleic acid (RNA) and deoxyribonucleic acid (DNA), each consisting of a chain of nucleotides. Each nucleotide contains a pentose sugar, a purine or pyrimidine base (see Chapter 3), and a phosphoric acid residue. The latter serves as a link between adjacent pentose units. The few estimates that have been made thus far indicate that no more than 3 percent of the organic P in soil occurs as nucleic acids or their derivatives.[22,33]

Phospholipids Phospholipids represent a group of biologically important organic compounds that are insoluble in water but soluble in fat solvents, such as benzene, chloroform, or ether. Included are the glycerophosphatides, such as phosphatidyl inositol, phosphatidyl choline, or lecithin **(VII)**, phosphatidyl serine **(VIII)**, and phosphatidyl ethanolamine **(IX)**, where RC=O is a long-chain fatty acyl group.

$$
\begin{array}{ccc}
\text{CH}_2\text{OCR} & \text{H}_2\text{COOCR} & \text{H}_2\text{COOCR} \\
| & | & | \\
\text{CHOCR'} & \text{R'COOCH} & \text{R'COOCH} \\
| & | & | \\
\text{CH}_2\text{OP=O} & \text{H}_2\text{C-O-P-OCH}_2\text{CHNH}_3^+ & \text{H}_2\text{C-O-P-OCH}_2\text{CH}_2\text{NH}_3^+ \\
| & \quad\;\; \text{O}^- \;\; \text{COO}^- & \quad\;\; \text{O}^- \\
\text{CH}_2\text{CH}_2\text{N(CH}_3)_3\text{OH} & & \\
\text{L-}\alpha\text{-Lecithin} & \text{Phosphatidyl serine} & \text{Phosphatidyl ethanolamine} \\
\text{VII} & \text{VIII} & \text{IX}
\end{array}
$$

The total quantity of phospholipids in soil is small, usually less than 5 µg P/g. Percentagewise, from 0.5 to 7.0 percent of the soil organic P occurs as phospholipids, with a mean value of 1 percent (see Table 5.3). The phospholipids in soil are undoubtedly of microbial origin.

The presence of large amounts of mineral matter may reduce the ability of organic solvents to remove phospholipids; thus, most estimates must be regarded as minimal. An improved procedure for recovering phospholipids from soil using sequential extraction with ethanol–benzene and methanol–chloroform has been described by Baker.[34]

Evidence for the presence of glycerophosphates in soil has come from the detection of glycerophosphate, choline, and ethanolamine in hydrolysates of extracted phospholipids.[2] Results obtained by Kowalenko and McKercher[35] indicates that phosphatidyl choline **(VII)** is the predominant soil phospholipid, followed by phosphatidyl ethanolamine **(IX)**.

Other Organic P Compounds Other organic P compounds present in small amounts in soil include metabolic phosphates. Robertson[36] detected two organic phosphates in citric acid extracts of soil, one of which corresponded closely to glucose-1-phosphate.

In a recent study, Anderson and Malcolm[37] identified a number of phosphate esters in alkaline extracts of soil. These included several mono-phosphorylated carboxylic acids, two esters containing glycerol, *myo*- and *chiro*-inositol, and an unidentified component.

Organic P of Humic Substances

As noted earlier, over one-half of the soil organic P cannot be accounted for in known compounds. Much of the unknown organic P may occur in association with humic and fulvic acids.[14,38-41]

The organic P associated with humic and fulvic acids may exist, at least in part, as complexes with simple phosphate esters (e.g., nucleotides or inositol phosphates) that modify the properties of the phosphate esters such that they are not estimated by conventional methods.

Evidence that inositol hexaphosphates are bound to other organic components in soil has been shown by the work of Anderson and Hance[42] and Cosgrove.[27] Other studies have shown that organic P compounds of several types are bound to high-molecular-weight organic colloids.[34,41] Model humic polymers containing organic P have been prepared and examined by Brannon and Sommers.[43]

Characterization of Organic P by ^{31}P-NMR Spectroscopy

Although present in low concentrations in soil organic matter, ^{31}P is a sensitive nucleus and great potential exists for characterization of soil organic P by ^{31}P-NMR spectroscopy. The method has been used to distinguish a range of organic and inorganic P in soil extracts, including inositol phosphates, teichoic acid P, and orthophosphate mono- and di-esters.[44-48] This work has shown that most of the organic P that is extractable from soils is present as monoesters and that a small amount of the P exists as phosphonates in which the P is bound to C rather than to O.

Biomass P

From 2 to 5 percent of the organic P of cultivated soils occurs as part of the biomass. Since the microbial biomass is relatively "labile" as compared to other organic matter fractions, the biomass P would be expected to be more available to plants than other organic P fractions. As one might expect, the amount of biomass P in soils follows closely the content of biomass C.

Biomass P has been estimated as follows: 1) from cell weights based on microbial counts plus estimates for the P content of microbial tissue, 2) from ATP measurements, and 3) from the inorganic P released during incubation of partially sterilized soil.[4,5]

In recent years, fumigation–extraction procedures (the third item) similar to those noted for biomass C and N (Chapters 1 and 3, respectively) have been used, in which case the fumigated and unfumigated soil is extracted with $0.5M$ NaHCO$_3$ at pH 8.5 (values corrected for phosphate fixation during extraction).[49-51]

Biomass P is calculated from the relationship:

Biomass P
$$= \frac{\text{P extracted from fumigated soil} - \text{P extracted from nonfumigated soil}}{0.4}$$

The assumption is made that 40 percent of the P in the biomass is rendered extractable to $0.5M$ $NaHCO_3$ by lysis and incubation. As one might expect, measurements of biomass P in soils that contain large amounts of $NaHCO_3^-$ extractable phosphate will be subject to considerable error.

As noted earlier, from 2 to 5 percent of the soil organic P resides in the biomass, although higher and lower values are not uncommon. Brookes et al.[50] accounted for from 5.0 to 24.3 percent of the organic P in some grassland soils as biomass P; values for some arable soils ranged from 2.5 to 3.4 percent. In agreement with other work, C/P ratios for the biomass (11.7 to 35.9) were substantially lower than for soil organic matter, which often exceed 200, as noted earlier. It appears evident, therefore, that the higher the percentage of the organic P as biomass P, the lower will be the C/organic P of the soil.

Chemical Fractionation of Organic P

In addition to identification of specific types or classes of organic compounds (discussed above), chemical fractionation methods have been used in attempts to partition organic P into pools based on availability to plants. In the method of Bowman and Cole[52] the following pools were recognized by sequential extraction of the soil with $0.3M$ $NaHCO_3$, dilute mineral acid, and $0.1M$ NaOH.

1. *Labile pool:* obtained by extraction of the soil with $0.5N$ $NaHCO_3$ at pH 8.5
2. *Moderately resistant pool:* acid-soluble organic P plus alkali-soluble inorganic P
3. *Moderately resistant pool:* fulvic acid P, obtained by extraction of the soil with $0.1M$ NaOH and removal of humic acid by acidification
4. *Highly resistant pool:* humic acid P

Recovery of the organic P in several grassland soils in the four pools is shown in Table 5.6. Less than 10 percent of the organic P occurred in labile forms, the fraction most readily available to plants.

A more complex fractionation scheme was used by Hedley et al.[53] Their approach included the following: 1) use of an anion exchange resin to remove inorganic PO_4^{3-}, 2) an extraction of the soil with $0.5M$ $NaHCO_3$ before and after treatment with chloroform to determine biomass N (discussed above), 3) extraction with $0.1N$ NaOH to remove inorganic and organic P held more strongly to Fe and Al components at soil surfaces, 4) ultrasonification of the

Table 5.6 Distribution of Organic P in Four Categories for Five Grassland Soils[a]

			Form of Organic P[b]			
Location of Soils	Total P (μg/g)	Organic P (μg/g)	Labile	Moderately Labile	Moderately Resistant	Highly Resistant
SW Montana	1200	744	9.9%	32.2%	48.1%	9.8%
SW South Dakota	600	298	4.5	33.8	47.6	14.1
NE Oklahoma	320	203	7.2	27.2	54.7	10.9
Texas Panhandle	240	211	5.0	33.1	49.6	12.3
NE Colorado	280	118	7.8	44.4	36.8	11.0

[a]From Bowman and Cole.[53]
[b]Labile = extractable with 0.5M NaHCO$_3$. Moderately labile = acid soluble P. Moderately resistant = fulvic acid-P. Highly resistant = humic acid-P.

soil residue in 0.1N NaOH to remove inorganic and organic P held at internal surfaces of soil aggregates, and 5) oxidation and acid digestion of the final soil residue to give stable organic P forms.

Role of Organic Matter on the Solubilization of Inorganic P

The availability of phosphate in soil is often limited by fixation reactions, which convert the monophosphate (H$_2$PO$_4^-$) to various insoluble forms. Insoluble Ca-phosphates predominate in calcareous soils while Fe- and Al-phosphates are formed in acidic soils. Adsorption by clay minerals can affect phosphate availability under neutral or slightly acid conditions.

Several studies indicate that the availability of phosphates in soil is enhanced by additions of decomposable organic matter. Several independent, but not necessarily exclusive, reactions may be involved, including the following:

1. Formation of chelate complexes with Ca, Fe, and Al with release of phosphate to water soluble forms, as illustrated in Fig. 5.5. The reader is referred to Chapter 16 for information as to the nature of the compounds that are involved and the conditions under which they are formed in soil.

   Reactions leading to the formulation of soluble phosphates are as follows:

   $$CaX_2 \cdot 3Ca(PO_4)_2 + \text{chelate} \rightarrow \text{soluble } HPO_4^{2-}$$
   $$+ \text{ Ca-chelate complex, where X = OH or F.}$$
   $$Al(Fe) \cdot (H_2O)_3(OH)_2H_2PO_4 + \text{chelate} \rightarrow \text{soluble } HPO_4^{2-}$$
   $$+ \text{ Al(Fe)-chelate complex}$$

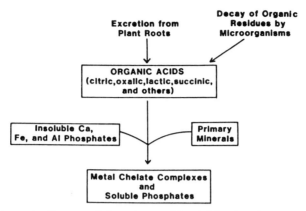

Fig. 5.5 Schematic diagram of the release of insoluble phosphates to soluble forms through the action of organic acids and other naturally occurring chelates

2. Competition between humates and phosphate ions for adsorbing surfaces, thereby preventing fixation of phosphate.
3. Formation of protective coatings over colloidal sesquioxides, with reduction in phosphate adsorption.
4. Formation of phospho–humic complexes through bridging with Fe and/ or Al.

Evidence that naturally occurring chelating agents enhance the availability of P to higher plants (item 1) is circumstantial and some investigators have questioned whether organic acids and other chelating agents are produced (or persist) in sufficient abundance to appreciably influence phosphate solubility. The effectiveness of these compounds may be greatest in unfertilized soils low in natural fertility and where most of the P is tied up as insoluble Ca-, Fe-, or Al-phosphates. A discussion of the role of phosphate-solubilizing microorganisms in enhancing the availability of P in soils has been given by Kucey et al.[54]

The competition of humates for adsorbing surfaces (item 2), together with the formation of protective coatings (item 3), is believed to be of some importance in allophanic soils in that fixation of phosphate is reduced. These soils are unique in that they fix appreciable amounts of phosphate due to reactions at Al-hydroxide surfaces. Through the interactions of humates with surface Al, fixation of phosphate can be reduced, thereby increasing its availability to plants.

With regard to phospho–humic complexes (item 4), the suggestion has been made that Al serves as a link between the phosphate anion and the negatively charged humate molecule,[55] such as in acid peats (**X**) and an allophanic soils (**XI**).

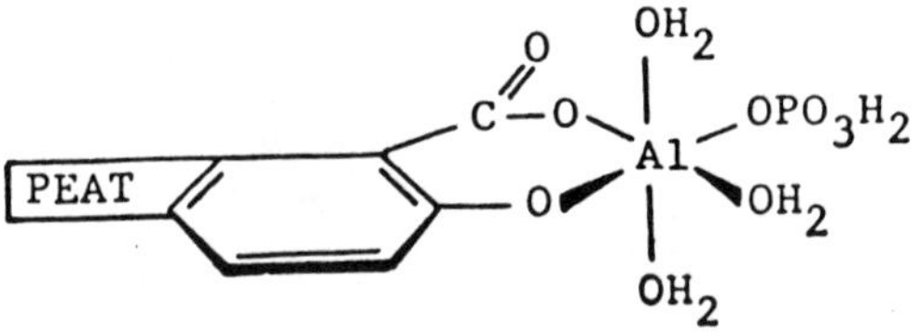

[Structures X and XI: PEAT–C(=O)–O–Al(OH₂)₂(OPO₃H₂)–O– aromatic ring; ALLOPHANE–Al(OH₂)₂–O– aromatic–HUMATE–aromatic –O–Al(OH₂)₂–OPO₃H₂]

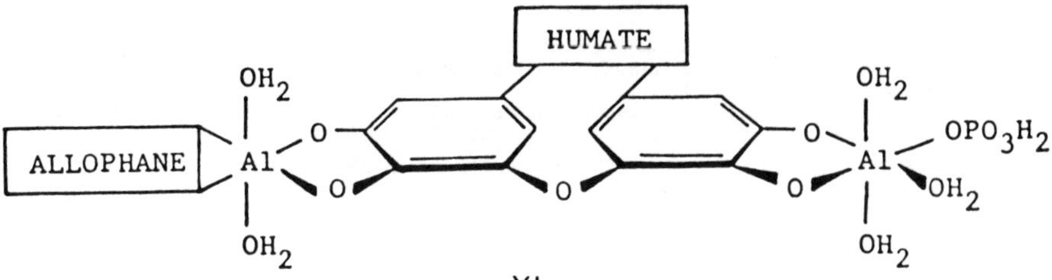

SOIL ORGANIC S

Sulfur has often been described as the neglected plant nutrient. The sporadic attention that has been given to this nutrient is partly caused by the fact that responses to applied S have been restricted to a few geographical areas, and to only a few crops. However, due to increased use of S-free fertilizers, reduction in the amount of S used as a pesticide, and higher crop yields, deficiencies are now being reported with increasing frequency and on a wider variety of crops.[56] A contributing factor has been a reduction in atmospheric levels of S gases through adoption of emission control systems designed to reduce the SO_2 content of the atmosphere through burning of fossil fuels.

Organic S in soil is an important link in the overall cycle of S in nature. When plant residues undergo decay in soils, part of the organic S reappears as SO_4^{2-} and part is incorporated into microbial tissue, and hence into humus. Unlike phosphate, SO_4^{2-} is subject to leaching; thus in highly leached soils, inorganic forms of S have been removed and only the S in organic forms remains.

Determination of Organic S

Organic S is usually taken as the difference between total soil S, as obtained by dry- or wet-ashing techniques, and inorganic S (SO_4^{2-} and sulfide) extracted from the untreated soil by reagents such as dilute HCl or $NaHCO_3$.[57]

$$\text{Organic S} = \text{total S} - \text{inorganic S}$$

The S content of digests or extracts is commonly determined by the Johnson and Nishita[58] methylene blue method. A discussion of methods for the determination of the various forms of soil S is given by Blanchar.[57]

Content of Organic S in Soil

Soils vary greatly in their content of S, being lowest in those developed from sands (~20 µg/g) and highest in those developed in tidal areas, where sulfides have accumulated (~35,000 µg/g). The normal range for soils of humid and semihumid regions of the earth is of the order of 80 to 140 µg/g, or 0.008 to 0.014 percent. The absolute amount of organic S in soils follows closely that for organic C and is related to those factors affecting organic matter content, as discussed in Chapter 2.

The fraction of the soil S that occurs in organic forms varies widely, depending on the nature of the soil (pH, drainage status, organic matter content, mineralogical composition) and depth in the soil profile. However, it is generally accepted that most of the S in soils of humid and semihumid regions occurs in organic forms, and this is attested by findings indicating that a close relationship exists between the amounts of C, N, and S that are present.[9-12,18,59]

Table 5.7 summarizes data obtained by Tabatabai and Bremner[60] for total and organic S in 37 Iowa surface soils (0 to 15 cm depth). Organic S accounted for from 95 to 99 percent of the total S and was significantly correlated with organic C and total N. Both the amounts of total S (55–618 µg/g) and the percentages of the soil S in organic forms are similar to values generally reported for agricultural soils (see review of Freney[6]). Very high percentages of the S in the subsurface soils examined by Tabatabai and Bremner[60,61] occurred in organic forms (84–99 percent).

Table 5.7 Amounts of Different Forms of S in Iowa Surface Soils[a]

	Amount of S in Form Specified			
Form of S	Range (µg/g)	Average (µg/g)	Range	Average
Total	56–619	292		100%
Inorganic	1–26	8	1–5%	3
Sulfate	1–26	8	1–5	3
Nonsulfate	0	0	0	0
Organic	55–604	382	95–99	97

[a]From Tabatabai and Bremner.[60] Total number of soils = 37.

Chemical Fractionation of Organic S

Much of the information regarding soil organic S has come from studies on the reactivity of organic S to reduction with hydriodic acid (HI).[18,61-67] A flow diagram is shown in Fig. 5.6. The soil is first leached with a phosphate solution to remove inorganic SO_4^{2-}, following which the leached soil is treated with HI. The S recovered as H_2S is believed to occur as ester sulfates; that which is not reducible is C-bonded S. Finally, the C-bonded S is separated into two fractions based on ability to be reduced to H_2S with Raney Ni.

Two main categories of organic S are as follows:

1. *Ester sulfate-S:* This S, which is readily reduced to H_2S by HI, occurs in compounds containing the C—O—S linkage, such as phenolic sulfates and sulfated polysaccharides. Evidence that most of the HI reducible S occurs as ester sulfates has been documented elsewhere.[5] Lowe[68] found that sulfated polysaccharides extracted from some Canadian soils accounted for <2 percent of the total S.
2. *Carbon-bonded S:* This S is directly attached to C through the S-C linkage and is taken as the difference between total organic S and ester sulfate S. Included are the S-containing amino acids, such as cysteine and methionine.

```
CH₂—SH          CH₂—CH₂—S—CH₃
|               |
CH—NH₂          CH—NH₂
|               |
COOH            COOH
Cysteine        Methionine
```

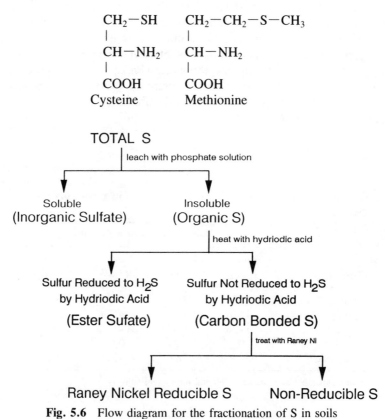

Fig. 5.6 Flow diagram for the fractionation of S in soils

As noted in Fig. 5.6, not all of the C-bonded S is reducible to H_2S by Raney Ni. This is not surprising because it is well-known that Raney Ni does not reduce the C-bonded S of some organic compounds, typical examples being cysteic acid and methionine sulfone. Furthermore, the S combined with humic and fulvic acids may not be reduced. Iron and other soil constituents can lead to low results using the Raney Ni method.[62]

For a range of Australian soils, Freney et al.[62] accounted for an average of 56 percent (range of 45–93 percent) of the C-bonded S by Raney Ni reduction. Good agreement has been obtained between Raney-Ni reducible S and the amount of S accounted for as amino acids.[69,70]

Freney[71] found that the average distribution of total S in 24 Australian soils was as follows:

$$\text{Ester sulfate-S} = 52 \text{ percent}$$

$$\text{C-bonded S} = 41 \text{ percent}$$

$$\text{Inorganic S} = 7 \text{ percent}$$

The average percentage of the total S as ester sulfate-S (52 percent) is similar to that reported in more recent studies by Bettany et al.[18] for the Ap layers of 54 Saskatchewan soils (36–50 percent) and by Tabatabai and Bremner[61] for 37 Iowa surface soils (50 percent). The fraction of the soil S as ester sulfates appears to be higher in grassland soils than in forest soils; for the latter, values as low as 18 percent have been observed.[72] Bettany et al.[73] found that the proportion of the total S as ester sulfate-S in some Canadian soils decreased along an environmental gradient of decreasing temperature and increasing rainfall. Their results are given in Table 5.8.

The percentage of the total S as ester sulfates may increase with depth in the soil profile whereas the percentage as C-bonded S decreases.[60,64] As can be seen from Fig. 5.7, ester sulfate S was the main organic form in some subsoils. The S which is reducible with HI occurs in both high and low molecular weight compounds of soil organic matter, but mostly in the former.[63]

Table 5.8 Effect of Environmental Gradient on the Proportion of the Total Soil S as Ester Sulfate-S[a]

Soil Type[b]	Total S ($\mu g/g$)	Ester Sulfate-S (% of S)
Brown Chernozem	270	52
Dark Brown Chernozem	379	51
Black Chernozem	450	48
Gray Luvisol	158	36

[a]Adapted from Bettany et al.[73]
[b]The four soils were a Aridic Haploboroll, a Typic Haploboroll, a Udic Haploboroll, and a Cyro Boralf, respectively. Temperature decreases and moisture increases in the order given.

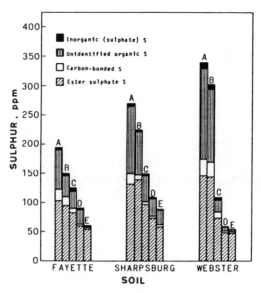

Fig. 5.7 Amounts of different forms of S in some Iowa soils. A = 0–15 cm, B = 15–30 cm, C = 30–60 cm, D = 60–90 cm, E = 90–120 cm. The unidentified fraction is the C-bonded S not reducible with Raney nickel. From Tabatabai and Bremner,[60] reproduced by permission of the Sulphur Institute, Washington, D.C.

McLaren and Swift[67] found that a high proportion (75 percent) of the S that is lost when soils are subject to long-term cultivation consisted of C-bonded forms. Ester sulfate-S was believed to be more transitory and of greater importance in the "short-term" mineralization of organic S. McLachlan and DeMarco[66] found that, on cropping, more S was withdrawn from C-bonded forms than from sulfate esters while Ghani et al.[74] found that C-bonded forms of S represented the major source of mineralizable S in soils. In other work, David et al.[72] observed an increase in ester sulfate-S, and a decrease in C-bonded S, by incubation of some forest soils.

S-Containing Amino Acids

The proportion of the soil S that occurs as amino acids is unknown. On the basis of published data for amino acids in soil hydrolysates, Whitehead[75] estimated that from 11 to 16 percent of the soil S occurs in this form. This estimate may be low, for the reason that cystine and methionine, the two main S-containing amino acids, are both known to undergo extensive destruction during acid hydrolysis. In an attempt to avoid this problem, Freney et al.[76] oxidized cystine and methionine to stable forms (cysteic acid and methionine sulfone) prior to acid hydrolysis. For two soils, an average of 26 percent of the soil S, equivalent to 46 percent of the C-bonded S, was estimated to be present as amino acids. By using a somewhat similar procedure, Scott et al.[77] found that from 11 to 15 percent of the total organic S (19 to 31 percent of the C-bonded S) in some Scottish soils occurred as amino acid-S.

Lipid S

The existence of sulfolipids in soil has been demonstrated from the occurrence of S in lipid extracts of soil.[78-80] Chae and Tabatabai[80] found that the amount of lipid S in 10 Iowa surface soils ranged from 0.87 to 2.63 μg/g and accounted for only 0.29 to 0.45 percent (average = 0.37 percent) of the total soil S. From 20 to 40 percent of the sulfolipid S was HI-reducible (i.e., as ester-type compounds). In other work, Chae and Lowe[78] found that lipid S accounted for from 0.5 to 3.5 percent of the total S in 27 soils of British Columbia.

The technique of column chromatography was used by Chae and Lowe[79] to fractionate soil lipid S into three lipid classes: polar, glycolipids, and less polar lipids. Glycolipids were the most dominant form (mean of 64 percent): polar lipids were the least abundant (mean of 4 percent).

Fractionations Based on Solubility Characteristics

Extractions and fractionations of soil organic S based on solubility characteristics (see Chapter 2) have shown that soil organic S occurs in all the humus fractions (i.e., humic acid, fulvic acid, and humin).[73,81]

By extraction with a $0.15M$ NaOH:$0.1M$ Na$_4$P$_2$O$_7$ solution at pH 15, Bettany et al.[73] recovered 63 to 72 percent of the total soil S from a series of soils along an environmental gradient in Canada. The approximate distribution of the extracted organic S was: humic acid (34 percent), fulvic acid "fraction" (39 percent), humin (26 percent).

Fractionations based on alkali extraction may lead to hydrolysis of extracted S, notably sulfate esters. To avoid this problem, milder extractants have been used.[63,82,83] They include dilute Na$_2$CO$_3$ and aqueous acetylacetone. In the latter case, extraction of the soil with $0.2M$ aqueous acetylacetone at pH 8, in combination with an ultrasonic treatment, has led to recovery of over 80 percent of the soil organic S.[83] Most of the extracted S occurred in molecular weight fractions >200,000 daltons, as estimated by gel permeation chromatography.

Sulfur of Humic and Fulvic Acids

Part of the C-bonded S, notably the fraction not reduced to H$_2$S with Raney Ni, is believed to occur as a structural component of humic and fulvic acids, namely, as complexes resulting from the reaction of thiol compounds with quinones and reducing sugars. The mechanisms involved are similar to those discussed in Chapter 9 for the formation of humic substances. For example, cysteine, thiourea, and glutathione are known to react with quinones (formed from polyphenols derived from lignin or synthesized by microorganisms) to form brown-colored pigments containing S.

Mason[84] pointed out that when there is a possibility for competition between thiol and NH_2 groups for the quinone, such as in the case of cysteine, the thiol will react first. For example, the first product of the reaction between cysteine and *p*-benzoquinone is 1,4-benzoquinone-2-cysteine, which subsequently undergoes an inner condensation between the quinone C=O group and the free NH_2 group to form a cyclic product. These reactions are shown in Fig. 5.8.

The resistance of soil organic S to attack by microorganisms may be partly caused by the existence of such C—S linkage in high-molecular-weight humic components. The reader is referred to Chapter 9 for additional information on the biochemistry of humus formation.

Physical Fractionation

Ultrasonic dispersion of the soil, followed by particle size separations (see Chapter 2), has been used for the physical fractionation of organic S.[85] In this work, consistent differences were observed for both the distribution and composition of S in the various size fractions (Table 5.9). Most of the S (70 percent) was recovered in the clay fractions, and most of the organic S occurred in HI-reducible forms. Carbon/S ratios were also lowest in the clay fractions. The bulk of the S in the coarse fractions occurred in C-bonded forms.

Biomass S

From 1 to 3 percent of the organic S in soils can be accounted for as part of the soil biomass.[86,87] The S content of most microorganisms lies between 0.1 and 1.0 percent on a dry weight basis.

The quantity of S in the microbial biomass has been estimated from the amount of inorganic SO_4^{2-} that is produced during incubation of soil fumigated with chloroform ($CHCl_3$), the extractant being $0.1 M$ $NaHCO_3$ or 10 mM $CaCl_2$.

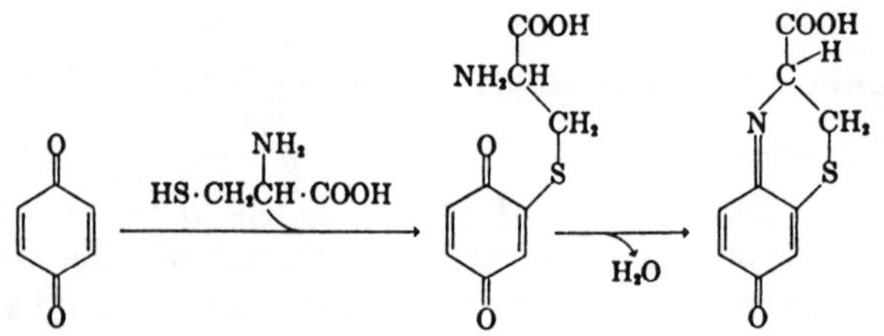

Fig. 5.8 Reactions between cysteine and *p*-benzoquinone to form complex structures containing C-S linkages

Table 5.9 Distribution of S and HI Reducible S (Ester Sulfate-S) in the Size Fractions of a Canadian Black Oxbow Soil (Udic Argiboroll)[a]

Fraction	Organic C µg/g	Total S µg/g	HI Reducible S % of S	C/Total S
Sand	1.26	11	27	115
Coarse silt	7.98	73	25	109
Fine silt	5.24	42	50	125
Coarse clay	12.03	140	63	86
Fine clay	5.45	151	77	36
Whole soil	33.20	418	62	79

[a]From Anderson et al.[85] Similar results were obtained for a Typic Argiboroll soil.

The equation is:

$$\text{Biomass S} = \frac{\text{S extracted from fumigated soil } - \text{ S extracted from nonfumigated soil}}{K_s}$$

where K_s is the fraction of the biomass S subject to release by the $CHCl_3$ treatment. Values of K_s for the two extractants were: $0.1 M$ $NaHCO_3$ = 0.41 and 10 mM $CaCl_2$ = 0.35.

Although only a small amount of the soil organic S resides in the biomass at any one time, this fraction is extremely labile and is the main driving force for S turnover in soil. The higher the amount of organic S in the biomass, the greater will be the availability of S to higher plants.

Soluble Forms of Organic S

In view of the broad synthetic activities of microorganisms, numerous S-containing biochemicals would be expected to be produced in soil. However, because these same compounds are themselves susceptible to further decomposition, they do not persist in the free form, and the low amounts found at any one time represent a balance between synthesis and destruction by microorganisms. In accordance with this concept, the concentration of uncombined organic S compounds in soil has been found to be low and variable. For example, only trace quantities of S-containing amino acids (cysteine, cystine, methionine, methionine sulfoxide, taurine, and cysteic acid) have been recovered from soil by extraction with water or such solvents as neutral ammonium acetate (see Chapter 3). The concentration of S-containing organics may be higher in rhizosphere soil than nonrhizosphere soil, since they are known to be excluded from plant roots.

Structures of some typical S-containing biochemicals that are known to occur in plants and microorganisms and that would be expected to occur in soils in trace amounts are given below.

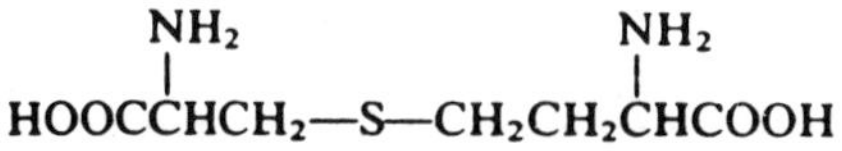

NH₂   NH₂
| |
HOOCCHCH₂—S—CH₂CH₂CHCOOH $(CH_3)_3\overset{+}{N}$—CH₂CH₂OSO₃⁻

Cystathionine   Choline sulfate

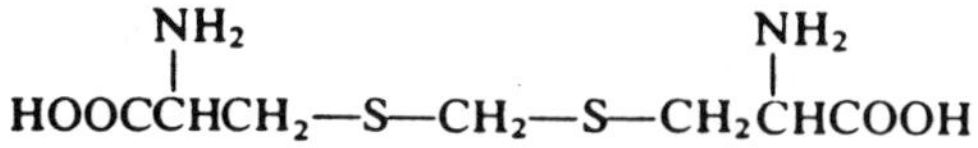

NH₂   NH₂
| |
HOOCCHCH₂—S—CH₂—S—CH₂CHCOOH   H₂N—CH₂CH₂—SO₃H

Djenkolic acid   Taurine

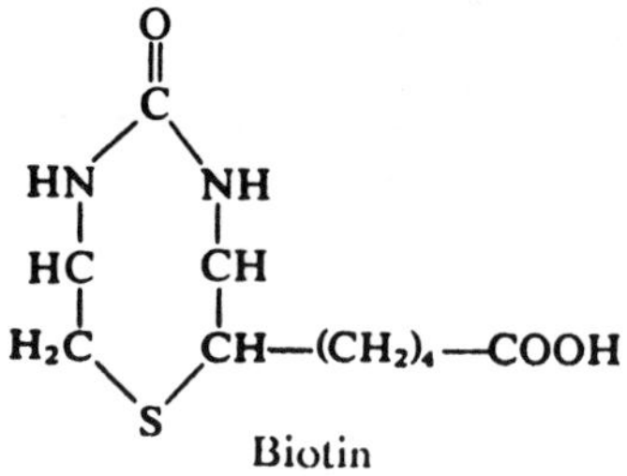

Biotin

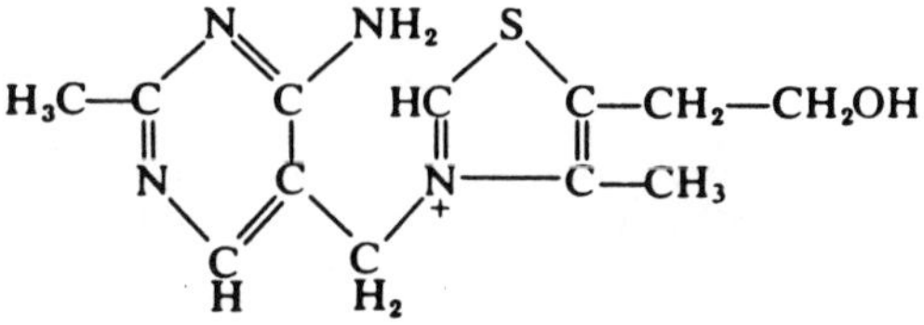

Thiamine

Volatile Organic S Compounds

In poorly drained soils, the decomposition of organic S compounds leads to the formation of trace amounts of mercaptans, alkyl sulfides, and other volatile organic S compounds, as well as H_2S.[88-91] Included with the volatile organic S compounds produced by microorganisms are the following:

CS_2	COS	CH_3—SH
Carbon disulfide	Carbonyl sulfide	Methyl mercaptan
CH_3—CH_2—S—CH_2—CH_3	CH_3—S—CH_3	CH_3—S—S—CH_3
Diethyl sulfide	Dimethyl sulfide	Dimethyldisulfide

Anaerobic conditions are especially favorable for the synthesis of volatile organic S compounds by microorganisms. Volatile S compounds in soil may be of importance as they may stimulate or suppress the growth of microorganisms and they may inhibit nitrification and other biochemical processes.

Much interest has been shown in recent years to the flux of S gases between soil and the atmosphere. These constituents are undesirable as atmospheric components because they have the potential for adversely affecting climate and the environment through their aerosol-forming properties.[92] In considering the soils' contributions to S gases in the atmosphere, one must keep in mind that most soils are able to adsorb S gases of various types.[93,94]

Use of ^{35}S For Following Organic S Transformations

As noted earlier, when fertilizer SO_4^{2-} is applied to the soil some of the S is converted to organic forms through immobilization by microorganisms. Use of ^{35}S-labeled SO_4^{2-} has provided valuable information on factors affecting incorporation of S into the various organic S forms and plant availabilities of C-bonded and ester sulfate forms.[83,95-98] A discussion of this work is beyond the scope of this book but has been reviewed elsewhere.[5,8]

SUMMARY

As much as two-thirds or more of the total P in soils occurs in organic forms, with the higher percentages being typical of peats, uncultivated forest soils, and tropical soils. An active P cycle exists in soil, in which turnover between organic and inorganic forms (mineralization–immobilization) plays a major role. Additions of decomposable organic matter to soil has been shown in laboratory and greenhouse studies to enhance the availability of insoluble Ca-, Fe-, and Al-phosphates to plants, but the importance of the process under conditions existing in the field is unknown.

Excellent progress has been made in determining the nature of organic P compounds in soils, but much of the organic P has yet to be identified in known compounds. As a general rule, less than one-half of the organic P can be accounted for. Approximate recoveries of organic P are: inositol phosphates, 2 to 50 percent; phospholipids, 1 to 5 percent; nucleic acids, 0.2 to 2.5 percent; phosphoproteins, trace; metabolic phosphates, trace. The technique of ^{31}P-NMR spectroscopy shows promise for the characterization of soil organic P.

Although the role of S in soil fertility has been neglected, sufficient information has accumulated to indicate that, in many soils, transformations involving organic forms are of utmost importance in the S mineral nutrition of plants. The total S content of soils is variable. For soils of humid and semi-humid regions, most of the S occurs in organic forms and the amounts found are directly related to organic matter content.

Little is known about the forms of organic S in soil. About 40 percent of the total S in soils of humid and semiarid regions occurs in C-bonded forms, only a fraction of which can be accounted for as amino acids. Another 50 percent of the total S occurs as unknown ester sulfates.

REFERENCES

1 S. K. Sanyal and S. K. De Datta, *Adv. Soil Sci.*, **16**, 1 (1991).
2 G. Anderson, "Other Organic Phosphorus Compounds," in J. E. Gieseking, Ed., *Soil Components: Vol. 1*, Springer-Verlag, New York, 1975, pp. 305–331.
3 R. C. Dalal, *Adv. Agron.*, **29**, 83 (1979).

4. R. L. Halsted and R. B. McKercher, "Biochemistry and Cycling of Phosphorus," in E. A. Paul and A. D. McLaren, Eds., *Soil Biochemistry, Vol. 4*, Marcel Dekker, New York, 1975, pp. 31–63.
5. F. J. Stevenson, *Cycles of Soil: Carbon, Nitrogen, Phosphorus, Sulfur, Micronutrients*, Wiley, New York, 1986.
6. J. R. Freney, "Forms and Reactions of Organic Sulfur Compounds in Soils," in M. A. Tabatabai, Ed., *Sulfur in Agriculture*, American Society of Agronomy, Madison, 1986, pp. 207–232.
7. J. R. Freney and F. J. Stevenson, *Soil Sci.*, **101,** 307 (1966).
8. J. J. Germida, M. Wainwright, and V. V. S. R. Gupta, "Biochemistry of Sulfur Cycling in Soil," in G. Stotzky and J.-M. Bollag, Eds., *Soil Biochemistry: Vol. 7.*, Marcel Dekker, New York, 1992, pp. 1–53.
9. A. M. L. Neptune, M. A. Tabatabai, and J. J. Hanway, *Soil Sci. Soc. Amer. Proc.*, **39,** 51 (1975).
10. T. W. Walker and A. F. R. Adams, *Soil Sci.*, **85,** 307 (1958).
11. C. H. Williams, E. G. Williams, and N. M. Scott, *J. Soil Sci.*, **11,** 334 (1960).
12. L. L. Somani and S. N. Saxena, *Anal. Edafol. Agrobiol.*, **37,** 809 (1978).
13. M. K. John, P. N. Sprout, and C. C. Kelley, *Can. J. Soil Sci.*, **45,** 87 (1964).
14. A. Kaila, *Soil Sci.*, **95,** 38 (1963).
15. K. M. Goh and M. R. Williams, *J. Soil Sci.*, **33,** 73 (1982).
16. E. T. Elliott, *Soil Sci. Soc. Amer. J.*, **50,** 627 (1982).
17. A. P. Uriyo and A. Kesseba, *Geoderma*, **13,** 201 (1975).
18. J. R. Bettany, J. W. B. Stewart, and E. H. Halstead, *Soil Sci. Soc. Amer. Proc.*, **37,** 915 (1973).
19. R. W. Pearson and R. W. Simonson, *Soil Sci. Soc. Amer. Proc.*, **4,** 162 (1939).
20. W. A. Dick and M. A. Tabatabai, *Soil Sci. Soc. Amer. J.*, **41,** 511 (1977).
21. J. H. Stewart and J. M. Oades, *J. Soil Sci.*, **23,** 38 (1972).
22. A. Islam and A. Ahmed, *J. Soil Sci.*, **24,** 193 (1973).
23. J. K. Martin, *New Zealand, J. Agric. Res.*, **7,** 723, 736, 750 (1964).
24. A. G. Caldwell and C. A. Black, *Soil Sci. Soc. Amer. Proc.*, **22,** 290, 293, 296 (1958).
25. D. H. Smith and F. E. Clark, *Soil Sci. Soc. Amer. Proc.*, **16,** 170 (1952).
26. R. Mandal and A. Islam, *Geoderma*, **22,** 315 (1979).
27. D. J. Cosgrove, *Aust. J. Soil Res.*, **1,** 203 (1963).
28. D. J. Cosgrove, *Soil Sci.*, **102,** 42 (1966).
29. D. J. Cosgrove, *Soil Biol. Biochem.*, **1,** 325 (1969).
30. D. J. Cosgrove, *Inositol Phosphates*, Elsevier, New York, 1980.
31. R. B. McKercher and G. Anderson, *J. Soil Sci.*, **19,** 47, 302 (1968).
32. T. I. Omotoso and A. Wild, *J. Soil Sci.*, **21,** 224 (1970).
33. G. Anderson, *J. Soil Sci.*, **21,** 96 (1970).
34. R. T. Baker, *J. Soil Sci.*, **26,** 432 (1975).
35. C. G. Kowalenko and R. B. McKercher, *Can. J. Soil Sci.*, **51,** 19 (1971).
36. G. Robertson, *J. Sci. Food Agr.*, **9,** 288 (1958).
37. G. Anderson and R. E. Malcolm, *J. Soil Sci.*, **25,** 282 (1974).

38 J. R. Moyer and R. L. Thomas, *Soil Sci. Soc. Amer. Proc.*, **34**, 80 (1970).
39 T. I. Omotoso and A. Wild, *J. Soil Sci.*, **21**, 216 (1970).
40 R. S. Swift and A. M. Posner, *J. Soil Sci.*, **23**, 50 (1972).
41 R. L. Veinot and R. L. Thomas, *Soil Sci. Soc. Amer. Proc.*, **36**, 71 (1972).
42 G. Anderson and R. J. Hance, *Plant Soil*, **19**, 296 (1963).
43 C. A. Brannon and L. E. Sommers, *Soil Biol. Biochem.*, **17**, 213, 221 (1985).
44 L. M. Condron, K. M. Goh, and R. H. Newman, *J. Soil Sci.*, **36**, 199 (1985).
45 L. M. Condron, E. Frossard, H. Tiessen, R. H. Newman, and J. W. B. Stewart, *J. Soil Sci.*, **41**, 41 (1990).
46 K. R. Tate and R. H. Newman, *Soil Biol. Biochem.*, **14**, 191 (1982).
47 G. E. Hawkes, D. S. Powlson, E. W. Randall, and K. R. Tate, *J. Soil Sci.*, **35**, 35 (1984).
48 F. Gil-Sotress, W. Zech, and H. G. Alt, *Soil Biol. Biochem.* **22**, 97 (1990).
49 P. C. Brookes, D. S. Powlson, and D. S. Jenkinson, *Soil Biol. Biochem.*, **14**, 319 (1982).
50 P. C. Brookes, D. S. Powlson, and D. S. Jenkinson, *Soil Biol. Biochem.*, **16**, 169 (1984).
51 M. J. Hedley and J. W. B. Stewart, *Soil Biol. Biochem.*, **14**, 337 (1982).
52 R. A. Bowman and C. V. Cole, *Soil Sci.*, **125**, 95 (1978).
53 M. J. Hedley, J. W. B. Stewart, and B. S. Chauhan, *Soil Sci. Soc. Amer. J.*, **46**, 970 (1982).
54 R. M. N. Kucey, H. H. Janzen, and M. E. Leggett, *Adv. Agron.*, **42**, 199 (1989).
55 F. J. Stevenson and G. F. Vance, "Naturally Occurring Aluminum–Organic Complexes," in G. Sposito, Ed., *The Environmental Chemistry of Aluminum*, CRC Press, Boca Raton, Florida, 1989, pp. 117–145.
56 R. Coleman, *Soil Sci.*, **101**, 230 (1966).
57 R. W. Blanchar, "Measurement of Sulfur in Soils and Plants," in M. A. Tabatabai, Ed., *Sulfur in Agriculture*, American Society of Agronomy, Madison, 1986, pp. 455–489.
58 C. M. Johnson and H. Nishita, *Anal. Chem.*, **24**, 736 (1952).
59 N. M. Scott and G. Anderson, *J. Soil Sci.*, **27**, 324 (1976).
60 M. A. Tabatabai and J. M. Bremner, *Sulphur Inst. J.*, **8**, 1 (1972).
61 M. A. Tabatabai and J. M. Bremner, *Soil Sci.*, **114**, 380 (1972).
62 J. R. Freney, G. E. Melville, and C. H. Williams, *Soil Sci.*, **109**, 310 (1970).
63 J. R. Freney, G. E. Melville, and C. H. Williams, *J. Sci. Food Agric.*, **20**, 440 (1969).
64 L. E. Lowe, *Can. J. Soil Sci.*, **45**, 297 (1965).
65 L. E. Lowe and W. A. deLong, *Can. J. Soil Sci.*, **43**, 151 (1963).
66 K. D. McLachlan and D. G. DeMarco, *Aust. J. Soil Res.*, **13**, 169 (1975).
67 R. G. McLaren and R. S. Swift, *J. Soil Sci.*, **28**, 445 (1977).
68 L. E. Lowe, *Can J. Soil Sci.*, **48**, 215 (1968).
69 J. R. Freney, G. E. Melville, and C. H. Williams, *Soil Biol. Biochem.*, **7**, 717 (1975).
70 N. M. Scott, W. Bick, and H. A. Anderson, *J. Sci. Food Agr.*, **32**, 21 (1981).

71 J. R. Freney, 1967. "Sulfur-Containing Organics," in A. D. McLaren and G. H. Peterson, Eds., *Soil Biochemistry*, Marcel Dekker, New York, 1967, pp. 229–259.
72 M. B. David, M. J. Mitchell, and J. P. Nakas, *Soil Sci. Soc. Amer. J.*, **46,** 847 (1982).
73 J. R. Bettany, J. W. B. Stewart, and S. Saggar, *Soil Sci. Soc. Amer. J.*, **43,** 981 (1979).
74 A. Ghani, R. G. McLaren, and R. S. Swift, *Biol. Fert. Soils*, **11,** 68 (1991).
75 D. C. Whitehead, *Soils Fertilizers*, **27,** 1 (1964).
76 J. R. Freney, F. J. Stevenson, and A. H. Beavers, *Soil Sci.*, **114,** 468 (1972).
77 N. M. Scott, W. Bick, and H. A. Anderson, *J. Sci. Food Agr.*, **32,** 21 (1981).
78 Y. M. Chae and L. E. Lowe, *Can. J. Soil Sci.*, **60,** 633 (1980).
79 Y. M. Chae and L. E. Lowe, *Soil Biol. Biochem.*, **13,** 257 (1981).
80 Y. M. Chae and M. A. Tabatabai, *Soil Sci. Soc. Amer. J.*, **45,** 20 (1981).
81 J. R. Bettany, S. Saggar, and J. W. B. Stewart, *Soil Sci. Soc. Amer. J.*, **44,** 70 (1980).
82 M. M. Scott and G. A. Anderson, *J. Soil Sci.*, **27,** 324 (1976).
83 J. I. Keer, R. G. McLaren, and R. S. Swift, *Soil Biol. Biochem.*, **22,** 97 (1990).
84 H. S. Mason, *Adv. Enzymol.*, **16,** 105 (1955).
85 D. W. Anderson, S. Saggar, J. R. Bettany, and J. W. B. Stewart, *Soil Sci. Soc. Amer. J.*, **45,** 767 (1981).
86 S. Saggar, J. R. Bettany and J. W. B. Stewart, *Soil Biol. Biochem.*, **13,** 493, 499 (1981).
87 J. E. Strick and J. P. Nakas, *Soil Biol. Biochem.*, **16,** 289 (1984).
88 W. L. Banwart and J. M. Bremner, *J. Environ. Qual.*, **4,** 363 (1975).
89 W. L. Banwart and J. M. Bremner, *Soil Biol. Biochem.*, **8,** 19, 329 (1976).
90 L. F. Elliott and T. A. Travis, *Soil Sci. Soc. Amer. Proc.*, **37,** 700 (1973).
91 J. A. Lewis and G. C. Papavizas, *Soil Biol. Biochem.*, **2,** 239 (1970).
92 W. W. Kellog, R. D. Cradle, E. R. Allen, A. L. Lazarus, and A. E. Martell, *Science*, **175,** 587 (1972).
93 J. M. Bremner and W. L. Banwart, *Soil Biol. Biochem.*, **8,** 79 (1975).
94 K. A. Smith, J. M. Bremner, and M. A. Tabatabai, *Soil Sci.*, **116,** 313 (1973).
95 J. R. Freney, G. E. Melville, and C. H. Williams, *Soil Biol. Biochem.*, **3,** 133 (1971).
96 J. R. Freney, G. E. Melville, and C. H. Williams, *Soil Biol. Biochem.*, **7,** 217 (1975).
97 K. M. Goh and P. E. H. Gregg, *New Zealand J. Sci.*, **25,** 135 (1982).
98 R. G. McLaren, J. I. Keer, and R. S. Swift, *Soil Biol. Biochem.*, **17,** 73 (1985).

6

SOIL CARBOHYDRATES

Carbohydrates have been estimated to constitute between 5 to 25 percent of the soil organic matter and thereby they are the second most abundant component of humus. Plant remains contribute carbohydrates in the form of simple sugars, hemicellulose, and cellulose but these are more or less decomposed by soil microorganisms, which in turn synthesize polysaccharides and other carbohydrates of their own. With the exception of the litter layer of the soil (see Chapter 1), it is the microbial products that make up the major part of the carbohydrate material in soil. Exudates of plant roots serve as a minor source of sugars, with a transient existence.

Results of hydrolysis procedures have given the following approximate ranges for the contribution of individual sugar types to soil organic matter. For the most part, they occur in complex forms; free sugars exist in soil in only trace amounts.

	Percent of Organic Matter
Amino sugars	2–6
Uronic acids	1–5
Hexose sugars	4–12
Pentose sugars	<5
Cellulose and cellulose derivatives	to 15
Others (e.g., methylated sugars)	trace

Several reviews can be consulted for more detailed information regarding soil carbohydrates.[1-5] The amino sugars, which constitute a significant fraction of the soil polysaccharide material, are discussed in Chapter 3 and will not be covered herein.

SIGNIFICANCE OF SOIL CARBOHYDRATES

The significance of carbohydrates in soil arises largely from the ability of complex polysaccharides to bind inorganic soil particles into stable aggregates (Chapter 18). Carbohydrates also form complexes with metal ions (Chapter 16), and they serve as building blocks for humus synthesis (Chapter 8). Mineralization of amino sugars and P-containing substances related to carbohydrates (phytin and nucleic acids) leads to the release of N and P for plant growth (Chapters 3 and 5). Some sugars may stimulate seed germination and root elongation. Other soil properties affected by polysaccharides include cation-exchange capacity (attributed to COOH groups of uronic acids), anion retention (occurrence of NH_2 groups), and biological activity (e.g., energy source for microorganisms). Neutral sugars generally constitute the most abundant form of organic matter in the soil solution and the content of water-soluble organic C has been shown to be a good index of the denitrification potential of the soil.[6]

In certain submerged soils, the production of microbial gums and nucilages may lead to an undesirable reduction in permeability due to blocking soil pores. The hydrophobic condition of certain sands may be due in part to waterproofing by gums and mucilages.

STRUCTURE AND CLASSIFICATION

Carbohydrates comprise one of the major groups of naturally occurring organic molecules. They can be divided into three subclasses: 1) monosaccharides, which are aldehyde and ketone derivatives of the higher polyhydric alcohols, 2) oligosaccharides, which comprise a large group of polymeric carbohydrates consisting of a relatively few monosaccharide units (Greek "oligos," meaning few), and 3) polysaccharides, which contain many monomeric units (eight or more) to give products high in molecular weight. Several excellent books can be consulted for detailed information on the chemistry and biochemistry of the carbohydrates.

The structures of sugars can be indicated in several ways, including the open-chain form, ring form (two models used), "chair" form, and "boat" form. The open-chain and ring forms of glucose are shown by structures I, II, and III. For the most part, structures shown by III will be used in the text.

STRUCTURE AND CLASSIFICATION 143

```
  HCO              HCOH ⎤
  |                |    |
  HCOH             HCOH |
  |                |    |
  HOCH             HOCH O
  |                |    |
  HCOH             HCOH |
  |                |    |
  HCOH             HC ──┘
  |                |
  CH₂OH            CH₂OH

    I                II
```

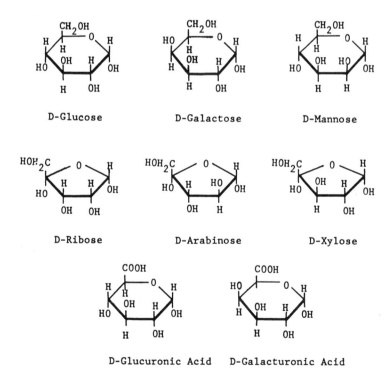

The ring form of some of the common hexose and pentose sugars are shown in Fig. 6.1. Also included are structures for several naturally occurring uronic acids. Sugars containing six-membered rings are called pyranoses; those with five-membered rings are furanoses.

Some typical polysaccharides occurring naturally in nature are shown in Fig. 6.2. Cellulose (A) is a linear polymer of β-(1 → 4)-D-glucopyranose units, amylose is a polymer of α(1 → 4)-D-glucopyranose units, and polygalacturonic acid is a pectic substance consisting of a polymer of α-(1 → 4) linked D-galacturonopyranose units in which the COOH groups are partially or fully methylated.

Fig. 6.1 Structural formulas (ring configuration) of some common sugars

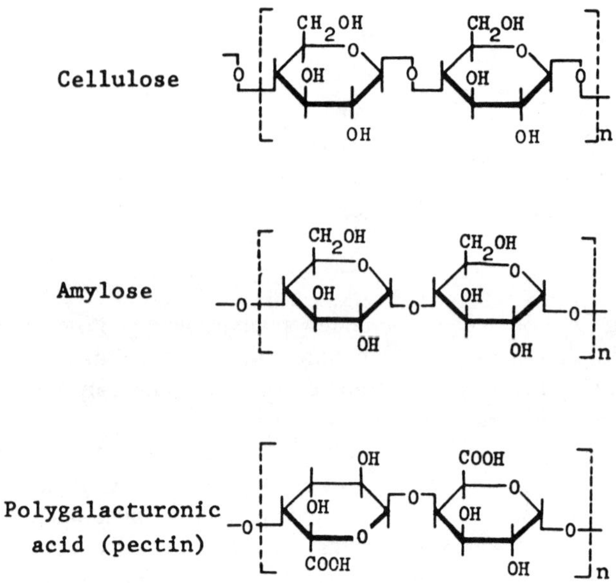

144 SOIL CARBOHYDRATES

Fig. 6.2 Structural formulas of some typical polysaccharides

Whereas the polysaccharides shown in Fig. 6.2 consist of repeating units of a single monosaccharide, many polysaccharides, including those of microbial origin, yield mixtures of monosaccharides and derived products by hydrolysis. Great variations in structure occur by virtue of the large number of monosaccharides that might be included in such substances and by variations in the nature of the glycosidic linkage (α or β, $1 \rightarrow 4$, $1 \rightarrow 6$, and others).

The complex polysaccharides synthesized by microorganisms have yet to be fully characterized but for the most part they appear to contain not more than three or four different monosaccharide types in a particular structure. They occur as: 1) extracellular surface polysaccharides, 2) cell-wall polysaccharides, and 3) somatic or intracellular polysaccharides. In many cases, the polysaccharides are bound to protein and lipid constituents. Polysaccharides containing uronic acids are usually acidic.

STATE IN THE SOIL

The carbohydrate material in soil occurs as: 1) free sugars in the soil solution, 2) complex polysaccharides that can be extracted and separated from other organic constituents, and 3) polymeric molecules of various sizes and shapes that are so strongly attached to clay and/or humic colloids that they cannot be easily isolated and purified. Stability of the polysaccharides is due to a combination of several factors, including structural complexity, which makes them resistant to enzymatic attack, adsorption on clay minerals and oxide surfaces, formation of insoluble salts or chelate complexes with polyvalent cations, and tanning by humic substances. In the latter case, the polysaccharides would occur as an integral component of humic and fulvic acids, being bound through an ester linkage (R_1COOR_2) and possibly other covalent bonds. Polysaccharides may be further protected when they occur within the micropores of soil aggregates.[7]

Martin[8] postulated that the polysaccharides in soil are formed by recombination of monomeric units derived from plant and microbial polysaccharides and that the products thus formed have more complex structures (e.g., are more highly branched) and are thus less susceptible to biodegradation. On the other hand, Cheshire et al.[9,10] concluded that the persistence of polysaccharides is related to inaccessibility caused by chemical combinations, complexing, or insolubility, but not to a biologically stable molecular structure.

In soils containing undecayed or partially decomposed plant remains, a portion of the carbohydrate will exist as cellulose, which is the main carbohydrate component of higher plants. Cellulose has been extracted from soils with Schweitzer's reagent, an ammoniacal solution of a cupric salt. The material recovered from soil is precipitated with alcohol and washed successively with hot water, $1N$ HCl, and water to give a product of high purity. For the soils examined by Gupta and Sowden,[11] from 20 to 40 percent of the total glucose could be accounted for as cellulose.

Soils rich in plant remains also contain relatively high amounts of the pentose sugars arabinose and xylose.[12] In studies using ^{14}C-labeled substrates, Cheshire and his coworkers[10,13] found that polysaccharides derived from the labeled material contained lower amounts of arabinose and xylose than the native polysaccharide, indicating that plants represent a major source of the pentoses in soil.

Free Sugars

Free carbohydrates are present in soils in such small quantities that they can only be detected using specialized microchemical techniques. This is not surprising in view of the ease with which simple sugars are utilized by microorganisms. The greatest quantities of free sugars are found in the litter layers of forest soils, notably those of cold climatic zones. The amounts found at any one time in mineral soils represent a delicate balance between synthesis and destruction by microorganisms, and are affected by those factors influencing microbial activity, such as moisture, temperature, and energy supply. The amounts and perhaps the kind of free sugars may be a reflection of methods used for collecting and handling the soil samples.[2]

Free sugars are most satisfactorily extracted from soils with water or 80 percent ethanol, the latter being the most desirable because soil dispersion is not a problem and the alcohol can easily be removed by evaporation *in vacuo* at a low temperature (40°C). The sugar residue is dissolved in water, centrifuged, clarified by various means, and analyzed for sugars by chromatographic procedures.

The free sugars reported by Gupta[2] for a variety of mineral and organic soils include glucose, galactose, mannose, fructose, arabinose, xylose, fucose, and ribose.

Complex Polysaccharides

Most polysaccharides in soil are believed to represent products of microbial metabolism. Evidence in support of this view is attested by their relatively low contents of xylose and glucose (a particularly abundant sugar in plant polysaccharides) but proportionally higher contents of other monosaccharides. Some of the sugars identified in hydrolysates of crude soil polysaccharides are not found to any extent in higher plants but are common constituents of exocellular and capsular polysaccharides of microorganisms. Chitin, a polysaccharide compose of N-acetylglucosamine units, is the main structural component of fungal cell walls.

Extraction and Preliminary Purification In order to characterize soil polysaccharides they must first be extracted from the soil. The ideal extractant is one which would: 1) be equally effective for all soils, 2) selectively extract polysaccharides or provide extracts from which impurities can easily be removed, 3) be nondestructive of labile polysaccharide components, and 4) give sufficiently complete extraction for the material recovered to be representative of the total.[1] It can safely be said that these objectives have yet to be fulfilled.

Extractants used for recovering polysaccharides from soil include water at elevated temperatures (70 to 100°C), aqueous buffers at pH 7, complexing agents (e.g., EDTA), organic solvents (formic acid), caustic alkalies, and dilute mineral acids.[14-24] The polysaccharide material, after removal of contaminants, has little or no color.

Quantitative data summarized in Table 6.1 are little more than a guide for the amounts extracted because not all of the extracted material may have occurred as polysaccharides and many contained high amounts of inorganic impurities.

Of the various extractants that have been used for recovering polysaccharides from soil, the highest yields have been obtained using hot formic acid (see Table 6.1). Yields equivalent to 3.5 to 11.5 percent of the soil organic matter have been recorded.[20] This procedure involves refluxing the soil twice for 30 min. with 98 percent formic acid containing $2.5N$ LiBr, after which the extracted organic matter is precipitated with diisopropyl ether. Partial purification of polysaccharides is accomplished by dissolution in aqueous $0.5N$ LiCl solution, followed by precipitation with cetavalon (cetyl trimethylammonium bromide). The major portion of the extracted material, amounting to 33 to 72 percent, was recovered as a neutral polysaccharide soluble in both solvents. Some loss of uronic acid material through decarboxylation may have occurred during extraction.

Table 6.1 **Typical Yields of Crude Polysaccharides From Soils Using Various Extractants**

Soil	Method	Yield % of Soil	Yield % of Organic Matter	Reference
Mineral soils				
Red-brown earth				
(a)	H_2O for 16 hours at 70°C	0.06	1.5	Swincer et al.[18]
(b)	$1N$ HCl for 16 hours at 20°C	0.04	1.0	Ibid.
(c)	$0.5N$ NaOH for 16 hours at 30°C	0.13	3.2	Ibid.
Five forest soils	pH 7 buffers with 5% KCl	0.01–0.04		Bernier[16]
Five British soils	Anhydrous HCOOH containing $0.2N$ LiBr, two hot extractions	0.4–5.6	4.0–11.5	Parsons and Tinsley[20]
Four Scottish soils	$0.5N$ NaOH for 48 hours at 15°C		1.3–1.9	Forsyth[17]
Saskatchewan soils	$0.5N$ NaOH for 3 hours at 20°C	0.14–0.78	3.2–14.0	Acton et al.[14]
Peat and muck soils				
Fen peat				
(a)	H_2O, conditions not specified	0.008	0.009	Barker et al.[15]
(b)	7% disodium EDTA	0.45	0.56	Ibid.
(c)	$0.6N$ H_2SO_4 for 24 hours at 30°C	0.03	0.04	Ibid.
(d)	Amberlite IRC resin, H^+-form	0.13	0.17	Ibid.
(e)	N-methyl-2-pyrrolidone	0.03	0.04	Ibid.
(f)	$8M$ urea	0.46	0.57	Ibid.
(g)	Dimethylformamide	3.0	3.8	Ibid.
Scottish peats	$0.09N$ HCl for 1 hour at 60–70°C	0.15	0.15	Black et al.[19]

Yields of polysaccharides recorded by Parsons and Tinsley[20] by formic acid extraction may be high for two reasons. First, uronic acids in the isolated polysaccharides were estimated by decarboxylation with 12 percent HCl, a procedure known to release CO_2 from humified organic matter. The second reason is that the reducing sugar content of the isolated polysaccharide material may have been overestimated due to interference by other reducing substances.[4]

Caustic alkali (i.e., 0.5N NaOH) has been a popular extractant for soil polysaccharides. This extractant also removes considerable quantities of true humic substances. The first step in recovering polysaccharides from the complex mixture is acidification to pH 2 to 3 to precipitate humic acids. However, a certain amount of polysaccharide material is coprecipitated or bound to the humic acid. One method used to reduce losses of polysaccharides through coprecipitation is to pass the alkali extract through the H^+-form of a cation exchange resin, which removes cations and aids in subsequent purification of polysaccharides.[18]

The approach used by Oades and Swincer[21] to isolate and partially purify soil polysaccharides appears to have been particularly successful and is as follows. Coarse plant fragments and incompletely humified materials are first removed from the soil by flotation, after which the carbohydrates are extracted with 0.2N NaOH. The suspension is centrifuged and the supernatant solution is passed upward through a column of a cation exchange resin (H^+-form) to remove humic acids and salts. The eluted polysaccharide material is subsequently clarified, or partly so, using Polyclar AT (polyvinyl pyrrolidone), which selectively removes colored material while leaving the carbohydrates in solution. Polysaccharides in the partially purified extract are freeze-dried prior to further fractionation by gel filtration (see next section).

A somewhat different extraction and purification approach has been utilized by Barker et al.[15,22] and Finch et al.[24] Recovery of polysaccharides was accomplished by extraction with 0.6N H_2SO_4 at a low temperature (3°C) for 24 h. The filtrate was neutralized, centrifuged, and the suspension was pressure filtered to remove all traces of suspended matter. Following dialysis, the nondialyzable material was freeze-dried to give a crude polysaccharide preparation. Additional polysaccharide was recovered from the precipitate obtained through neutralization of the extract by dissolving the residue in 0.3N HCl and repeating the dialysis procedure.

Since no single trial has given complete extraction of soil polysaccharide, multiple treatments are required for maximum yields. Swincer et al.[4] reported that a double treatment with 1N HCl and 0.5N NaOH generally removed more than 50 percent of the soil polysaccharide and a third treatment using acetic anhydride in 2.5 percent H_2SO_4 at 60°C for 2 h. further increased the yield to 80 percent. The sequential extraction procedure is shown in Fig. 6.3.

A flow sheet illustrating the various approaches for obtaining crude polysaccharide preparations from soil is shown in Fig. 6.4. The main methods for recovering polysaccharides from soil extracts are described below.

Dialysis: This procedure eliminates salts and low-molecular-weight components. Only the higher molecular weight polysaccharides are recovered. Some work has shown that up to one-half of the carbohydrate material is sufficiently low in molecular weight to be removed in the dialysis water, presumably as small fragments (oligosaccharides) rather than simple monosaccharides. Also, losses of polysaccharides can occur through adsorption on the dialysis tubing.[18] The colloid material may contain sub-

STATE IN THE SOIL 149

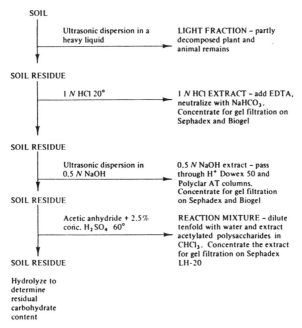

Fig. 6.3 Sequential extraction procedure for soil polysaccharides. From Swincer et al.,[4] reproduced by permission of Academic Press, Inc.

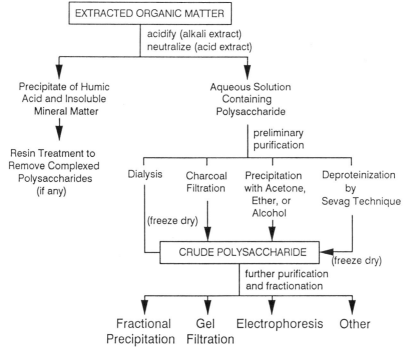

Fig. 6.4 Approaches for recovering crude polysaccharides from soil extracts

stances other than carbohydrates that must be removed by further purification.

Charcoal filtration: The first successful attempt to isolate polysaccharides from soil was made by Forsyth,[17] who passed a fulvic acid extract through a pad of activated charcoal and subsequently recovered the polysaccharide by elution with water. Subsequent studies have shown that recovery is incomplete, due possibly to association of polysaccharide material with dark-colored humic substances, which are adsorbed by the activated charcoal.

Precipitation: Many polysaccharides can be precipitated from aqueous solution by addition of acetone, ether, or alcohol at concentrations of about 80 percent (v/v). With soil polysaccharides, precipitation is incomplete and the final product usually contains noncarbohydrate impurities.

Deproteinization: Proteins in complex mixtures are often recovered by the Sevag procedure, which involves solvent-solvent extraction in the presence of a detergent. Proteins are removed at the interface, leaving polysaccharides in the solution phase. As with other procedures, complete separations are not achieved.

Fractionation of Soil Polysaccharides The extraction and preparation procedures described in the previous section nearly always yield crude polysaccharide preparations that require further treatment in order to obtain products suitable for characterization studies. Methods used for further fractionation include fractional precipitation with ethanol and quaternary ammonium compounds,[16] and with cetavalon.[20] However, the most promising results have been achieved by filtration on Sephadex gels[15, 18] and by combined chromatography on charged cellulose and Sephedex.[23]

Complete deproteinization of soil polysaccharides has apparently not yet been achieved, which suggests that part of the polysaccharide material occurs as complexes with peptides (glycopeptides) and/or proteins (glycoproteins).

For a discussion of the principles involved in gel filtration, the reader is referred to Chapter 14. Results obtained using this approach can be illustrated by data reported by Barker et al.[22] A crude soil polysaccharide preparation was subject to large-scale gel filtration by gradient elution from DEAE Sephadex A-50 (Fig. 6.5) after which part of the high-molecular-weight component (A) was rechromatographed on the same gel. The heterogeneous nature of the material was shown by the broad, ill-defined band obtained by the preliminary separation (Fig. 6.5) and by detection of at least four polysaccharides when the high-molecular-weight component (A) was rechromatographed (Fig. 6.6).

A procedure for the extraction and characterization of soil polysaccharides by whole soil methylation has been described by Cheshire et al.[25] The methylated product was soluble in chloroform and had an infrared spectrum typical of a methylated polysaccharide. The preparation was hydrolyzed with $2M$ trifluoroacetic acid and characterized by gas chromatography-mass spectrometry.

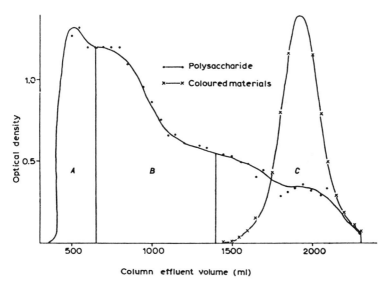

Fig. 6.5 Preliminary gel filtration of a crude polysaccharide on Sephadex G-100. The component marked "A" was rechromatographed on DEAE Sephadex A-50 as shown in Fig. 6.6. From Barker et al.,[22] reproduced by permission of Elsevier Scientific Publishing Co.

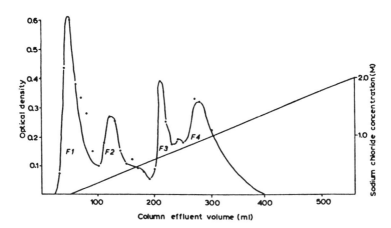

Fig. 6.6 Gradient elution on DEAE Sephadex A-50 of a high-molecular-weight polysaccharide material (Fraction A, Fig. 6.5). The effluent consisted of pH 6 phosphate buffer containing variable amounts of NaCl. From Barker et al.,[22] reproduced by permission of Elsevier Scientific Publishing Co.

Table 6.2 Characterization of Crude Polysaccharides from Various Sites

	Forsyth[17]	Black et al.[19]	Parsons and Tinsley[20]			
Soil type	Scottish agricultural soil	Scottish peat	Sandy loam	Meadow soil	Peat	Spodosol
Extractant	0.5N NaOH	0.9N HCl	Anhydrous formic acid containing LiBr			
Purification	Charcoal filtration	Ethanol precipitation	Cetavalon precipitation			
Ash (%)	ND	3.5	11.7	6.0	4.4	5.7
Carbohydrate (%)	80	89.8	26.3	37.8	22.9	24.7
Nitrogen (%)	0.34	—	3.16	2.65	3.67	3.02
Uronic anhydride (%)	15.8	5.6	15.8	17.8	16.0	16.3
Amino sugars (%)	0	—	5.2	5.0	3.3	3.1
Composition[a]						
Glucose	20.8	66.0	33.6	37.7	26.8	33.6
Galactose	20.0	8.2	19.1	20.1	20.8	23.1
Mannose	21.9	15.0	18.0	18.2	15.8	17.9
Arabinose	11.7	—	7.9	7.6	7.9	6.7
Xylose	23.6	9.8	7.9	8.2	8.9	7.5
Ribose	1.5	—	—	—	—	—
Rhamnose	0	+	13.5	8.2	18.8	11.2

[a]Individual monosaccharides are expressed as a percentage of total neutral sugars in the preparation.

Derivatives corresponding to (1 → 4) linked sugars predominated for both hexose and pentose sugars but there was also an abundance amount of (1 → 3) linkages.

Nature of Soil Polysaccharides Properties of several select polysaccharides as reported in the literature are recorded in Table 6.2. Irrespective of origin, about the same kinds of sugars have been found, the most common being glucose, galactose, mannose, xylose, ribose, rhamnose, and fructose.[17,19,20] The structures of most of these sugars are shown in Table 6.1. Several other sugars have also been identified, including amino sugars (see Chapter 3). Two methylated sugars, 2-0-methyl-L-rhamnose (**IV**) and 4-*O*-methyl-*D*-galactose (**V**), were identified in hydrolysates of soil polysaccharides by Duff.[26] Two methylated pentose sugars, 2-0-methylxylose and 3-0-methylxylose, have been reported in an acid hydrolysate of peat.[27]

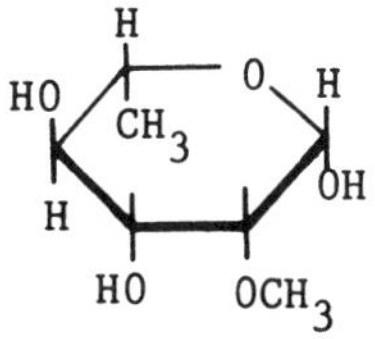

2-O-Methyl-L-rhamnose
IV

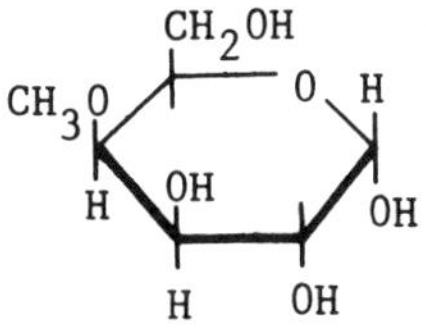

4-O-Methyl-D-galactose
V

Studies with the ultracentrifuge show that polysaccharides, as normally recovered from soil, are extremely polydisperse, thereby making molecular weights difficult to calculate. Mean molecular weights of 124,000 and 130,000 were obtained by Bernier[16] and Ogston,[28] respectively, using sedimentation and viscometry procedures. Martin[8] suggested that partial degradation products of plant and microbial polysaccharides "serve as structural units for the formation of new polysaccharides characteristic of the soil environment." According to this concept, synthesizing enzymes at cell surfaces, or released during autolysis of microbial cells, link the units together, thereby forming very heterogeneous polymers that become further resistant to decomposition through their association with clay, metal ions, or humic colloids. Failure to obtain pure or uniform polysaccharide fractions by various fractionation methods was attributed to their structural complexity.

On the basis of an exhaustive study of the nature of soil carbohydrates and their associations with humic substances, Cheshire et al.[23] concluded that the polysaccharide fraction of soil is composed of two types: 1) primary polysaccharides that consist of chains of sugar residues which are not combined with other substrates, and 2) secondary pseudopolysaccharides that have an elemental analysis corresponding to a carbohydrate (CH_2O) but that do not yield sugars upon hydrolysis.

Results of Sephadex gel filtration of soil polysaccharides indicate a continuum of molecular sizes ranging from molecules of oligosaccharide size to components with molecular weights approaching 200,000 (see previous section). Infrared studies indicate the occurrence of COOH groups, thereby confirming the presence of uronic acids (see Table 6.2).

Bound Carbohydrates As noted in the previous section, complete recovery of polysaccharides from soil has not been achieved, even using rather elaborate extraction procedures (see Figures 6.3 and 6.4). This has led to the conclusion that part of the carbohydrate material is covalently bound to humic and fulvic acids, an observation in accord with results of ^{13}C-NMR investigations (see Chapter 11). Some polysaccharides may be irreversibly adsorbed to fine-grained mineral matter.

QUANTITATIVE DETERMINATION OF SOIL CARBOHYDRATES

Methods for estimating the total carbohydrate content of soils are based on an initial hydrolysis to liberate individual sugars from polysaccharides and other polymeric components to which they may be attached. Individual sugars are then estimated by several methods as outlined in subsequent sections.

Sulfuric acid has been used universally for hydrolysis. This acid can be eliminated from the hydrolysate by precipitation as the highly insoluble $BaSO_4$. Colored impurities are not removed completely and some sugars are lost with the precipitate. Neutralization of the acid with $NaHCO_3$ and extraction of sugars with methanol from the dried Na_2SO_4 precipitate has also been used.[12]

A major problem is encountered in selecting proper hydrolysis conditions. Pretreatment with 72 percent H_2SO_4 results in increased extraction of sugars by subsequent heating with dilute H_2SO_4 ($1N$, $3N$, or $5N$ solutions usually used). Losses of sugars during hydrolysis can be appreciable, particularly pentoses and uronic acids. The stabilities of the various classes of monosaccharides follow the approximate order: hexoses > deoxyhexoses > pentoses > uronic acids.

The effect of several different hydrolysis conditions on the quantity of sugars recovered from a silt loam soil is given in Fig. 6.7. Of the procedures tested, highest yields of sugars were obtained by soaking the soil for 16 h. in $26N$ H_2SO_4 followed by refluxing with $1N$ H_2SO_4. Oades et al.[29] adopted a combination of procedures in which the soil was initially refluxed with $5N$ H_2SO_4 for 20 min. to recover easily hydrolyzable components. The soil residue was then soaked in $26N$ H_2SO_4 for 16 h. and refluxed with $1N$ H_2SO_4 for 5 h. The material most difficult to hydrolyze was found to be rich in glucose and presumed to consist partly of cellulosic material of plant origin.

Total Reducing Sugars

A variety of methods have been used to estimate total reducing sugars in soil hydrolysates (often recorded as total carbohydrates). Most workers have used

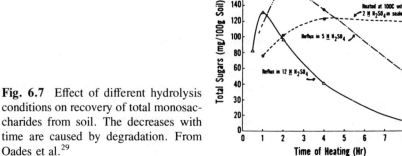

Fig. 6.7 Effect of different hydrolysis conditions on recovery of total monosaccharides from soil. The decreases with time are caused by degradation. From Oades et al.[29]

the classical Fehling's reagent, which consists of a copper sulfate solution (reagent A) and an alkaline tartrate solution (reagent B). The basis for the method is that when an alkaline solution of cupric hydroxide is heated in the presence of a reducing agent, the cupric hydroxide is reduced to the insoluble yellow or red cuprous oxide.

$$2\ Cu(OH)_2 \xrightarrow[\text{sugars}]{\text{reducing}} Cu_2O + H_2O + \text{oxidized form of sugar (R-COOH)}$$

The quantity of reducing sugars is obtained by titrating the unused reagent with a standard glucose solution, using methylene blue as the indicator. The role of tartrate in the reaction is to stabilize the $Cu(OH)_2$ through formation of cupric tartrate. Reducing compounds other than sugars are invariably present in soil hydrolysates; thus, results obtained by the method are probably high.

Another approach that has been used for total neutral sugars is the phenol–concentrated H_2SO_4 method. This reagent reacts with reducing sugars and their derivatives to give a yellow color. When applied to soil hydrolysates by McGrath,[30] values for reducing sugars were lowered by about 30 percent by treatment on anion- and cation-exchange resins, and attributed to removal of nonsugar contaminants. Subsequent studies[31,32] have confirmed that the phenol–H_2SO_4 method can be used for analysis of a wide variety of soil types provided the extracts are treated on anion- and cation-exchange resins.

A relatively precise and rapid colorimetric method using 3-methyl-2-benzothiazolinone hydrazone chloride was used for analysis of dissolved monosaccharides in seawater.[33] The method, which gives nearly equivalent color with hexoses, pentoses, and sugar acids, does not employ concentrated mineral acid and can be coupled with acid hydrolysis to give polymeric carbohydrates by difference.[34] The method does not seem to have been applied to soil organic matter.

Results obtained for total reducing sugars in soil hydrolysates serve as a basis for the conclusion that from 5 to 25 percent of the soil organic matter occurs in the form of carbohydrates. However, because of the nonspecificity of the methods, confirmation by other approaches is required. Total carbohydrates have also been obtained by summation of the results for individual sugar types (see next section). A more satisfactory way may be by summation of individual sugars as estimated by chromatographic procedures.

Colorimetric Determination of Specific Types of Monosaccharides

Colorimetric procedures are convenient because they are rapid, simple, and do not require elaborate instrumentation.

Many reactions are known that give rise to colored products with monosaccharides. For the most part, they involve an initial conversion of the sugars to hydroxymethylfurfural (hexoses) and furfural (pentoses, uronic acids) by treatment with mineral acids. The reactions are as follows:

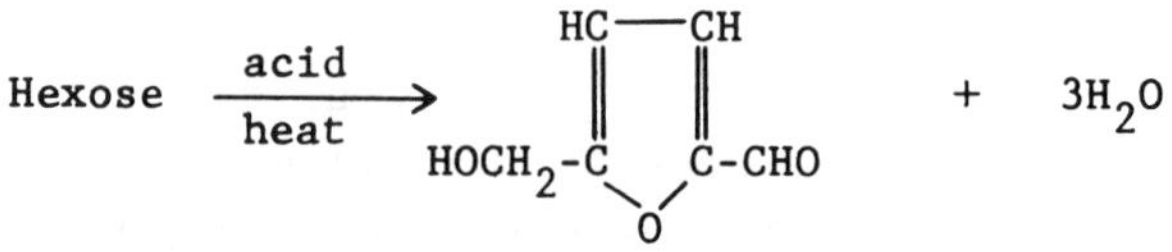

Hexose $\xrightarrow[\text{heat}]{\text{acid}}$ HOCH$_2$-C(HC=CH)C-CHO (O ring) + 3H$_2$O

Hydroxymethylfurfural

Pentose $\xrightarrow[\text{heat}]{\text{acid}}$ HC(HC=CH)C-CHO (O ring) + 3H$_2$O

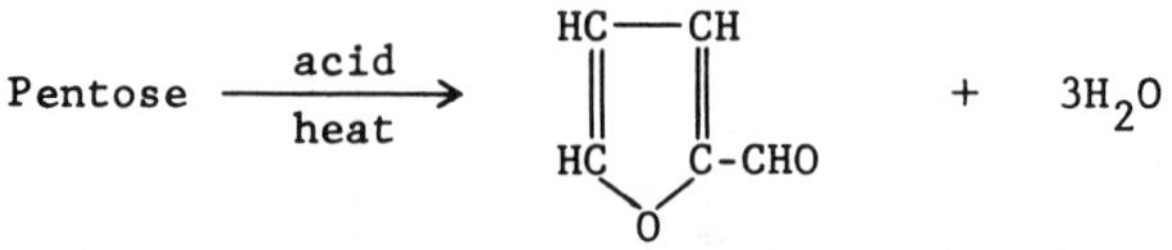

Furfural

The furfurals are highly reactive compounds that form colored complexes of unknown composition with such reagents as anthrone, carbazole, and orcinol (3,5-dihydroxytoluene). Because of differences in the ease with which the reactions occur and the colors produced, conditions can be selected for the quantitative determination of each class of monosaccharides.

As Greenland and Oades[1] correctly point out, colorimetric procedures for estimating hexoses, pentoses, and uronic acids cannot be considered accurate for two reasons: 1) none of the procedures give the same color intensity with equimolar concentrations of different sugars, particularly the anthrone method for hexoses and the carbazole method for uronic acids, and 2) hydrolysis conditions are not entirely satisfactory and represent a compromise between good glycosidic cleavage and destruction of monomers. Moreover, organic and inorganic constituents in soil hydrolysates can interfere with the analyses.[30-32]

A summary of the quantities of hexoses, pentoses, and uronic acids in soil organic matter as determined by colorimetric methods[35,36] is recorded in Table 6.3.

Hexose Sugars A popular method for determining hexose sugars is by the well-known anthrone procedure. The hexoses are converted to derivatives of hydroxymethylfurfural through the action of sulfuric acid, which subsequently form green-colored complexes with anthrone (maximum absorbance near 625 nm).

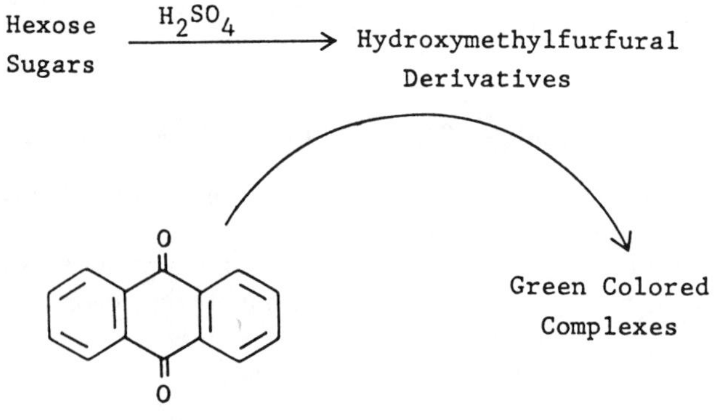

Hexose Sugars $\xrightarrow{\text{H}_2\text{SO}_4}$ Hydroxymethylfurfural Derivatives

Anthrone → Green Colored Complexes

Table 6.3 Content of Hexoses, Pentoses, and Uronic Acids in Soil Organic Matter as Estimated by Colorimetric Procedures (in %)[a]

Soil[a]	Organic Matter	Hexoses	Pentoses	Uronic Acids	Total
Solodized solonetz					
A_1 Horizon	9.8	9.6	4.8	1.8	16.2
A_2 Horizon	2.1	12.4	—	3.3	—
B_2 Horizon	1.4	19.0	—	3.0	—
B_{Sa} Horizon	0.5	20.0	—	2.3	—
Solod					
A_1 Horizon	6.9	10.6	5.3	2.0	17.9
A_2 Horizon	1.7	15.3	6.0	5.1	26.4
B_2 Horizon	1.6	14.7	4.8	5.1	24.6
B_3 Horizon	1.0	17.5	—	4.7	—
Orthic					
A_1 Horizon	10.5	9.5	4.2	1.3	15.0
B_2 Horizon	1.7	17.0	4.0	3.4	24.4
B_3 Horizon	1.4	16.0	—	1.8	—
Brown forest	14.9	8.3	2.2	1.8	12.3
Podzol	18.7	8.0	2.2	1.9	12.1
Dark Brown	8.6	12.7	4.4	1.9	19.0
Black	10.6	7.2	2.7	1.3	11.2

[a]Top portion from Graveland and Lynch[35] (hexoses, anthrone; pentoses, orcinol–$FeCl_3$; uronic acids, carbazole); bottom portion from Ivarson and Sowden[36] (hexoses, anthrone; pentoses, aniline acetate; uronic acids, carbazole).

Colored complexes are also formed with furfural derivatives derived from pentoses and uronic acids but the colors fade rapidly and equimolar intensities are small compared with hexoses. As with other colorimetric methods, individual hexoses do not give equal color intensities. For example, the color equivalent for galactose is only slightly more than one-half of that given by glucose. Results obtained by the method are usually expressed in terms of glucose equivalents.

The anthrone procedure cannot be applied directly to acid hydrolysates of soil but appears to be suitable for extracts that have been neutralized with $CaCO_3$. However, the work of Oades[12] indicates that losses of sugars can occur through neutralization, a problem that was overcome by removing interfering substances in the hydrolysate through adsorption on activated charcoal. Sugars were recovered by elution with 8.5N acetic acid and estimated colorimetrically with anthrone. Other workers have removed interfering substances by treatment of the hydrolysate with anion- and cation-exchange resins.[30]

Hexoses in soil hydrolysates have also been estimated by the chromotrophic acid procedure, which is based on the violet color produced through the action of this reagent in acid (H_2SO_4) solution. The reaction depends upon the conversion of hexoses to 5-hydroxymethylfurfural and splitting off of the methylol group to form formaldehyde, which reacts with the chromotrophic acid. Sugars that form furfural do not split off formaldehyde and thus do not react. Ivarson and Sowden[36] found that the method gave consistently higher results than with the anthrone procedure.

The results of numerous studies (reviewed by Greenland and Oades[1]) indicate that from 4 to 12 percent of the soil organic matter occurs in the form of hexose sugars. Since the kinds of monosaccharides may be reasonably constant, the anthrone method may provide a good index of the relative content of total carbohydrates in different soil types, as well as in soils under different management practices.

Pentose Sugars As noted earlier, methods for analysis of pentose sugars involve transformation of the sugars to furfural through the action of mineral acids. Several methods are available for analysis of the furfural thus produced, the most popular being the orcinol–ferric chloride procedure.

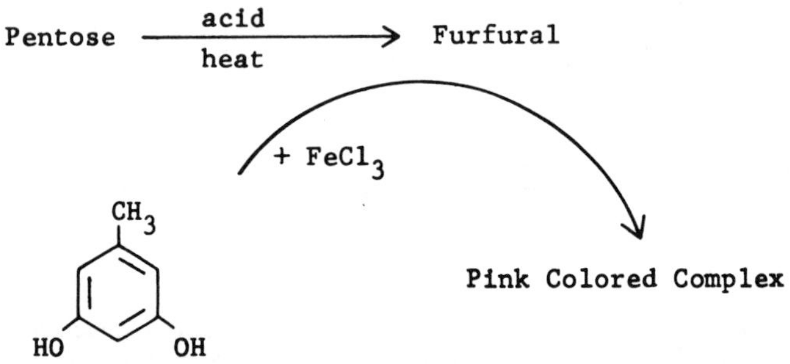

$$\text{Pentose} \xrightarrow[\text{heat}]{\text{acid}} \text{Furfural} \xrightarrow{+ \text{FeCl}_3 \text{, Orcinol}} \text{Pink Colored Complex}$$

Carbohydrates other than pentoses can be converted to furfural and a significant error can be introduced from the furfural derived from uronic acids, which are decarboxylated during the acid treatment. Interference from uronic acids has been overcome by removing them from the hydrolysate through adsorption on an anion-exchange resin.[37]

Uronic Acids The uronic acids can be defined as carbohydrate derivatives possessing both the aldehyde (CHO) and carboxyl (COOH) groups. These compounds are easily decarboxylated by hot mineral acids to form furfural and CO_2. The basis of the usual colorimetric estimation is the formation of pink-colored complexes by reaction with carbazole.

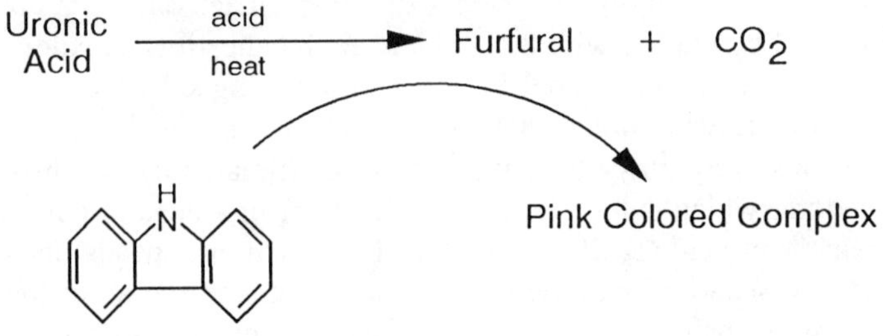

$$\text{Uronic Acid} \xrightarrow[\text{heat}]{\text{acid}} \text{Furfural} + CO_2 \xrightarrow{\text{Carbazole}} \text{Pink Colored Complex}$$

Ferric iron, pentoses, and other carbohydrates interfere with the estimation. Interference from Fe^{3+} has been avoided by reduction with stannous chloride.[38] Neutral sugars can be separated from uronic acids by adsorption of the latter on an anion exchange resin.[36,39] After the neutral sugars have been washed from the resin column with distilled water, the uronic acids are recovered by elution with $0.25N$ NH_4Cl at pH 10 and analyzed by the carbazole method. Benzing-Purdie and Nikiforuk[40] found that a colorimetric procedure using m-hydroxydiphenyl as reactant gave a fast and reliable estimate of uronic acids in soil hydrolysates.

A major problem in the analysis of uronic acids in soil is that hydrolysis losses are high, often exceeding 50 percent when uronic acids are added to hydrolysates in known amounts.

In some early work, attempts were made to estimate uronic acids by measuring the quantity of CO_2 produced by hydrolysis of soil with 12 percent HCl (the Lefevre–Tollens decarboxylation method). Whereas the method gives good results with plant material, impossibly high values are obtained for uronic acids in soil, often exceeding over 40 percent of the total C. The decarboxylation method has since been abandoned as a suitable method for the determination of uronic acids in soils because most of the CO_2 is derived from colored humic substances.

Results obtained by the carbazole method indicate that from 1 to 5 percent of the soil organic matter occurs as uronic acids. Greenland and Oades[1] believe that these estimates are far from satisfactory and represent minimal amounts.

Separation of Monosaccharides by Chromatographic Procedures

Chromatographic techniques are ideally suited for the separation, detection, and identification of monosaccharides. The technique of paper chromatography was applied in initial studies of soil monosaccharides but this method has been replaced by gas–liquid chromatography (GLC) and high performance anion-exchange chromatography, both of which are well-suited for the precise quantitative analysis of very small samples containing a complex mixture of sugars.[12,29-32,40-43]

For the determination of monosaccharides, the soil is first hydrolyzed with a mineral acid (H_2SO_4) to liberate individual monomers (see Fig. 6.7), following which the acid is removed from the hydrolysate using $Ba(OH)_2$ or $SrCO_3$ to precipitate SO_4^{2-} and to raise the pH.[29] Alternately, the extract can be passed successively through an anion- and a cation-exchange resin, as done by McGrath.[30]

For separations by GLC, the sugars are next converted to derivatives that are volatile at temperatures below about 300°C. Types of sugar derivatives that have been found suitable for analysis by GLC include methyl esters, trimethylsilyl ethers, and acetates. For complex mixtures of monosaccharides, such as would be expected in hydrolysates of soil, the acetate form appear to be the most suitable.[12,29]

In this procedure, the carbonyl group of the sugar is first reduced to the alditol form prior to acetylation. The reactions are:

$$\begin{array}{c} H \\ C{=}O \\ | \\ H\text{-}C\text{-}OH \\ | \end{array} \xrightarrow[\text{(NaBH}_4\text{)}]{\substack{\text{Reduction with} \\ \text{sodium} \\ \text{borohydride}}} \begin{array}{c} H \\ H\text{-}C\text{-}OH \\ | \\ H\text{-}C\text{-}OH \\ | \end{array} \xrightarrow[\text{anhydride}]{\substack{\text{Acetylation with} \\ \text{acetic}}} \begin{array}{c} H \\ H\text{-}COOCH_2CH_3 \\ | \\ H\text{-}COOCH_2CH_3 \\ | \end{array}$$

A flow diagram for the preparation of alditol acetates is shown in Fig. 6.8; a diagram for the separations obtained by GLC is given in Fig. 6.9. For the soil examined, 98 percent of the peak area was represented by rhamnose, fucose, arabinose, xylose, galactose, mannose, and glucose.

A troublesome feature of GLC is the preparation of volatile derivatives, a drawback that can be avoided using the technique of high-performance anion-exchange chromatography.[31,43] A chromatogram obtained by Martens and Frankenberger[31] is shown in Fig. 6.10. The monosaccharides detected are the same as those noted in Fig. 6.9, where GLC was used. However, the relative distribution of monosaccharides would appear to be dissimilar (e.g., note prominence of arabinose band in Fig. 6.10, as compared to Fig. 6.9).

Result obtained by Oades et al.[29] for the monosaccharide composition of three soil types are given in Table 6.4. For all soils, glucose was by far the most abundant sugar. A similar result was obtained by Cheshire and Ander-

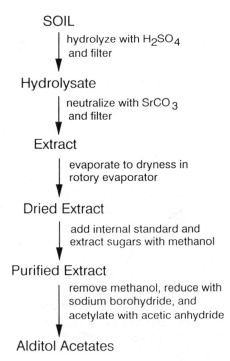

Fig. 6.8 Flow diagram for the preparations of alditol acetates during the gas–liquid separation of monosaccharides in soil hydrolysates

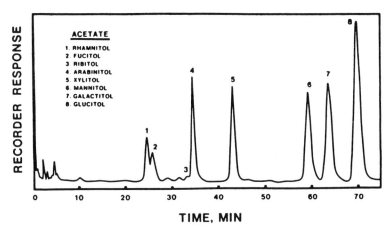

Fig. 6.9 Gas–liquid chromatograph of alditol acetates derived from soil sugars. From Oades et al.[29]

son,[44] who also found that the absolute amount of all sugars was lower in cultivated versus uncultivated soils, with only minor differences in the relative proportion of individual sugars. Their data are shown in Table 6.5.

Considerable differences have been observed in the relative distribution of individual monosaccharides in hydrolysates of soil and soil organic matter preparations. The order of decreasing abundance, as recorded in several studies, is shown below. The diversity of results is illustrated by the connecting lines for mannose.

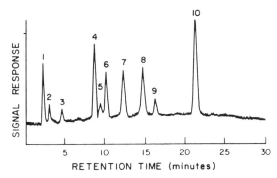

Fig. 6.10 Chromatogram of H_2SO_4 (80°C) extract of soil saccharides as detected by high-performance anion-exchange chromatography coupled with pulsed amperometric detection: 1 = inositol; 2 = ribitol; 3 = fucose; 4 = arabinose; 5 = rhamnose; 6 = galactose; 7 = glucose; 8 = xylose; 9 = mannose; 10 = lactose (internal standard). From Martens and Frankenberger,[31] reproduced by permission of Elsevier Scientific Publication Co.

Table 6.4 Monosaccharide Composition of Soil and Percentage of the Organic Matter Recovered in the Sugars Identified[a]

Sugars	Mexico Silt Loam (mg/100 g)	Summit Silty Clay (mg/100 g)	Muck (mg/100 g)
Glucose	124	257	664
Galactose	37	77	319
Mannose	40	105	282
Xylose	36	53	222
Arabinose	39	99	326
Rhamnose and fucose	21	56	187
Total sugars	297	647	2000
Organic matter in soil (%)	2.1	4.4	59.0
Organic matter as sugars (%)	14.1	14.7	3.4

[a] From Oades et al.[29]

1. Glucose > galactose > mannose > arabinose > xylose > rhamnose/fucose
2. Glucose > galactose > arabinose > xylose > mannose > rhamnose/fucose
3. Glucose > arabinose > galactose > mannose > xylose > rhamnose/fucose
4. Xylose > glucose > galactose > mannose > arabinose/ribose > rhamnose/fucose
5. Glucose > mannose > galactose > rhamnose > arabinose > xylose > fucose > ribose

Table 6.5 Effect of Cultivation on the Sugar Composition of Two Soils at the Rothamsted Experimental Station in England[a]

Sugar Composition	Highfield Cultivated	Highfield Fallow	Fosters Cultivated	Fosters Fallow
Relative proportions of sugar in the hydrolysate:				
Glucose	100	100	100	100
Galactose	40	46	46	52
Mannose	37	50	47	54
Arabinose	45	53	67	74
Xylose	52	42	62	49
Rhamnose/fucose	28	23	21	31
Pentose, %	32	30	38	34
Reducing sugar as glucose (mg/g soil)	13.5	6.8	5.2	4.2
Pentose as xylose (mg/g soil)	5.8	2.5	2.1	1.8

[a] From Cheshire and Anderson.[44]

Sources for the above sequences are: 1) average for the soils of Table 6.4, 2) average for the soils of Table 6.5; 3) results obtained by Doutre et al.[32] for a Canadian soil; 4) McGrath's[30] average for 36 Irish soils, and 5) values obtained by Schnitzer and Preston[45] for organic matter recovered from an Aquoll soil by supercritical gas extraction with ethanol.

Reasons for the variable distribution patterns are unknown but they attest to the complex nature of the polysaccharide material in soils of diverse sources. On the basis of separations obtained by GLC, Folsom et al.[41] found that some prairie soils contained proportionally more pentoses but less hexoses than some forest soils. Also, the proportion of the carbohydrate material as mannose increased with increasing depth in the soil profile, indicating that polysaccharides containing this sugar are more resistant to decomposition than other polysaccharides. Another general trend was that the proportion of each individual monosaccharide to organic C increased with depth in the soil profile. In other work, Murayama[46] found that the heavy fraction of soil contained relatively higher amounts of galactose, mannose, rhamnose, and ribose than the light fraction.

Both GLC and high-performance anion-exchange chromatography have been used for the separation of uronic acids in soil hydrolysates. By GLC, Mundie[47] found that the uronic acid contents of some Irish soils were near the high values obtained by colorimetric methods. Galacturonic acid occurred in greater amounts than glucuronic acid. In contrast, Martin and Frankenberger,[48] using high-performance anion-exchanger chromatography, found a 1.83-fold higher content of glucuronic acid relative to galacturonic acid in the soil they examined.

SUMMARY

Carbohydrates constitute 5 to 25 percent of the organic matter in most soils. A wide array of compounds are present, include pentoses, hexoses, uronic acids, amino sugars, and deoxy- and O-methyl sugars. A major portion of the carbohydrate material occurs in polymeric molecules of varying degrees of complexity, some of which are associated with clay and humus colloids.

A number of methods have been used for the determination of total carbohydrates in soil, all of which involve hydrolysis of the soil with mineral acids to release monosaccharides from polymeric molecules. Colorimetric procedures have been developed for estimating hexoses, pentoses, uronic acids, and amino sugars but they cannot be regarded as accurate for two reasons: 1) considerable breakdown of monomers occurs during hydrolysis (particularly pentoses and uronic acids), and 2) none of the colorimetric methods give the same color intensity with equimolar concentrations of different sugars. Estimates for the kinds and amounts of individual sugars have also been obtained by application of GLC and high-performance anion-exchange chromatography.

Excluding amino sugars (see Chapter 3), the contribution of individual sugar types to soil organic matter follows the order: hexose sugars (4–12 percent) > uronic acids (1–5 percent) > pentose sugars > methylated sugars (trace).

REFERENCES

1. D. J. Greenland, and J. M. Oades, "Saccharides," in J. E. Gieseking, Ed., *Soil Components: Vol. 1. Organic Components*, Springer-Verlag, New York, 1975, pp. 213-261.
2. U. C. Gupta, "Carbohydrates," in A. D. McLaren and G. H. Peterson, Eds., *Soil Biochemistry*, Arnold, London, 1967, pp. 91-118.
3. L. E. Lowe, "Carbohydrates in Soil," in M. Schnitzer and S. U. Khan, Eds., *Soil Organic Matter*, Elsevier, New York, 1978, pp. 65-93.
4. G. D. Swincer, J. M. Oades, and D. J. Greenland, *Adv. Agron.*, **21**, 195 (1969).
5. M. V. Cheshire, *Nature and Origin of Carbohydrates in Soils*, Academic Press, New York, 1979.
6. J. R. Burford and J. M. Bremner, *Soil Biol. Biochem.*, **7**, 389 (1975).
7. J. K. Adu and J. M. Oades, *Soil Biol. Biochem.*, **10**, 109 (1978).
8. J. P. Martin, *Soil Biol. Biochem.*, **3**, 33 (1971).
9. M. V. Cheshire, *J. Soil Sci.*, **28**, 1 (1977).
10. M. V. Cheshire, M. P. Greaves, and C. M. Mundie, *J. Soil Sci.*, **25**, 483 (1974).
11. U. C. Gupta and F. J. Sowden, *Soil Sci.*, **97**, 328 (1964).
12. J. M. Oades, *Aust. J. Soil Res.*, **5**, 103 (1967).
13. M. V. Cheshire, C. M. Mundie, and H. Shepherd, *J. Soil Sci.*, **24**, 54 (1973).
14. C. J. Acton, E. A. Paul, and D. A. Rennie, *Can. J. Soil Sci.*, **43**, 141 (1963).
15. S. A. Barker, P. Finch, M. H. B. Hayes, R. G. Simmonds, and M. Stacey, *Nature*, **205**, 68 (1965).
16. B. Bernier, *Biochem. J.*, **70**, 790 (1958).
17. W. G. C. Forsyth, *Biochem. J.*, **46**, 141 (1950).
18. G. D. Swincer, J. M. Oades, and D. J. Greenland, *Aust. J. Soil Res.*, **6**, 211, 225 (1968).
19. W. A. P. Black, W. J. Cornhill, and F. N. Woodward, *J. Appl. Chem. Lond.*, **5**, 484 (1955).
20. J. W. Parsons and J. Tinsley, *Soil Sci.*, **92**, 46 (1961).
21. J. M. Oades and G. D. Swincer, *Trans. 19th Intern. Congr. Soil Sci.*, **3**, 183 (1968).
22. S. A. Barker, M. H. B. Hayes, R. G. Simmonds, and M. Stacey, *Carbohyd. Res.*, **5**, 13 (1967).
23. M. V. Cheshire et al., *J. Soil Sci.*, **43**, 359 (1992).
24. P. Finch, M. H. B. Hayes, and M. Stacey, *Trans. 9th Intern. Congr. Soil Sci.*, **3**, 193 (1968).
25. M. V. Cheshire, C. M. Mundie, J. M. Bracewell, G. W. Robertson, J. D. Russell, and A. R. Fraser, *J. Soil Sci.*, **34**, 539 (1983).
26. R. B. Duff, *J. Sci. Food Agr.*, **12**, 826 (1961).
27. J.-F. Bouhours and M. V. Cheshire, *Soil Biol. Biochem.*, **1**, 185 (1969).
28. A. G. Ogston, *Biochem. J.*, **70**, 598 (1958).
29. J. M. Oades, M. A. Kirkman, and G. H. Wagner, *Soil Sci. Soc. Amer. Proc.*, **34**, 230 (1970).

30 D. McGrath, *Geoderma*, **10**, 227 (1973).
31 D. A. Martens and W. T. Frankenberger, Jr., *Soil Biol. Biochem.*, **22**, 1173 (1990).
32 D. A. Doutre, G. W. Hay, A. Hood, and G. W. VanLoon, *Soil Biol. Biochem.*, **10**, 457 (1978).
33 K. M. Johnson and J. McN. Sieburth, *Marine Chem.*, **5**, 1 (1977).
34 C. M. Burney and J. McN. Sieburth, *Marine Chem.*, **5**, 15 (1977).
35 D. N. Graveland and D. L. Lynch, *Soil Sci.*, **91**, 162 (1961).
36 K. C. Ivarson and F. J. Sowden, *Soil Sci.*, **94**, 245 (1962).
37 U. C. Gupta and F. J. Sowden, *Can. J. Soil Sci.*, **45**, 237 (1965).
38 D. L. Lynch, E. E. Hearns, and L. J. Cotnoir, *Soil Sci. Soc. Amer. Proc.*, **21**, 160 (1957).
39 R. L. Thomas and D. L. Lynch, *Soil Sci.*, **91**, 312 (1961).
40 L. M. Benzing-Purdie and J. H. Nikiforuk, *Soil Sci.*, **145**, 264 (1988).
41 B. L. Folsom, G. H. Wagner, and C. L. Scrivner, *Soil Sci. Soc. Amer. Proc.*, **38**, 305 (1974).
42 G. Ogner, *Geoderma*, **23**, 1 (1980).
43 D. A. Martens and W. T. Frankenberger, Jr., *Chromatographia*, **29**, 7 (1990).
44 M. V. Cheshire and G. Anderson, *Soil Sci.*, **119**, 356 (1975).
45 M. Schnitzer and C. M. Preston, *Soil Sci. Soc. Amer. J.*, **51**, 639 (1987).
46 S. Murayama, *Soil Sci. Plant Nutr.*, **23**, 479 (1977).
47 C. M. Mundie, *J. Soil Sci.*, **27**, 331 (1976).
48 D. A. Martens and W. T. Frankenberger Jr., *Chromatographia*, **30**, 249 (1990).

7

SOIL LIPIDS

The class of organic compounds designated as lipids represents a convenient analytical group rather than a specific type of compound, the common property being their solubility in various organic solvents (benzene, methanol, ethanol, acetone, chloroform, ether, etc.) or organic solvent mixtures. They represent a diverse group of materials ranging from relatively simple compounds such as fatty acids to more complex substances such as the sterols, terpenes, polynuclear hydrocarbons, chlorophyll, fats, waxes, and resins. Many lipid constituents, such as the low-molecular-weight organic acids and sterols, are present in the soil in extremely low amounts. The bulk of the soil lipids occurs as the so-called fats, waxes, and resins. The resins, being more polar than fats and waxes, are more soluble in methanol or ethanol, a property that has been used to separate the group.

Lipids have been found to be widely distributed throughout soils of the world, ranging from the highly weathered Oxisols of the humid tropics to the weakly developed tundras of the arctic zone. In normal aerobic soil, the lipids probably exist largely as remnants of plant and microbial tissues; low and variable quantities of these constituents may be associated with undecomposed plant residues and the bodies of living and dead microfaunal organisms. As shown later, from 2 to 6 percent of soil humus occurs as fats, waxes, and resins, although considerably higher percentages are not uncommon.

Detailed reviews of soil lipids include those of Braids and Miller,[1] Morrison,[2] and Stevenson.[3]

LIPID CONTENT OF SOIL HUMUS

The proportion of the humus that occurs as lipids has traditionally been determined using Waksman's proximate method of soil analysis (see Chapter 2). In this method, lipids are determined by sequential extraction with ether (fats and waxes) and alcohol (resins). Distribution of ether- and alcohol-soluble material in the humus of a series of brown and black prairie soils as reported by Waksman and Stevens[4] is summarized in Table 7.1.

The range given in Table 7.1 for total fats, waxes, and resins (1.2–6.3 percent) is characteristic for most agriculturally important soils of the world. However, higher values have been reported, as illustrated in Fig. 7.1 for a series of mineral and organic soils. For the mineral soils, the highest values (up to 16 percent of the humus) occurred in the podzolized soils (i.e., Spodosols). In the case of the organic soils, the highmoor peats contained larger amounts than the lowmoor and sedimentary peats.

The variations noted in the lipid content of humus (see Fig. 7.1) can be explained by differences in vegetation, pH, or a combination of these two factors. Spodosols are developed on well-drained sites where mor is the characteristic type of humus on the forest floor—this humus typically occurs under coniferous forests growing on soils low in available Ca. On the other hand, woodland soils (i.e., Alfisols) are formed on moderately well-drained sites where mull is the characteristic humus type on the forest floor—this humus typically occurs under deciduous forests growing on soils well supplied with Ca. Spodosols are generally more acidic than Alfisols, and, on the average, a higher proportion of the humus occurs in the form of lipids. Of the soils shown in Fig. 7.1, the prairie soils (i.e., Mollisols) have the highest pH values, and the humus in these soils is lowest in fats, waxes and resins.

A pH relationship similar to that noted for the mineral soils appears to hold for organic soils. Highmoor peats are formed in waters which are low in Ca (and other nutrients); lowmoor and sedimentary peats are formed in places where waters are relatively rich in Ca. Thus, highmoor peats are generally more acidic than lowmoor and sedimentary peats, and, on the average, they contain substantially higher amounts of lipid material. As much as 20 percent

Table 7.1 Quantity of Fats and Waxes (Ether Soluble) and Resins (Alcohol Soluble) Materials in Various Mineral Soils (in % of humus)[a]

Soil	Fats and Waxes	Resins	Total
Summit soil (Missouri), A horizon	3.56	0.58	4.14
Grassland soil (Kansas), A horizon	4.71	1.53	6.24
Grassland soil (Alberta, Canada) (1–25 cm)	0.80	0.82	1.62
Grassland soil (Manitoba, Canada) (1–20 cm)	0.46	0.84	1.30
Grassland soil (Manitoba, Canada) (25–50 cm)	0.52	0.63	1.15
Brown soil (Saskatchewan, Canada) (1–20 cm)	1.02	0.88	1.90
Prairie soil, dark-colored, A horizon	0.62	0.61	1.23
Alpine humus, Pikes Peak (13,800 ft elevation)	0.94	3.10	4.04

[a]From Waksman and Stevens.[4]

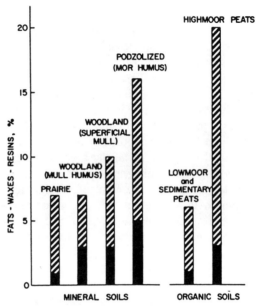

Fig. 7.1 Lipid content of the humus in several mineral and organic soils. The broken portion of the bars indicate the range of values reported. From Stevenson[3]

of the organic matter in individual layers of highmoor peats may exist in lipid or lipid-like constituents.

Information on the ether- and alcohol-extractable substances in a number of peat profiles has been summarized by Stevenson.[3] As can be seen from Fig. 7.2, the highest amounts (12–20 percent) occurred in the highly acidic Scottish sphagnum peat profile (pH 3.4–4.3); the lowest values (<1–2 percent) occurred in the Florida sedimentary peat profile (pH 7.1–7.5).

It is not known whether the high lipid content of the humus in acidic soils results from inability of microorganisms to decompose completely the lipids occurring in plant remains, or if larger quantities of lipids are synthesized by microorganisms. As an average, bacterial cells contain from 5 to 10 percent lipids; fungi usually contain more (from 10 to 25 percent). It is of some interest that fungi are the predominant type of microorganisms in highly acidic soils. Still another explanation is that more of the lipid material in acidic soils occurs as bituminous material not related to the true lipids; alternately, in acidic soils, humic-like substances may be partially solubilized by virtue of the relatively low pH of the extracting solution.

Sources of Soil Lipids

Little information is available concerning the rates at which plant lipids decompose in soils. Compounds resistant to microbial decomposition, such as

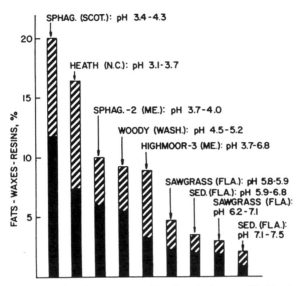

Fig. 7.2 Lipid content of various peat profiles in relation to pH. The broken portions of the bars indicate the range of values reported. From Stevenson[3]

certain higher alcohols and paraffins, should persist for longer periods than those that are attacked readily by microorganisms, such as true fats and phospholipids. Waxes, protecting the surfaces of leaves, needles, trunks, and fruit of many higher plants are particularly resistant to decomposition; these should be capable of surviving essentially unchanged over long periods. Plants belonging to the Coniferaceae and Myrtaceae produce rather large quantities of relatively stable terpenoid compounds.

Because of their resistance to microbial decomposition, the cuticle waxes of plants may represent a major source of soil waxes, long-chain acids, alcohols, and alkanes. It is also probable that the steroids and terpenoids that occur in soil are of plant origin. On the other hand, microorganisms are probably the main source of the glyceride and phosphatide components.

Except for some early research (reviewed by Stevenson[3]), no report has been published showing a significant accumulation of difficultly decomposable lipids in productive agricultural soil, even when large annual increments of plant residues and manures have been applied. Thus, most soils probably contain sufficient numbers of the proper kinds of microorganisms to decompose completely the lipid material contained in plant and animal residues. It should be noted in this respect that past studies on lipid transformations have been concerned with gross chemical changes accompanying decomposition, and that no attempt has been made to distinguish between lipids remaining as a remnant of the original material and those newly synthesized by microorganisms active in the decay process.

Environmental factors undoubtedly have a profound influence on the persistence of lipids in the soil. Most compounds are rapidly decomposed in aerobic soils, but not in waterlogged soil. Some compounds, such as the sterols, may resist decomposition in highly acidic soils.

Soil organisms are also able to degrade abiotic lipid-like constituents introduced into soil, such as from gas well leaks and oil spills. One compound of special interest is 3,4-benzopyrene, a carcinogenic compound occurring as an air pollutant through the combustion of fossil fuels. Shcherbak[5] found that soil collected near an industrial plant known to be emitting 3,4-benzopyrene contained as much as 8.35 mg of the compound per kilogram of soil. A contaminant found in soils near munitions works/arsenals is TNT (2,4,6-trinitrotoluene). This compound can persist in soils for long periods of time.[6]

Extraction of Lipids

Recovery of lipids from soil is not as easily accomplished as might be supposed. Some lipids may be associated with inorganic material or other organic constituents in such a way as to be insoluble in the usual lipid solvents. On the other hand, many solvents remove dark-colored products that may be of a nonlipid nature. Morrison[2] suggested the term "soil bitumen" to describe crude lipid preparations from soil.

Isolated humic and fulvic acids have been found to contain lipid constituents of various types[7,8] and recent studies indicate that organic matter associated with the fine clay fraction may be enriched with such substances as alkanes and long-chain fatty acids.[9]

Treatment of the soil with an acid mixture containing HF to destroy hydrated clay minerals may enhance extraction of lipids from mineral soils. An HF:HCl pretreatment was adopted by Wang et al.,[10] after which the lipids were recovered by extraction with chloroform–methanol (2:1, v/v). In other work, Meinschein and Kenny[11] used a ball milling technique to extract lipids from some subsurface soils. Efficiency of extraction might also be increased by using modern ultrasonic vibration techniques.

Supercritical gas extraction with n-pentane (temperature of 250°C, pressure of 11 MPa, time of 2 h.) has been reported to be selective for recovery of alkanes and alkanoic acids from soil[9,12,13] Extracts obtained in this way were essentially free of proteinaceous materials, carbohydrates, and humic substances.[9] Because of their high densities, high diffusion coefficients, and low viscosities, supercritical gasses have exceptional power to penetrate complex matrices and solubilize components otherwise inaccessible in the solvent used (see Chapter 2).

Rice and MacCarthy[14] have proposed a unique procedure for extraction of lipids from soil using methyl–isobutylketone (MIBK) as the water-immiscible phase in a two-phase solvent system. In their procedure, the soil is shaken with an aqueous NaOH solution, MIBK is added to the suspension, and the sample is acidified by addition of acid. The fulvic acid fraction (aqueous phase) is discharged and the remaining mixture (MIBK layer, humin, and the humic acid precipitated during acidification) is made alkaline with NaOH. The aqueous layer (containing the humic acid) is subsequently discharged and the lipid (MIBK) layer is separated from the soil residue. The extractable lipid was referred to as "bitumen." Extraction was incomplete and the lipid material retained in the soil residue was referred to as "bound lipid."

Isolation of Components

A wide range of analytical techniques are available for the isolation of components from complex lipid mixtures. The techniques used in organic geochemistry studies have been outlined by Eglinton and Murphy.[15]

Adsorption chromatography would appear to have wide application for separation of classes of substances from soil lipid mixtures. Both column and layer techniques can be used, with the former being more suitable for fractionating large quantities of material. A generally useful adsorbent is silica gel; other materials (e.g., neutral alumina) are useful for specific separations.

Meinschein and Kenny[11] fractionated crude lipid extracts on silica gel, eluting successively with n-heptane, carbon tetrachloride (CCl_4), benzene, and methanol. About 20 percent of the material was eluted with the first three solvents, about 70 percent with methanol, and 12 percent was not recovered. The heptane and CCl_4 effluents consisted mainly of saturated hydrocarbons while benzene eluted mostly ester waxes. Polar lipids, such as alcohols and long-chain fatty acids, were not separated from the contaminating material.

Figure 7.3 illustrates an example of a procedure for the isolation and fractionation of major lipid components from soil. The soil is extracted with chloroform ($CHCl_3$) and shaken with aqueous NaOH to remove organic acids, which are esterified and further separated into saturated and unsaturated fatty acids by thin-layer chromatography. Alkanes, aromatic hydrocarbons, and alcohols are recovered from the $CHCl_3$ layer by thin-layer chromatography. All subfractions are subjected to chromatographic analysis for isolation of individual components, such as by gas–liquid chromatography (GLC), thin-layer chromatography, or high-pressure liquid chromatography.

The importance of confirming identifications by two or more independent methods cannot be overemphasized. A gas chromatographic separation combined with high resolution mass spectrometry is highly recommended when appropriate.

Other fractionation techniques include urea adduction or use of molecular sieves for separating straight- from branched-chain and/or cyclic compounds prior to application of adsorption chromatography.

FUNCTION IN SOIL

Interest in soil lipids arises from the fact that many of these constituents are physiologically active. Some compounds have a depressing effect on plant

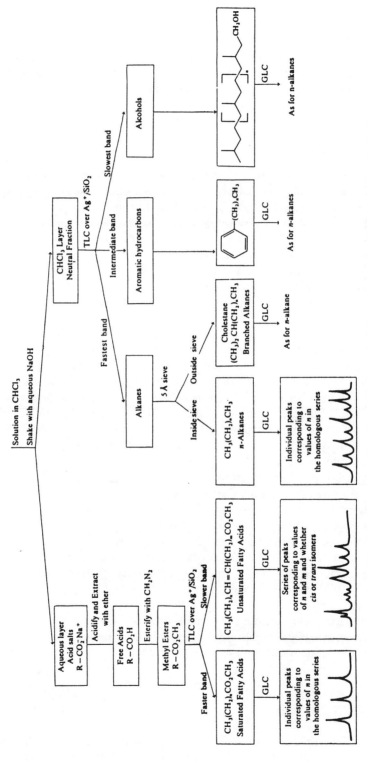

Fig. 7.3 Separation procedure for preliminary isolation of components from a complex lipid mixture. TLC = thin layer chromatography; GLC = gas liquid chromatography. Modified from Eglinton and Murphy,[15] used with permission of Springer-Verlag

growth whereas others act as growth hormones. Waxes and similar materials may be responsible for the water repellent condition of certain soils.

Phytotoxic Substances

Accumulation of lipids was considered by many early workers as one of the causes of soil fatigue. Greig-Smith,[16] for example, explained soil exhaustion on the accumulation of fats and waxes which he termed "agroire." In a series of articles from 1907 to 1911, Schreiner and Shorey placed considerable emphasis on the accumulation of organic toxins in soil (for references see Stevenson[3]). Particular emphasis was given to the accumulation in soil of dihydroxystearic acid, $CH_3(CH_2)_7CHOHCHOH(CH_2)_7COOH$, which was believed to be one of the compounds responsible for the low fertility of worn out soils. Many of the claims of Schreiner and Shorey have not been substantiated.

In more recent times, Whitehead[17] concluded that the concentration of phenolic acids in the aqueous phase of some English soils (from 0.07×10^{-5} to 4.9×10^{-5} *M*) was sufficiently high to adversely affect plant growth. His results are given in Table 7.2. The extensive studies of Wang and his associates[18-20] suggested that many agricultural soils of Taiwan contained inhibitory substances.

Toxic amounts of ethylene and other volatile organic products may be produced by decay of organic residues under anaerobic soil conditions.[21-23] Ethylene is of particular importance because very low concentrations in the soil atmosphere (>1 μL/L) can adversely affect plant growth. Both ethylene and methane have been shown to be produced in the organic horizons of forest soils.[24]

Lipids of various types are believed to be responsible for the low yields of wheat under the stubble-mulch system of farming in semiarid regions of the world. At present, it is not known whether the phytotoxic substances are derived from the residues or if they are produced by microorganisms active in the decay process. The review of McCalla and Haskins[25] suggests that a dual source is likely.

Table 7.2 Distribution of Phenolic Acids in Four Contrasting Soil Types[a]

	Soil Type (concentration, moles/liter $\times 10^{-5}$)			
Phenolic acid	Calcareous Loam	Sand	Clay With Flint	Clay Loam
p-Hydroxybenzoic	1.9	3.9	3.2	0.8
Vanillic	1.4	4.9	1.7	0.7
p-Coumaric	3.8	4.2	2.0	0.9
Ferulic	3.2	0.4	0.07	0.3

[a]From Whitehead.[12]

The alleopathic effect in higher plants, which can be defined as the suppression of the growth of one plant by chemicals released from another plant, is known to be due to a number of secondary plant substances, including phenolic acids, flavonoids, terpenoids, steroids, alkaloids, and organic cyanides. These alleopathic materials enter the soil by excretion or exudation from roots, by rain wash or fog drip from leaf surfaces, by volatilization from leaves, and through decay of aboveground or belowground plant parts. A wide array of phytotoxic compounds have been isolated from soil, root excretions, leaf washings, plant residues, and microbial decomposition products of organic residues.[25-27] Included are a variety of aldehydes, acids, ketones, courmarins, and glycosides. Documentation of this work is beyond the scope of the present volume but the papers mentioned above cite innumerable examples of phytotoxicity from compounds emitted or derived from the various sources.

Although many lipids are known to be phytotoxic at very low concentrations, little is known about their effectiveness in normal agricultural soils. In most cultivated soils, any toxins produced during decomposition of organic residues, or secreted from plant roots, would themselves undergo rapid microbial decomposition. Accumulations are possible, however, under conditions that are unfavorable for incomplete decomposition, such as in heavy clay soils and in those that are poorly aerated.

Growth-Promoting Substances

Growth-promoting compounds can be considered in the same category as antagonisms in that they can affect plant growth at very low concentrations. It is generally recognized that B vitamins and other growth factors are normal constituents of fertile soils, their presence being due to release from organic residues, liberation from plant roots, and synthesis by microorganisms. Products derived from lignin during decay by microorganisms are known to be physiologically active.[28] Another class of compounds known to stimulate plant growth are the phytochelates, a subject discussed in Chapter 16.

In summarizing existing information on the subject, Flaig[28] concluded that the action of physiologically active compounds in soil is dependent upon environmental conditions, such as temperature, light, water, oxygen tension, and availability of nutrients. Physiologically active substances were believed to exert their greatest effect on increasing yields when conditions were suboptimum for plant growth.

As was the case noted in the previous section, an extensive review of the role of growth-promoting substances in soil is beyond the scope of the present chapter. Microbial aspects of growth-promoting substances in soil are discussed by Lochhead.[29]

Influence on Soil Physical Properties

Lipids are hydrophobic and thus have the potential for altering such soil physical properties as degree of wetting and aggregate stability. As early as 1910,

Schreiner and Shorey (see Chapter 2) observed that a California soil could not be readily wetted by rain, irrigation, or movement of water through the subsoil, thus making it unfit for agriculture. The problem was attributed to waxes present in the soil organic matter.

The condition of water-repellency or nonwettability is a common problem in citrus groves, burned-over areas of forest soil, and turf. This subject is covered in Chapter 18.

COMPOSITION

Rather extensive use has been made of modern analytical methods for characterizing organic constituents in soil. The result has been a marked increase in our knowledge of the nature of soil nitrogen, saccharides, and the various humic functions. Until recently, few examples could be cited where the newer methods have been applied to soil lipids. Soils undoubtedly contain a wide variety of lipids, ranging from the rather stable paraffin hydrocarbons to ephemeral chlorophyll degradation products.

Waxes

Typical waxes are mainly esters of long-chain fatty acids and the higher aliphatic alcohols, although cyclic alcohols may also be present. The acid and alcohol components are usually unbranched and saturated, with C numbers ranging from C_{12} to C_{34}. Compounds with an even number of C atoms predominate.

A significant part of the lipids in soil may consist of mixtures of waxes. However, individual waxes per se have not been separated and most characterization studies have been based on the identification of degradation products.

Nearly all research on soil waxes has been carried out using material extracted from peat. The vast majority of papers are dated 1940 or earlier and most of the others appeared between 1940 and 1960 (see reviews[1-3]). It is apparent, therefore, that modern analytical techniques have not been adequately used to characterize soil waxes.

The most extensive studies on the nature of the wax material in mineral soils appears to be those of Meinschein and Kenny[11] and Butler et al.[30] In the study by Meinschein and Kenny, the soil was extracted with a benzene–alcohol mixture and the extract was placed on a silica gel column prewet with *n*-heptane. Four fractions were recovered by successive elution with *n*-heptane, carbon tetrachloride, benzene, and methanol. The materials recovered with *n*-heptane and carbon tetrachloride consisted largely of saturated hydrocarbons; only the wax recovered with benzene, representing about 5 percent of the extracted material, was examined in detail. The types of acids and alcohols in the wax esters were determined by converting the esters to saturated hydrocarbons by high-pressure hydrogenation and analyzing them mass spectrometrically. Even C-numbered (C-even) waxes were present in considerably higher amounts than their odd C-number (C-odd) homologs (90 versus 10 percent). The waxes ranged from C_{36} to C_{52}, and the C-even waxes were formed from C-even normal aliphatic acids and normal primary aliphatic alcohols. The waxes in urea-nonadduct samples were chiefly esters of cyclic alcohols and normal aliphatic acids, a result that is in agreement with the observation that cyclic alcohols are more prevalent than cyclic acids in natural waxes.

Butler et al.[30] utilized gas chromatography to characterize the acids and alcohols produced by saponification of a neutral wax obtained from a green-colored Australian soil. The acids (analyzed as the esters) ranged from C_{12} to C_{30}, with the even numbers predominating. The acids present in greatest amounts were C_{22} (13 percent), C_{24} (22 percent), and C_{26} (21 percent). The crude wax also contained a complex mixture of hydrocarbons. A pigment identified as a hexachloropolynuclear quinone of the dihydroxyperylenequinone or dihydroxydinapthylquinone type (chemical formula of $C_{20}H_4O_5Cl_6$) was isolated from the original extract containing the wax; this material was believe to be responsible for the green color of the soil.

Morrison and Bick[31] saponified the neutral fraction of the lipids from a garden soil and by adsorption chromatography of the methyl esters obtained a fraction consisting of long-chain fatty acids. Analysis by gas–liquid chromatography showed that acids from n-C_{18} to n-C_{34} were present and that about 80 percent were even numbered. A wide variety of compounds, including some benzenepolycarboxylic acid polymethyl esters, were identified by Ogner[32] by permanganate oxidation of the hexane and ether extracts of raw humus. These findings indicate the presence of substituted aromatic rings in soil lipids.

The general observation that soil waxes are chiefly esters of the higher members of the homologous series of n-acids and n-alcohols, both having a preponderance of C-even atoms, is in agreement with the observation that animal and plant waxes are, for the most part, saturated straight-chain compounds containing primarily C-even acids and alcohols.

Organic Acids

More than a dozen organic acids have been isolated from soils, many of high molecular weight. In early work (reviewed by Stevenson[3]), the following long-chain fatty acids were identified: α-hydroxystearic (**I**), cerotic (**II**), triacosanoic (**III**), dihydroxystearic (**IV**), lignoceric (**V**), and pentacosanoic (**VI**); others included crotonic, $CH_3CH=CHCOOH$, and two compounds identified as "agroceric," $C_{21}H_{42}O_3$, and "paraffinic," $C_{24}H_{48}O_2$.

$CH_3(CH_2)_{15}CHOHCOOH$ $CH_3(CH_2)_{24}COOH$

α-Hydroxystearic   Cerotic

I   **II**

$CH_3(CH_2)_{21}COOH$ $CH_3(CH_2)_7CHOH-CHOH(CH_2)_7COOH$

Triacosanoic   Dihydroxystearic

III   IV

$CH_3(CH_2)_{22}COOH$ $CH_3(CH_2)_{23}COOH$

Lignoceric   Pentacosanoic

V   VI

In more modern times, Morrison and Bick[31] obtained a purified lipid preparation from a garden soil which contained 20 percent free acids, 55 percent of which consisted of a mixture of fatty acids of n-C_{20} to n-C_{34}, with 80 percent of even-numbered C atoms. The even-numbered acids of n-C_{24} to n-C_{32} made up 75 percent of the fraction. Gallopini and Raffladi[33] found that fatty acids and unsaponifiable constituents were the major components of the ether extracts of some forest soils. Numerous long-chain fatty acids, some of which could not be identified, were indicated in the study by Wang et al.[10,20] Soils planted with sugarcane contained components not found in soil planted with paddy rice, banana, or pineapple.

As noted earlier, supercritical extraction of the soil with n-pentane has been found to be effective for recovery of fatty acids (as well as alkanes).[9,12,13] The major acids identified by Schnitzer et al.[9] were n-fatty acids (n-C_7–n-C_{29}) accompanied by lesser amounts of branched fatty acids (C_{12}–C_{19}), unsaturated fatty acids (n-C_{16}, n-C_{18}, and n-C_{20}), hydroxy fatty acids (C_{12}–C_{16}), and α,ω-diacids (C_{15}–C_{25}). For the n-fatty acids, a bimodel distribution pattern was observed, with maximum at C_{16} and C_{24}. In as much as C-even fatty acids in the C_4–C_{26} range predominate in microorganisms, while C-even acids in the C_{26}–C_{38} range are abundant in the waxes of plants and insects, the n-fatty acids were considered to be primarily of microbial origin.

As far as isolated soil components are concerned, Schnitzer and Neyroud[7] found that fatty acids, as well as alkanes, occur in association with humic and fulvic acids. The organic acids were believed to be present in two different forms: physically adsorbed and chemically bonded, with the latter being the most important.

Low-molecular-weight organic acids (e.g., formic, acetic, oxalic, and butyric), are also normal constituents of mineral soils. These acids may be important agents in the mobilization and transport of metals, the weathering of rocks, and the solubilization of plant nutrients, as discussed in Chapters 16 and 20.

Organic acids of various types occur in small quantities in the rhizosphere of plant roots, and in the raw humus layers of forest soils. In a recent study, Fox and Commerford[34] examined the low-molecular-weight aliphatic organic acids in water extracts from the O, A, Bh, and Bt horizons of a series of forest soils (Ultisols, Entisols, Spodosols) of the southeastern United States. Oxalic acid was generally present in the highest concentration (range of 25-1000 μM in the soil solution) and was greater in the Bh and Bt horizons than in the A horizon. High concentrations of formic acids were also observed (5-174 μM in the soil solution). Other organic acids present in trace amounts were citric, acetic, maleic, aconitic, and succinic.

In aerobic soils, organic acids would be expected to have a transitory existence as long as moisture and temperature remain favorable for biological activity. Anaerobic conditions are particularly suitable for the preservation of organic acids; accumulations of butyric acid in rice fields is believed to contribute to the "Aki-och" condition of rice (see review of Stevenson[3]).

Hydrocarbons

Normal paraffins in the C_{16}-C_{32} range have been found in the nonsaponifiable fraction of ether, alcohol, and benzene-alcohol extracts of soil. In an early study by Stevens et al.,[35] alkanes in the C_{23}-C_{31} range were observed, among which n-nonacosane, $C_{29}H_{60}$, and n-hentriacontane, $C_{31}H_{64}$, predominated. There was a preponderance of C-odd to C-even atoms. Somewhat similar results were obtained by Morrison and Bick[31] for n-alkanes in a garden soil and a peat; main alkanes in the garden soil were n-C_{29} (21 percent), n-C_{31} (31 percent), and n-C_{33} (15 percent).

In contrast to the above, a preponderance of C-even to C-odd atom ratios were observed by Schulten and Schnitzer[12] for n-alkanes recovered from the fine clay fractions of several soils. Their results for an Armadale soil, together with distribution patterns reported for n-alkanes (C_{23}-C_{33} range) in cattle manure, a Gulf of Mexico sediment, and the soil examined by Stevens et al.,[35] are illustrated in Fig. 7.4. For the soil and the marine sediment, C_{29} and C_{31} were the most abundant alkanes; for the clay fraction of the Armadale soil, C_{24} and C_{26} were the most abundant. The cattle manure contained a relatively large amount of the C_{33} hydrocarbon (n-tritriacontane), which was attributed to the presence of plant species in the cattle's diet that contained high concentrations of this hydrocarbon.[36]

The variations observed for the relative abundance of C-even and C-odd n-alkanes in soil and soil organic matter preparations may reflect different origins for the n-alkanes (i.e., microorganisms or higher plants). The n-alkanes of higher plants are reported to have a predominance of C-odd to C-even atoms; n-nonacosane and n-hentriacontane are often major components. In contrast, for bacteria and fungi, the average ratio of C-odd to C-even atoms appears to be near unity.[37] Thus, in those soils where the n-alkanes are primarily of microbial origin, the C-odd to C-even ratio should be closer to that of the biomass (i.e., near unity) than to higher plants. Additional research is required on factors affecting the distribution of n-alkanes in soil.

Alkanes have been identified in isolated humic and fulvic acids.[7,8] Ogner and Schnitzer[8] found that up to 0.16 percent of the weight of a soil fulvic acid consisted of normal plus branched-cyclic alkanes, most of which could be

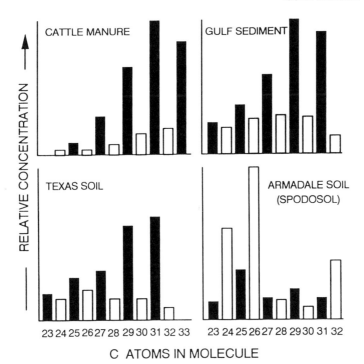

Fig. 7.4 Relative amounts of n-alkanes in soils, sediments, and cattle manure. Adapted from Stevenson.[3] Results for the Armadale soil are from Schulten and Schnitzer[12]

extracted only after methylation of the fulvic acid to reduce H-bonding. Weight distribution of the n-alkanes exhibited two well-defined regions, the first consisting of C_{14} to C_{23} n-alkanes and the second to C_{24} to C_{36} n-alkanes. The two groups were believed to be of microbial and plant origin, respectively. Ratios of C-odd to C-even n-alkanes were near unity.

Simonart and Batistic[38] identified the following aromatic hydrocarbons in soil: benzene, toluene, ethylbenzene, p/m-xylene, o-xylene, and naphthalene.

Polycyclic Hydrocarbons

Polynuclear aromatic hydrocarbons have been detected in small quantities in soils.[39-43] The term polynuclear (or "polycyclic") is used in an operational sense and implies C—H compounds containing benzenoid structures; the remainder of the molecule may consist of straight or branched chains of olefinic groups or of saturated rings. Many of these compounds are carcinogenic and are of interest because they occur as air pollutants in combustion products where fossil fuels are burned, such as the exhaust of motor vehicles and industrial flue gases. In addition, a complex assemblage of polynuclear aromatic hydrocarbons is formed in natural fires, from which they are dispersed, mixed by air transport, and eventually deposited into surface sediments.[42,43] The work of Hites et al.,[39] and LaFlamme and Hites[40] indicates that the major sources of polycyclic aromatic hydrocarbons in soils and sediments are anthropogenic (e.g., resulting from combustion) rather than natural.

Polycyclic hydrocarbons are typically determined by gas chromatography–mass spectrometry of methanol and methanol–benzene extracts of soils. A typical gas chromatogram is shown in Fig. 7.5. Some of the polynuclear aromatic hydrocarbons reported in soils are listed in Table 7.3.

Blumer[41] found that 3,4-benzpyrene and 1,2-benzpyrene were common constituents of soils in rural areas distant from major highways and industries (40–1300 μg/kg soil). Also detected were phenanthrene, fluoranthene, pyrene, chrysene, perylene, anthanthrene, benzanthracene, anthracene, and others (see Table 7.3 for typical structural formulas). It is not known whether any of these compounds are indigenous to the soil or if they are derived exclusively from atmospheric fallout.

LaFlamme and Hites[40] postulated that part of the polycyclic hydrocarbons in recent sediments may be of natural origin. Perylene, for example, may be formed from extended quinone pigments, as illustrated in Fig. 7.6. Swan[44] found that an Alaskan forest soil contained a significant amount of dehydroabietin, possibly derived from a conifer resin.

Glycerides and Phospholipids

True fats have yet to be isolated from soil, although glycerol has been detected in products obtained by saponification.

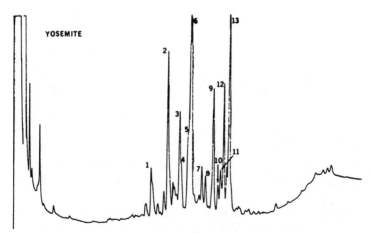

Fig. 7.5 Gas chromatography separation of polycyclic hydrocarbons from a soil taken from Yosemite National Park. Peaks 1 and 2 are fatty acid methyl esters. From LaFlamme and Hites,[40] reproduced by permission of Pergamon Press Ltd.

Table 7.3 Some Polynuclear Aromatic Hydrocarbons in Soils

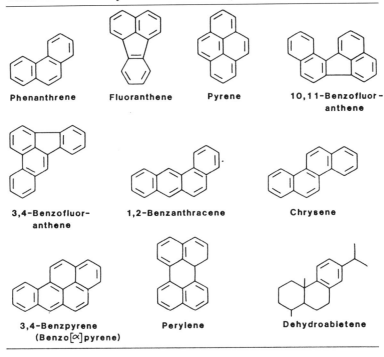

Considerable attention has been given from time to time to the occurrence of phospholipids in soil, because these compounds are potential sources of P for plant growth. This work is discussed in Chapter 5.

Steroids and Terpenoids

The group of compounds called steroids can be defined as derivatives of a fused, reduced-ring system, perhydrocyclopentano–phenanthrene, composed of three fused cyclohexane rings (A, B, and C) in the nonlinear or phenanthrene arrangement, and a terminal cyclopentane ring (D).

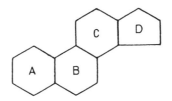

These compounds would be expected to occur in soils because they are widely distributed in plant and animal tissues. The compounds β-sitosterol and

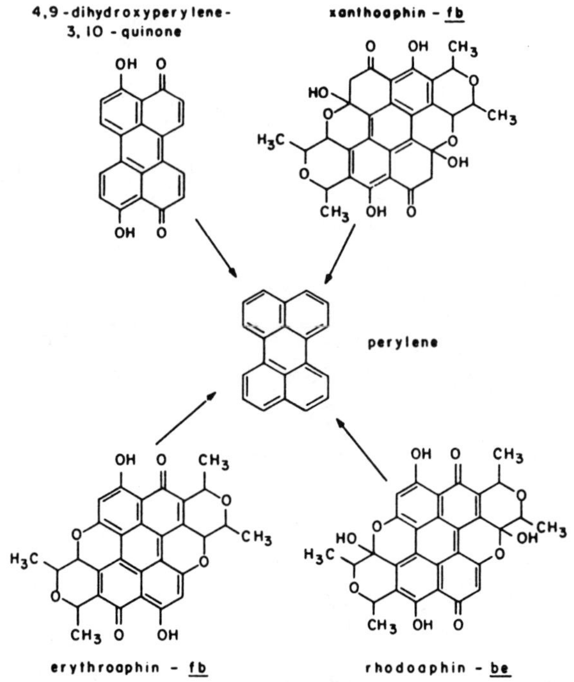

Fig. 7.6 Extended quinone pigments as possible sources of perylene in soils and sediments. From LaFlamme and Hites,[40] reproduced by permission of Pergamon Press Ltd.

dehydro-β-sitosterol have both been reported in peat or soil.[1,2] The former is a sterol commonly found in higher plants. The structures of these and other terpenoids identified in soils are shown in Table 7.4.

Meinschein and Kenny[11] detected sterols and penta- and hexacyclic compounds (tentatively identified as triterpenoids) in the benzene extracts of several mineral subsoils.

Carotenoids

These compounds are of interest because they are the orange pigment of many plants and are associated with chlorophyll. The structure of α-carotene is as follows:

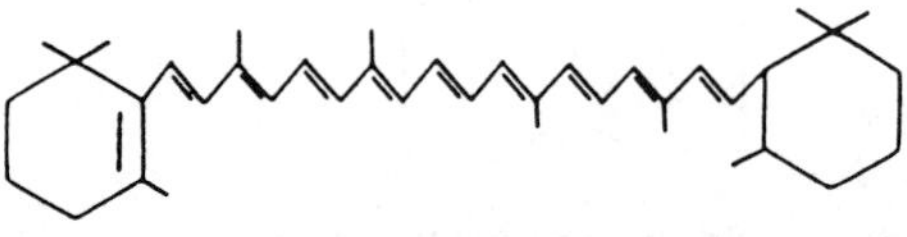

Carotenoids are sensitive to heat and light, and they are readily oxidized in the presence of air; thus, the conditions in most soils are suitable for their rapid

Table 7.4 Some Sterols and Terpenoids Isolated from Soil

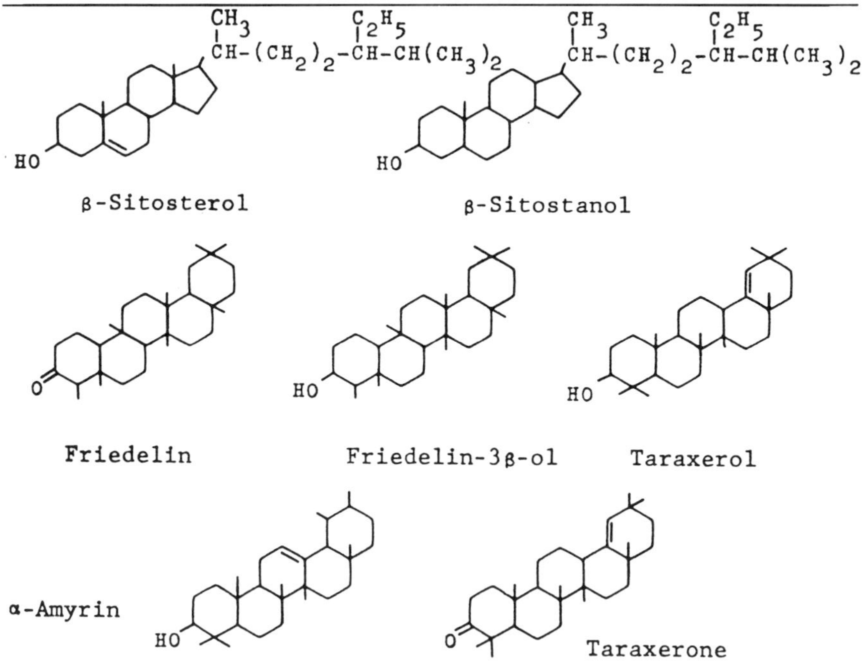

CH3 C2H5
CH-(CH2)2-CH-CH(CH3)2
HO
β-Sitosterol
CH3 C2H5
CH-(CH2)2-CH-CH(CH3)2
HO
β-Sitostanol
O
Friedelin
HO
Friedelin-3β-ol
HO
Taraxerol
α-Amyrin
HO
O
Taraxerone

decomposition. No report seems to have been published on the occurrence of carotenoids in agricultural soils, although they are undoubtedly present in trace amounts, particularly in poorly drained soils. About 20 different carotenoids have been found in wet sediments, only a few of which have been identified.[15]

Porphyrins

The occurrence of porphyrin pigments in freshwater and marine sediments, and in petroleum, has been recognized for many years. The major porphyrins are regarded as being derived from chlorophyll; they have been referred to as "chlorophyll derivatives" and "sedimentary chlorophyll degradation products." Chopra[45] found that the distribution pattern of plant pigments in soil was determined to a large extent by soil type.

The geochemical significance of chlorophyll has been discussed by Orr et al.[46] and Hodgson et al.[47] Changes produced in chlorophyll A during attack by microorganisms involves loss of Mg to form pheophytin A and then the phytol group to form pheophorbide A (Fig. 7.7). In sediments, the latter is transformed to a variety of porphyrin-type compounds.[46] The occurrence of pheophytins and other chlorophyll-type compounds in soil has been reported by Chopra.[45]

Data obtained by Gorham[48] showed that, per unit of C, the chlorophyll content of woodland soils was an order of magnitude lower than that observed

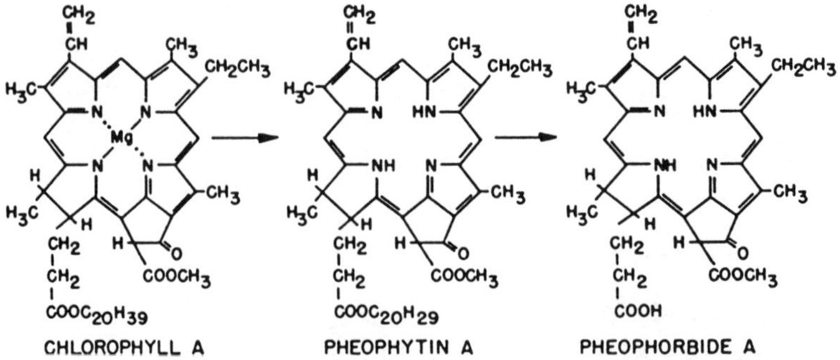

Fig. 7.7 Initial changes in chlorophyll during decay in soil

for lake muds. This result can be explained on the basis that deficiency of oxygen in the latter inhibits chlorophyll destruction. Poorly drained soils may contain porphyrins in amounts approximating that detected in freshwater sediments. Additional information on chlorophyll degradation products is given in Chapter 3, where their importance as nitrogen sources is discussed.

Pristane (2,6,10,14-tetramethylpentadecane) and phytane (2,6,10,14-tetramethylhexadecane) have been observed in ancient sedimentary rocks, including the one-billion-year-old Precambrian chert from the Gunflint iron formation. According to Oró et al.,[36] these compounds are probably derived from phytol, the alcohol component of chlorophyll. As yet, no attempt has been made to determine if hydrocarbons of chlorophyll origin occur in terrestrial soil.

Other Compounds

Many other substances of a lipid nature are likely to be present in soils, but in very small quantities. They include the lower aliphatic alcohols,[18] phenolic acids,[20] polyphenols,[49] ketones,[50] and complex quinones.[51] A number of alcohols, aldehydes, esters, and organic acids have been found to be evolved from soils when incubated with glucose under anaerobic conditions.[52]

A number of long-chain methyl ketones were isolated by Morrison and Bick[50] from a garden soil and peat. Analysis by gas–liquid chromatography showed that the components from peat ranged from n-C_{17} to n-C_{33}, with 92 percent having odd-C numbers. Compounds of n-C_{25} (26 percent) and n-C_{27} (37 percent) were especially abundant. For the garden soil, the range was from n-C_{19} to n-C_{35}, with 81 percent odd-C numbers. This appears to be the first report of methyl ketones of this C range in natural products.

Phthalates have been reported to occur as natural products in soil.[53,54] However, as Mathur[55] pointed out, phthalic acid esters of various types are used extensively in plastics and the possibility of environmental contamination cannot be eliminated. Phthalate esters have the general structure shown below, where R_1 and R_2 indicate alcohol radicals.

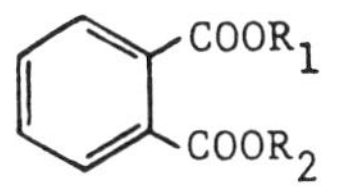

$$\text{benzene-1,2-dicarboxylate: } C_6H_4(COOR_1)(COOR_2)$$

SUMMARY

Although often considered inert and unimportant, soil lipids in many cases have a direct influence on soil properties and plant growth.

Evidence for organic toxins in some soils is conclusive but little is known of the circumstances under which they are formed and persist. Their effects on plant growth are influenced by such variables as the nature of the soil, inherent toxicity of the compound or compounds involved, and resistance of the toxin(s) to microbial decomposition. Growth-promoting substances of various types are also common constituents of soils.

Because of the diverse nature of soil and the complexity of the organic constituents contained therein, characterization of the lipid component will require the skill and patience of scientists working in several disciplines of science. Many of the separation and analytical techniques developed by lipid chemists can be applied directly. Gas–liquid chromatography in conjunction with mass spectroscopy has been extended to a wide variety of lipids, and, with conventional apparatus and sensitive detectors, the technique is suitable for the analysis of nanogram amounts of fatty acids, steroids, hydrocarbons, and other compounds. Of further interest and general applicability are liquid-column chromatographic techniques. The rapidly accumulating body of reliable knowledge on the lipid composition of plants, animals, and microorganisms will undoubtedly be of great assistance to those working on the chemistry of lipids in soil.

REFERENCES

1 O. C. Braids and R. H. Miller, "Fats, Waxes, and Resins in Soil," in J. E. Gieseking, Ed., *Soil Components: Vol. 1. Organic Components*, Springer-Verlag, New York, 1975, pp. 343–368.
2 R. I. Morrison, "Soil Lipids," in G. Eglinton and M. T. J. Murphy, Eds., *Organic Geochemistry*, Pergamon, New York, 1969, pp. 559–575.
3 F. J. Stevenson, *J. Amer. Oil Chemist's Soc.*, **43**, 203 (1966).
4 S. A. Waksman and K. R. Stevens, *Soil Sci.*, **30**, 97 (1930).
5 N. P. Shcherbak, *Chem. Abstr.*, **71**, 48877 (1969).
6 W. L. Banwart, personal communication.
7 M. Schnitzer and J. A. Neyroud, *Fuel*, **54**, 17 (1975).
8 G. Ogner and M. Schnitzer, *Geochim. Cosmochim. Acta*, **34**, 921 (1970).
9 M. Schnitzer, C. A. Hindle, and M. Meglic, *Soil Sci. Soc. Amer. J.*, **50**, 913 (1986).

10 T. S. C. Wang, Y-C. Liang, and W-C. Shen, *Soil Sci.*, **107**, 181 (1969).
11 W. G. Meinschein and G. S. Kenny, *Anal. Chem.*, **29**, 1153 (1957).
12 H.-R. Schulten and M. Schnitzer, *Soil Sci. Soc. Amer. J.*, **54**, 98 (1990).
13 M. Schnitzer and C. M. Preston, *Soil Sci. Soc. Amer. J.*, **51**, 639 (1987).
14 J. A. Rice and P. MacCarthy, *Sci. Total Environ.*, **81/82**, 61 (1982).
15 G. Eglinton and M. T. J. Murphy, *Organic Geochemistry: Methods and Results*, Springer-Verlag, New York, 1969.
16 R. Greig-Smith, *Proc. Linnean Soc. N. S. Wales*, **35**, 808 (1910).
17 D. C. Whitehead, *Nature*, **202**, 417 (1963).
18 T. S. C. Wang and T-T. Chuang, *Soil Sci.*, **104**, 40 (1967).
19 T. S. C. Wang, S-Y. Cheng, and H. Tung, *Soil Sci.*, **104**, 138 (1967).
20 T. S. C. Wang, P-T. Hwang, and C-Y. Chen, *Soil Sci. Soc. Amer. Proc.*, **35**, 584 (1971).
21 G. Goodlass and K. A. Smith, *Soil Biol. Biochem.*, **10**, 201 (1978).
22 K. A. Smith and S. W. F. Restall, *J. Soil Sci.*, **22**, 430 (1971).
23 K. A. Smith and R. S. Russell, *Nature*, **222**, 769 (1969).
24 A. J. Sexstone and C. N. Mains, *Soil Biol. Biochem.*, **22**, 1315 (1990).
25 T. M. McCalla and F. A. Haskins, *Bact. Rev.*, **28**, 181 (1964).
26 W. Mojé, "Organic Soil Toxins," in H. D. Chapman, Ed., *Diagnostic Criterion for Plants and Soils*, University of California Press, Berkeley, 1966, pp. 533–569.
27 Z. A. Patrick, *Soil Sci.*, **111**, 13 (1971).
28 W. Flaig, "Effect of Lignin Degradation Products on Plant Growth," in *Isotopes and Radiation in Soil Organic Matter Studies*, International Atomic Energy Agency, Vienna, 1965, pp. 3–19.
29 A. G. Lochhead, *Bact. Rev.*, **22**, 145 (1958).
30 J. H. A. Butler, D. T. Downing, and R. J. Swaby, *Aust. J. Chem.*, **17**, 817 (1964).
31 R. I. Morrison and W. Bick, *J. Sci. Food Agr.*, **18**, 351 (1967).
32 G. Ogner, *Soil Sci.*, **120**, 25 (1975).
33 C. Gallopini and R. Riffladi, *Agrochimica*, **13**, 207 (1969).
34 T. R. Fox and N. B. Comerford, *Soil Sci. Soc. Amer. J.*, **54**, 1139 (1990).
35 N. P. Stevens, E. E. Bray, and E. D. Evans, *Bull. Amer. Assoc. Petrol. Geol.*, **40**, 975 (1956).
36 J. Oró, D. W. Nooner, and S. A. Wikström, *Science*, **147**, 870 (1965).
37 J. G. Jones, *J. Gen. Microbiol.*, **59**, 145 (1969).
38 P. Simonart and L. Batistic, *Nature*, **212**, 1461 (1966).
39 R. A. Hites, R. E. LaFlamme, and J. W. Farrington, *Science*, **189**, 829 (1977).
40 R. E. LaFlamme and R. A. Hites, *Geochim. Cosmochim. Acta*, **42**, 289 (1978).
41 M. Blumer, *Science*, **134**, 474 (1961).
42 M. Blumer and W. W. Youngblood, *Science*, **188**, 53 (1975).
43 W. W. Youngblood and M. Blumer, *Geochim. Cosmochim Acta*, **39**, 1303 (1975).
44 E. P. Swan, *Forest Prod. J.*, **15**, 272 (1965).
45 N. M. Chopra, *Soil Sci.*, **121**, 103 (1976).

46 W. L. Orr, K. O. Emery, and J. R. Grady, *Bull. Amer. Assoc. Petrol. Geol.*, **42**, 925 (1958).
47 G. W. Hodgson, B. Hitchon, K. Taguchi, B. L. Baker, and E. Peake, *Geochim. Cosmochim. Acta*, **32**, 737 (1968).
48 E. Gorham, *Soil Sci.*, **87**, 258 (1959).
49 H. Morita, *Geoderma*, **13**, 163 (1975).
50 R. I. Morrison and W. Bick, *Chem. Ind. (London)*, **1966**, 596 (1966).
51 E. N. Lambert, C. E. Seaforth, and N. Ahmed, *Soil Sci. Soc. Amer. Proc.*, **35**, 463 (1971).
52 J. D. Adamson, A. J. Francis, J. M. Duxbury, and M. Alexander, *Soil Biol. Biochem.*, **7**, 45 (1975).
53 C. F. Cifrulak, *Soil Sci.*, **107**, 63 (1969).
54 S. U. Khan and M. Schnitzer, *Soil Sci.*, **112**, 231 (1971).
55 S. P. Mathur, *J. Environ. Quality*, **3**, 189 (1974).

8

BIOCHEMISTRY OF THE FORMATION OF HUMIC SUBSTANCES

The biochemistry of the formation of humic substances is one of the least understood aspects of humus chemistry and one of the most intriguing. Continued research on this subject can be justified on theoretical and practical grounds. In the first case, information as to how humic substances are formed would provide valuable clues as to their structures. Practically speaking, an understanding of the pathways of humus synthesis would result in greater comprehension of the carbon (C) cycle and of changes that occur when plant residues and organic wastes undergo decay by soil microorganisms.

For detailed information on the subject, several excellent reviews can be consulted.[1-7]

MAJOR PATHWAYS OF HUMUS SYNTHESIS

Several pathways exist for the formation of humic substances during the decay of plant and animal remains in soil, the main ones being shown in Fig. 8.1. The classical theory, popularized by Waksman,[8] is that humic substances represent modified lignins (pathway 4) but the majority of present-day investigators favor a mechanism involving quinones (pathways 2 and 3). In practice, all four pathways must be considered as likely mechanisms for the synthesis of humic and fulvic acids in nature, including sugar-amine condensation (pathway 1).

In this chapter, the four main theories are discussed at some length but first a brief résumé of each will be given.

1. For many years it was thought that humic substances were derived from lignin (pathway 4 of Fig. 8.1). According to this theory, lignin is incom-

MAJOR PATHWAYS OF HUMUS SYNTHESIS 189

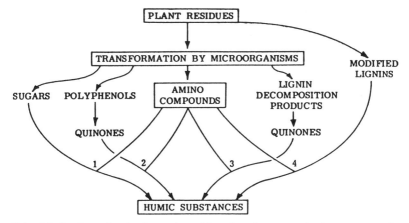

Fig. 8.1 Mechanisms for the formation of humic substances. Amino compounds synthesized by microorganisms are seen to react with modified lignins (pathway 4), quinones (pathways 2 and 3), and reducing sugars (pathway 1) to form complex dark-colored polymers

pletely utilized by microorganisms and the residuum becomes part of the soil humus. Modifications in lignin include loss of methoxyl (OCH_3) groups with the generation of *o*-hydroxyphenols and oxidation of aliphatic side chains to form COOH groups. This pathway, illustrated further in Fig. 8.2, is exemplified by Waksman's lignin–protein theory.

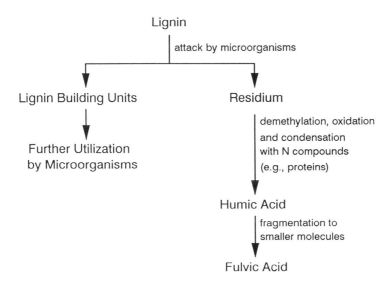

Fig. 8.2 Schematic representation of the lignin theory of humus formation

Assuming that humic substances represent a system of polymers, the initial product would be components of humin; further oxidation and fragmentation would yield first humic acids and then fulvic acids.

2. In pathway 3, lignin still plays an important role in humus synthesis, but in a different way. In this case, phenolic aldehydes and acids released from lignin during microbiological attack undergo enzymatic conversion to quinones, which polymerize in the presence or absence of amino compounds to form humic-like macromolecules.
3. Pathway 2 is somewhat similar to pathway 3 except that the polyphenols are synthesized by microorganisms from nonlignin C sources (e.g., cellulose). The polyphenols are then enzymatically oxidized to quinones and converted to humic substances as per pathway 3.
4. The notion that humus is formed from sugars (pathway 1) dates back to the early days of humus chemistry. According to this concept, reducing sugars and amino acids, formed as by-products of microbial metabolism, undergo nonenzymatic polymerization to form brown nitrogenous polymers of the type produced during dehydration of certain food products.

Pathways 2 and 3 form the basis of the now popular polyphenol theory. Unlike the lignin theory, the starting material consists of low molecular weight organic compounds, from which large molecules are formed through condensation and polymerization. A schematic representation of the polyphenol theory is shown in Fig. 8.3.

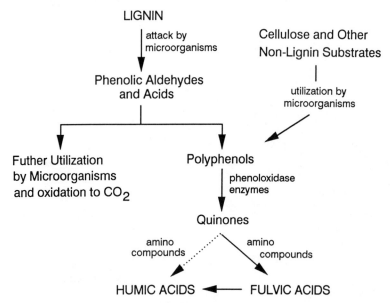

Fig. 8.3 Schematic representation of the polyphenol theory of humus formation

In considering the origin of humic substances, it should be emphasized that a completely satisfactory scheme for explaining the nature and occurrence of humic and fulvic acids in diverse environments has not yet been developed. The four pathways suggested by Fig. 8.1 may operate in all soils, but not to the same extent or in the same order of importance. A lignin pathway may predominate in poorly drained soils and wet sediments (swamps, etc.) whereas synthesis from polyphenols in leachates from leaf litter may be of considerable importance in certain forest soils. The frequent and sharp fluctuations in temperature, moisture, and irradiation in terrestrial surface soils under a harsh continental climate may favor humus synthesis by sugar–amine condensation.

One consideration that has not received adequate attention is that, in any given soil, not all humic components may be formed by the same mechanism. Humic acids, for example, may originate from polyphenols of plant or microbial origin whereas fulvic acids may consist of products arising from the condensation of sugars and amines, sometimes called the Maillard reaction.

A major difficulty in deciphering environmental effects on humus formation is that, more often than not, the humic material is derived from several sources, such as in sediments and natural waters, where a portion is formed *in situ* and the remaining is of terrestrial origin.

As noted later, the lignin and polyphenol models fail to account for the formation of humic substances in environments where little if any aromatic constituents are produced, such as in seawater and ocean sediments. In the case of seawater, still another mechanism has been proposed, namely, that humic substances are formed from unsaturated lipids produced by plankton.[6,7] The reactions involved will be shown later in Chapter 12.

THE LIGNIN THEORY

For many years, the view was held that lignin was the source of humic substances. The theory was popularized by Waksman,[8] who concluded that the nitrogen contained in humic acids resulted from the condensation of modified lignin with protein, the latter being a product of microbial synthesis. Stabilization of the protein was believed to occur through formation of a Schiff base.

(Modified lignin)–CHO + RNH_2 → (Modified lignin)–CH=NR + H_2O

The following evidence was cited by Waksman in support of the lignin theory of humic acid formation.

1. Both lignin and humic acid are decomposed with considerable difficulty by the great majority of fungi and bacteria.

2. Both lignin and humic acid are soluble in alkali and precipitated by acids. Both are partly soluble in alcohol and pyridine and they both contain OCH_3 groups; the OCH_3 content diminishes with stage of decomposition.
3. Both lignin and humic acid are acidic in nature, both are able to combine with bases, and both are characterized by their capacity to undergo base exchange, although to a different quantitative extent.
4. When lignins are warmed with aqueous alkali, they are transformed into methoxyl-containing humic acids. Humic acids have many properties in common with oxidized lignins.

Modifications have been made in the lignin–protein theory to include reactions with NH_3, which is produced during decay of nitrogenous organic substances by microorganisms. Reactions involved in the combination of NH_3 with lignin and lignin-like substances are discussed in Chapter 4.

Although lignin is less easily attacked by microorganisms than other plant components, mechanisms exist in nature for its complete aerobic decomposition. Otherwise, undecomposed plant remains would accumulate on the soil surface and the organic matter content of the soil would gradually increase until depletion of CO_2 from the atmosphere. The ability of soil organisms to degrade lignin has been underestimated in some quarters and its contribution to humus has been exaggerated.

In normally aerobic soils, lignin may be broken down into low-molecular-weight products prior to humus synthesis (see next section). On the other hand, it is known that oxygen is required for the microbial depolymerization of lignin; furthermore, the fungi that degrade lignin are not normally found in excessively wet sediments. Accordingly, it seems logical to assume that modified lignins will make a major contribution to the humus of poorly drained soils, peat, and lake sediments. The importance of lignin as a source of humic substances has been emphasized in recent times by Ertel and Hedges[9,10] and Hatcher and Spiker.[5]

The nature of the products obtained by modification of lignin can best be understood in terms of the biochemistry of lignin formation and its chemistry. These subjects are discussed briefly in the following sections.

Biosynthesis and Structure of Lignin

Evidence is rather conclusive that lignins are made up of phenylpropane units (C_6-C_3) of the types represented by coniferyl alcohol (**I**), *p*-hydroxycinnanyl alcohol (**II**), and sinapyl alcohol (**III**), the contribution of each being dependent upon the nature of the plant. The lignins of conifers, for example, contain relatively high amounts of sinapyl alcohol; those of herbaceous plants contain *p*-hydroxycinnamyl units as well as coniferyl and sinapyl alcohols.

Coniferyl alcohol
I

p-Hydroxycinnamyl alcohol
II

Sinapyl alcohol
III

Steps involved in the synthesis of lignins by plants are outlined briefly in the following sections. For more detailed information, the reviews of Crawford,[11] Pearl,[12] and Schubert[13] are recommended.

Synthesis of Phenylpropane Units An overall scheme for the formation of lignin building blocks is shown in Fig. 8.4. Glucose, produced during photo-

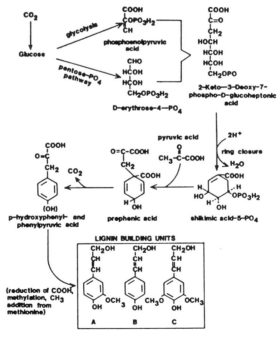

Fig. 8.4 Pathways for the synthesis of lignin-building blocks—A: Coniferyl alcohol, B: *p*-hydroxycinnamyl alcohol, C: sinapyl alcohol

synthesis, is converted to phosphoenolpyruvic acid through glycolysis and to D-erythrose-4-phosphate via a pentose–phosphate pathway. A combination of the two yields a 7-C compound (2-keto-3-deoxy-7-phospho-D-glucoheptonic acid), which by ring closure and addition of pyruvic acid yields shikimic acid-5-phosphate and prephenic acid. Decarboxylation of the latter gives p-hydroxyphenyl- and phenylpyruvic acids, from which the lignin-building units are ultimately formed.

Several steps are involved in the formation of primary building units from prephenic acid. In the grasses, prephenic acid is first converted to p-hydroxyphenylpyruvic acid (see Fig. 8.4); in most other plants the ring OH is eliminated to form phenylpyruvic acid. Present evidence suggests that these two aromatic acids are converted to the corresponding amino acids (phenylalanine and tyrosine), both of which are converted to p-coumaric acid and ultimately to the three primary lignin-building blocks (i.e., coniferyl alcohol, p-hydroxycinnamyl alcohol, and sinapyl alcohol).

Formation of Mesomeric Free Radicals The formation of larger molecules from the individual building blocks (structures I–III) can best be explained by a pairing mechanism involving mesomeric free radicals, such as those shown below for coniferyl alcohol.

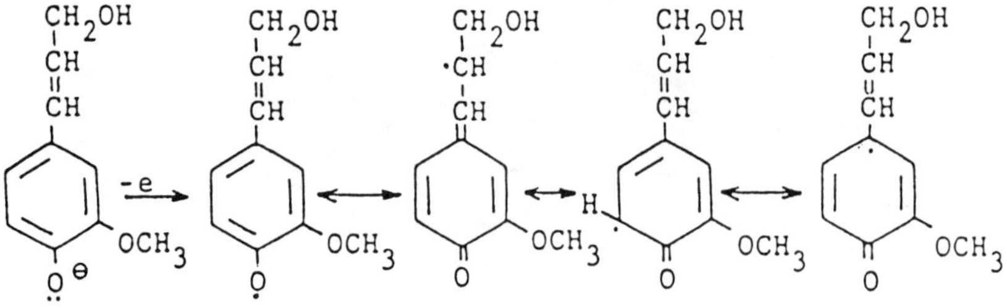

Thus, the enzymatic dehydrogenation of coniferyl alcohol leads to the formation of the phenoxy radical **(IV)**, which occurs in equilibrium with its mesomeric forms (e.g., **V, VI,** and **VII**). The latter process is initiated with the anion remaining after ionization of the free phenolic OH group of coniferyl alcohol. Unpaired electrons are shown as large dots.

Intermediates of Lignin Formation The next step in lignin formation is believed to be the pairing of mesomeric free radicals in various combinations to yield dilignols in which covalent bonds link the two monomers together. For example, a combination of the free radicals shown by VI and VII lead to the dilignol known as dehydrodiconiferyl alcohol (IX); other important pairings include VI + VI to form pinoresinol (X) and VI + V to form guaiacylglycerol-β-coniferyl ether (XI).

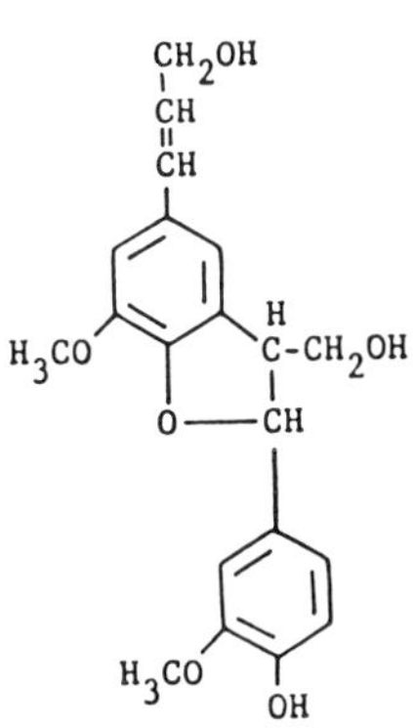

Dehydrodiconiferyl alcohol
IX

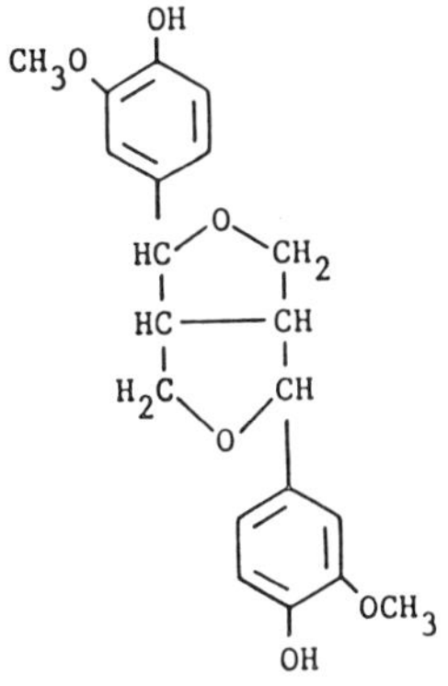

Pinoresinol
X

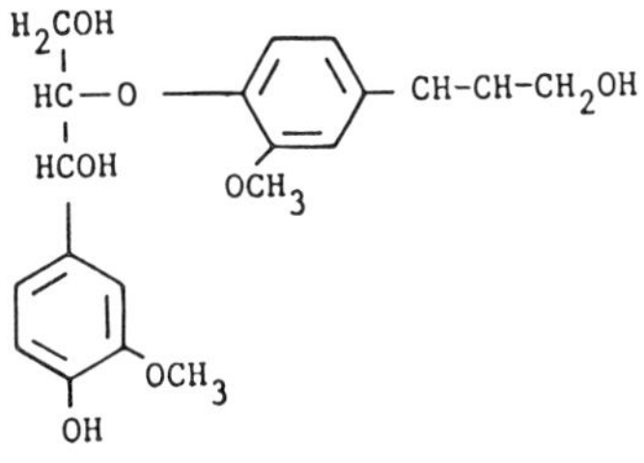

Guaiacylglcerol-β-coniferyl ether
XI

Other dilignols are known, but the important ones are those shown above. Many lignols have been identified in plant tissues, including some tetra- and hexalignols.

Structure of Lignin According to present concepts, lignification is an extension of the coupling reactions noted above to produce higher polymers. Numerous possibilities exist for the formation of extensively branched structures. Whereas the illustrations shown in the previous section involve coniferyl alcohol, other free radicals also participate in polymerization reaction, notably those originating from *p*-hydroxycinnamyl and sinapyl alcohols. A diagrammatic representation of the structure of spruce lignin is shown in Fig. 8.5.

Conversion of Lignin to Humic Substances

On the assumption that humic substances represent modified lignins, the most obvious changes in lignin would be a loss of OCH_3 groups, with exposure of phenolic OH groups, and oxidation of terminal side chain to form COOH groups.[4,14,15] A large number of bacteria are apparently able to demethylate lignin without further degrading the polymer.[1] As can be seen from Fig. 8.6, demethylation and oxidation of side chains lead to the formation of products enriched in acidic functional groups (COOH and phenolic OH) and that have lower C but higher oxygen contents than the original lignin. It should be noted, however, that COOH groups arising from the side chains of lignin would appear to be insufficient to account for the high COOH content of humic and fulvic acids. Additional COOH groups may arise from ring cleavage of aromatic components of lignin.

In view of the large size of the lignin macromolecule and its insolubility in water, the modifications outlined above would necessarily have to be brought about by exocellular enzymes. The *o*-dihydroxy-benzene units resulting from

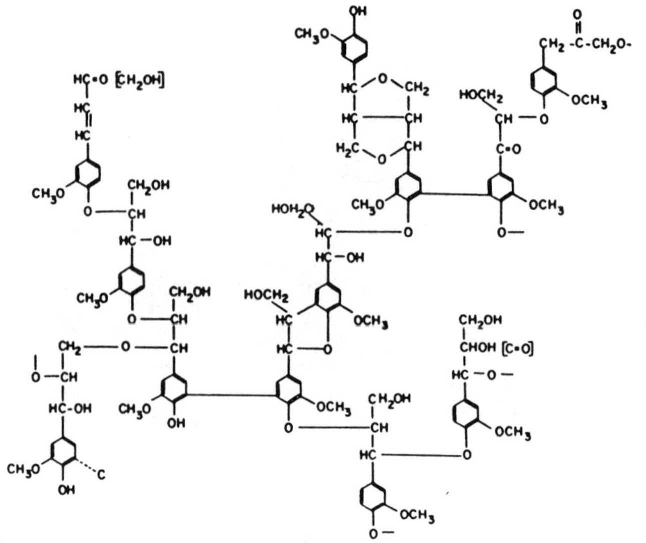

Fig. 8.5 Diagrammatic representation of spruce lignin. From Pearl,[12] reproduced by permission of the American Chemical Society, Washington, D.C.

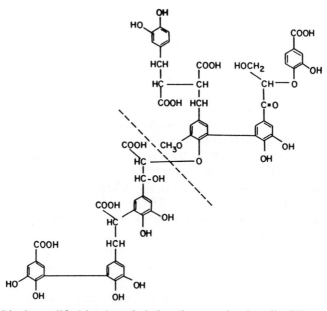

Fig. 8.6 Lignin modified by demethylation (increase in phenolic OH) and oxidation of side chains (increase in COOH)

demethylation of lignin would be subject to oxidation to quinones capable of undergoing condensation reactions with NH_3 and amino compounds, as shown below.

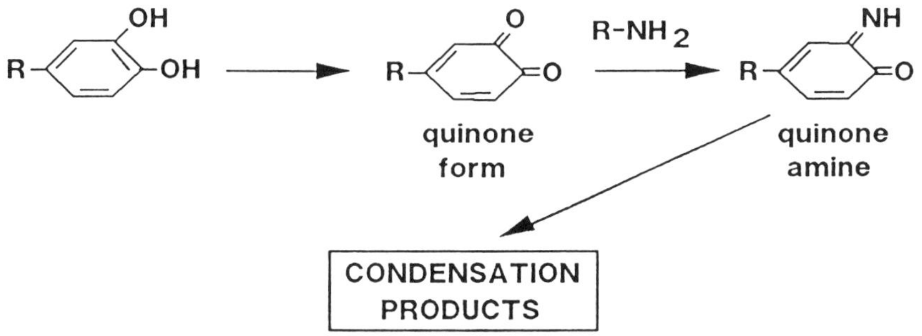

$$\underset{}{R-\!\!\bigcirc\!\!\!\!\overset{OH}{\underset{OH}{}}} \longrightarrow \underset{\text{quinone form}}{R-\!\!\bigcirc\!\!\!\!\overset{O}{=}\!\!O} \xrightarrow{R-NH_2} \underset{\text{quinone amine}}{R-\!\!\bigcirc\!\!\!\!\overset{NH}{=}\!\!O}$$

$$\boxed{\text{CONDENSATION PRODUCTS}}$$

A modified degradative model for humification has been advanced by Hatcher and Spiker.[5] In their model, resistant plant (lignin, cutin, suberin) and microbial (melanins, paraffinic macromolecules) biopolymers are the precursors from which humic substances evolve. The refractory biopolymers are selectivity preserved and become part of what is operationally defined as humin (alkali insoluble component of humus). Increasing degradation leads to the formation of macromolecules that have somewhat similar molecular weights but that are enriched in oxygen-containing functional groups (i.e., COOH, C=O, OH). The increased concentration of acidic groups promotes increased solubility in alkali and the evolution, first, of humic acids and then fulvic acids, the latter being regarded as the most humified fraction of humic substances.

Essentially, the degradative model of Hatcher and Spiker[5] is a modification of Waksman's lignin–protein theory, extended to include refractive macromolecules other than lignin. It is noteworthy in this respect that refractory paraffinic structures have been detected in the humin and clay fractions of soil by ^{13}C–NMR spectroscopy (see Chapter 11).

THE POLYPHENOL THEORY

As noted earlier, the classical lignin–protein theory of Waksman is now considered obsolete by many investigators. According to current concepts, quinones of lignin origin, together with those synthesized by microorganisms, are the major building blocks from which humic substances are formed. In this model, the first step consists of the breakdown of all plant biopolymers (including lignin) into their monomeric structural units, some of which polymerize enzymatically (or spontaneously under certain circumstances) to produce humic molecules of increasing complexity. The order of formation of humic substances would thereby follow the order: fulvic acid → humic acid → components of humin.

The formation of brown-colored substances by reactions involving quinones is not a rare event but is a well-known phenomenon that takes place in melanin formation, such as in the flesh of ripe fruits and vegetables following mechanical injury and during seed coat formation.[16]

Sources of Polyphenols

Possible sources of phenols for humus synthesis include lignin, microorganisms, uncombined phenols in plants, glycosides, and tannins. Of these, only the first two have received serious attention.

Flaig's[14,15] concept of humus formation is as follows:

1. Lignin, freed of its linkage with cellulose during decomposition of plant residues, is subjected to oxidative splitting with the formation of primary structural units (i.e., derivatives of phenylpropane).
2. The side-chains of the lignin-building units are oxidized, demethylation occurs, and the resulting polyphenols are converted to quinones by polyphenoloxidase enzymes.
3. Quinones arising from the lignin (as well as from other sources) react with N-containing compounds to form dark-colored polymers.

The role of microorganisms as sources of polyphenols was emphasized in early work by Kononova,[2] who gives a detailed account of research in which histological microscopic techniques and chemical methods were used to study the decomposition of plant residues. She concluded that humic substances were being formed by cellulose-decomposing myxobacteria prior to lignin decomposition. The stages leading to the formation of humic substances were postulated to be as follows:

Stage 1: Fungi attack simple carbohydrates and parts of the protein and cellulose in the medullary rays, cambium, and cortex of plant residues.

Stage 2: Cellulose of the xylem is decomposed by aerobic myxobacteria. Polyphenols synthesized by the myxobacteria are oxidized to quinones by polyphenoloxidase enzymes, and the quinones subsequently react with N compounds to form brown humic substances.

Stage 3: Lignin is decomposed. Phenols released during decay also serve as source materials for humus synthesis.

The relative importance of lignins and microorganisms as sources of phenolic units for humus synthesis is unknown—this may depend upon environmental conditions in the soil. Because lignins are relatively resistant to microbial decomposition and a major plant constituent, they are sometimes considered to be the major, if not the primary source, of phenolic units. However, as will be shown later, some of the microscopic fungi that decompose lignin in soil produce humic acid-like substances in which the phenolic units originate from both lignin and through biosynthesis.

Indirect evidence that both lignins and microorganisms contribute to humic formation has come from studies of products produced by oxidative and reductive cleavage of humic acids, where components typical of both lignins (syringyl and guaiacyl derivatives) and microorganisms (flavonoids) have been identified (see Chapter 10).

Formation of Polyphenols from Lignin during Biodegradation Fungi are undoubtedly the most important group of organisms responsible for cleavage of lignin. Greatest attention has been given to certain basidiomycetes called the "white-rot fungi," which use lignin as a preferred C and energy source. However, the ability of these organisms to synthesize humic substances has been questioned by Martin and Haider[3] on the basis that the "white rot fungi," when grown on lignified plant material, produce only small amounts of dark-colored, humic-like substances even though the lignin is decomposed and small quantities of phenols appear in the growth medium. In some basidiomycetes, humic-like substances are produced in fruiting bodies, presumably by transport of phenolic lignin degradation products into the fruiting bodies and polymerization following conversion to quinones by phenyloxidase enzymes.

According to Martin and Haider,[3] microscopic fungi of the *Imperfecti* group play a significant role in the synthesis of humic substances in soil. Their studies have shown that such fungi as *Hendersonula toruloidea*, *Epicoccum nigrum*, *Stachybotrys atra*, *S. chartarum*, and *Aspergillus sydowi* degrade lignin as well as cellulose or other organic constituents and in the process synthesize appreciable amounts of humic acid-like polymers. Phenolic units making up the polymer originated from the lignin, as well as through synthesis by the fungi (see next section).

To some extent, the initial degradation of lignin by fungi can be regarded as the reverse of lignin synthesis, as described earlier. Thus, the initial step involves release of the dilignol component (guaiacylglycerol-β-coniferyl ether, pinoresinol, and dehydrodiconiferyl alcohol) and the formation of primary phenylpropane (C_6—C_3) units. For example, the decomposition of guaiacyl-β-coniferyl ether leads to the formation of guaiacylglycerol, coniferyl alcohol, coniferaldehyde, ferulic acid, and other structurally related compounds as shown in Fig. 8.7.

Transformation by this means is believed to occur within or on exposed edges of intact molecules by extracellular enzymes produced by the fungi. The C_6—C_3 units then undergo oxidation in the side–chain position to yield a variety of low molecular weight aromatic acids and aldehydes, including vanillin and vanillic acid (see Fig. 8.7); other common degradation products are syringaldehyde (XII), syringic acid (XIII), *p*-hydroxybenzaldehyde (XIV), *p*-hydroxybenzoic acid (XV), protocatechuic acid (XVI), and gallic acid (XVII).

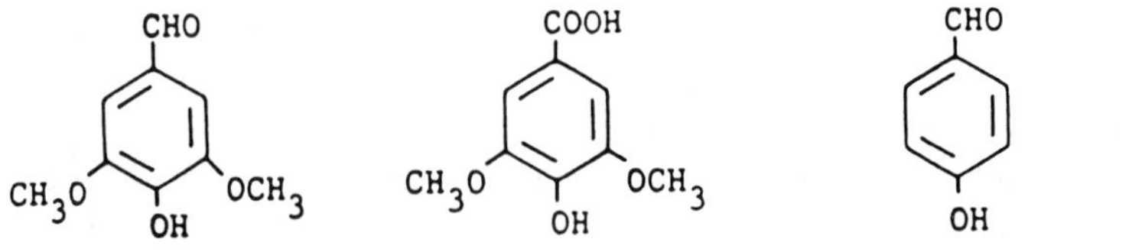

Syringaldehyde XII | Syringic acid XIII | p-Hydroxybenzaldehyde XIV

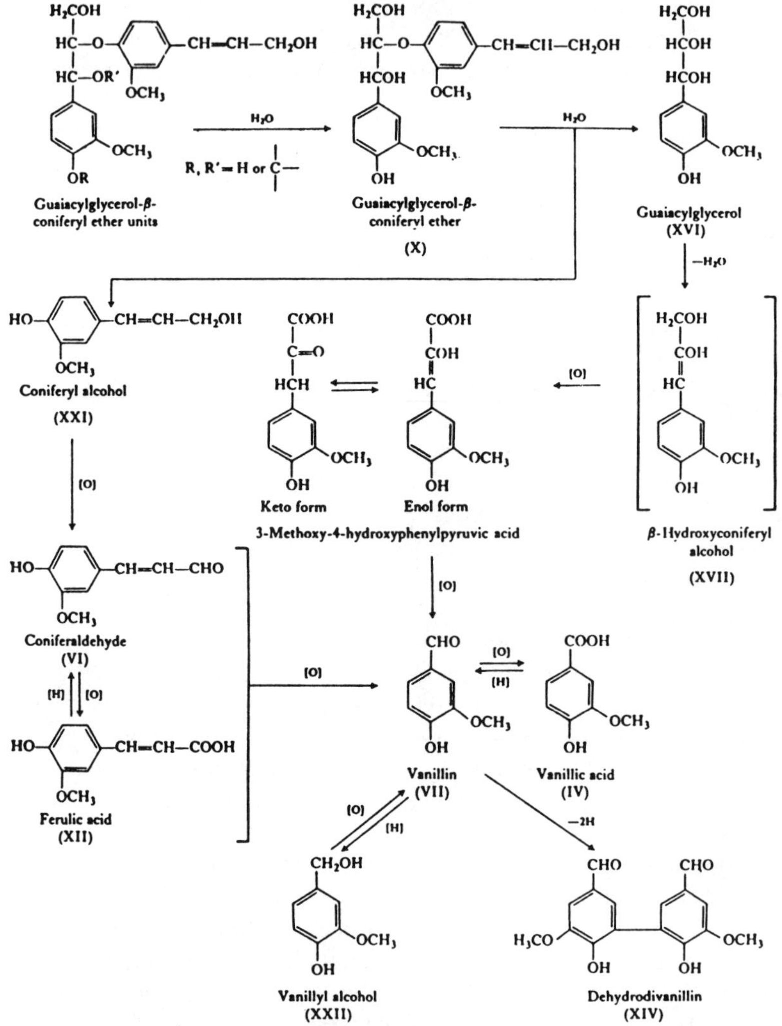

Fig. 8.7 Pathways of lignin decomposition by fungi. From Schubert,[13] reproduced by permission of Academic Press, Inc.

p-Hydroxybenzoic acid
XV

Protocatechuic acid
XVI

Gallic acid
XVII

Transformation of coniferaldehyde and p-hydroxycinnamaldehyde by three fungi of the *Imperfecti* group as recorded by Martin and Haider[3] are shown in Fig. 8.8. Additional OH groups are introduced on the aromatic rings; in addition, terminal chains are oxidized and decarboxylation may occur. Martin and Haider[3] pointed out that although numerous phenolic compounds are formed by these transformations, some of the fungi involved have a limited capacity for cleavage of aromatic rings.

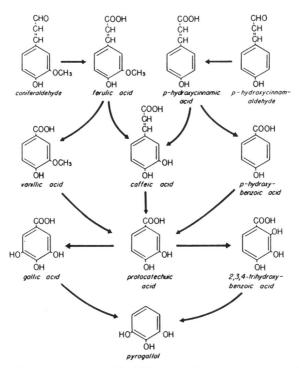

Fig. 8.8 Transformations of coniferaldehyde and p-hydroxycinnamaldehyde by *E. nigrum*, *S. atra*, and/or *A. sydowi*. From Martin and Haider,[3] reproduced by permission of Williams and Wilkins Co.

Synthesis by Microorganisms While there can be little doubt that lignins are important sources of polyphenols for humus formation, synthesis by microorganisms is also believed to be of importance. In some environments, lignin may make a major contribution; in other cases not at all.

Evidence for the formation of humic substances without the participation of lignin has been provided by the observation that humus can accumulate in soils and sediments without lignin being present. In the great plains of the United States, peat-like deposits of lichen origin have been described beneath translucent pebbles. In cold, wet regions humus is apparently formed from lower plants that do not contain lignin (e.g., the mosses). Buildup of organic matter has also been observed in gas-saturated soils, such as near natural gas wells or above gas main leaks. In this case, the organic matter has originated from the tissues of microorganisms that utilize hydrocarbon gases as sources of energy for cell synthesis. In the geological literature, peat deposits originating from the remains of algae (algae oozes) have been described. The likelihood also exists that part of the humic material in the B horizon of the Spodosol has been formed from polyphenols originating in the surface litter and washed down the profile, where polymerization and condensation occurred. Unsaturated hydrocarbons may be involved in the formation of humic substances in seawater.[6,7]

Numerous phenolic and hydroxy aromatic acids are synthesized from nonaromatic C sources by microorganisms. This ability is more a characteristic of the actinomycetes and fungi than of bacteria, although not exclusively so. Documentation of this work will not be attempted and the reader is referred to the reviews mentioned earlier. In brief, many species of streptomycetes and fungi known to be natural soil inhabitants have been demonstrated to synthesize dark-colored, humic-like substances in solution culture. These polymers have many of the properties of soil humic acids, including resistance to microbial decomposition, exchange capacity, total acidity, elemental analysis, molecular weight, and compounds released by various degradation procedures.

Several studies,[3,17-20] have been concerned with the synthesis of humic acid-like substances by fungi of the *Imperfecti* group. As noted earlier, these microscopic fungi degrade cellulose and other organic constituents besides lignin and in the process form dark-colored macromolecules (i.e., melanins) from phenols synthesized by the fungi.

Considerable variation exists in the kinds and amounts of structural units in fungal melanins. The synthesis and possible transformations of polyphenols by a typical soil fungi, *E. nigrum*, is illustrated in Fig. 8.9. Orsellinic and 2-methyl-3,5-dihydroxybenzoic acids are formed first, and, with time, other phenols are formed from the introduction of OH groups, though decarboxylation, and by oxidation of methyl groups. A noteworthy feature is the occurrence of resorcinol and resorcinol-type constituents (e.g., 3,5-dihydroxybenzoic acid), which are not found in lignin transformation products (see Fig. 8.7). For additional information of the synthesis of polyphenols by microscopic fungi, the review of Martin and Haider[3] can be consulted.

The quantities of humic acid synthesized by fungi can be appreciable. For

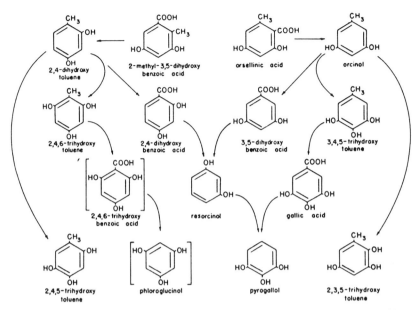

Fig. 8.9 Synthesis and transformation of phenols by *E. Nigrum*. From Haider and Martin[17]

example, Martin et al.[19] found that as much as one-third of the substances synthesized by *H. toruloidea*, including the biomass, consisted of humic acid. Furthermore, a humic acid-type polymer could be recovered from the mycelium tissue by extraction with 0.5N NaOH.

The production of humic substances by microorganisms may be partly an extracellular process. Following synthesis, the polyphenols are secreted into the external solution where they are enzymatically oxidized to quinones, which subsequently combine with other metabolites (e.g., amino acids and peptides) to form humic polymers. The study of Saiz-Jimenez et al.[20] suggests that anthraquinones, as well as polyphenols, serve as building blocks for humus synthesis.

Formation of Quinones

Polyphenols arising in soil by decomposition of lignin, or by microbial synthesis, are not stable but are subject to further decomposition by bacteria, actinomycetes, and fungi. Alternately, they may undergo recombination, either alone or with other organic molecules, after conversion of quinones. While phenols can be spontaneously oxidized to quinones (e.g., in alkaline media), the conversion is more likely carried out by polyphenoloxidase enzymes.

Numerous mono-, di-, and trihydroxy phenols and aromatic acids serve as building units for the synthesis of humic substances, and the number of derivatives formed is equally numerous. For the products derived from lignin, oxidation to quinones and consequent polymerization to humic substances must be preceded by demethylation to account for the low OCH_3 but high phenolic OH contents of humic and fulvic acids; other changes include oxidation of aldehyde components ($R-CHO \rightarrow R-COOH$), decarboxylation and hydroxylation, and coupling of intermediates. These transformations are illustrated for vanillin in Fig. 8.10. Oxidation products of other lignin-derived compounds (syringic and gallic acids) have been given by Flaig et al.[15]

Reactions of Quinones with N-Containing Compounds

Although self-condensation of quinones or free radicals occurs under conditions similar to those existing in soil, the reactions are greatly enhanced in the presence of amino compounds, such as amino acids. Enzymatic oxidation of phenolic compounds in the presence of amino acids, peptides, and proteins yields nitrogenous polymers having many of the properties of natural humic acids.

Reactions postulated to occur between quinones and amino acids are shown in Fig. 8.11. An interesting aspect of chemical investigations into the nature of synthetic humic acids formed from quinones and amino acids is that only a portion of the amino acid–N can be recovered by acid hydrolysis; part of the amino acid–N becomes incorporated into the polymer. As can be seen from

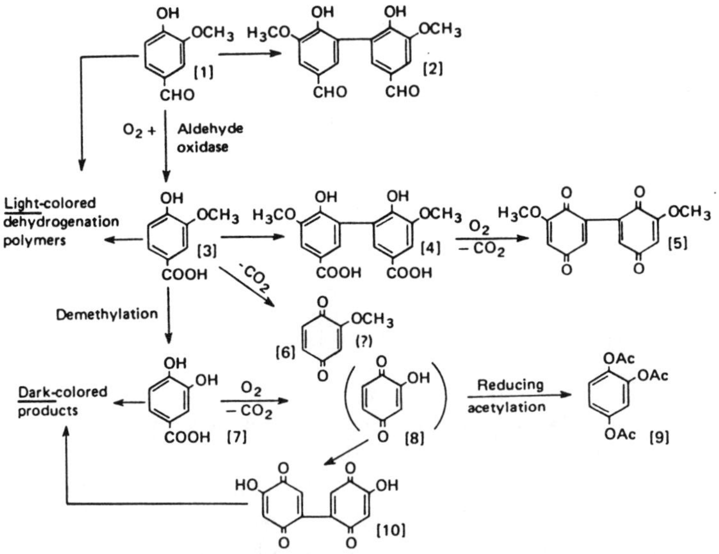

Fig. 8.10 Formation of quinones from lignin degradation products (vanillin and vanillic acid) in cultures of microorganisms. From Flaig et al.,[15] reproduced by permission of Springer-Verlag

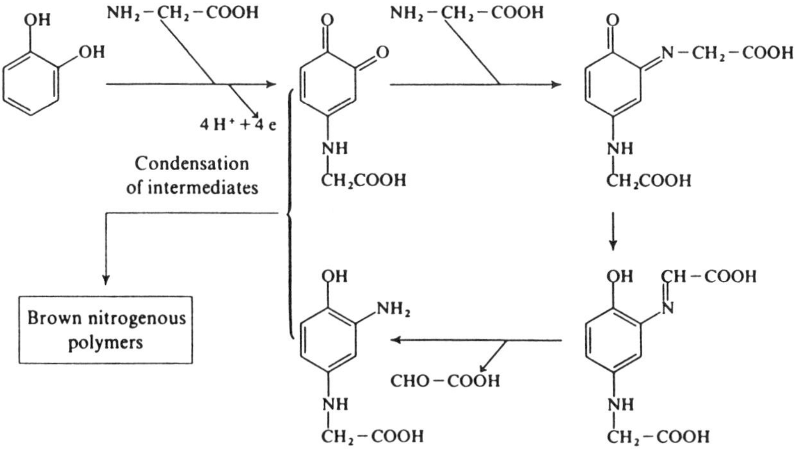

Fig. 8.11 Formation of humic substances from quinones and amino acids, as illustrated by the reaction between catechol and glycine

Fig. 8.11, reaction of the amino acid with quinone C=O leads to degradation of the amino acid and the formation of an aryl amine.

Amino sugars are ubiquitous to microorganisms and soil (Chapter 3), and they have the potential for reacting with phenols in much the same manner as an amino acids,[21,22]. A scheme for the stabilization of amino sugars through oxidative polymerization with phenols is shown in Fig. 8.12.

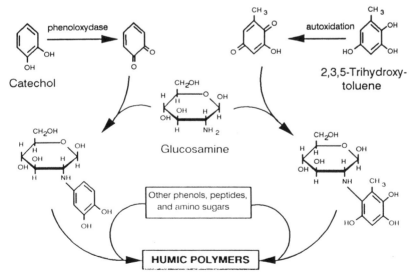

Fig. 8.12 Oxidative polymerization of phenyl derivatives involving amino sugars units. Modified from Martin et al.[22]

SUGAR–AMINE CONDENSATION

The production of brown nitrogenous polymers by condensation of reducing sugars and amines (e.g., amino acids) occurs extensively in the dehydration of food products at moderate temperatures and the reaction has been postulated to be of importance in the formation of humic substances in soil. The theory was initially proposed in 1911 by Maillard[23] and has been periodically endorsed since then, including more recent times.[24-27]

One attractive feature of the sugar–amine condensation theory is that the reactants (sugars, amino acids, etc.) are produced in abundance in soil through the activities of microorganisms. A second attractive feature, is that the theory affords an explanation for the formation of humic substances in environments where lignin and lignin-degradation products are not prevalent. An argument in opposition to the theory is that the reaction proceeds rather slowly at the temperatures found under normal soil conditions. However, drastic and frequent changes in the soil environment (freezing and thawing, wetting and drying), together with the intermixing of reactants with mineral material having catalytic properties, may facilitate condensation. Hedges[24] found that melanins have chemical properties characteristic of humic substances formed in marine environments.

The initial reaction in sugar–amine condensation involves addition of the amine to the aldehyde group of the sugar to form a Schiff base and the n-substituted glycosylamine.[16]

$$\underset{\text{ALDOSE IN ALDEHYDE FORM}}{\begin{array}{c} H-C=O \\ | \\ (CHOH)_N \\ | \\ CH_2OH \end{array}} \xrightarrow{RNH_2} \underset{\text{ADDITION COMPOUND}}{\begin{array}{c} R-N-H \\ | \\ H-C-OH \\ | \\ (CHOH)_N \\ | \\ CH_2OH \end{array}} \xrightarrow{H_2O} \underset{\text{SCHIFF BASE}}{\begin{array}{c} R-N \\ \| \\ CH \\ | \\ (CHOH)_N \\ | \\ CH_2OH \end{array}} \longrightarrow \underset{\text{N-SUBSTITUTED GLYCOSYLAMINE}}{\begin{array}{c} R-N-H \\ | \\ HC \\ | \\ (CHOH)_{N-1} \quad O \\ | \\ HC \\ | \\ CH_2OH \end{array}}$$

The glycosylamine subsequently undergoes the Amadori rearrangement to form the N-substituted-1-amino-1-deoxy-2-ketose.

$$\underset{\text{N-SUBSTITUTED GLYCOSYLAMINE}}{\begin{array}{c} R-N-H \\ | \\ HC \\ | \\ (HCOH)_N \quad O \\ | \\ HC \\ | \\ CH_2OH \end{array}} \xrightarrow{+H^+} \underset{\text{CATION OF SCHIFF BASE}}{\left[\begin{array}{c} R-N-H \\ \| \\ CH \\ | \\ HCOH \\ | \\ (HCOH)_N \\ | \\ CH_2OH \end{array}\right]^+} \xrightarrow{-H^+} \underset{\text{ENOL FORM}}{\begin{array}{c} R-N-H \\ | \\ CH \\ \| \\ COH \\ | \\ (HCOH)_N \\ | \\ CH_2OH \end{array}} \longrightarrow \underset{\substack{\text{N-SUBSTITUTED} \\ \text{1-AMINO-1-} \\ \text{DEOXY-2-KETOSE,} \\ \text{KETO FORM}}}{\begin{array}{c} R-N-H \\ | \\ CH_2 \\ | \\ C=O \\ | \\ (HCOH)_N \\ | \\ CH_2OH \end{array}}$$

The product of Amadori rearrangement is subject to: 1) fragmentation, with the liberation of the amine and formation of 3-C chain aldehydes and ketones, such as acetol, glyceraldehyde, and dihydroxyacetone, 2) loss of two water molecules to form reductones, and 3) loss of three water molecules to form hydroxymethyl furfural.[16] All of these compounds are highly reactive and readily polymerize in the presence of amino compounds to form brown-colored products of unknown composition. The overall reactions are illustrated in Fig. 8.13.

Although not shown in Fig. 8.13, aromatic and furanoid units are also known to be formed from carbohydrates.[27,28] A scheme for the formation of aromatic compounds is shown in Fig 8.14.

The extent to which browning reactions participate in the formation of humic substances is unknown; future research may show that the process is of particular importance in the synthesis of components generally found in the fulvic acid fraction of soil organic matter. Enders,[29] who did much work on the subject, concluded that humic acids are formed by reactions of amino acids and methyl glyoxal (CH_3COCHO). He postulated that when metabolic processes of microorganisms are disturbed by autolysis, the products of microbial decomposition of carbohydrates (methyl glyoxal) and proteins (amino acids)

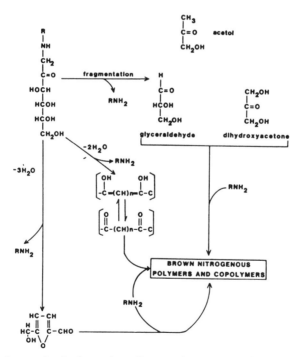

Fig. 8.13 Pathways for the formation of brown nitrogenous polymers from the product of Amadori rearrangement. Three pathways are illustrated: fragmentation to yield 3-carbon aldehydes and ketones, loss of two water molecules to form reductones, and loss of three water molecules to form hydroxymethyl furfural

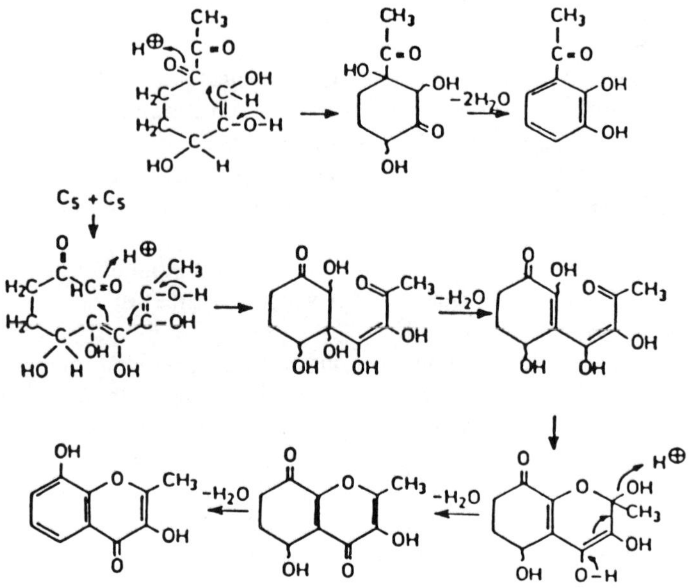

Fig. 8.14 Reactions leading to the formation of phenols and quinones from carbohydrates. From Popoff and Theander[28]

condense to form humic substances, in much the same way as melanoidins are formed during the production of beer. The disappearance of amino acids in deep-sea sediments has been attributed to reactions involving carbohydrates.[30]

SCHEME FOR THE FORMATION OF HUMIC SUBSTANCES

Humic substances in soil may be formed by all of the mechanisms mentioned in this chapter. As noted earlier, most workers favor a mechanism based on condensation of phenolic compounds or quinones. The number of precursor molecules is large (see Figs. 8.7–8.11) and the number of combinations in which they react is astronomical. The possibility that a given suite of compounds will combine in exactly the same way to form two identical molecules is so remote that it is probably safe to say that few, if any, humic molecules will be precisely the same.

In some respects, the synthetic process can be thought of in terms of card playing, where each structural unit represents a separate card in the deck and a given "hand" represents the combinations in which the structural units combine to form a humic molecule.

A humic acid molecule could conceivably be formed from among the structures shown in Figs. 8.7–8.10 but numerous other compounds may also be involved. A given molecule could include a "dimer" formed by the coupling

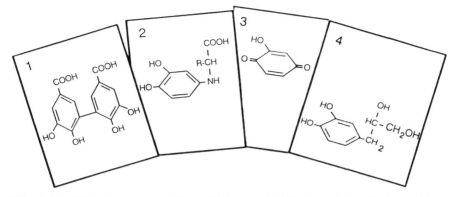

Fig. 8.15 Suite of structures from which a typical humic acid molecule could be formed

of two lignin-derived oxidation products (see Figs. 8.7 and 8.10), a "phenol-amino acid complex" (see Fig. 8.11), a hydroxyquinone, and a C_6-C_3 "structural unit of lignin." This sequence of four building blocks is illustrated as "cards" from a "deck" of structural units in Fig. 8.15; their combination leads to the "core" structure shown in Fig. 8.16. Incorporation of additional structural units and attachment of aliphatic constituents leads to the "type structure" for humic acid as shown in Fig. 12.4 of Chapter 12.

SUMMARY

Many different types of reactions can lead to the production of dark colored pigments (e.g., humic and fulvic acids) in soil. Although a multiple origin is

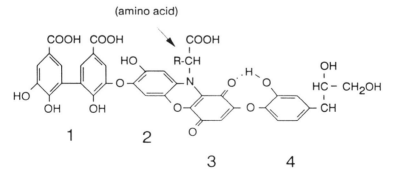

Fig. 8.16 Hypothetical structural unit for a soil humic acid as obtained by combination of the four building blocks shown in Fig. 8.15. In the natural state, the molecule would contain additional aromatic and aliphatic constituents (see Chapter 12)

suspect, the major pathway in most soils appears to be through condensation reactions involving polyphenols and quinones. According to present-day concepts, polyphenols derived from lignin, or synthesized by microorganisms, are enzymatically converted to quinones, which undergo self-condensation or combine with amino compounds to form N-containing polymers. The number of precursor molecules is large and the number of ways in which they combine is astronomical, thereby accounting for the heterogeneous nature of the humic material in any given soil.

REFERENCES

1. K. Haider, J. P. Martin, and Z. Filip, "Humus Biochemistry," in E. A. Paul and A. D. McLaren, Eds, *Soil Biochemistry, Vol. 4*, Marcel Dekker, New York, 1975, pp. 195-244.
2. M. M. Kononova, *Soil Organic Matter*, Pergamon, Oxford, 1966.
3. J. P. Martin and K. Haider, *Soil Sci.*, **111**, 54 (1971).
4. W. Flaig, "Generation of Model Chemical Precursors," in F. H. Frimmel and R. F. Christman, Eds., *Humic Substances and their Role in the Environment*, Wiley, New York, 1988, pp. 75-92.
5. P. G. Hatcher and E. C. Spiker, "Selective Degradation of Plant Biomolecules," in F. H. Frimmel and R. F. Christman, Eds., *Humic Substances and their Role in the Environment*, Wiley, New York, 1988, pp. 59-74.
6. J. I. Hedges, "Polymerization of Humic Substances in Natural Environments," in F. H. Frimmel and R. F. Christman, Eds., *Humic Substances and their Role in the Environment*, Wiley, New York, 1988, pp. 45-58.
7. R. Francois, *Reviews Aquatic Sciences*, **3**, 41 (1990).
8. S. A. Waksman, *Humus*, Williams and Wilkins, Baltimore, 1932.
9. J. R. Ertel and J. I. Hedges, *Geochim. Cosmochim. Acta*, **48**, 2065 (1984).
10. J. R. Ertel and J. I. Hedges, *Geochim. Cosmochim. Acta*, **49**, 2097 (1985).
11. H. A. Crawford, *Lignin Biodegradation and Transformation*, Wiley, New York, 1981.
12. I. A. Pearl, "Lignin Chemistry," in *Chemistry and Engineering News*, July 6, 1964, pp. 81-92.
13. W. J. Schubert, *Lignin Biochemistry*, Academic, New York, 1965.
14. W. Flaig, "The Chemistry of Humic Substances," in *The Use of Isotopes in Soil Organic Matter Studies*, Report of FAO/IAEA Technical Meeting, Pergamon, New York, 1966, pp. 103-127.
15. W. Flaig, H. Beutelspacher, and E. Rietz, "Chemical Composition and Physical Properties of Humic Substances," in J. E. Gieseking, Ed., *Soil Components: Vol. 1. Organic Components*, Springer-Verlag, New York, 1975, pp. 1-211.
16. J. E. Hodge, *J. Agr. Food Chem.*, **1**, 928 (1953).
17. K. Haider and J. P. Martin, *Soil Sci. Soc. Amer. Proc.*, **31**, 766 (1967).
18. K. Haider and J. P. Martin, *Soil Biol. Biochem.*, **2**, 145 (1970).
19. J. P. Martin, K. Haider, and D. Wolf, *Soil Sci. Soc. Amer. Proc.*, **36**, 311 (1972).

20 C. Saiz-Jimenez, K. Haider, and J. P. Martin, *Soil Sci. Soc. Amer. Proc.*, **39,** 649 (1975).
21 E. Bondietti, J. P. Martin, and K. Haider, *Soil Sci. Soc. Amer. Proc.*, **36,** 597 (1972).
22 J. P. Martin, K. Haider, and E. Bondietti, "Properties of Model Humic Acids Synthesized by Phenoloxidase and Autoxidation of Phenols and Other Compounds Formed by Soil Fungi," in *Proc. Intern. Meeting Humic Substances, Nieuwersluis (1972)*, Pudoc, Wageningen, 1975, pp. 171–186.
23 L. C. Maillard, *C. R. Acad. Sci.*, **156,** 148 (1913).
24 J. I. Hedges, *Geochim. Cosmochim. Acta*, **42,** 69 (1978).
25 R. Ikan, Y. Rubinsztain, P. Ioselis, Z. Aizenshtat, R. Pugmire, L. L. Anderson, and W. R. Woolfenden, *Org. Geochem.*, **9,** 199 (1986).
26 R. Ishiwatari, S. Morinaga, S. Yamamoto, T. Machihara, Y. Rubinsztain, P. Ioselis, Z. Aizenshtat, and R. Ikan, *Org. Geochem.*, **9,** 11 (1986).
27 Y. Rubinsztain, P. Ioselis, R. Ikan, and Z. Aizenshtat, *Org. Geochem.*, **6,** 791 (1984).
28 T. Popoff and O. Theander, *Acta Chem. Scand.*, **B30,** 705 (1976).
29 C. Enders, *Biochem. Z.*, **318,** 44 (1947).
30 F. J. Stevenson, "Nonbiological Transformations of Amino Acids in Soils and Sediments," in B. Tissot and F. Bienner, Eds., *Advances in Organic Geochemistry*, Editions Technip, Paris, 1975, pp. 701–714.

9

REACTIVE FUNCTIONAL GROUPS

The chemistry of the complex organic colloids in soil is undoubtedly the least understood field of soil science, and the most perplexing. Many of the more important functions of humus, including its effect on soil structure, chelation of heavy metals, and adsorption of pesticides and other toxic pollutants, will remain obscure until more is known about the structural chemistry of humic and fulvic acids.

A variety of functional groups, including COOH, phenolic OH, enolic OH, quinone, hydroxyquinone, lactone, ether, and alcoholic OH, have been reported in humic substances. For reasons outlined in this chapter, considerable disagreement exists as to the exact amounts present; in some cases, proof of existence is lacking. Methods for the determination of functional groups and their distribution in humic and fulvic acids are provided in several reviews.[1-5]

Some of the more important structural groups of organic molecules are shown in Table 9.1.

ELEMENTAL CONTENT

As shown below (Table 9.2) the major elements in humic and fulvic acids are carbon (C) and oxygen. The C content of humic acids generally ranges from 53.8 to 58.7 percent; oxygen content varies from 32.8 to 38.3 percent. Fulvic acids have lower C (usual range of 40.7–50.6 percent) but higher oxygen (39.7–49.8 percent) contents. Percentages of hydrogen, nitrogen (N), and sulfur (S) vary from 3.2 to 7.0, 0.8 to 4.3, and 0.1 to 3.6 percent, respectively. Disregarding S, the average chemical formula for humic acid would be $C_{10}H_{12}O_5N$; for fulvic acid, the formula would be $C_{12}H_{12}O_9N$.[6]

Table 9.1 Some Important Structural Groups of Organic Molecules

Amino	$-NH_2$	Anhydride	$R-\overset{\overset{O}{\|}}{C}-O-\overset{\overset{O}{\|}}{C}-R'$
Amine	$R-\overset{\overset{H}{\|}}{\underset{\underset{H}{\|}}{C}}-NH_2$	Imine	$R-\overset{\overset{H}{\|}}{C}=NH$, $R-CHNH$
Amide	$R-\overset{\overset{O}{\|\|}}{C}-NH_2$	Imino	$=NH$
Alcohol	$R-CH_2OH$	Ether	$R-CH_2-O-CH_2-R'$
Aldehyde	$R-\overset{\overset{H}{\|}}{C}=O$, $R-CHO$	Ester	$R-\overset{\overset{O}{\|\|}}{C}-O-R'$, $R-COOR'$
Carboxyl	$R-\overset{\overset{O}{\|\|}}{C}-OH$, $R-COOH$	Quinone	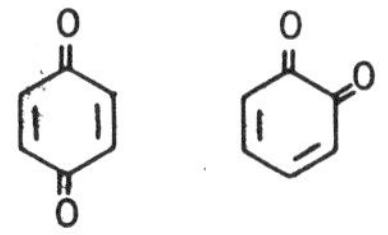 (quinone structures)
Carboxylate ion	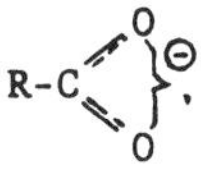 $R-C\overset{O^{\ominus}}{\underset{O}{\diagdown\!\!\diagup}}$, $R-COO^-$		
Enol	$R-CH=CH-OH$	Hydroxyquinone	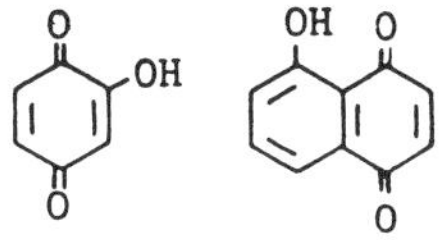 (hydroxyquinone structures)
Ketone	$R-\overset{\overset{O}{\|\|}}{C}-R'$, $R-CO-R'$		
Keto acid	$R-\overset{\overset{O}{\|\|}}{C}-COOH$		
		Peptide	(peptide structure)
Unsaturated carbonyl	$-\overset{\overset{H}{\|}}{C}=\overset{\overset{H}{\|}}{C}-\overset{\overset{H}{\|}}{C}=O$		

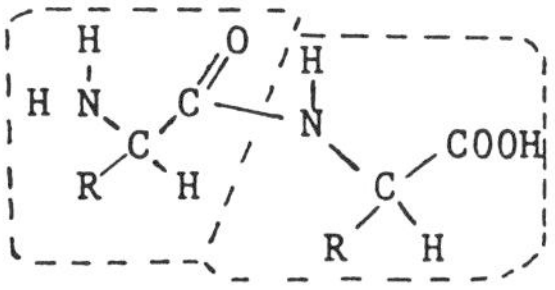

As will be noted later, fulvic acids have somewhat higher contents of COOH groups than humic acids, and a higher percentage of the oxygen occurs in this form. Accordingly, on a COOH-free basis, the C contents of humic and fulvic acids should be about the same, and this has been found to be the case. Values calculated by Varadachari and Ghosh[5] for 10 humic acids gave C values of

Table 9.2 Usual Range for the Elemental Composition of Humic Substances[a]

Element	Humic Acids	Fulvic Acids
Carbon	53.8–58.7%	40.7–50.6%
Oxygen	32.8–38.3	39.7–49.8
Hydrogen	3.2–6.2	3.8–7.0
Nitrogen	0.8–4.3	0.9–3.3
Sulfur	0.1–1.5	0.1–3.6

[a] From Steelink.[6]

from 59.0 to 64.6 percent; for six fulvic acids, the range was from 57.6 to 63.5 percent.

One way to express information about the elemental composition of humic substances is to use atomic ratios, as applied to humic and fulvic acids by Visser.[7] Some generalizations regarding O/C and H/C ratios are as follows.[6]

1. Ratios of O/C for soil humic acids cluster around 0.5; for fulvic acids, the ratios center near 0.7. The difference may reflect higher amounts of COOH and/or carbohydrates in the fulvic acids.
2. Ratios of H/C for soil humic and fulvic acids are clustered around 1.0. Lake and marine sedimentary humic substances have somewhat higher ratios. For terrestrial humic acids, the H/C ratio is directly correlated to the E_4/E_6 ratio (see Chapter 13). The latter is inversely proportional to "aromaticity" or degree of condensation; E_4/E_6 ratios above 1.3, corresponding to low H/C ratios, indicate that much of the material is aliphatic in nature.

The elemental content of humic substances would be expected to be affected by such factors as pH, parent material, vegetation, and age of the soil. For lake sediments, the C content of humic acids has been found to increase with increasing depth (see Steelink[6]).

METHODS OF FUNCTIONAL GROUP ANALYSIS

Emphasis is given in this chapter to the determination of the common oxygen-containing functional groups by wet chemical methods, several of which involve the formation of derivatives through methylation or acetylation. Techniques in which isotopically enriched reagents have been used for the formation of derivatives, followed by NMR spectroscopic analysis, have been described by Leenheer and Noyes[8]; this work is described in Chapter 11.

Most methods for determining reactive groups in humic materials have been based on the acidic properties of the group involved. Because of the complex nature of humic substances, the acidities of the various groups may overlap. For this reason, results obtained by methods dependent on ion-exchange or pK values must be interpreted with caution. Polycarboxylic acids, for example, exhibit a whole series of dissociation constants that decrease as successive protons dissociate. On the other hand, substituted phenols are often more strongly dissociated than the unsubstituted compound. Some groups may be unreactive because of H-bonding or through stearic hinderance.

Other problems in the quantitative determination of functional groups include: 1) insolubility of the material (particularly humic acids) in water and most organic solvents, 2) oxidation and reduction, 3) interactions with reagents used for forming derivatives, and 4) the nonstoichiometric nature of the reactions.

Total Acidity (COOH Plus Phenolic- and/or Enolic-OH)

Barium Hydroxide (Baryta Absorption) Method The barium hydroxide [$Ba(OH)_2$] approach is an indirect potentiometric method in that it is based on release of H^+ from acidic groups that ionization up to the pH of the reaction solution (i.e., $0.1M$ $Ba(OH)_2$) with a pH > 13). Due to its simplicity, this has been the most popular method for determining total acidity of humic substances.

Briefly, the sample is allowed to react with excess $Ba(OH)_2$, following which the unused reagent is titrated with standard acid. The overall reaction is

$$2RH + Ba(OH)_2 \rightarrow R_2Ba + 2H_2O$$

where R is the macromolecule and H is the proton of a COOH or acidic OH group.

The method used by Schnitzer[2] is as follows:

To between 50 and 100 mg of humic preparation in a 125-ml ground-glass stoppered Erlenmeyer flask, add 20 ml of $0.2N$ $Ba(OH)_2$ solution. Simultaneously, set up a blank consisting of 20 ml of $0.1M$ $Ba(OH)_2$ only. Displace the air in each flask by N_2, stopper flask carefully and shake the system for 24 h at room temperature. Following this, filter the suspension, wash the residue thoroughly with CO_2-free distilled water and titrate the filtrate plus washing potentiometrically (glass–calomel electrodes) with standard $0.5M$ HCl solution to pH 8.4.

Total acidity (in cmole/kg) is calculated by the equation:

$$\text{Total acidity} = \frac{(V_b - V_s) \times N \times 10^6}{\text{milligrams of sample}}$$

where V_b and V_s represent the volumes of standard acid used for the blank and sample, respectively, and N is the normality of the acid.

The main advantage of the method is its simplicity: unfortunately, arbitrary values are probably secured. The assumption is made that all of the humic material is precipitated and removed as the insoluble Ba–salt during filtration, but this has not always been the case.[9] A second difficulty is that humic acids, as normally prepared by precipitation under acidic conditions, may contain residual mineral acid (usually HCl) not removed by washing with distilled water (see Chapter 15). Reactions carried out in the presence of O_2 lead to high results; low results are obtained for OH groups in certain substituted phenols.

Methylation Procedures A second approach involves an estimate of OCH_3 formed in samples after methylation with diazomethane (CH_2N_2).

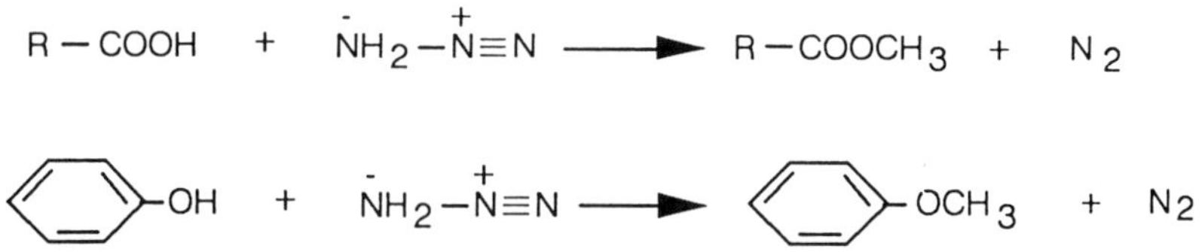

$$R-COOH + NH_2-\overset{+}{N}\equiv N \longrightarrow R-COOCH_3 + N_2$$

$$\text{C}_6\text{H}_5-OH + NH_2-\overset{+}{N}\equiv N \longrightarrow \text{C}_6\text{H}_5-OCH_3 + N_2$$

Diazomethane reacts with acidic-H of a wide variety of structures, including COOH, phenolic OH, enolic compounds, N—H acidic groups and certain aliphatic alcohols. On the other hand, H-bonded phenolic OH groups may not react. Side reactions include the formation of polymethylene under the catalytic action of heavy metals, as observed by Farmer and Morrison.[10]

Following methylation, the samples are analyzed for OCH_3 by the well-known Zeisel method.[11] In this technique, the CH_3 group is split off with HI in phenol as solvent and the methyl iodide (CH_3I) thus formed is removed in a stream of CO_2-free air and collected in a bromine–water solution, where oxidation to periodic acid (HIO_3) occurs. The HIO_3 is then allowed to react with potassium iodide (KI) in acidic solution to form I_2, which is titrated with standard sodium thiosulfate ($Na_2S_2O_3$) using starch as an indicator. The pertinent reactions are

$$R-OCH_3 + HI \longrightarrow ROH + CH_3I$$

$$CH_3I + 6Br_2 + 5H_2O \longrightarrow HIO_3 + 12HBr + CO_2$$

$$2HIO_3 + 10KI + 5H_2SO_4 \longrightarrow 6I_2 + 6H_2O + 5K_2SO_4$$

$$I_2 + 2Na_2S_2O_3 \xrightarrow[\text{indicator}]{\text{starch}} 2NaI + Na_2S_4O_6$$

Methylation leads to an increase in weight and it is necessary to convert percentage values for the methylated product to percentage of acidic OH in the original sample. This conversion is made as shown below.[12]

$$\text{OH percentage that can be methylated} = \frac{1700(Y_2 - Y_1)}{3100 - 14Y_2}$$

where Y_1 is the percentage of OCH_3 in the original sample and Y_2 is percentage OCH_3 in the methylated product.

One difficulty with this method is that of obtaining complete methylation. A three-step methylation procedure has been proposed by Wershaw and Pinckney.[13] In the first step, COOH and some of the OH groups are methylated by CH_2N_2 in dimethylformamide; in the second step, additional OH groups are methylated with methyl iodide (CH_3I) and dimethyl sulfinyl carbanion in dimethylformamide; in the third step, any acidic groups hydrolyzed in earlier steps are methylated with CH_2N_2 in methylene chloride.

Reaction with Diborane Dubach et al.[14] and Martin et al.[15] used still another approach to determine active H in humic substances. They employed diborane, B_2H_6, which is believed to react with stearically hindered active H. The method, which involves an estimate of the H_2 produced, is believed to be independent of pK values. For a soil fulvic acid, good agreement was obtained between total acidity determined by this method and the $Ba(OH)_2$ procedure.

Reactions with Lithium Aluminum Hydride This reagent ($LiAlH_4$) has been used for determining active H in coals and may have application for analysis of humic substances. The reaction is

$$4ROH + LiAlH_4 \rightarrow LiAl(OR)_4 + 4H_2$$

Carboxyl Groups

The Ca-Acetate Method Like the $Ba(OH_2)$ procedure for total acidity, the Ca-acetate method is based on ion exchange and can be regarded as an indirect potentiometric titration approach. Extensive use has been made of the method for determining COOH groups. The reaction is

$$2R-COOH + Ca(CH_3COO)_2 \rightarrow (R-COO)_2Ca + 2CH_3COOH$$

The acetic acid (CH_3COOH) liberated during the reaction is titrated with a standardized NaOH solution. The procedure is as follows[2]:

To between 50 and 100 mg of humic preparation in a 125-ml ground-glass stoppered Erlenmeyer flask, add 10 ml of a $1N$ $Ca(CH_3COO)_2$ solution and 40 ml of CO_2-free distilled water. Set up a blank simultaneously, consisting of 10 ml of a $0.5M$ $Ca(CH_3COO)_2$ solution and 40 ml of CO_2-free distilled water only. After shaking for 24 h at room temperature, filter the suspension, wash the residue with CO_2-free distilled water, combine the filtrate and the washing and titrate potentiometrically (glass–calomel electrodes) with standard $0.1M$ NaOH solution of pH 9.8.

The content of COOH groups (in cmole/kg) is obtained by the equation:

$$COOH = \frac{(V_s - V_b) \times N \times 10^6}{\text{milligrams of sample}}$$

where V_s and V_b represent the volumes of standard base used for the sample and blank, respectively, and N is the normality of the base.

Because the procedure involves ion exchange, and since humic substances may contain acidic OH groups that ionize below the pH of the $Ca(CH_3COO)_2$ reaction mixture, strictly quantitative values may not be achieved. Dubach et al.,[14] and Martin et al.,[15] for a series of humic preparations, found that the discrepancy between values obtained for COOH groups by chemisorption and those calculated from active H (diborane) and total OH measurements became increasingly serious as the OH contents of the preparations increased.

Still another difficulty is that of incomplete removal of humic matter during filtration. A modified procedure involving recovery of CH_3COOH by steam distillation has been given by Holtzclaw and Sposito.[16] Other work has shown that the CH_3COOH can be selectively separated from the reaction mixture by ultrafiltration.[17]

Finally, it should be noted that binding of Ca^{2+} to humic and fulvic acids may lead to release of protons from sites other than a COOH group. Perdue et al.[17] found that the binding of Ca^{2+} by humic substances resulted in the displacement of protons that are not released when sodium acetate is used as the exchange base.

Methylation Procedures Another method of estimating COOH groups is by methylation and subsequent saponification of the resulting methyl esters. Several methylation procedures have been used, including CH_2N_2 and CH_3OH in dry HCl. Farmer and Morrison[10] found that the latter procedure gave incomplete esterification.

Several saponification procedures have been used for the analysis of methyl esters, including distillation of the liberated CH_3OH and determination of the change in OCH_3 content accompanying saponification (hydrolysis of esters). A disadvantage of the latter procedure is that quantitative recovery of material following saponification is difficult. In the former case, the OCH_3 is split off as CH_3OH by heating with alkali, following which the CH_3OH is recovered by distillation and oxidized to formaldehyde with $KMnO_4$. The formaldehyde is then determined by a sensitive colorimetric procedure.

The Iodometric Method Wright and Schnitzer[18] employed the iodometric method to determine COOH groups in some humic preparations. This technique, which is based on ion exchange, gave higher values than the Ca–acetate method.

Quinoline Decarboxylation Aromatic acids are decarboxylated when heated with quinoline in the presence of a suitable catalyst.

$$R-\text{C}_6\text{H}_4-COOH \xrightarrow{\text{Quinoline}} R-\text{C}_6\text{H}_5 + CO_2$$

Application of this method to soil humic substances has given results comparable to the Ca–acetate method,[18] indicating that most of the COOH groups are attached to aromatic rings. However, CO_2 can also be released from α-hydroxy aliphatic acids.

Indirect Methods Indirect methods for estimating COOH groups involve subtraction of acidic OH groups (to be discussed) from total acidity values. Obviously, results secured by this practice include accumulated errors of both determinations.

Total OH

The two methods used most frequently to determine total OH content of humic substances are the following: 1) methylation with dimethyl sulfate, $(CH_3)_2SO_4$, and 2) acetylation with acetic anhydride.

Methylation with Dimethyl Sulfate In this method, the sample is treated repeatedly with $(CH_3)_2SO_4$ in an alkaline solution, after which the resulting precipitate is analyzed for OCH_3 by the Zeisel method. Only phenolic- and alcoholic OH groups, but not COOH, are assumed to be methylated. The reaction is referred to as a synchronous substitution and proceeds as follows:

$$R-OH + NaOH \rightleftharpoons RO^- + Na^+ + H_2O$$

$$RO^- + CH_3-O-SO_2-O-CH_3 \rightarrow R-OCH_3 + CH_3-O-SO_2-O^-$$

Dimethylsulfate is capable of reacting with phenolic OH groups that are too weakly acidic to react with CH_2N_2, and this has been used as the basis for determining H-bonded OH groups in humic acids. Unfortunately, results obtained by $(CH_3)_2SO_4$ are difficult to interpret because of possible side reactions in the strongly alkaline solution. A less drastic procedure has been used in which the humic material is refluxed with $(CH_3)_2SO_4$ over anhydrous K_2CO_3 in acetone.[19] A further modification was made by Leenheer and Noyes,[8] who used methanol as the solvent.

Acetylation A second method (acetylation with acetic anhydride to form acetate esters) has been used more extensively. The procedure is as follows:[2]

Reflux 50 to 100 mg samples with a mixture of equal parts of pyridine and acetic anhydride (5 ml) for 2 to 3 h under nitrogen. Cool the mixture and pour into water. Collect the solid by filtration, wash thoroughly with water and dry under vacuum over P_2O_5. The acetylated sample (50 mg) is then refluxed with 25 ml of 3N aqueous NaOH solution for 2 h under nitrogen. Add 25 ml 6N H_2SO_4 and 25 ml distilled water and distill the mixture through a splash head, and titrate the distillate with 0.1N NaOH using phenolphthalein as indicator. Maintain the volume of distillation mixture by adding distilled water in 25 ml portions. Determine the reagent blank and continue the distillation until aliquots of the sample and reagent blank distillations titrate equally.

Acetyl content (in cmole/kg) is calculated as follows:

$$\text{Acetyl content} = \frac{(V_s - V_b) \times N \times 10^6}{\text{milligrams of sample}}$$

where V_s and V_b represents the volumes of standard base required for the sample and blank, respectively, and N is the normality of the base.

Total OH (in cmole/kg) is given by

$$\text{OH content} = \frac{\text{acetyl content}}{1 - 0.042 \times \text{acetyl content}}$$

Leenheer and Noyes[8] found that when the above procedure was applied to an aquatic fulvic acid, the sample darkened and partially precipitated. They proposed a procedure in which COOH groups were methylated prior to acetylation. Acetylation was carried out under the conditions shown below.

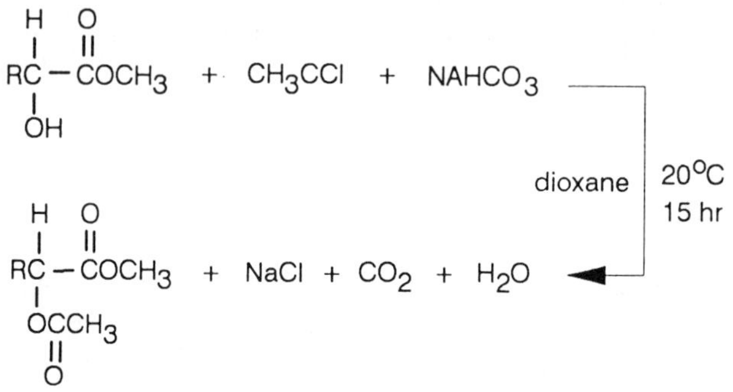

$$\underset{\underset{OH}{|}}{\overset{\overset{H}{|}}{RC}} - \overset{\overset{O}{\|}}{C}OCH_3 + CH_3CCl + NAHCO_3 \xrightarrow[\text{dioxane}]{20°C, 15\text{ hr}} \underset{\underset{OCCH_3 \; (\|O)}{|}}{\overset{\overset{H}{|}}{RC}} - \overset{\overset{O}{\|}}{C}OCH_3 + NaCl + CO_2 + H_2O$$

The findings of Wagner and Stevenson[20] and others indicate that many early investigators probably used too short a reaction period. For reasons which are not clear, the acetyl content of humic acids methylated with CH_2N_2 and saponified prior to acetylation is higher than that of unmethylated samples.[20]

Since COOH groups are also acetylated with acetic anhydride (formation of mixed anhydrides), a correction has to be made for COOH as determined by an independent method. The possibility that cyclic anhydrides can be formed during acetylation[20,21] does not seem to have been taken into account in evaluating the method.

The overall reactions of acetic anhydride with OH groups (formation of esters) and COOH groups (formation of mixed and cyclic anhydrides) are as follows:

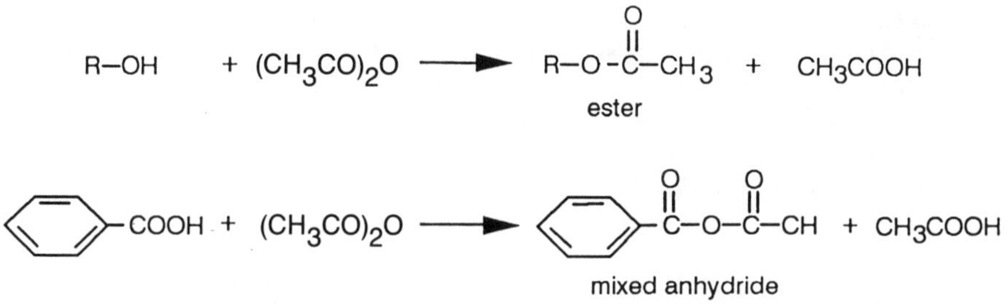

$$R-OH + (CH_3CO)_2O \longrightarrow \underset{\text{ester}}{R-O-\overset{\overset{O}{\|}}{C}-CH_3} + CH_3COOH$$

$$\text{Ph-COOH} + (CH_3CO)_2O \longrightarrow \underset{\text{mixed anhydride}}{\text{Ph}-\overset{\overset{O}{\|}}{C}-O-\overset{\overset{O}{\|}}{C}-CH} + CH_3COOH$$

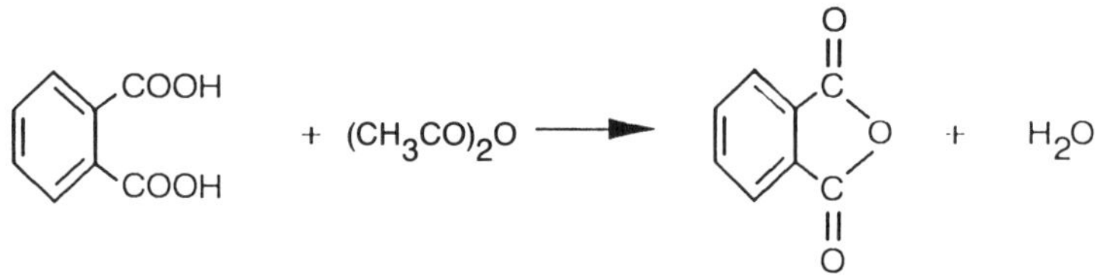

[Scheme: benzene-1,2-dicarboxylic acid + $(CH_3CO)_2O \rightarrow$ cyclic anhydride + H_2O]

cyclic anhydride

Primary and secondary amines, as well as sulfhydryl groups, also react with acetic anhydride, but interference from these compounds may not be serious. Acetylations conducted under reducing conditions (detection of hindered phenols) have given inconclusive results.[22]

A quantitative estimate of OH content by acetylation is obtained either by hydrolysis of the excess reagent and titration of the resulting acetic acid, or by hydrolysis of the derivative with alkali. Martin et al.[15] concluded that determination of the unreacted acetic anhydride gave unreliable results.

Phenolic OH

By Difference (Total Acidity—COOH)   The quantity of phenolic OH (or more correctly, acidic OH) is calculated as the difference between total acidity (in cmole/kg) and COOH content (in cmole/kg).

$$\text{Phenolic OH} = \text{total acidity} - \text{COOH}$$

Only those phenolic OH groups sufficiently acidic to react with $Ba(OH)_2$ or CH_2N_2 (see section on total acidity) will be measured by this approach. In the case of methylation with CH_2N_2, hydroxyquinones of the general types **I, II,** and **III** may not be completely methylated, because of the high stabilities of the ring structures formed through H-bonding.

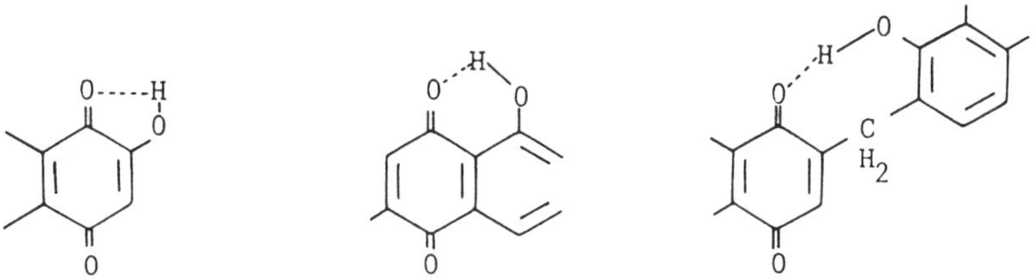

[Structures I, II, III showing hydroxyquinones with intramolecular H-bonding]

I   II   III

Until the question of hydroxyquinones in humic substances is resolved, values obtained by analysis of methylated derivatives, or those calculated from total acidity and COOH estimations, should not be regarded as accurate values for phenolic OH content. The term "acidic OH" may be more appropriate.

The Ubaldini Method In this method, K-salts of both COOH and phenolic OH groups are formed by heating the material with alcoholic KOH. Carbon dioxide is then bubbled through the solution and the K^+ released as K_2CO_3, presumably originating from phenolic OH groups, is estimated by titration. As shown below, no K_2CO_3 is formed from K^+-salts of COOH groups.

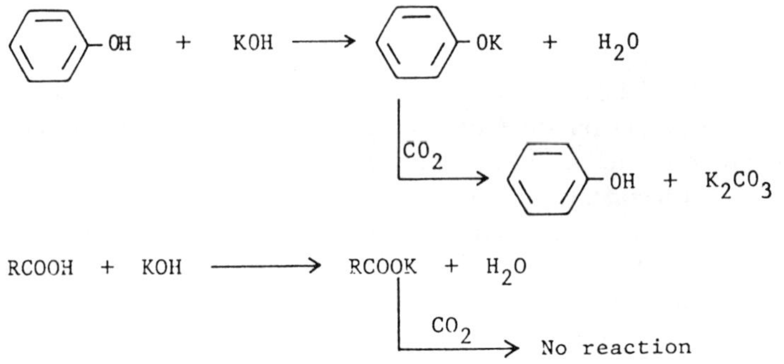

$$\text{Ph-OH} + \text{KOH} \longrightarrow \text{Ph-OK} + H_2O$$
$$\xrightarrow{CO_2} \text{Ph-OH} + K_2CO_3$$

$$\text{RCOOH} + \text{KOH} \longrightarrow \text{RCOOK} + H_2O$$
$$\xrightarrow{CO_2} \text{No reaction}$$

Serious objections include the nonspecificity of the reaction and the danger of hydrolysis during the treatment with hot alcoholic KOH. No information is available as to how results obtained for humic and fulvic acids compare with other methods.

Alcoholic OH

Estimates for alcoholic OH groups are obtained from the difference between total OH (in cmole/kg) and phenolic OH (in cmole/kg).

$$\text{Alcoholic OH} = \text{total OH} - \text{phenolic OH}$$

For reasons discussed above, estimates for alcoholic OH groups must be accepted with reservation. Schnitzer and Gupta[23] reported significant amounts of "alcoholic OH" in humic and fulvic acids by subtracting "phenolic OH," also obtained by difference (total acidity minus COOH by the Ca–acetate method), from total OH as determined by acetylation. On the other hand, Dubach et al.[14] and Martin et al.[15] failed to detect alcoholic OH groups from the difference between active-H (diborane) and total acidity by baryta absorption.

Carbonyl (C=O)

Formation of Oximes, Hydrazones, and Carbazones Most methods that have been used for determining C=O groups in humic substances are based on the formation of derivatives by reaction with such reagents as hydroxylamine, phenylhydrazine and 2,4-dinitrophenylhydrazine. Typical reactions are as follows:

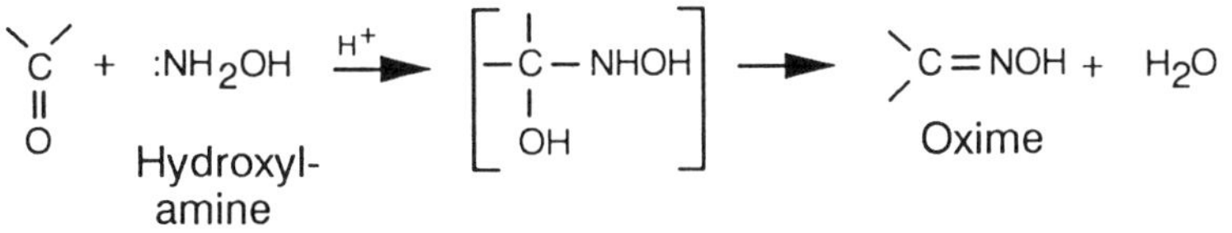

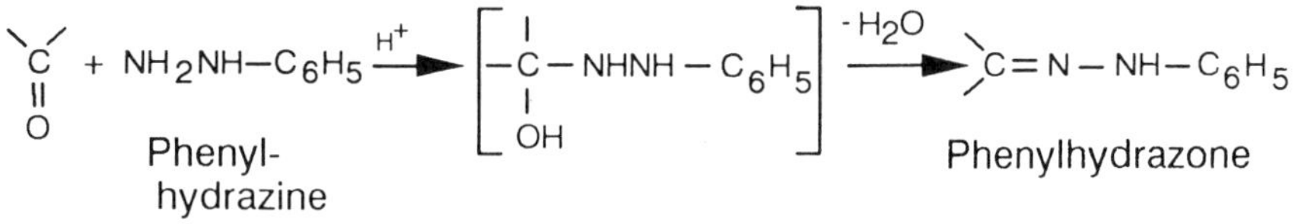

$$\underset{O}{\overset{\diagdown\diagup}{\underset{\|}{C}}} + :NH_2OH \xrightarrow{H^+} \left[\underset{OH}{\overset{|}{\underset{|}{-C-NHOH}}} \right] \longrightarrow \underset{Oxime}{\overset{\diagdown}{\diagup}C=NOH} + H_2O$$

Hydroxyl-
amine

$$\underset{O}{\overset{\diagdown\diagup}{\underset{\|}{C}}} + NH_2NH-C_6H_5 \xrightarrow{H^+} \left[\underset{OH}{\overset{|}{\underset{|}{-C-NHNH-C_6H_5}}} \right] \xrightarrow{-H_2O} \underset{Phenylhydrazone}{\overset{\diagdown}{\diagup}C=N-NH-C_6H_5}$$

Phenyl-
hydrazine

The reactions can be followed in several ways, including analysis of the unused reagent and measurement of the increase in N content of the derivative.[22] In the case of hydroxylamine, the unused reagent can be analyzed by back titration with perchloric acid,[18] or polarographically.[24]

The procedure recommended by Schnitzer[2] is as follows:

To 50 mg of humic material in a 50-ml ground-glass stoppered Erlenmeyer flask add 5 ml of 0.25M 2-dimethylaminoethanol solution plus 6.3 ml 0.4M hydroxylammonium chloride solution. Heat the system on steam bath for 15 min; cool and back-titrate potentiometrically (glass–calomel) the excess of hydroxylammonium chloride with standard perchloric acid solution. Determine the end point by plotting mV vs ml of acid. Set up a blank simultaneously consisting of 5 ml of 0.25M 2-dimethylaminoethanol and 6.3 ml 0.4M hydroxylammonium chloride solution only.

Carbonyl (C=O) content (in cmole/kg) is given by

$$C{=}O = \frac{(V_b - V_s) \times N \times 10^6}{\text{milligrams of sample}}$$

where V_b and V_s represent the volumes of standard acid for the blank and sample, respectively, and N is the normality of the acid.

Procedures based on the formation of derivatives can be criticized on the grounds that the reagents employed may react without the participation of C=O groups. Humic acids, for example, are known to combine with NH_3 to form compounds that are nonhydrolyzable. Also, hydroxyquinones of the types shown by structures **I, II,** and **III** will not always react with these reagents. Evidence for the presence of aromatic ketone groups in aquatic humic substances has been given by Leenheer et al.[25]

Reduction with Sodium Borohydride (NaBH₄) Another method for estimating >C=O is based on reduction to $-CH_2OH$ with $NaBH_4$.

$$4R_2C{=}O + NaBH_4 + 2NaOH + H_2O \rightarrow Na_3BO_3 + 4R_2CHOH$$

224 REACTIVE FUNCTIONAL GROUPS

Reduction is carried out in an alkaline solution and the unused $NaBH_4$ is estimated manometrically after decomposition with HCl.

$$NaBH_4 + HCl + 3H_2O = H_3BO_3 + NaCl + 4H_2\uparrow$$

The method is reported to be highly specific for C=O groups. Martin et al.[15] obtained values for some humic and fulvic acids that compared favorably with those reported using the methods mentioned above.

Quinones

The approaches described in this section represent attempts to estimate the C=O group of quinones. Methods outlined in the previous section are non-specific in that all reactive C=O groups are measured, including those of ketones (R_2C=O) and quinones.

Estimates for quinone groups in humic acids have been based on selective reduction by 1) $SnCl_2$ in acid solution, 2) $SnCl_2$ in alkaline solution, and 3) Fe^{2+} in alkaline triethanolamine solution.

Reduction with $SnCl_2$ The first attempt to determine quinone groups was by reduction with $SnCl_2$ in acid solution.[26] The humic material is heated for 4 h. in a sealed tube with an acid solution of $SnCl_2$. The residue is filtered and Sn^{2+} in the filtrate is titrated with standard I_2, using starch as an indicator.

The above method has been criticized on the grounds that the experimental conditions are drastic, that adsorption or complexing of the reducing agent (Sn^{2+}) occurs, and that air oxidation may occur during filtration and titration. The net result is excessively high values for quinones.[27-29]

In attempts to overcome the above difficulties, Vasilyevskaya et al.[29] recommended reduction under alkaline rather than acid conditions. The humic material, dissolved in 0.1M NaOH under N_2, is treated with $SnCl_2$ for 1 h. in the absence of O_2, after which excess Sn^{2+} is back titrated potentiometrically with standard $K_2Cr_2O_7$ solution, using a platinum–calomel electrode.

Reduction with Ferrous Iron Still another method, proposed by Glebko et al.,[30] is based on reduction of quinone groups by Fe^{2+} in alkaline triethanolamine. The excess reductant is back-titrated amperometrically with standard chromate solution.

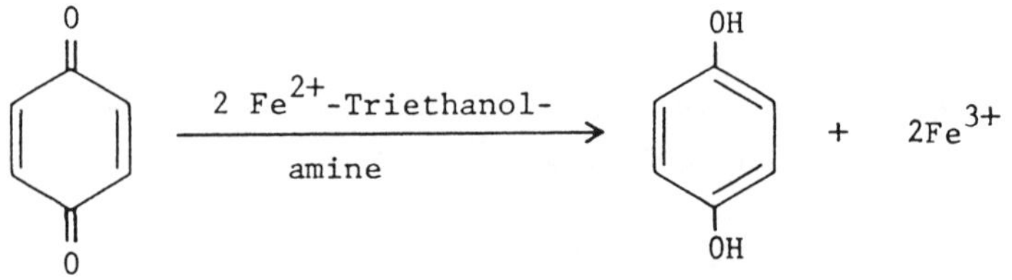

In a comparison of methods for determining quinone groups in humic substances, Schnitzer and Riffaldi[28] obtained somewhat higher results using the acid $SnCl_2$ approach (250 to 510 cmole/kg versus 9 to 220 cmole/kg with the Fe^{2+}-triethanolamine method). There was excellent agreement between results obtained by reduction with $SnCl_2$ and Fe^{2+}-triethanolamine in alkaline solutions. Evidence for specificity of the Fe^{2+}-triethanolamine reduction method has been given by Maximov and Glevko.[27]

Ether Linkages

Practically all of the humic and fulvic acids examined thus far by the Zeisel method have been found to contain low but variable OCH_3 contents.

The oxygen not accounted for in COOH, OH, C=O, and OCH_3 groups has often been recorded as existing in unknown ether linkages. However, some of the unaccounted-for oxygen exists in extremely stable quinones and lactones.

Free Amino (NH_2) Groups

Inasmuch as nearly all humic substances contain N, the determination of free NH_2 groups has been of considerable interest. In proteins, free NH_2 groups have been determined by the nitrous acid method and by reaction of the NH_2 group with such reagents as fluorodinitrobenzene and phenylisocyanate.

Reaction with Nitrous Acid The nitrous acid method is based on the following reaction:

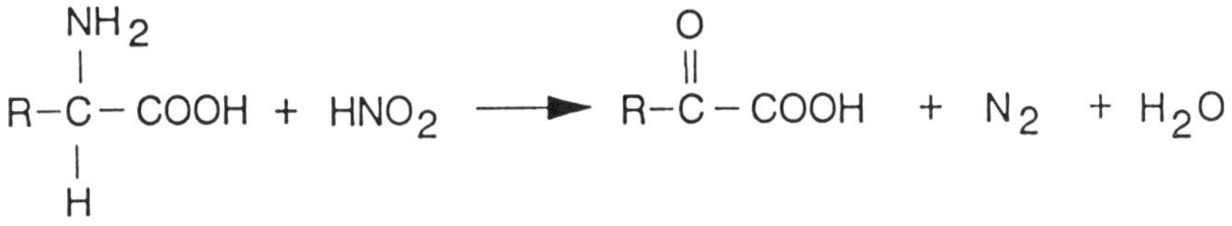

$$\underset{H}{\underset{|}{R-\overset{\overset{NH_2}{|}}{C}-COOH}} + HNO_2 \longrightarrow R-\overset{\overset{O}{\|}}{C}-COOH + N_2 + H_2O$$

A significant fraction of the N in humic acids (to 30 percent) can be accounted for as "free amino groups" by the nitrous acid method. However, lignin and other phenolic compounds interfere with the determination; consequently, the results cannot be considered valid. Reactions between nitrous acid and humic substances are discussed in Chapter 3.

Formation of Organic Derivatives The fluorodinitrobenzene method involves the formation of a yellow-colored dinitrophenyl derivative of the free amino group.

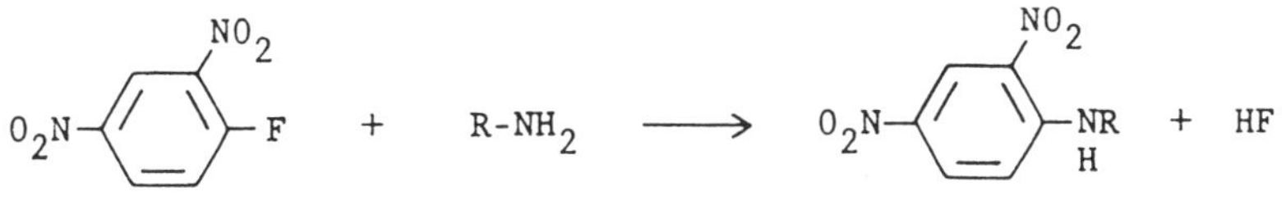

$$O_2N-\underset{}{\bigcirc}^{NO_2}-F + R-NH_2 \longrightarrow O_2N-\underset{}{\bigcirc}^{NO_2}-\underset{H}{NR} + HF$$

Sowden and Parker[31] and others have applied this technique to soil organic matter fractions, but were not successful in detecting any free amino groups.
The phenylisocyanate method is based on the following reactions:

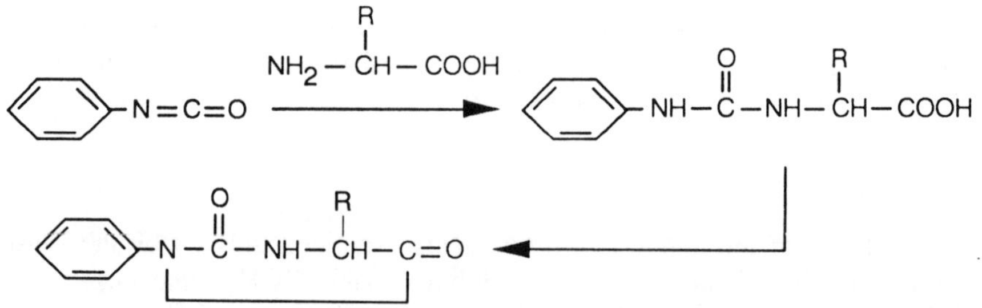

$$\text{Ph-N=C=O} + \text{NH}_2-\underset{R}{\text{CH}}-\text{COOH} \longrightarrow \text{Ph-NH-}\underset{\text{O}}{\overset{\|}{\text{C}}}-\text{NH}-\underset{R}{\text{CH}}-\text{COOH}$$

$$\text{Ph-N-}\underset{\text{O}}{\overset{\|}{\text{C}}}-\text{NH}-\underset{R}{\text{CH}}-\text{C=O} \longleftarrow$$

Sowden[32] detected a small amount of free amino groups in humic acids using this approach.

DISTRIBUTION OF OXYGEN-CONTAINING FUNCTIONAL GROUPS

As might be apparent from the discussion of the previous section on methods, data reported in the literature for functional groups in humic and fulvic acids are difficult to interpret. Not only has there been a lack of standardized methods for extraction, fractionation, and purification, but many of the analytical techniques have lacked specificity. Reasons for contradictory results include:

1. Divergent origins and purity (e.g., ash contents) of humic substances
2. Incomplete reactions and adsorption of reagents
3. Variable molecular weights, leading to fractionation during manipulations
4. Proximity of many and different reactive functional groups, which influence both the reactivity of the groups and the specificity of the reagents used for their detection and measurement

Relative Distribution of Functional Groups in Humic and Fulvic Acids

Notwithstanding difficulties in estimating functional groups, sufficient information has now accumulated to permit tentative conclusions to be drawn regarding the nature and relative abundance of oxygen-containing functional groups in humic substances of various types.

Because of the variability in reactivity of OH groups, values reported for "phenolic OH" and "alcoholic OH" groups should be accepted with reservation. Based on wet-chemical methods, more of the oxygen in humic substances would appear to exist as "phenolic OH" than has been recorded, because very weak and hindered phenols are difficult to estimate by existing methods. For reasons that are not clear, "phenolic OH" groups, as determined by wet-chemical methods, are somewhat lower than those estimated by the technique of ^{13}C-NMR spectroscopy, as will be noted later in Chapter 11.

In this section, the term "acidic OH" will be used rather than "phenolic OH" to refer to the fraction of the total acidity not accounted for as COOH groups. Other OH groups will be referred to as "weakly acidic- plus alcoholic-OH."

Schnitzer's[33] summary of the distribution of oxygen-containing functional groups in humic and fulvic acids from soils of widely different climatic zones is summarized in Table 9.3. Although a considerable range of values is apparent for humic material from within a group of soils, certain trends are evident. Total acidities of the fulvic acids (from 640 to 1420 cmole/kg) are unmistakably higher than those of the humic acids (from 560 to 890 cmole/kg). Both COOH and acidic OH groups (presumed to be phenolic OH) contribute to the acidic nature of these substances, with COOH being the most important. Of particular interest is the very high COOH content of fulvic acids.[18,23,32-36]

On the assumption that the core structure of soil humic substances (see Chapter 12) consist primarily of C_6-ring units held together by $-O-$, $-CH_2-$, $-NH-$, and $-S-$linkages, the conclusion would have to be made that multiple substitution of the ring has occurred. Even so, considerable difficulty is encountered in accounting for all acidic functional groups. A discussion of constraints on assignment of COOH, acidic OH, and C=O groups has been given by Perdue.[1]

This author is of the opinion that the exceptionally high values recorded for total acidity of humic substances, often approaching 1400 meq/100 g, are in error due to failure to removal mineral acids from the preparations during isolation/purification (see Chapter 15).

Tsutsuki and Kuwatsuka,[37] in an examination of over 30 humic acids from a wide variety of soils, found that COOH and C=O groups increased in amounts during humification while alcoholic OH, phenolic OH, and OCH$_3$ decreased. Thus, the ratio of the oxygen in COOH and C=O groups to total oxygen increased during humification, in some cases accounting for 80 percent of the total oxygen. A direct relationship was observed between COOH and C=O, as shown in Fig. 9.1.

Methoxyl and C=O groups seem to be universally present in humic substances. On the other hand, not all samples are reported to contain weakly acidic- or alcoholic-OH groups. For example, Dubach et al.[14] and Martin et al.[15] found that active-H, as determined by diborane, agreed well with total acidity measurements by reaction with Ba(OH)$_2$, a finding that excludes the presence of alcoholic OH groups.

A major difference between the functional group content of humic and fulvic acids is that a smaller fraction of the oxygen in the former can be accounted for in COOH, OH, and C=O groups. Table 9.4 shows that whereas essentially all of the oxygen in some soil fulvic acids has been recovered in these groups, somewhat less than 75 percent of that in most humic acids was similarly distributed.

Table 9.3 Distribution of Oxygen-Containing Functional Groups in Humic and Fulvic Acids Isolated from Soils of Widely Different Climatic Zones (in cmole/kg)[a]

Functional Group	Climatic Zone						
	Arctic	Cool, Temperate Acid Soils	Cool, Temperate Neutral Soils	Subtropical	Tropical	Range	Average
Humic acids							
Total acidity	560	570–890	620–660	630–770	620–750	560–890	670
COOH	320	150–570	390–450	420–520	380–450	150–570	360
Acidic OH	240	320–570	210–250	210–250	220–300	210–570	390
Weakly acidic + alcoholic OH	490	270–350	240–320		20–160	20–490	260
Quinone C=O	230	{10–180	{450–560	290		{10–560	{290
Ketonic C=O	170			{80–150	{30–140		
OCH$_3$	40	40	30	30–50	60–80	30–80	60
Fulvic acids							
Total acidity	1100	890–1420		640–1230	820–1030	640–1420	1030
COOH	880	610–850		520–960	720–1120	520–1120	820
Acidic OH	220	280–570		120–270	30–250	30–570	300
Weakly acidic + alcoholic OH	380	340–460		690–950	260–520	260–950	610
Quinone C=O	200	{170–310		{120–260	30–150	{120–420	{270
Ketonic C=O	200				160–270		
OCH$_3$	60	30–40		80–90	90–120	30–120	80

[a]From Schnitzer.[33]

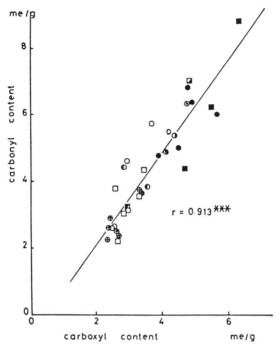

Fig. 9.1 Relationship between COOH and C=O contents of humic acids. From Tsutsuki and Kuwatsuka[37]

The COOH content of humic substances appears to be inversely related to molecular weights. It was pointed out earlier that fulvic acids have lower molecular weights than humic acids. The results given in Tables. 9.3 and 9.4 show rather conclusively that the proportion of the oxygen that occurs in the form of COOH is highest for the fulvic acids.

Rashid[38] reported unusually high contents of C=O groups in the humic and fulvic acids of marine environments. His C=O values were in excess of 500 cmole/kg and were often higher than COOH and "phenolic" OH combined. Some aspects of this work have been discussed by Mathur.[39] Finally, rather high values for C=O (to 570 cmole/kg) were obtained by Piccolo et al.[40] for humic substances isolated from organic waste-amended soils.

Quinone C=O Groups in Humic and Fulvic Acids

Recent studies permit further breakdown of total C=O content into ketonic C=O and quinone C=O. Table 9.5 gives the results for a variety of humic and fulvic acids as recorded in the literature.[27,34-36,41] Whereas a wide range of values have been obtained for the two subfractions, the results indicate that the quinone C=O content of humic acids is generally higher than for fulvic

Table 9.4 Distribution of Oxygen in Reactive Functional Groups of Humic and Fulvic Acids (All Values as Percentage of Oxygen)

	Number of Samples	COOH	Acidic OH	Weakly Acidic plus Alcoholic OH	C=O	OCH$_3$	Oxygen accounted for
Humic acids[a]							
A	6	34.0–50.0	7.0–14.0	1.0–8.0	15.0–30.0	2.0–4.0	71.0–95.0
B	5	39.0–46.0	9.0–11.0	0.0–13.0	4.0–11.0	ND[b]	61.0–74.0
C	1	31.0	12.0	24.0	19.0	2.0	88.0
Fulvic acids							
A	6	57.0–75.0	1.0–10.0	9.0–20.0	11.0–17.0	3.0–5.0	90.0–103.0
B	3	39.0–64.0	5.0–9.0	24.0–35.0	4.0–10.0	ND	89.0–101.0
C	1	64.0	8.0	14.0	14.0	2.0	102.0

[a]Sources were: A = Griffith and Schnitzer[34]; B = Ortiz de Serra and Schnitzer[35]; C = Schnitzer and Vendette.[36]
[b]Not determined.

Table 9.5 Quinone- and Ketonic C=O Contents of Humic and Fulvic Acids (in cmole/kg)[a]

	Number of Samples	Quinone C=O	Ketonic C=O	Total C=O	Ketonic C=O / Total C=O
Humic acids					
A	6	140–264	6–177	173–394	55.1–96.7
B	7	140–210	80–420	270–560	25.0–70.4
C	1	230	170	400	57.5
D	1	350	—	—	—
E	2	105–126	78	204	53.0–61.8
Fulvic acids					
A	6	28–145	104–336	220–383	9.4–52.7
B	1	60	250	310	19.4
C	1	200	200	400	50.0
D	1	420	—	—	—

[a]Sources were: A = Griffith and Schnitzer[34]; B = Schnitzer and Riffaldi[28]; C = Schnitzer and Vendette[36]; D = Mathur[39]; and E = Maximov and Glebko.[27]

acids; furthermore, a higher percentage of the total C=O in humic acids occurs in quinone linkages than is the case for fulvic acids.

The possible presence of terminal-ring 1,4-quinones (30–110 cmole/kg) has been indicated by an increase in N content following methylation with CH_2N_2.[4] This reagent is known to combine with terminal-ring quinones to form the pyrazoline ring.

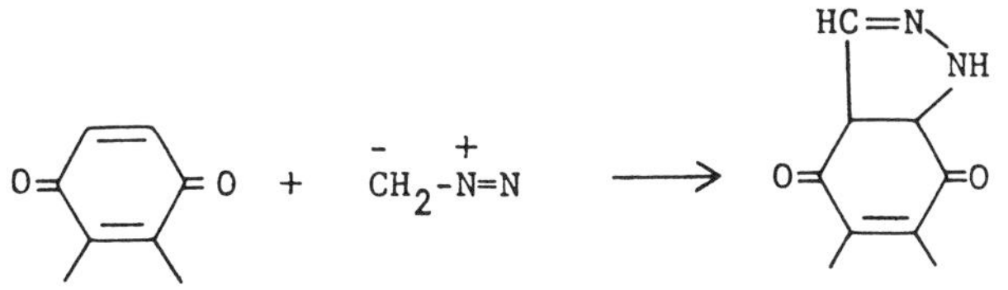

An increase in N does not constitute unequivocal proof for 1,4-quinones, because CH_2N_2 can also combine with other groups, such as olefinic double bonds.

DIAGENETIC TRANSFORMATIONS

In natural soil, the balance of humus is maintained by the continued synthesis of new material as part of the old is mineralized; consequently, the chemical properties of the humic and fulvic acids will remain essentially constant from one year to the next. It should be noted, however, that the humus of each soil will have its own characteristic "equilibrium composition," both with regard to chemical nature and composition.

Highly organic soils (Histosols) vary greatly in the extent to which plant remains have undergone decomposition. Increase humification has been found to be associated with increased solubility in dilute pyrophosphate solution, a property that serves as a practical test for ascertaining the "degree of humification" in organic soils. Schnitzer and Desjardins[42] also observed increases in COOH and C=O groups but decreases in alcoholic OH groups. Higher subsidence rates of cultivated Histosols in the Everglades of southern Florida were found by Volk and Schnitzer[43] to be associated with the following changes in the chemistry of their humic acids: 1) increases in COOH, phenolic OH, quinone, and ketonic C=O groups, 2) decreases in alcoholic OH groups, and 3) decreases in molecular complexity as indicated by E_4/E_6 ratios. (*Note:* the subject of E_4/E_6 ratios is discussed in Chapter 13).

When the source of plant raw material for humus synthesis is cut off, such as is the case following burial, the humus is exposed to successive cycles of biological attack, with concomitant changes in chemical composition. Easily decomposable compounds, such as proteins and carbohydrates, are attacked first, with the result that these constituents are eliminated at the expense of resistant molecules, such as the core structures of humic and fulvic acids. Further modifications are brought about through chemical processes. An extensive discussion of the transformations of organic substances in geological environments is beyond the scope of this paper. For more complete details, reference should be made to recent books on organic geochemistry.

During diagenesis, humic substances are either slowly mineralized by microorganisms or altered to kerogen- or coal-like products, depending upon environmental conditions in the sediment.[4,44] Carbonization results from losses of weakly attached polysaccharide and proteinaceous residues, peripheral side chains, and oxygen-containing functional groups, particularly COOH and OH. These changes are accompanied by an increase in C content and a decrease in oxygen content. Other modifications include a loss of N and greater insolubility in alkali. All humic and fulvic acids surviving biological decomposition will eventually be diagenetically altered to kerogen or coal-like substances.

Blom et al.[45] prepared a diagram illustrating the distribution of oxygen-containing functional groups in lignites and coals as related to their C and oxygen contents. From published data on the C, oxygen, and function group content of humic and fulvic acids it has been possible to extend their diagram to include pigments having higher oxygen but lower C contents.[4] The modification is given in Fig. 9.2. The greatest change is in the percentage of the oxygen as COOH groups. It is apparent that if humic and fulvic acids are involved in the coalification process, COOH groups disappear first, followed by OCH_3 and C=O groups.

Maximov and Glebko[27] found that the quinone content of humic acids from various sources followed the order: weathered bituminous > coal > weathered brown coal > brown coal > peat > woodland soil. For this sequence, quinone C=O ranged from 337 cmole/kg for the humic acid from the weathered bituminous coal to 105 cmole/kg for the soil humic acid. Ketonic C=O varied within a narrow range (69–102 cmole/kg).

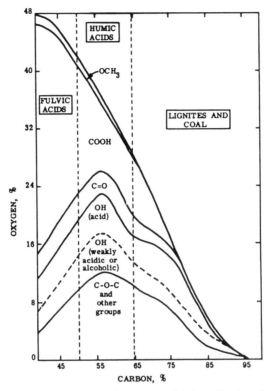

Fig. 9.2 Oxygen containing functional groups in fulvic acids, humic acids, coal, and lignite as related to carbon and oxygen contents. From Stevenson and Butler[4] as modified from Bloom et al.[45]

At this time, interrelationships between humic and fulvic acids and other natural products are not clear (see Chapter 8). Proposed relationships are illustrated in the following diagram.

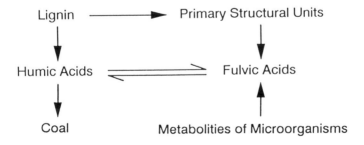

Recent studies of the chemical characteristics of paleosol humic acids include those of Sakai et al.[46] and Schnitzer and Calderone.[47] Changes brought about in oxygen-containing functional groups during early diagenesis have been examined by Pillon et al.[48]

SUMMARY

The reactivity of humic substances is due in large measure to their high content of oxygen-containing functional groups, including COOH, phenolic-and/or enolic-OH, alcoholic OH, and the C=O of quinones, hydroxyquinones, and possibly α,β-unsaturated ketones. Because of lack of specificity, absolute values reported for the various groups must be accepted with reservation. The results obtained thus far indicate that total acidities of fulvic acids (640–1420 cmole/kg) are much higher than for humic acids (560 to 890 cmole/kg), and that more of the oxygen in the latter occurs in unknown ether linkages. The content of oxygen-containing functional groups in fulvic acids appears to be substantially higher than for any other naturally occurring organic polymer.

REFERENCES

1 E. M. Perdue, "Acidic Functional Groups in Humic Substances," in G. R. Aiken, D. M. McKnight, R. L. Wershaw, and P. MacCarthy, Eds., *Humic Substances in Soil, Sediment, and Water*, John Wiley, New York, 1985, pp. 493–526.
2 M. Schnitzer, "Chemical, Spectroscopic, and Thermal Methods for the Classification and Characterization of Humic Substances," in *Proc. Intern. Meetings on Humic Substances*, Pudoc, Wageningen, 1972, pp. 293–310.
3 M. Schnitzer and S. U. Khan, *Humic Substances in the Environment*, Marcel Dekker, New York, 1972.
4 F. J. Stevenson and J. H. A. Butler, "Chemistry of Humic Acids and Related Pigments," in G. Eglinton and M. T. J. Murphy, Eds., *Organic Geochemistry*, Springer-Verlag, New York, 1969, pp. 534–557.
5 C. Varadachari and K. Ghosh, *Plant Soil*, **77**, 305 (1984).
6 C. Steelink, "Elemental Characteristics of Humic Substances," in G. R. Aiken, D. M. McKnight, R. L. Wershaw, and P. MacCarthy, Eds., *Humic Substances in Soil, Sediment, and Water*, John Wiley, New York, 1985, pp. 457–476.
7 S. A. Visser, *Water Res.*, **17**, 1393 (1983).
8 J. A. Leenheer and T. I. Noyes, "Derivatization of Humic Substances for Structural Studies," in M. H. B. Hayes, P. MacCarthy, R. L. Malcolm, and R. S. Swift, Eds., *Humic Substances II: In Search of Structure*, John Wiley, New York, 1989, pp. 257–289.
9 J. A. Davis, *Geochim. Cosmochim. Acta*, **46**, 2381 (1981).
10 V. C. Farmer and R. I. Morrison, *Sci. Proc. Roy. Dublin Soc., Ser. A.*, **1**, 85 (1960).
11 Association of Official Agricultural Chemists (AOAC), *Methods of Analysis*, Banta, Menasha, Wisconsin, 1955.
12 W. G. C. Forsyth, *J. Agr. Sci.*, **37**, 132 (1947).
13 R. L. Wershaw and D. J. Pinckney, *Science*, **199**, 906 (1978).
14 P. Dubach, N. C. Mehta, T. Jakab, F. Martin, and N. Roulet, *Geochim. Cosmochim. Acta*, **28**, 1567 (1964).

15 F. Martin, P. Dubach, N. C. Mehta, and H. Deuel, *Z. Pfl. Dung. Bodenk*, **183,** 27 (1963).
16 K. M. Holtzclaw and G. Sposito, *Soil Sci. Soc. Amer. J.*, **43,** 318 (1979).
17 E. M. Perdue, J. H. Reuter, and M. Ghosal, *Geochim. Cosmochim. Acta*, **44,** 1841 (1980).
18 J. R. Wright and M. Schnitzer, *Trans. 7th Intern. Congr. Soil Sci.*, **2,** 120 (1960).
19 M. Schnitzer, *Soil Sci.*, **117,** 94 (1974).
20 G. H. Wagner and F. J. Stevenson, *Soil Sci. Soc. Amer. Proc.*, **29,** 43 (1965).
21 J. C. Wood, S. E. Moschopedes, and W. den Hertog, *Fuel*, **40,** 491 (1961).
22 M. Schnitzer and S. I. M. Skinner, *Soil Sci. Soc. Amer. Proc.*, **29,** 400 (1965).
23 M. Schnitzer and U. C. Gupta, *Soil Sci. Soc. Amer. Proc.*, **28,** 374 (1964).
24 M. Schnitzer and S. I. M. Skinner, *Soil Sci.*, **101,** 120 (1966).
25 J. A. Leenheer, M. A. Wison, and R. L. Malcolm, *Org. Geochem.*, **11,** 123 (1987).
26 T. A. Kukharenko and L. N. Yekaterinina, *Soviet Soil Sci.*, **7,** 933 (1967).
27 O. B. Maximov and L. I. Glebko, *Geoderma*, **11,** 17 (1974).
28 M. Schnitzer and R. Riffaldi, *Soil Sci. Soc. Amer. Proc.*, **36,** 772 (1972).
29 N. A. Vasilyevskaya, L. I. Glebko, and O. B. Maximov, *Soviet Soil Sci.*, **3,** 224 (1971).
30 L. I. Glebko, Zh. I. Ulkina, and O. B. Maximov, *Mikrochim. Acta*, **1970,** 1247 (1970).
31 F. J. Sowden and D. I. Parker, *Soil Sci.*, **76,** 201 (1953).
32 F. J. Sowden, *Can. J. Soil Sci.*, **37,** 143 (1957).
33 M. Schnitzer, "Recent Findings on the Characterization of Humic Substances Extracted from Soils from Widely Differing Climatic Zones," in *Proc. Symposium on Soil Organic Matter Studies*, Braunschweig, International Atomic Energy Agency, Vienna, 1977, pp. 117–131.
34 S. M. Griffith and M. Schnitzer, *Soil Sci. Soc. Amer. Proc.*, **39,** 861 (1975).
35 M. I. Ortiz de Serra and M. Schnitzer, *Soil Biol. Biochem.*, **5,** 281 (1973).
36 M. Schnitzer and E. Vendette, *Can. J. Soil Sci.*, **55,** 93 (1975).
37 K. Tsutsuki and S. Kuwatsuka, *Soil Sci. Plant Nutr.*, **24,** 547 (1978).
38 M. A. Rashid, *Soil Sci.*, **113,** 181 (1972).
39 S. P. Mathur, *Soil Sci. Soc. Amer. Proc.*, **37,** 486 (1973).
40 A. Piccolo, P. Zaccheo, and P. G. Genevini, *Bioresource Tech.*, **40,** 275 (1992).
41 S. P. Mathur, *Soil Sci. Soc. Amer. Proc.*, **36,** 175 (1972).
42 M. Schnitzer and J. G. Desjardins, *Can. J. Soil Sci.*, **46,** 237 (1966).
43 B. G. Volk and M. Schnitzer, *Soil Sci. Soc. Amer. Proc.*, **37,** 886 (1973).
44 F. J. Stevenson, *Soil Sci.*, **107,** 470 (1969).
45 L. Blom, L. Edelhausen, and D. W. van Krevelen, *Fuel*, **36,** 135 (1957).
46 C. Sakai, K. Sakagami, R. Hamada, and T. Kurobe, *Soil Sci. Plant Nutr.*, **28,** 37 (1982).
47 M. Schnitzer and G. Calderone, *Chem. Geol.*, **53,** 175 (1985).
48 O. Pillon, J. M. Portal, B. Gerard, P. Jeanson, and L. Jocteur-Monrozier, *Org. Geochem.*, **9,** 313 (1986).

10

STRUCTURAL COMPONENTS OF HUMIC AND FULVIC ACIDS AS REVEALED BY DEGRADATION METHODS

A valuable approach for characterizing humic and fulvic acids is through degradation to individual monomers.[1-6] The primary objective of such studies is to produce simple compounds representative of main structural units in the humic macromolecule. Ideally, the products obtained would provide information from which "type" structures and formulas could be devised. Since both plant and microbial products serve as precursors for humus synthesis, the array of compounds produced should reflect the conditions of climate and vegetation under which the humus was formed. Chromatographic patterns of degradation products from humic substances of different origins may provide a "fingerprint" technique for their classification.[7]

Breakdown products of humic substances can also provide additional worthwhile information. Thus, the occurrence of phenolic aldehydes in degradation products has been used as an indicator of terrestrial organic matter in ocean sediments; the presence of resorcinol derivatives (e.g., 3,5-dihydroxybenzoic acid) has been taken as evidence that microorganisms contribute polyphenols for humus synthesis.

EXPERIMENTAL APPROACHES

A variety of degradation methods have been applied to soil humic substances, as shown in Table 10.1. Each approach has its own set of advantages and limitations. With mild degradation procedures (i.e., acid or base hydrolysis), yields of identifiable products are extremely low, usually of the order of a few percent or less of the starting material. On the other hand, with drastic methods (i.e., oxidation), much of the material is broken down into such small fragments

Table 10.1 Degradative Approaches for Identifying Structural Units of Humic Substances

Hydrolysis	Reduction	Oxidation	Other
Water	Na-amalgam	Cu-NaOH	Na-sulfide degradation
Base	Zn dust distillation	Alkaline nitrobenzene	Depolymerization
Acid		Alkaline KMnO$_4$	
		Nitric acid	
		Peracetic acid	
		Sodium hypochlorite	
		Hydrogen peroxide	

(e.g., CO$_2$, H$_2$O, acetic acid) that any resemblance to the original structure is lost. On occasion, high yields of ether-soluble products have been reported but the nature of much of the material was not established.

The number of kinds (class) of organic degradation products depends on the method chosen for degradation. Typical degradation products are as follows:

Na-amalgam reduction: Phenols and phenolic acids, including lignin-derived components (e.g., vanillin)

Zinc dust distillation: Complex polycyclic hydrocarbons

Permanganate and sodium hypochlorite oxidation: Aromatic di-, tri-, tetra-, penta-, and hexapolycarboxylic acids

Nitric acid oxidation: Nitrophenolic acids

Sodium sulfide degradation: Aliphatic and aromatic mono- and dicarboxylic acids; aromatics containing aliphatic side chains.

An additional difficulty in interpreting results of degradation studies is that yields of identified compounds have often been recorded in terms of their methylated analogous (acids as esters; phenolics as methylated ethers). Also, some of the separation techniques have been semiquantitative at best. An even more serious problem is that humic substances, as normally prepared, nearly always contain appreciable quantities of adsorbed and coadsorbed impurities, both aliphatic and aromatic. These constituents are partially recovered and included with products derived from the so-called "aromatic core" of the humic molecule. For humic acids, Riffaldi and Schnitzer[8] recommended hydrolysis with 6*N* HCl to remove N-containing compounds, carbohydrates, and other adsorbed materials, thereby providing a more homogeneous starting material for subsequent structural investigations.

In principle, the ideal method should be mild but would at the same time yield high amounts of degradative products of moderate molecular complexities while minimizing the formation of artifacts. A combination of degradative procedures with increasing severity will undoubtedly be required, as done by Neyroud and Schnitzer[9] and Schnitzer and Ortiz de Serra.[10] According to these investigators, humic acids contain an easily degradable portion that consists of quaiacyl and syringyl units and that may be lignin-derived (about 10 percent). This material was believed to be released by $CuO-CaOH$ oxidation and by Na-amalgam reduction. The bulk of the humic acid structure was believed to consist of a more chemically condensed "core" that degrades into complex phenolic and benzenecarboxylic acids with more drastic oxidation. The drastic conditions of Zn dust distillation and fusion leads to destruction of phenolic constituents, the main degradation product being polycyclic hydrocarbons.

Thus far, well over 300 different compounds have been identified in degradation products of humic substances but not all of them can be regarded as being derived from "core" structures of the macromolecule. In many instances, a constituent initially released by degradation is transformed to other products or modified during residence in the reaction mixture. In any event, no given compound or class of compounds has been shown to represent more than a few percent of the initial starting material.

A partial list of phenols and phenolic acids in degradation products of humic substances is shown in Table 10.2; other compounds include benzenecarboxylic acids (alkaline-KMnO oxidation), nitrophenols (nitric acid oxidation), polycyclic hydrocarbons (Zn dust distillation), and aromatics with aliphatic side chains (sodium sulfide degradation).

In addition to aromatic compounds, most degradation procedures produce aliphatic substances (e.g., alkanes, fatty acids, carboxylic acids). With drastic degradation, a substantial portion of the aliphatics arise from destruction of aromatic constituents; other sources include side chains on aromatic rings.

The general order that will be followed in discussing degradation procedures will be from mild hydrolysis to the more drastic conditions of reduction, oxidation, and pyrolysis.

HYDROLYSIS METHODS

Hydrolytic procedures are effective in removing protein, carbohydrate, and other aliphatic constituents from crude humic acid preparations and have been used primarily to "purify" humic acids prior to further chemical and physicochemical characterization. The residue from hydrolysis is sometimes termed the "backbone" or "core" of the humic macromolecule. Although there is little evidence to indicate that basic structural units of the "core" are affected by hydrolysis, some loss of humic material occurs during removal of the hydrolyzed products. With acid hydrolysis, considerable amounts of CO_2 are liberated due to decarboxylation of structures where OH or C=O groups are in α or β positions to a COOH group.

Table 10.2 Partial List of Organic Compounds Identified as Chemical Degradation Products of Humic and Fulvic Acids

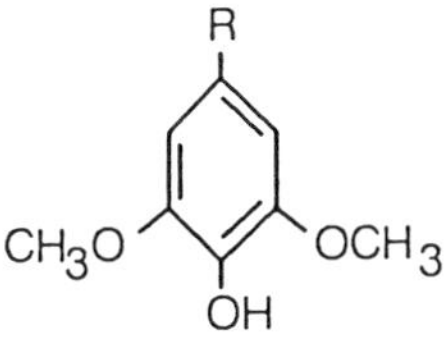

Structures (R-substituted benzene rings):
- I, II, III: 4-R-2,6-dimethoxyphenol (CH$_3$O, OCH$_3$, OH)
- IV, V, VI: 4-R-2-methoxyphenol (OCH$_3$, OH)
- VII, VIII: 4-R-phenol (OH)
- IX, X: 4-R-catechol (OH, OH)
- XI, XII: 4-R-1,2,4-trihydroxybenzene (HO, OH, OH)
- XIII, XIV, XV: 3,5-dihydroxy-R-benzene (R, OH, OH)
- XVI, XVII: 3,5-dihydroxybenzoic acid derivative (R, OH, COOH)

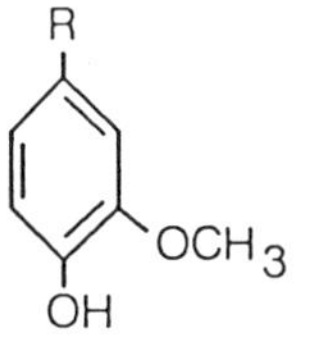

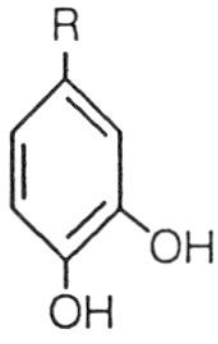

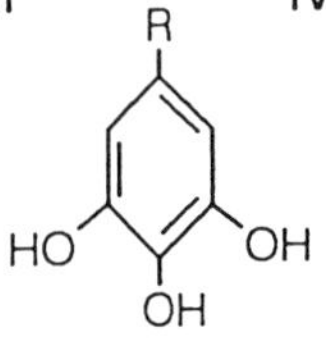

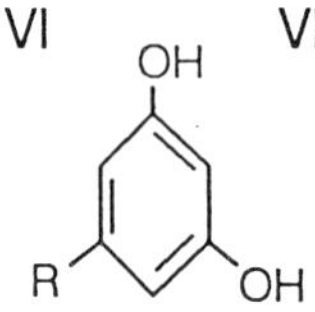

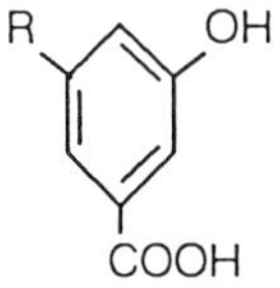

	R	Compound
I	R = CHO	Syringaldehyde
II	R = COOH	Syringic acid
III	R = CH$_2$-CH$_2$-COOH	Syringylpropionic acid
IV	R = CHO	Vanillin
V	R = COOH	Vanillic acid
VI	R = CH$_2$-CH$_2$-COOH	Guaiacylpropionic acid
VII	R = COOH	p-Hydroxybenzoic acid
VIII	R = CHO	p-Hydroxybenzaldehyde
IX	R = H	Catechol
X	R = COOH	Protocatechuic acid
XI	R = H	Pyrogallol
XII	R = COOH	2,3,4-Trihydroxybenzoic acid
XIII	R = H	Resorcinol
XIV	R = COOH	2,4-Dihydroxybenzoic acid
XV	R = OH	Phloroglucinol
XVI	R = H	m-Hydroxybenzoic acid
XVII	R = OH	Same as XIV

Others
Benzene carboxylic and hydroxycarboxylic acids
Nitrophenols (nitric acid oxidation)
Polycyclic ring compounds

Water

Boiling humic and fulvic acids with H$_2$O has been observed to extract polysaccharides, polypeptides, and small quantities of relatively simple phenolic acids and aldehydes.[11] Neyroud and Schnitzer[12] heated humic and fulvic acids for 3 h. at 170°C under N$_2$ gas, after which the supernatant solution was adjusted to pH 2 and extracted with ethyl acetate. Yields of ethyl acetate-soluble products ranged from 64 to 81 mg/100 g of humic acid and from 197 to 401 mg/100 g of fulvic acid. The dried ethyl acetate extracts were then extracted with several solvents of increasing polarity: n-hexane, benzene, ethyl acetate, and methanol. Most of the material was recovered in the ethyl acetate and methanol extracts but no attempt was made to characterize these materials. The main products in the n-hexane + benzene extracts of the humic acid were n-fatty acids and n-alkanes; some furan derivatives, a benzene carboxylic acid, and two phenolic compounds were also identified. Substantially greater yields of phenolic compounds (five identified) and benzene carboxylic acids (seven identified) were recovered from the fulvic acid, equivalent to about 4 percent of the initial weight.

Acid Hydrolysis

Acid hydrolysis of humic materials involves suspending the substance in an aqueous solution of the acid and heating the mixture for a time. In addition to carbohydrate and protein derivatives (i.e., sugars and amino acids), the hydrolysates have been found to contain substances extractable with ether. In the work of Jakab et al.,[11] the ether extracts represented from 0.5 to 2.5 percent of the organic material, from which protocatechuic acid (**X**), p-hydroxybenzoic acid (**VII**), vanillic acid (**V**), and vanillin (**IV**) were identified (see Table 10.1 for structures). The conclusion was reached that these products represented lignin impurities in the humic material.

Alkaline Hydrolysis

Hydrolysis with dilute aqueous alkaline solutions is carried out in much the same way as acid hydrolysis except that care must be taken to prevent concurrent oxidation of the degradation products by excluding O_2. In addition to removal of adsorbed aliphatics and N-compounds (e.g., proteinaceous constituents), a variety of phenolic substances are liberated, presumably without degrading the central "core" of the molecule.

Jakab et al.[11] hydrolyzed a humic acid with $5N$ NaOH at 170 and 250°C with and without $CuSO_4$ additions. Phenolic compounds (30 detected) accounted for 6 percent of the total organic matter for reactions at 250°C but only 2 percent at 170°C.

REDUCTIVE CLEAVAGE

Reductive methods result in saturation of C=C bonds, a change in oxygen-containing functional groups (reduction of COOH), and cleavage of ether linkages. The latter point is of interest because of the possibility that ether linkages are a prominent structural feature of humic substances.

Sodium Amalgam Reduction

Sodium amalgam is a relatively mild degradation procedure that leads to digest products containing phenols and phenolic acids thought to be released through cleavage of ether linkages.

In a typical procedure, a given quantity of humic material, usually about 1 g, is added to a three-neck, round-bottomed flask equipped with a sealed stirrer, a condenser, and an N_2 gas inlet tube. The material is dissolved in dilute NaOH solution, air in the flask is displaced with N_2 gas, 5 percent Na-amalgam is added, and the contents of the flask are heated to boiling while under continuous stirring with N_2 gas passing through the system. Heating is continued at 100 to 110°C for several hours, during which time the dark color of the solution changes to a translucent green, orange, or gray color depending upon the sample used. The solution is cooled, acidified, and filtered, following which the supernatant solution is extracted with diethyl ether to remove phenolic compounds and other low-molecular-weight degradation products. Chromatographic procedures are then used to identify the degradation products, the most popular method being thin-layer chromatography (TLC).

Mechanism of Bond Cleavage The action of Na-amalgam results from attack of atomic or "nascent" hydrogen (H^+) on electron-rich areas of the substrate molecule. Diphenyl ether (**XVIII**), for example, is cleaved at either bond between the oxygen atom and the aromatic ring to give low yields (2 percent) of phenol. Introduction of an OH group into the aromatic rings (**XIX**) directs cleavage towards the substituted ring and increases greatly the yield of degradation products (to 30 percent).

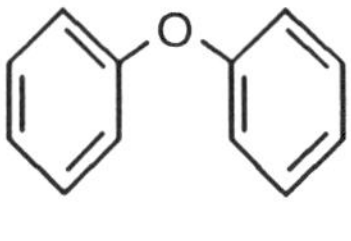

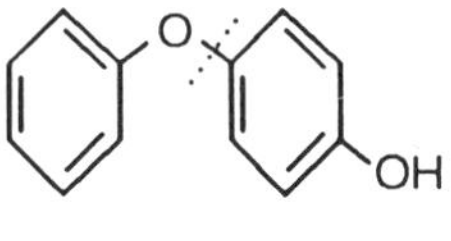

Results with known flavonoids provide an indication of the ways in which cleavage occurs. These constituent have structures of the type illustrated below, where the numbers indicate the accepted numbering system for the **A**- and **B**-ring components.

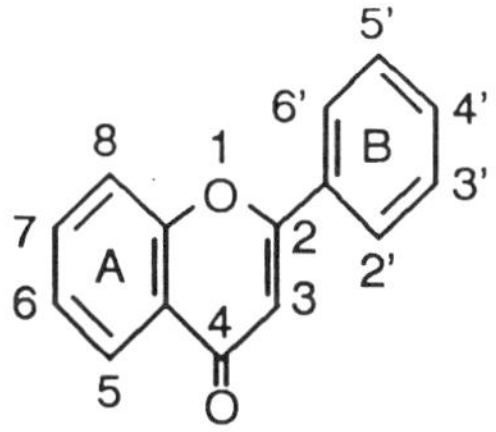

Products formed by Na-amalgam reduction of cyanidin (**XXI**) are shown in Fig. 10.1. The initial reduction leads to the formation of protocatechuic acid (**X**), phloroglucinol (**XV**), 2,4,6-trihydroxy toluene (**XXII**), and 2,4,6-trihydroxyphenylethanoic acid (**XXIII**), which are themselves transformed through dehydroxylation to other products, such as resorcinol (**XIII**), 2,6-dihydroxytoluene (**XXIV**), and 2,4-dihydroxytoluene (**XXV**).

Limitations of Sodium Amalgam Reduction Like most other approaches, the Na-amalgam reduction method has defects that must be taken into account in interpreting the results. In addition to transformations of phenolic constituents through dehydroxylation (see Fig. 10.1), some compounds are modified drastically, both during release from the macromolecule and during residence in the reaction mixture. Accordingly, any given phenolic compound recovered in the reaction mixture might originate from a variety of structure units.

Another troublesome feature of the Na-amalgam reduction method is that some of the products are labile and easily reoxidized in the presence of molecular O_2. Also, some phenolic compounds are further modified so that the final products are not typical of those occurring in the original macromolecule.[13] In general, degradation is enhanced in the presence of a COOH side group, or with more than two OHs on the aromatic ring. Attempts have been made to minimize adverse side effects through methylation of the reduction products soon after recovery.[10,14]

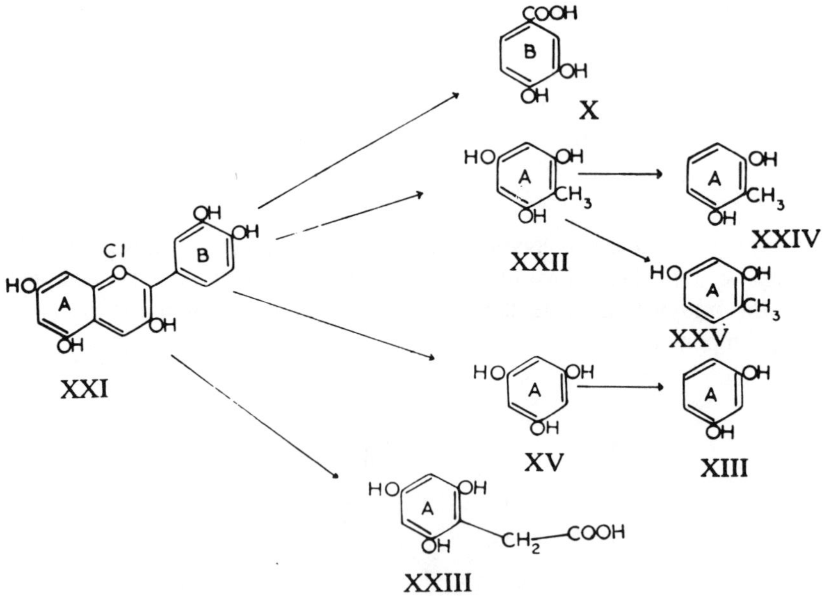

Fig. 10.1 Products from the reductive cleavage of cyanidin (**XXI**) using Na-amalgam under alkaline conditions. From Burges et al.[7]

Table 10.3 Yields of Products from Reduction of Soil Humic Acids with Na-Amalgam

Reference	Percentage of Humic Acid as:		Number of Samples Examined	Number of Compounds Detected
	Ether-Soluble Materials	Phenolic Compounds		
Burges et al.[7]	—	30–35	6	9–12
Stevenson and Mendez[20]	12	trace	1	3
Martin and Haider[15]	15–22	2–6	3	13–18
Dormaar[18]	—	—	6	9–14
Piper and Posner[19]	—	4–20[a]	13	7–16
Schnitzer and Ortiz de Serra[10]	9–16	3–4	2	9–10
Tate and Goh[21]	—	5–8[a]	3	6–16
Martin et al.[13]	8–24	3–5	5	18
Matsui and Kumada[16]	8–22	2–3	8	5

[a]May include aliphatic material.

Yields of Reduction Products As applied to humic acids, good results have been obtained by some investigators,[7,13,15–17] but not by others.[10,18–20] From Table 10.3, it can be seen that yields of reduction products (as ether-soluble material) have ranged from as much as 35 percent to less than 10 percent. Unfortunately, much of the ether-soluble material may consist of aliphatic constituents. In most studies, yields of identifiable compounds have been less than 5 percent of the starting material.

Martin et al.[13] obtained higher yields of both ether-soluble material and phenolic compounds from digests of synthetic and fungal humic acids than from soil and peat humic acids (Table 10.4). In other work, Piper and Posner[17]

Table 10.4 Comparison of Ether-Soluble Materials and Phenolic Compounds in Digests after Reduction with Na-Amalgam of Synthetic and Fungal "Humic Acids" (HAs) and of Soil and Peat Humic Acids[a]

Source	Percentage of HA Recovered as:		Number of Samples Analyzed
	Ether-Soluble Materials	Phenolic Compounds	
Synthetic HAs (enzymatic)	15–44	4–18	9
Synthetic HAs (autoxidation)	22–25	7–23	4
Fungal HAs	12–59	4–32	10
Soil and peat HAs	8–24	4–6	5

[a]From Martin et al.[13] The synthetic "humic acids" were prepared from various mixtures of hydroxyphenols, toluenes, and benzoic acids.

found that yields of phenolic materials could be increased substantially (to 30 percent) by keeping the amount of humic acid below an optimum level of 25 to 50 mg/30 g of 5 percent Na-amalgam.

Reduction Products Although up to 35 different phenolic compounds have been identified from digests of Na-amalgam degradation of humic acids, the maximum for any study is 18. The compounds can be classified with regard to type and source, such as those with origins in lignin or synthesized by microorganisms (Table 10.5). Compounds of lignin origin include those containing methoxy substituents (**II–V**). The resorcinol-type phenols (**XIII, XIV, XV**) and the hydroxytoluenes (**XXVI–XXVIII**) are believed to be of microbial

Table 10.5 Compounds Recovered from Digests after Reduction of Humic Acids with Na-Amalgam

Aromatics of Lignin Origin			
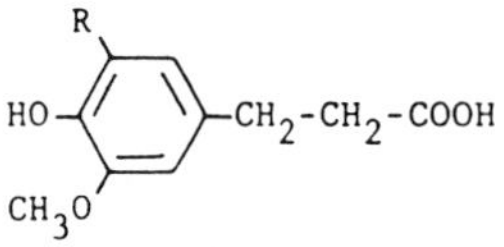 [structure: HO-, CH_3O-, R- substituted benzene with $-CH_2-CH_2-COOH$]	VI III	R = H R = OCH_3	Guiacylpropionic acid Syringylpropionic acid
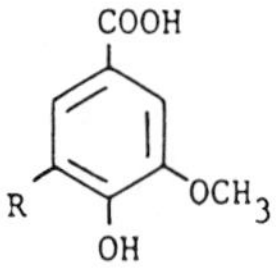 [structure: benzene with COOH, OCH_3, OH, R substituents]	V II	R = H R = OCH_3	Vanillic acid Syringic acid
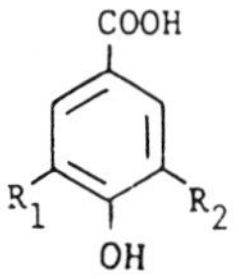 [structure: benzene with COOH, OH, R_1, R_2 substituents]	VII X XII	$R_1 = R_2 = H$ $R_1 = H, R_2 = OH$ $R_1 = R_2 = OH$	p-Hydroxybenzoic acid Protocatechuic acid Gallic acid
Phenols and Toluenes Based on Resorcinol-Type Structures			
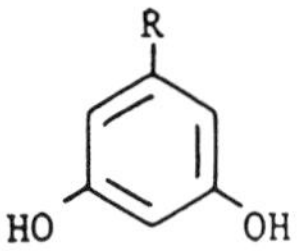 [structure: HO-, OH, R-substituted benzene (resorcinol)]	XIII XV XIV	R = H R = OH R = COOH	Resorcinol Phloroglucinol 2,4-Dihydroxybenzoic acid
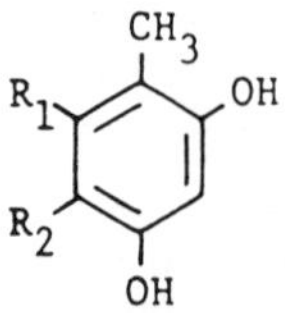 [structure: benzene with CH_3, OH, OH, R_1, R_2 substituents]	XXVI XXVII XXVIII	$R_1 = R_2 = H$ $R_1 = H, R_2 = OH$ $R_1 = OH, R_2 = H$	2,4-Dihydroxytoluene 2,4,5-Trihydroxytoluene Methylphloroglucinol

origin, because such compounds are not found in lignin but are known to be synthesized by fungi (see Chapter 8). Additional phenolic compounds recovered by reduction of humic substances with Na-amalgam include catechol (IX), pyrogallol (XI), 3-hydroxybenzoic acid, 2,4-dihydroxybenzoic acid, p-coumaric acid, ferulic acid, 4-hydroxytoluene 2,6-dihydroxytoluene, and 2,3,5-trihydroxytoluene.

Considerable variation has been observed in the number and kind of phenolic compounds that are released by Na-amalgam reduction of humic acids (Table 10.6), indicating that there may be intrinsic differences in the nature of humic acids. However, the possibility that chromatographic patterns (TLC) of degradation products might provide a "fingerprint" technique for identifying a particular type of humic acid has not been substantiated. Piper and Posner[17] concluded that the Na-amalgam method could be employed to estimate "degree of transformation" but not to "finger print" humic acids. Martin et al.[13] found that the compositions of some fungal humic acid-like compounds were affected by the substrates on which the fungi were cultured; for example, vanillic acid (V) and syringic acid (II) were found only when the organisms were cultured on lignaceous crop residues.

Zinc Dust Distillation and Fusion

Zinc dust distillation and Zn dust fusion are drastic procedures that lead to extensive degradation of oxygen-containing compounds, including polyphenols. The main application of these methods has been for structural analysis of alkaloids and other complex organic molecules that give rise, under the distillation and fusion conditions, to heteroaromatic products, notably polycyclic aromatic hydrocarbons. The methods have been used in attempts to obtain information regarding the central "core" or "nuclei" of humic substances.

Experimental Approaches

General methods for Zn dust distillation and Zn dust fusion of humic substances are described elsewhere.[22-25] In the approach of Hansen and Schnitzer,[24] a mixture of humic acid (0.5 g) and Zn dust (12.5 g) were inserted into a Pyrex glass test tube, covered with a further layer of Zn dust (5 g), and heated at 510–530°C for about 15 min. while a stream of inert gas was passed through the tube. Polycyclic hydrocarbons are sublimed into the cooler part of the tube, recovered by extraction with benzene, purified by vacuum sublimation, and separated by chromatographic procedures (TLC). Identification of reaction products was accomplished by ultraviolet and spectrofluorimetry. A somewhat similar procedure is followed by Zn dust fusion, except that heating is done at a lower temperature and the reduction products are recovered by solvent extraction from the reaction mixture.

Degradation Products The main compounds released from soil humic and fulvic acids by Zn dust distillation and Zn dust fusion are polycyclic aromatic

Table 10.6 Types of Structural Units Observed in Digests from Na-Amalgam Degradation of Humic Acids

Reference	Number of Samples Studied	Lignin-Derived Units[a]			Toluenes[a]	Miscellaneous Flavonoid Units[a]	Other Units[a]	Total Compounds Identified
		C_6-C_3	C_6					
Burges et al.[7]	6	2	4		2	3	1	12
Stevenson and Mendez[20]	1	0	2		0	0	1	3
Martin and Haider[15]	3	2	5		4	4	3	18
Dormaar[18]	6	1	4		0	2	—	7
Piper and Posner[17]	13	2	4		4	3	3	16
Schnitzer and Ortiz de Serra[10]	2	0	4		0	2	4	10
Tate and Goh[21]	3	0	3		0	3	2	8
Matsui and Kumada[16]	8	0	4		0	1	0	5

[a]Values refer to number of compounds identified in each category. Unidentified compounds were observed in some studies.

REDUCTIVE CLEAVAGE   247

hydrocarbons containing from two to five rings. Major compounds are listed in Table 10.7; other products include fluoranthene, 1,2-benzanthracene, 1,2-benzofluorene, 2,3-benzofluorene, 1,2-benzopyrene, 3,4-benzopyrene, naphthalene, chrysene, 1,12-benzoperylene, and coronene (see Chapter 7 for typical structures).[22]

Yields of Degradation Products Yields of polycyclic aromatic hydrocarbons by Zn dust distillation and fusion of humic and fulvic acids are invariably low and often account for less than 1 percent of the starting material (see Table 10.7). However, similar low yields have been obtained for alkaloids and other polycyclic organics of known structures. Hansen and Schnitzer[24] assumed a recovery of 10 percent from humic substances and calculated, on a functional group-free basis, that polycyclic aromatic hydrocarbons might account for from 12 to 25 percent of the mass of the humic material. Cheshire et al.[22] proposed that humic substances are built around a polynuclear aromatic "core" to which other constituents are attached, such as by H-bonding.

A word of caution should be mentioned with respect to the interpretation of results obtained by Zn dust distillation and fusion. Polycyclic aromatic hydrocarbons, believed to be derived from natural fires, have been detected in lipid

Table 10.7 Major Products of Zn Dust Distillation and Fusion of Humic Acids. Yields are Given as a Percentage of the Initial Starting Material. From Hansen and Schnitzer[24]

Structure		Compound	Yield, %
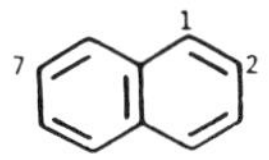 (1,2,7-trimethylnaphthalene structure)	XXIX	1,2,7-Trimethylnaphthalene	0.06
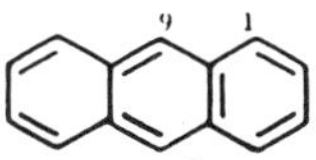 (anthracene structure)	XXX	Anthracene	
	XXXI	1-Methylanthracene	0.06-0.07
	XXXII	9-Methylanthracene	
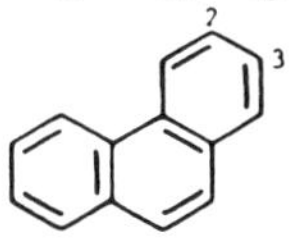 (phenanthrene structure)	XXXIII	2-Methylphenanthrene	0.06-0.07
	XXXIV	3-Methylphenanthrene	
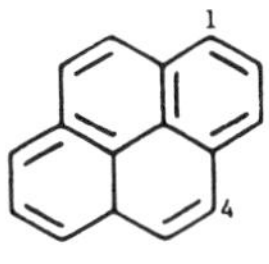 (pyrene structure)	XXXV	Pyrene	
	XXXVI	1-Methylpyrene	0.12-0.13
	XXXVII	4-Methylpyrene	
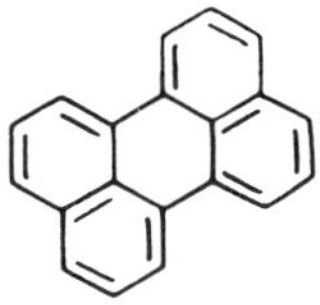 (perylene structure)	XXXVIII	Perylene	0.15-0.20

extracts of soil (see Chapter 7). Accordingly, it is possible that a portion of those found in humic and fulvic acids may occur as contaminants rather than as structural components. It should further be noted that the harsh condition of Zn dust distillation and fusion can lead to formation of polycyclic aromatic artifacts from aliphatic substances (i.e., cellulose).

Other Reduction Methods

Several other reductive procedures have been applied to humic substances, but with limited success. These include sodium and liquid ammonia, phosphorus and hydriodic acid, and hydrogenation–hydrogenolysis.

OXIDATIVE METHODS

Several oxidation techniques have been applied to soil humic substances with variable success. Much of the early work was disappointing but later research using improved isolation techniques was more fruitful, especially when labile oxygen-containing functional groups were protected by methylation prior to chemical treatment.

A general procedure for the separation and identification of products in digests obtained by oxidative degradation of humic substances is shown in Fig. 10.2.

Alkaline Cupric Oxide Oxidation

This is a relatively mild oxidation that has been used rather extensively in lignin chemistry. According to Schnitzer,[26] CuO−NaOH oxidation releases primarily phenolic compounds; aromatic structures bound through C−C bonds are not readily attacked. The main reaction governing alkaline CuO–oxidation may be alkaline hydrolysis.[27]

Green and Steelink[28] obtained a yield of 20 percent ether-soluble products by CuO−NaOH oxidation of humic acid but only 10 percent of this material, or 2 percent of the original humic acid, could be accounted for. Resorcinol-derived (e.g., **XIII–XV** of Table 10.1) and guaiacyl-type compounds (**I** to **VI** of Table 10.2) were found in about equal amounts.

Schnitzer[26] applied the technique to a fulvic acid previously methylated with diazomethane (CH_2N_2) and was able to account for about 30 percent of the material in degradation products. Slightly more than one-half of the degradation products, representing 18 percent of the original methylated fulvic acid, was identified in known compounds. Two-thirds of the identified compounds were phenolics, mainly phenolic acids; benzenecarboxylic acids accounted for about 15 percent. The remainder occurred in such compounds as alkanes, fatty acid methyl esters, and aliphatic dicarboxylic acid esters. A substantial quantity of dioctyl adipate was also found.

Griffith and Schnitzer,[29] on the basis of major products produced by alkaline–

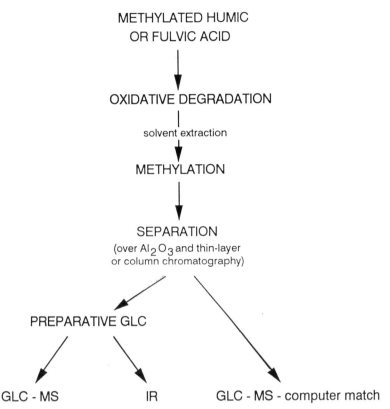

Fig. 10.2 General procedure for the separation and identification of products in digests from the oxidative degradation of humic substances. From Griffith and Schnitzer[2]

CuO oxidation, concluded that humic and fulvic acids of tropical volcanic soils did not differ significantly from those formed in other pedological environments.

Alkaline Nitrobenzene Oxidation

The technique of alkaline nitrobenzene oxidation, which has been of immense value for determining the structure of lignin, has been applied to soil organic matter by Morrison.[30] Yields of lignin-derived compounds were extremely low (about 1 percent). The method would seem to be inferior to the CuO−NaOH oxidation procedure mentioned above.

Alkaline Permanganate Oxidation

A useful method for characterizing humic substances is by alkaline-$KMnO_4$ oxidation. Although rather drastic, a number of compounds are produced that are not found by other degradation methods (e.g., benzenecarboxylic acids). Compounds containing benzene rings substituted with OH groups are subject to cleavage but this problem can be partially solved by prior methylation with diazomethane (CH_2N_2).[31-37] Results summarized by Schnitzer[31] show that yields of benzenecarboxylic acids were somewhat similar for methylated and unmethylated fulvic acid while yields of methoxyl compounds were higher for the methylated sample.

The approach used by Schnitzer and his associates[9, 10, 31-36] consists of oxidizing a 1-g sample of humic acid with 250 mL of 4 percent (w/v) aqueous $KMnO_4$ solution at pH 10, extraction of the products into ethyl acetate, re-methylation, and separation of components by preparative gas chromatography. Individual compounds are then identified by matching their mass and infrared spectra with those of standards (see Fig. 10.2). A gas chromatogram showing the separation of products resulting from the oxidation of a soil humic acid is given in Fig. 10.3.

Major products identified by alkaline–$KMnO_4$ oxidation of methylated humic substances include benzenecarboxylic acids, phenolic acids, and aliphatic dicarboxylic acids. The benzenecarboxylic acids include an array of di-, tri-, and tetra-benzenecarboxylic acids (Table 10.8) and are generally thought to be formed from oxidation of aliphatic side chains, some of which might be involved in the linking together of aromatic units.[2]

Yields of degradation products reported by Schnitzer,[31] and recorded in Table 10.9, are higher than others have reported using similar techniques. It should be noted in this respect that data obtained by gas–liquid chromatography is semiquantitative at best, especially when baseline drift is excessive (see Fig. 10.3).

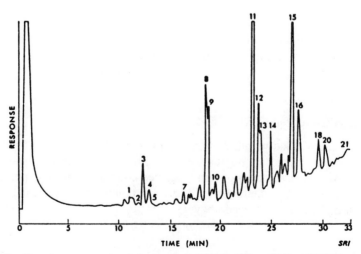

Fig. 10.3 Gas chromatogram of the products obtained by alkaline–$KMnO_4$ oxidation of a methylated humic acid. From Ortiz de Serra and Schnitzer,[35] reproduced by permission of Pergamon Press, Inc.

OXIDATIVE METHODS

Table 10.8 Benzene–Carboxylic Acids Identified in Digests from the Oxidative Degradation of Humic Substances with Alkaline KMnO$_4$. From Schnitzer[31]

$$R_1, R_2, R_3, R_4, R_5, R_6 \text{ on benzene ring}$$

XXXIX	$R_1 = R_2 = COOH$	1,2-Benzenedicarboxylic acid
XL	$R_1 = R_3 = COOH$	1,3-Benzenedicarboxylic acid
XLI	$R_1 = R_4 = COOH$	1,4-Benzenedicarboxylic acid
XLII	$R_1 = R_2 = R_3 = COOH$	1,2,3-Benzenetricarboxylic acid
XLIII	$R_1 = R_2 = R_4 = COOH$	1,2,4-Benzenetricarboxylic acid
XLIV	$R_1 = R_3 = R_5 = COOH$	1,3,5-Benzenetricarboxylic acid
XLV	$R_1 = R_2 = R_3 = R_4 = COOH$	1,2,3-4-Benzenetetracarboxylic acid
XLVI	$R_1 = R_2 = R_4 = R_5 = COOH$	1,2,4,5-Benzenetetracarboxylic acid
XLVII	$R_1 = R_2 = R_3 = R_5 = COOH$	1,2,3,5-Benzenetetracarboxylic acid
XLVIII	$R_1 = R_2 = R_3 = R_4 = R_5 = COOH$	Benzenepentacarboxylic acid
XLIX	All R's = COOH	Benzenehexacarboxylic acid

Problems encountered in interpreting results of KMnO$_4$ oxidations have been discussed by Maximov et al.,[38] who concluded that yields of identifiable oxidation products from humic acids are probably too high. Ogner and Gronneberg[37] accounted for only 6–7 percent of the weight of methylated humic and fulvic acids as degradation products, about one-half of which existed as benzenecarboxylic acids and their methylated derivatives (15 compounds identified). Most of the remaining degradation products consisted of C-4–C-18 dicarboxylic acids, presumed to be derived from the aliphatic part of the humic structure. In other work, Riffaldi and Schnitzer[8] found that removal of carbohydrates and other adsorbed materials from humic acids by acid hydrolysis (weight loss of 46 percent) did not greatly change the kind and amount of degradation products. However, products from the acid-treated samples contained higher amounts of phenolic constituents but relatively less benzenecarboxylic acids.

Considerable variation has been shown to exist in the kinds and amounts of oxidation products obtained by alkaline–KMnO$_4$ oxidation of humic and fulvic acids. Results recorded by Schnitzer[31] for some acid and neutral soils are summarized in Table 10.10. The most abundant oxidation products for typical samples were: 1) humic acids from acid soils; benzene tetra-, penta-, tri-, and hexacarboxylic acids and hydroxybenzenepentacarboxylic acid; 2) humic acids

Table 10.9 Yields of Oxidation Products Resulting From the Alkaline–KMnO$_4$ Oxidation of Methylated Humic Substances (in Milligrams/Gram)[a]

Substance	Aliphatic Dicarboxylic Acids	Benzenecarboxylic Acids	Phenolic Acids
Spodosol humic acid	41.74	58.98	70.70
Other soil humic acids	17.63	190.14	86.96
Fungal humic acid	4.15	30.15	4.94
Spodosol fulvic acid	0.00	114.89	89.47
Other fulvic acids	1.40	59.35	76.61

[a] The compounds were isolated and identified in their fully methylated forms (from Schnitzer[31]).

from neutral soils; benzenecarboxylic acids followed in order by hydroxybenzenepentacarboxylic acid and benzene tri- and pentacarboxylic acids; 3) fulvic acids from acid soils; 5-hydroxy-1, 2,4-benzenetetracarboxylic acid and benzene penta-, tetra, and hydroxybenzene–pentacarboxylic acids; 4) fulvic acids from neural soils; 2-hydroxy-1, 3, 5-benzenetricarboxylic acid, followed by benzene tetra-, penta-, and tricarboxylic acids. It should be emphasized that samples from a wide variety of soils and sediments will need to be analyzed before any generalizations can be made.

The aliphatic carboxylic acids recovered from humic substances by alkaline–KMnO$_4$ oxidation may arise from oxidation of straight-chain compounds or labile ring systems. A logical source of the benzenecarboxylic acids is from

Table 10.10 Summary of Major Types of Products Resulting from Alkaline–KMnO$_4$ Oxidation of Humic and Fulvic Acids from Contrasting Soil Types (in Milligrams)[a]

Types of Product	Humic Acids		Fulvic Acids	
	Acid Soils	Neutral Soils	Acid Soils	Neutral Soils
Aliphatic acid esters	53.80	7.22	1.00	6.03
Benzenecarboxylic acid esters	155.10	170.42	114.89	84.94
Phenolic acid esters and ethers	66.22	74.23	89.47	41.26
Total	275.12	251.87	204.36	132.23
Weight ratio, $\frac{\text{benzenecarboxylic}}{\text{aliphatic}}$	2.9	23.6	114.9	14.1
Weight ratio, $\frac{\text{phenolic}}{\text{aliphatic}}$	1.2	10.3	89.5	6.8
Weight ratio, $\frac{\text{benzenecarboxylic}}{\text{phenolic}}$	2.3	3.0	1.3	2.1

[a] From Schnitzer.[31]

oxidation of side chains associated with polycyclic aromatic compounds and/or destruction of aromatic rings containing oxygen.

Nitric Acid Oxidation

Only limited use has been made of the HNO_3 oxidation method to examine soil humic substances. Hansen and Schnitzer[39] subjected organic matter from the illuvial horizon of a Spodosol to oxidation and were successful in identifying a variety of aliphatic dicarboxylic acids (to 0.7 percent of the organic matter), benzenecarboxylic acids (to 3.8 percent), hydroxybenzoic acids (to 0.6 percent), and nitro compounds (to 5.5 percent). The nitro compounds included o-, m-, and p-nitrophenols, o-, m-, and p-nitrobenzoic acids, some di- and trinitrophenols, and some mono- and dinitrosalicyclic acids.

Peracetic Acid

With the exception of HNO_3, most oxidations are conducted under alkaline conditions, which may result in both serious alteration of the original material and the formation of undesirable by-products. In an attempt to avoid these problems, Schnitzer and Skinner[40,41] investigated the use of peracetic acid as an oxidizing agent under acidic conditions. Since the magnitude and extent of chemical alterations under acidic conditions would be minimized, the products obtained should be more representative of components associated with the structures of humic substances. Yields of major oxidation products from a humic acid were phenolic acids (4.3 percent) and benzenecarboxylic acids (15.2 percent). For a fulvic acid, the corresponding values were 4.1 and 7.3 percent, respectively. In addition, small amounts of fatty acids and aliphatic dicarboxylic acids were isolated. The peracetic acid oxidation of unmethylated humic substances produced essentially similar compounds as alkaline $KMnO_4$ and Cu−NaOH oxidations of methylated humic substances. However, total yields were somewhat different, particularly from fulvic acids, where peracetic acid yielded greater amounts of aliphatic but smaller amounts of phenolic compounds than did the other methods. The compounds produced by peracetic oxidation resembled qualitatively those obtained by HNO_3 oxidation, except that no nitro compounds were formed.

Sodium Hypochlorite

Oxidation of a soil humic acid with this reagent gave a variety of benzene-polycarboxylic acids (7 compound identified), along with relatively high amounts of aliphatic carboxylic acids.[42] Phenolic structures were absent, indicating cleavage of aromatic rings. Over 50 percent of the total C appeared as CO_2.

Oxidation with Hydrogen Peroxide (H_2O_2)

Principal products are CO_2 and H_2O with a maximum of 5 percent of the original material soluble in ether. The approach would appear to be of limited value as applied to soil organic matter.[43]

MISCELLANEOUS CHEMICAL APPROACHES

Degradation with Sodium Sulfide

Several investigators have used sodium sulfide solutions (10 percent) at 250°C under autoclave conditions to degrade humic acids.[44,45] This work, reviewed by Hayes and O'Callaghan,[3] shows that aromatic structures in the digests are significantly different from those obtained by oxidative procedures. Among the compounds identified were several aliphatic dicarboxylic acids and some long-chain alcohols and fatty acids. Only one benzene–polycarboxylic acid (benzene-1-2-dicarboxylic acid) was identified, and only in trace amounts. A result of particular interest is that several aromatic structures were identified that contained aliphatic side chains, as exemplified by structures **XLIX–LII**. Hayes and O'Callaghan[3] concluded that some of the compounds identified had origins in phenolpropane structures of the types associated with lignins.

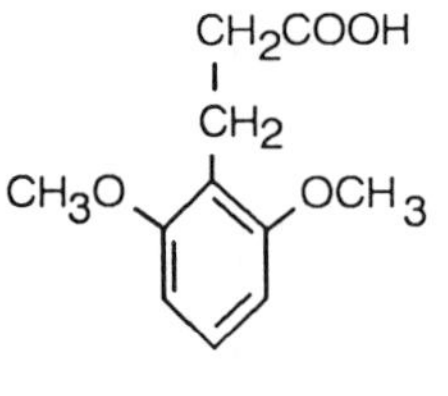

XLIX 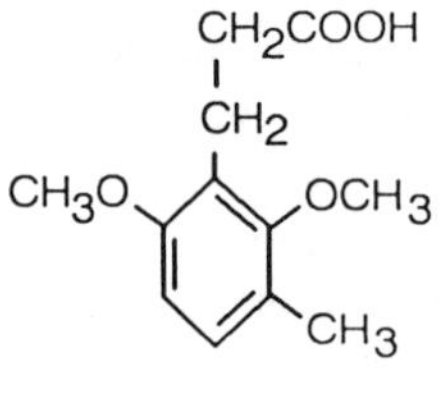   **L** 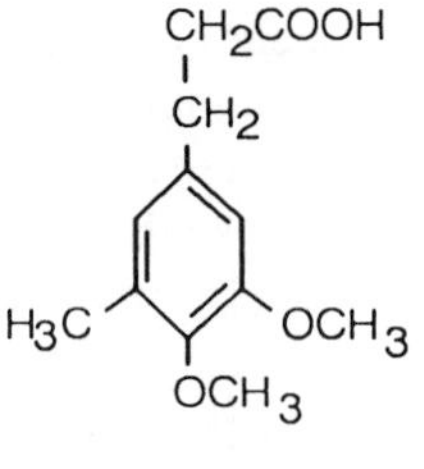   **LI** 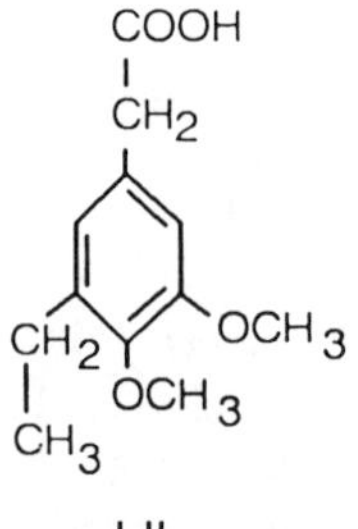   **LII**

Degradations carried out with sodium sulfide would cleave methyl ethers and their occurrence in the degradation products arises from the fact that phenolic OH groups had been methylated (using diazomethane) prior to separations by gas chromatography (GC) and identification by GC–mass spectrometry (GC–MS).

Depolymerization with Phenol

One useful approach for determining the nature of interaromatic linkages in coal is through depolymerization with phenol in the presence of a catalyst (*p*-toluenesulphonic acid or boron trifluoride). The mechanism involved is believed to be replacement of aromatic units by the phenol, yielding new compounds from which the original interaromatic linkage can be identified. The technique was applied by Jackson et al.[46] to a humic acid from a well-humified organic soil.

THERMAL METHODS

Thermogravimetric Analysis: Estimation for "Aromaticity"

Another tool for characterization soil organic matter fractions is by thermogravimetric analysis. Both differential thermal analysis (DTA) and isothermal heating have been used. For humic acids, plots of rate of weight loss versus temperature (DTG curves) show two principal peaks, one at a low temperature (about 280°C) and one at a higher temperature (>400°C). In contrast, curves for fulvic acid fail to exhibit a well-defined low temperature peak. Schnitzer[31] concluded that the main reactions governing the pyrolysis of humic and fulvic acids were: 1) dehydration up to 200°C, 2) elimination of functional groups between 250 and 280°C, and 3) decomposition of the "nuclei" at a high-temperature maximum (>400°C).

A differential thermogravimetric (DTG) curve for the organic matter from the Ao horizon of a Spodosol (temperature range of 127 to 577°C, or 400 to 800°K) is given in Fig. 10.4. At least two different peaks are evident in addition to the main ones shown as A and D. Turner and Schnitzer[47] concluded that reaction A signaled the breakup of aliphatic and/or alicyclic structures whereas D was due to decomposition of aromatic structures. Both reactions B and C

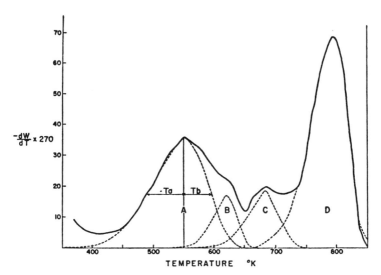

Fig. 10.4 DTG curve of a soil organic matter preparation. From Turner and Schnitzer,[47] reproduced by permission of Williams and Wilkins Co.

had high activation energies and specific rate constants and were also thought to be due to breakup of aromatic structures.

Since weight loss below 400°C (673°K) may be due to elimination of functional groups and aliphatic substance, an estimate for "degree of aromaticity" can be obtained by measuring weight loss above this temperature. For the results depicted in Fig. 10.4, approximately 60 percent of the weight was lost at temperatures below 400°C, leaving a "nucleus" of about 40 percent by weight. Ishiwatari[48] concluded that this approach greatly overestimated apparent aromaticity and that corrections would be required for residues left from carbohydrate and protein impurities in order to obtain true aromaticity.

Pyrolysis/Mass Spectrometry (Py/MS)

Analysis of products released by pyrolysis has emerged as a valuable approach for characterization of soil organic matter.[1] This subject is discussed in Chapter 10.

SUMMARY

Degradation methods show great promise for characterizing humic substances but additional research will be required before the full potential of the various approaches can be realized.

The nature and kind of degradation products depends on the method used for degradation. In interpreting results, it is essential to understand mechanisms of bond cleavages and to be able to relate the products to structures in the parent macromolecule. Some general observations are as follows:

1. Reductive degradation with Na-amalgam leads to cleavage of phenolic ether and biphenyl structures bearing activating functional groups (i.e., an OH group in ortho or para position to C linking the biphenol). By this approach, phenolics are obtained from humic substances that can be identified as to source (i.e., lignin or synthesized by microorganisms).
2. Polycyclic aromatic structures have been detected in digests obtained by Zn dust distillation and fusion and this has led to the conclusion that the central "core" of humic substances consists of polycyclic hydrocarbons to which other constituents (saccharides, peptides, phenols) are attached. Due to the harsh conditions of such degradation, some of the compounds that have been identified may be artifacts.
3. Oxidative degradations with $KMnO_4$ produce aliphatic substances (alkanes, fatty acids, and aliphatic carboxylic acids) in addition to phenolic acids and benzenecarboxylic acids. Whereas similar quantities of aliphatic compounds are obtained from humic and fulvic acids, lower amounts of phenolic acids, but higher amounts of benzenecarboxylic acids, are obtained from humic acids.

4. Products obtained from degradation of humic acids with sodium sulfide are unique in that many of the products contain aliphatic side chains; benzenecarboxylic acids are generally absent. As is the case for Na-amalgam reduction, results of sodium sulfide degradation allow predictions to be made of the origins of the compounds identified.
5. Chemical degradation studies have provided information from which possible "core" structures for humic substances can be derived (see Chapter 12).

REFERENCES

1 J. M. Bracewell, K. Haider, S. R. Larter, and H.-R. Schulten, "Thermal Degradation Relevant to Structural Studies of Humic Substances," in M. H. B. Hayes, P. MacCarthy, R. L. Malcolm, and R. S. Swift, Eds., *Humic Substances II: In Search of Structure*, Wiley, New York, 1989, pp. 181–222.
2 S. M. Griffith and M. Schnitzer, "Oxidative Degradation of Soil Humic Substances," in M. H. B. Hayes, P. MacCarthy, R. L. Malcolm, and R. S. Swift, Eds., *Humic Substances II: In Search of Structure*, Wiley, New York, 1989, pp. 69–98.
3 M. H. B. Hayes and M. R. O'Callaghan, "Degradations with Sodium Sulfide and with Phenol," in M. H. B. Hayes, P. MacCarthy, R. L. Malcolm, and R. S. Swift, Eds., *Humic Substances II: In Search of Structure*, Wiley, New York, 1989, pp. 143–180.
4 M. H. B. Hayes and R. S. Swift, "The Chemistry of Soil Organic Colloids," in D. J. Greenland and M. H. B. Hayes, Eds., *The Chemistry of Soil Constituents*, Wiley, New York, 1978, pp. 179–230.
5 J. W. Parsons, "Hydrolytic Degradations of Humic Substances," in M. H. B. Hayes, P. MacCarthy, R. L. Malcolm, and R. S. Swift, Eds., *Humic Substances II: In Search of Structure*, Wiley, New York, 1989, pp. 99–120.
6 F. J. Stevenson, "Reductive Cleavage of Humic Substances," in M. H. B. Hayes, P. MacCarthy, R. L. Malcolm, and R. S. Swift, Eds., *Humic Substances II: In Search of Structure*, Wiley, New York, 1989, pp. 121–142.
7 N. A. Burges, H. M. Hurst, and B. Walkden, *Geochim. et Cosmochim. Acta*, **28,** 1547 (1964).
8 R. Riffaldi and M. Schnitzer, *Soil Sci.*, **115,** 349 (1973).
9 J. A. Neyroud and M. Schnitzer, *Geoderma*, **13,** 171 (1975).
10 M. Schnitzer and M. I. Ortiz de Serra, *Geoderma*, **9,** 119 (1973).
11 T. Jakab, P. Dubach, N. C. Mehta, and H. Deuel, *Z. Pflnähr Dung. Bodenk.*, **102,** 8 (1963).
12 J. A. Neyroud and M. Schnitzer, *Agrochemica*, **19,** 116 (1975).
13 J. P. Martin, K. Haider, and C. Saiz-Jimenez, *Soil Sci. Soc. Amer. Proc.*, **38,** 760 (1974).
14 M. Schnitzer, M. I. Ortiz de Serra, and K. Ivarson, *Soil Sci. Soc. Amer. Proc.*, **37,** 229 (1973).
15 J. P. Martin and K. Haider, *Soil Sci.*, **107,** 260 (1969).

16 Y. Matsui and K. Kumada, *Soil Sci. Plant Nutr.*, **23**, 341, 491 (1977).
17 T. J. Piper and A. M. Posner, *Soil Biol. Biochem.*, **4**, 513, 525 (1972).
18 J. F. Dormaar, *Plant and Soil*, **31**, 182 (1969).
19 J. Mendez and F. J. Stevenson, *Soil Sci.*, **102**, 85 (1966).
20 F. J. Stevenson and J. Mendez, *Soil Sci.*, **103**, 383 (1967).
21 K. R. Tate and K. M. Goh, *New Zealand J. Sci.*, **16**, 59 (1973).
22 M. V. Cheshire, P. A. Cranwell, C. P. Falshaw, and R. D. Haworth, *Tetrahedron*, **23**, 1669 (1967).
23 M. V. Cheshire, P. A. Cranwell, and R. D. Haworth, *Tetrahedron*, **24**, 5155 (1968).
24 E. H. Hansen and M. Schnitzer, *Soil Sci. Soc. Amer. Proc.*, **33**, 29 (1969).
25 E. H. Hansen and M. Schnitzer, *Fuel*, **48**, 41 (1969).
26 M. Schnitzer, *Soil Biol. Biochem.*, **6**, 1 (1974).
27 J. A. Neyroud and M. Schnitzer, *Soil Sci. Soc. Amer. Proc.*, **38**, 907 (1974).
28 G. Green and C. Steelink, *J. Org. Chem.*, **27**, 170 (1962).
29 S. M. Griffith and M. Schnitzer, *Soil Sci.*, **122**, 191 (1976).
30 R. I. Morrison, *J. Soil Sci.*, **14**, 201 (1963).
31 M. Schnitzer, "Chemical, Spectroscopic, and Thermal Methods for the Classification and Characterization of Humic Substances," *Proc. Intern. Meetings on Humic Substances*, Pudoc, Wageningen, 1972, pp. 293–310.
32 S. U. Khan and M. Schnitzer, *Geoderma.*, **7**, 113 (1972).
33 S. U. Khan and M. Schnitzer, *Can. J. Soil Sci.*, **52**, 43 (1972).
34 K. Matsuda and M. Schnitzer, *Soil Sci.*, **114**, 185 (1972).
35 M. I. Ortiz de Serra and M. Schnitzer, *Soil Biol. Biochem.*, **5**, 287 (1973).
36 M. Schnitzer and J. G. Desjardins, *Soil Sci. Soc. Amer. Proc.*, **34**, 77 (1970).
37 G. Ogner and T. Gronneberg, *Geoderma*, **19**, 237 (1977).
38 O. B. Maximov, T. V. Shvets, and Yu. N. Elkin, *Geoderma*, **19**, 63 (1977).
39 E. H. Hansen and M. Schnitzer, *Soil Sci. Soc. Amer. Proc.*, **31**, 79 (1967).
40 M. Schnitzer and S. I. M. Skinner, *Soil Sci.*, **118**, 322 (1974).
41 M. Schnitzer and S. I. M. Skinner, *Can. J. Chem.*, **52**, 1072 (1974).
42 S. K. Chakbarartty, H. O. Kretschmer, and S. Cherwonka, *Soil Sci.*, **117**, 318 (1974).
43 S. M. Savage and F. J. Stevenson, *Soil Sci. Soc. Amer. Proc.*, **25**, 35 (1961).
44 J. D. Craggs, M. H. B. Hayes, and M. Stacey, *Trans. 10th Int. Congr. Soil Sci. (Moscow)*, **2**, 318 (1974).
45 M. H. B. Hayes, M. Stacey, and R. S. Swift, *Fuel*, **51**, 211 (1972).
46 M. P. Jackson, R. S. Swift, A. M. Posner, and J. R. Knox, *Soil Sci.*, **114**, 75 (1972).
47 R. C. Turner and M. Schnitzer, *Soil Sci.*, **93**, 225 (1962).
48 R. Ishiwatari, *Soil Sci.*, **107**, 53 (1969).

11

CHARACTERIZATION OF SOIL ORGANIC MATTER BY NMR SPECTROSCOPY AND ANALYTICAL PYROLYSIS

The new instrumental techniques described in this chapter (nuclear magnetic resonance spectroscopy, NMR, and analytical pyrolysis) have led to major advances in studies on the nature and chemical composition of soil organic matter. Both techniques have the potential for characterizing organic matter in the intact soil and its size fractions without the need for extraction and fractionation; furthermore, they can provide information on compositional changes in crop residues, peat, and the litter of forest soils during biodegradation and humification.

Of the new approaches, ^{13}C–NMR has been the most frequently used and the most widely accepted; accordingly, greatest attention will be given to this subject. Pyrolysis techniques have enormous potential but they have been used only sparingly because of high instrument costs and the sparsity of operating equipment.

THEORY OF NUCLEAR MAGNETIC RESONANCE SPECTROSCOPY (NMR)

Nuclear magnetic resonance (NMR) spectroscopy is a form of absorption spectroscopy akin to infrared and ultraviolet spectroscopy, as discussed later in Chapter 13. Like other spectroscopic methods, NMR spectroscopy depends upon the interaction of electromagnetic radiation with nuclear, atomic, or molecular species. Both ^{13}C–NMR and ^{1}H–NMR spectroscopy have been applied to soil organic matter, with considerable success. A few applications have been made of ^{15}N–NMR and ^{31}P–NMR spectroscopy.

The subject of NMR spectroscopy is complex and a knowledge of advanced physics, chemistry, and NMR instrumentation is required for a thorough comprehension of the subject. However, appreciation of the method can be achieved without a full understanding of all aspects of NMR theory and technology. Due to space considerations, only a cursory examination can be given herein.

The topic of NMR has been well-documented and there are several books and reviews on application of ^{13}C-NMR and ^{1}H-NMR to soil organic matter.[1-8] The theory and practice of NMR spectroscopy has been covered by Silverstein et al.[9]

All nuclei carry a charge. In some nuclei (e.g., ^{13}C and ^{1}H), this charge spins about the nuclear axis and thereby generates a magnetic dipole along the axis. Under the influence of an external magnetic field, H_0, the nuclei precess about the magnetic field in much the same manner as a gyroscope precesses under the influence of gravity, as illustrated in Fig. 11.1.

The angular momentum of the spinning charge is described in terms of spin number, I, which specifies the number of orientations that a nucleus may assume in an external magnetic field, in accordance with the formula $2I + 1$. For both ^{13}C and ^{1}H, $I = 1/2$; thus, each have two energy levels. When electromagnetic radiation is applied to generate a second small alternating field, H_1, at a right angle to the external magnetic field, H_0, the molecule absorbs energy. At the point where the frequency matches the precession frequency, the nuclei flip over or resonate, thereby inducing a voltage change (resonance signal), which is amplified and recorded.

The fundamental equation relating the frequency of electromagnetic radiation, ν, to the magnetic field strength, H_0, is given by

$$\nu = \frac{\gamma H_0}{2\pi}$$

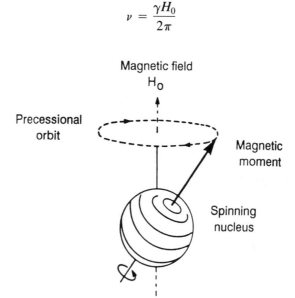

Fig. 11.1 Precession of nuclear spin about a magnetic field. Adapted from Wilson,[7] reproduced by permission of John Wiley & Sons, Inc.

where γ, the gyromagnetic ratio, is constant for a particular nuclear type. Of the elements useful in soil organic matter research, ^1H has the most favorable gyromagnetic ratio, followed in order by ^{31}P, ^{13}C, and ^{15}N. With ^1H assigned a value of unity, the sensitivities for the four elements are 1, 0.405, 0.251, and 0.092, respectively.

The value obtained for ν is dependent on the strength of the magnetic field. To facilitate comparisons, results of an NMR experiment are expressed in terms of "chemical shifts" with respect to a reference standard. The chemical shift, δ, is given by

$$\delta = \frac{\nu_{example} - \nu_{reference}}{\nu_{reference}} \times 10^6$$

For both $^{13}C-$ and ^1H–NMR spectroscopy, the most common reference standard is an aqueous solution of tetramethylsilane, $Si(CH_3)_4$, TMS.

In NMR spectroscopy, the experimental sample is placed in a uniform magnetic field and a second magnetic field is introduced, as illustrated in Fig. 11.2. Either the magnetic field or the frequency of the oscillating field is varied until the condition of resonance is reached. A spectrum is subsequently produced relating the amount of energy absorbed from the oscillating field to chemical shift δ.

Essential components of an NMR spectrometer are: 1) a strong magnet with a homogeneous field that can be varied continuously and precisely over a relatively narrow range, 2) a radiofrequency transmitter and a radiofrequency

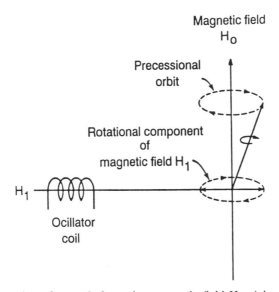

Fig. 11.2 Generation of second alternating magnetic field H_1. Adapted from Silverstein et al.[9]

receiver, 3) a sample holder that positions the sample relative to the main magnetic field, the transmitter coil, and the receiver coil, and 4) a V/F converter and a computer for recording the data.

The sample is placed in the sample holder of the spectrometer and the magnetic field H_0 is increased, thereby causing the precession rate of the nuclei to increase. When the frequency of the alternating magnetic field H_1 is matched by the precession frequency, the nuclei flip over and a voltage is induced in the detector coil. This resonance signal is subsequently amplified and recorded. From the spectrum that is produced, information is provided regarding the chemical environment of C or H atoms in the sample.

For ^{1}H-NMR, adequate spectra can be achieved by continuous wave measurements, in which the irradiation frequency is fixed and the magnetic field strength is slowly and continuously changed. However, routine ^{13}C-NMR spectra could not be obtained until the development of Fourier transform spectrometers, which allows for the rapid acquisition and averaging of many individual spectra to produce an average spectrum with a high signal-to-noise ratio.

As indicated earlier, chemical shift δ is the parameter from which structural information is obtained. The basis for the analysis is that the frequency at which nuclei resonate (e.g., ^{13}C and ^{1}H) is governed by the chemical environment of the nuclei (e.g., C and H) in the sample. To use one example, the chemical shift δ of ^{13}C in an aromatic ring is different from that of a COOH group; thus, the two can be separated by ^{13}C-NMR spectroscopy.

Chemical shifts δ of ^{13}C in structures of importance in the analysis of humic substances are shown in Fig. 11.3.

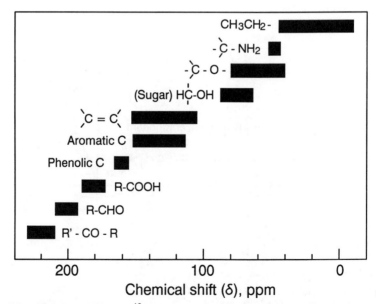

Fig. 11.3 Chemical shifts for ^{13}C in select functional groups relative to tetramethylsilane. Adapted from Malcolm[1] and Wilson[8]

TECHNIQUES FOR ^{13}C–NMR SPECTROSCOPY

The technique of ^{13}C–NMR spectroscopy is currently the most widely used and definitive instrumental technique for the characterization of soil organic matter. Two general approaches have been used: liquid-state ^{13}C–NMR spectroscopy and solid-state ^{13}C–NMR spectroscopy. Because of the insolubility of some components of soil organic matter (i.e., humin but also humic acids), greater use has been made of solid-state ^{13}C–NMR spectroscopy. Also, solid-state ^{13}C–NMR spectroscopy is suitable for examination of soil organic matter *in situ*.

Limitations of ^{13}C–NMR Spectroscopy

Some limitations of ^{13}C–NMR spectroscopy are listed below.

1. The natural abundance of ^{13}C is very low (~1.1 percent of total organic C). The most abundant isotope (^{12}C) does not exhibit a magnetic spin momentum and thus cannot be observed by NMR.
2. It is the inherent nature of the ^{13}C–NMR experiment that a very large number of signals must be collected and averaged before an adequate response can be measured and recorded. Accordingly, long analysis times are required.
3. For solids, line broadening and other effects reduce resolution, as discussed in the section on solid-state ^{13}C–NMR spectroscopy.
4. Costs for the purchase and maintenance of a modern ^{13}C–NMR spectrometer are beyond the means of most laboratories where research on soil organic matter is underway.
5. As applied to humic substances, results of ^{13}C–NMR spectroscopy must be interpreted with caution. While it is usual to analyze the data by dividing the spectrum into regions based on results obtained for known compounds, these "chemical-shift" regions are not completely exclusive or specific. For soil organic matter, processes of decomposition and humification may produce chemical structures that are no longer amenable to conventional interpretations. As noted later, results obtained by ^{13}C–NMR spectroscopy have not been in complete agreement with results obtained by chemical approaches, a result that may be due to failure of ^{13}C in the various groups to follow precisely the chemical shifts obtained for well-defined organic molecules.

Advantages and Disadvantages of Liquid-State ^{13}C–NMR Spectroscopy

Most liquid-state ^{13}C–NMR spectra of humic substances have been obtained in an alkaline solution (i.e., 0.1M NaOH) with sample concentrations of 50 to 100 mg/mL (100–200 mg in sample analyzed) on NMR spectrometers with resonant frequencies ranging from 200 to 400 MHz. The spectrometer time required to obtain a spectrum of satisfactory signal-to-noise ratio depends on the type of experiment but will normally be of the order of 6 to 12 h.

Liquid-state ^{13}C–NMR has both advantages and disadvantages. One advantage is decreased line broadening due to dipolar interactions and chemical shift anisotropy (see next section). Some disadvantages of liquid-state ^{13}C–NMR are as follows:[1]

1. Low solubility of humic substances in suitable NMR solvents. The differential solubility of various components in a mixture can result in sample fractionation
2. Formation of micelles and colloidal suspensions
3. Long relaxation times and proton coupling effects
4. Solvent effects on chemical shifts. Chemical shifts of the sample can also be obscured by chemical shifts due to the solvent
5. Inability to cool the sample to low temperatures and still maintain a liquid phase

Limitations of Solid-State ^{13}C–NMR Spectroscopy

An undesirable feature of solid-state ^{13}C–NMR is that of line-broadening due to: 1) dipolar interactions between ^{13}C and ^{1}H nuclei, 2) chemical shift anisotropy of highly anisotropic Cs (primary aromatic and carboxyl), and 3) a poor signal-to-noise ratio. These problems are not normally encountered by liquid-state ^{13}C–NMR. One advantage of solid-state ^{13}C–NMR spectroscopy is that the sample can be recovered in an unaltered state.

The utility of solid-state ^{13}C–NMR has been greatly enhanced in recent years by using a combination of the following techniques to improve spectral quality:

High-power proton decoupling: Humic substances contain two interacting spin species—^{13}C and ^{1}H. Much of the ^{13}C line broadening arises from interactions between the two. Through high-power proton decoupling, the magnetic influence of the proton nuclei on a neighboring ^{13}C nuclei is eliminated or minimized, thereby reducing line broadening.

Cross-polarization (CP): A second technique for reducing line broadening is cross-polarization or proton-enhanced nuclear induction spectroscopy. Essentially, this technique results in the transfer of net magnetization from the abundant ^{1}H spins to the less abundant ^{13}C spins. The procedure also overcomes the problem of dipolar line broadening; in addition, resolution is enhanced by an increase in net ^{13}C magnetization.

Magic-angle spinning (MAS): This is a technique for further decreasing line broadening by eliminating the remaining vestiges of dipolar ^{13}C–^{1}H interactions and chemical shift anisotropy effects by rapidly rotating the sample at the so called "magic-angle" with respect to the applied magnetic field. This magic angle is 54.7°.

All three techniques must be used simultaneously in order to obtain optimum results; further improvements are possible using "dephasing," a variation of the basic approach in which an extra delay period is used to enhance signal responses. The symbol CPMAS is used to designate application of cross-polarization and magic-angle spinning in the solid-state ^{13}C–NMR analysis of solids (i.e., **CPMAS ^{13}C–NMR**).

In addition to soil humic and fulvic acids,[10-22] CPMAS ^{13}C–NMR spectroscopy has been used for examination of the humin fraction of soil organic matter,[23,24] mineral soils and their size and density fractions,[25-31] Histosols,[32-34] forest litter,[35-38] and humic materials in sewage sludges.[39] The technique has also been used to determine cultivation effects on soil organic matter,[28] to monitor chemical changes associated with incorporation of ^{13}C-labeled substrates into components of soil organic matter,[40-42] for examining pathways of humus synthesis,[43,44] for evaluating purification procedures for humic acids,[16] and to follow chemical changes in composed organic matter.[45] The literature is extensive and additional references can be found in comprehensive reviews on the subject.[1-8]

^{13}C–NMR SPECTROSCOPIC ANALYSIS OF HUMIC AND FULVIC ACIDS

Use has been made of both liquid-state and solid-state ^{13}C–NMR spectroscopy for characterization of humic substances. Malcolm[1] concluded that almost identical ^{13}C–NMR spectra can be obtained by both procedures provided proper standardization procedures are used. Schnitzer and Preston,[20] on the other hand, found that liquid-state NMR gave higher resolution, and provided more detailed information, than solid-state NMR.

A CPMAS ^{13}C–NMR spectrum of a humic acid from a Mollisol is given in Fig. 11.4. Unlike simple organic molecules where sharp, well-separated peaks are the rule, spectra of humic substances invariably show broad and diffuse bands. This result can be explained by the complex and heterogeneous nature of humic substances.

A ^{13}C–NMR spectrum of a humic or fulvic acid provides an inventory of the different components of which the material is composed. It has been common practice to divide the spectra into regions corresponding to specific chemical classes, as listed below:

Unsubstituted aliphatic C (e.g., alkanes, fatty acids)	0–50 ppm
N-Alkyl (e.g., amino acid-, peptide, and protein-C) + methoxyl C	50–60 ppm
Aliphatic C—O (notably carbohydrates)	60–110 ppm
Aromatic C, consisting of unsubstituted and alkyl substituted aromatic C (110–150 ppm) and phenolic C (150–160 ppm). Alkenes also resonate in this region	110–160 ppm
Carboxyl C (includes the carboxylate ion, COO$^-$)	160–190 ppm
Ketonic C=O of esters and amides	190–200 ppm

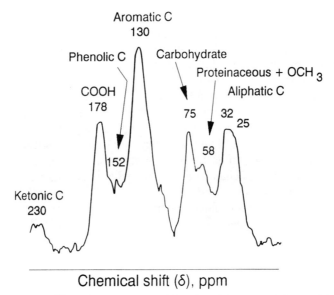

Fig. 11.4 CPMAS ^{13}C-NMR spectrum of humic acid extracted from a Mollisol soil. Adapted from Schnitzer[4]

The above chemical shift zones are approximations in that published spectra are somewhat variable, depending on operating parameters for the spectrometer. Also, the chemical shifts are in reference to the commonly used tetramethylsilane and some investigators have used other standards. For quantitative analysis, areas of the spectrum corresponding to the various chemical shift zones are measured, such as with an integrator.

In some studies, the first two chemical shift zones (0–60 ppm) have been combined and labeled as alkyl C; in other cases, the 50–60 ppm zone has been assigned exclusively to methoxyl groups. The regions corresponding to 60–110 ppm (C—O resonance) is normally attributed to carbohydrates but other compounds may contribute to resonance in this region.

The aromatic region (110–160 ppm) consists of two main groups of resonances, one centered near 130 ppm and the other near 150 ppm. The latter is characteristic of C next to an oxygen atom and has been assigned to phenolic C. It should be noted that alkenes (aliphatic compounds containing the double bond) may also contribute to resonance in the 110–160 ppm region.

In Fig. 11.5, spectra for the fulvic acid "fractions" of three soils are compared with those for humic acids from the same source. The most obvious difference in the spectra for the fulvic acid "fractions" is the pronounced peak for carbohydrates near 70 ppm. As can be seen from Fig. 11.6, a large portion of the carbohydrate material in the fulvic acid "fraction" is removed when the colored component ("generic" fulvic acid) is recovered through sorption-desorption on an XAD-8 resin column (see Chapter 2 for a discussion of methods for recovering "generic" fulvic acids).

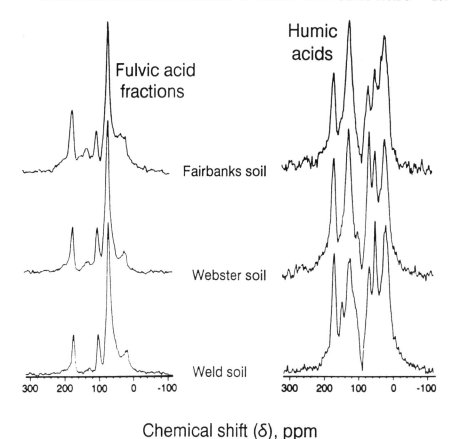

Fig. 11.5 CPMAS ^{13}C–NMR spectra of the fulvic acid *"fractions"* and the humic acids from three soil types. Adapted from Malcolm[2]

An examination of the published literature (see reviews[1-8]) shows that considerable dissimilarities have been reported in the composition of humic substances from diverse soil types through application of ^{13}C–NMR spectroscopy, indicating wide differences in the composition of humic and fulvic acids from different sources. However, a word of caution must be noted in that comparison of results from different laboratories are fraught with difficulties. Not all work has been done with acquisition parameters optimized for quantitative analysis, and there has been a lack of standardized procedures for sample treatment and instrumental operation.[1] Another major problem is that humic acids, as normally prepared, are not completely ash-free (see Chapter 2) and the presence of paramagnetic inorganic species (notably Fe^{3+}) can mask responses of the ^{13}C nuclei. It should also be noted that spectra have been interpreted by correlating the various resonance signals with those of known chemical structures; with complex macromolecules, such as humic and fulvic acids, the ^{13}C nuclei

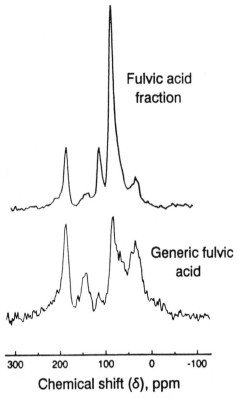

Fig. 11.6 Comparisons of the CPMAS ^{13}C-NMR spectrum for the fulvic acid "*fraction*" (top) and the "*generic*" fulvic acid (bottom) from a Webster soil. Adapted from Malcolm[2]

can exist in a broad array of chemical environments, thereby producing a wide variety of chemical shifts.

Special uncertainties are encountered in interpreting ^{13}C-NMR spectra of soil fulvic acids, namely: 1) most spectra are for characterization of the fulvic acid "fraction" rather than "generic" fulvic acids, 2) the tendency of soil researchers has been to emphasize the humic acid component and to discard the fulvic acid fraction without investigation, and 3) a diversity of extractants have been used and methods for recovery of "generic" fulvic acids have not been standardized.[2]

Results obtained by Malcolm[2] for the distribution of C in the humic acids and fulvic acid "fractions" of several diverse soil types are shown in Fig. 11.7; results for a "generic" fulvic acid are shown in Fig. 11.8. Similar data for the C composition of some selected stream and groundwater humic and fulvic acids are shown in Fig. 11.9.

All samples are shown to be more aliphatic than aromatic, especially the

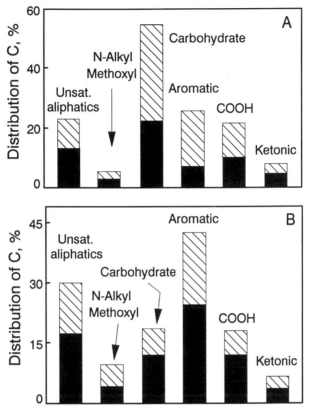

Fig. 11.7 Distribution of organic C in the fulvic acid *fractions* (A) and the humic acids (B) from eight soil types. The broken portion of the bars indicate the range of values obtained. Prepared from data published by Malcolm[2]

fulvic acid "fractions." The content of total aliphatic C (i.e., unsubstituted aliphatic C + N-alkyl C + carbohydrate C) is somewhat higher for the stream and groundwater humic and fulvic acids than for the soil humic acids and fulvic acid "fractions" (compare Fig. 11.9 with Figs. 11.7 and 11.8). As expected, the content of carbohydrate C in the "generic" fulvic acid was somewhat lower than for the fulvic acid "fractions."

The percentage of the C accounted for as aromatic C (f_{a1}) has been referred to as the "aromaticity" of soil organic matter, although in some work the C of COOH groups has been omitted (f_{a2}) when making the calculations. The equations are

$$f_{a_1} = \frac{\text{Aromatic peak area (110–160 ppm)}}{\text{Total peak area (0–230 ppm)}} \times 100$$

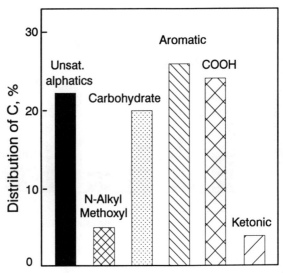

Fig. 11.8 Composition of organic C in the "*generic*" fulvic acid from an Elliott silt loam soil. Prepared from data published by Malcolm[2]

and

$$f_{a_2} = \frac{\text{Aromatic peak area (110–160 ppm)}}{\text{Total peak area–COOH peak area (160–190 ppm)}} \times 100$$

Aromatics for the humic substances used in the above-mentioned studies are given in Table 11.1. For the soil humic acids, f_{a1} values ranged from 25 to 42 percent, with an average of 29 percent; aromaticity with exclusion of the C of COOH averaged 37 percent. The range in aromaticities for the soil humic acids were of the same order as those generally recorded for humic acids.[10–13,15,16] However, somewhat higher aromaticities (to 80 percent) were obtained by Schnitzer and Preston[20] and Schnitzer et al.[22] Newman and Tate[14] examined the ^{13}C-NMR characteristics of humic acids from soils of a developing sequence and concluded that the aromatic nature of humic acids decreased with increasing soil development.

Results of ^{13}C-NMR spectroscopic analyses of humic substances can be summarized as follows[3]:

1. The most significant result of ^{13}C-NMR spectral studies of humic and fulvic acids is that these materials have been shown to be more aliphatic in nature than previously thought. Recognition of the rather high aliphatic content of humic substances has changed previous concepts as to their chemical nature and composition.
2. Data obtained by ^{13}C-NMR spectroscopy support findings obtained with other techniques (described in earlier chapters) indicating that humic acids are very different from fulvic acids in composition and reactivity.

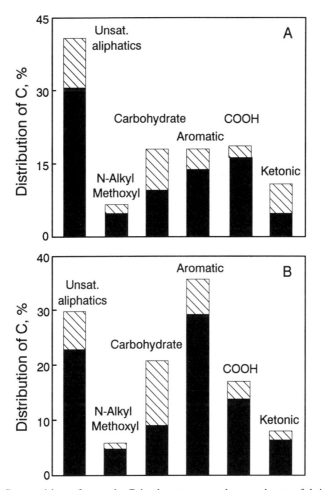

Fig. 11.9 Composition of organic C in the stream and groundwater fulvic acids (A) and humic acids (B) from four different sources. The broken portion of the bars indicate the range of values obtained. Prepared from data published by Malcolm[2]

Table 11.1 Ranges in Aromaticities (f_{a1} and f_{a2}) for the Humic Samples Depicted in Figs. 11.7, 11.8, and 11.9. Adapted from Malcolm[2]

Type	Number of Samples	f_{a1}	f_{a2}
Humic acids			
Soil	7	25–42	30–53
Stream	4	30–36	38–47
Soil fulvic acid "*fraction*"	6	6–26	7–37
"*Generic*" fulvic acids			
Soil	1	26	36
Stream	4	14–18	18–25

3. Fulvic acids are less aromatic than humic acids. For the former, aromatic structures account for only 25 percent or so of their molecular masses. "Generic" fulvic acids have lower carbohydrate (and possibly peptide) contents than the fulvic acid "fraction," although "generic" fulvic acids may contain compounds sometimes classed as pseudopolysaccharides.
4. Contrary to previous views, results of ^{13}C-NMR studies have shown that fulvic acids from diverse environments (soils, sediments, natural waters) are not identical. Differences also exist among humic acids from different environments.
5. In comparison to chemical methods (see Chapter 9), results of ^{13}C-NMR studies indicate a lower phenolic but a higher COOH content of humic substances. One explanation that has been given for this effect is that the signals for phenolic C may be shifted downfield into the region of COOH groups and that resonances for the two overlap.[4,20,21] Still another anomaly is that estimates for carbohydrates by ^{13}C-NMR are higher than expected based on chemical methods for monosaccharides.[4,11] In the studies of Schnitzer and Preston,[21] the carbohydrate content of the organic matter in some soil extracts, as estimated by ^{13}C-NMR spectroscopy, was 2.7 to 57 times higher than that measured by gas–liquid chromatography (see Chapter 5). Methods and problems in the quantification of ^{13}C-NMR spectral data have been discussed by Wilson.[7,8]
6. As a general rule, the aromaticity of humic acids is of the order of 35 percent (i.e., one out of three C atoms are aromatic) and follows the order: peat humic acids > soil humic acids > marine humic acids. The order of fulvic acids is: soil fulvic acids > sediment fulvic acids > stream fulvic acids.

The technique of ^{13}C-NMR has opened a new area for the determination of functional groups in humic substances through methylation with ^{13}C-enriched methylating agents.[46,47] Good agreement has been obtained for COOH and phenolic OH groups by ^{13}C-NMR analysis of the enriched material and potentiometric titration.[46]

^{13}C-NMR SPECTROSCOPIC ANALYSIS OF SOIL ORGANIC MATTER IN SITU

CPMAS ^{13}C-NMR spectroscopy has been found to be a valuable tool for the direct analysis of organic matter in soils. By this technique, estimates can be obtained for the major types of organic C in soils, namely, aliphatic C of alkanes and fatty acids, N-alkyl (proteinaceous substances) + methoxyl, carbohydrates, aromatic C, and COOH.

There are several limitations in application of CPMAS ^{13}C-NMR spectroscopy to mineral soils, including the following:

1. The magnitude of the signal, and the quality of the resulting spectra, are dependent upon the amount of ^{13}C in the soil. The low C content of most mineral soils, along with the low natural abundance of ^{13}C (1.1 percent of the total organic C), makes the acquisition of high-quality spectra extremely difficult unless excessively long scan periods are used. Physical fractionations have been used to concentrate C by removal of the coarser particle-size fractions and chemical procedures have been used to selectively remove inorganic components.[24,25]

2. The presence of paramagnetic species containing unpaired electrons can reduce the efficiency of signal acquisition in that ^{13}C nuclei near the paramagnetics may be rendered NMR "invisible." The major paramagnetic component in mineral soils is Fe^{3+} but other paramagnetic species may also be present. Reduction of Fe^{3+} to the nonparamagnetic Fe^{2+} has been used to improve signal resolution.[25]

3. A solid-state ^{13}C-NMR spectrum represents the sum of major organic structures in the sample, but it does not distinguish between humic substances that vary in molecular weight or of structures that are associated with inorganic soil components. According to Wilson,[8] there is no reason to believe that "all" of the C in a soil is being observed by CPMAS ^{13}C-NMR.

A CPMAS ^{13}C-NMR spectrum for a Mapourika soil from New Zealand is shown in Fig. 11.10. As has been the case for many other soils,[25-31] the most

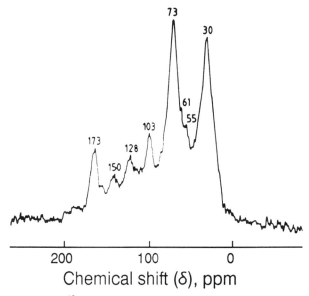

Fig. 11.10 CPMAS ^{13}C-NMR spectrum of a Mapourika soil. From Wilson[8]

dominant peaks are for the C—O of carbohydrates and related substances (73 ppm) and the aliphatic C of alkanes (30 ppm). As for humic and fulvic acids, variations in the magnitude of the aromatic C signals (110–160 ppm) have been used to demonstrate differences in the aromaticity of soil organic matter. As a rule, from 15 to 20 percent of the organic C in mineral soils occurs as aromatic C. Zech et al.[31] found that cultivation led to an increase in the aromaticity of humus in some Vertisol soils of Mexico.

In agreement with findings for humic and fulvic acids (discussed earlier), results of ^{13}C–NMR studies indicate a higher proportion of O—alkyl structures (carbohydrate C) in soil organic matter than can be accounted for by conventional analyses for monosaccharides by acid hydrolysis.[27] This result suggests that C—O structures other than those of carbohydrates contribute to resonance in the 60–110 region of the spectrum. It should also be noted that many spectra indicate a higher proportion of alkyl or lipid-like C (resonance peak at 30 ppm) that is not extractable by organic solvents. The humin fraction of soil organic matter has been shown to be enriched in alkyl C.[23,24]

The relative proportion of each type of C in the size fractions of some Australian soils is given in Fig. 11.11. A unique feature of the results is the relatively high content of aliphatic (alkyl) C in the fine clay fraction, particularly for the Urrbrae soil. The high content of aliphatic C in the fine clay was thought to be due to accumulation of recalcitrant plant waxes strongly associated with soil clays.[26]

In addition to analysis of organic matter in the intact soil and its size fractions, the technique has been used to follow the humification process in forest soils.[38] Among the changes are the following: 1) polysaccharides of plant litter are extensively decomposed and substituted by microbial polysaccharides, 2) lignin is partly decomposed and the remnant molecule is transformed through side-chain oxidation and ring cleavage, 3) further modifications of aromatic C lead to an increase of C-substituted aromatic C and a concomitant loss of phenolic structures in the humic acid fraction during humification, 4) the refractory alkyl C moieties in humified forest soil organic matter do not result from selective preservation of plant-derived biomacromolecules, 5) aliphatic biopolymers (i.e., cutin and suberin) do not accumulate but there may be an increase in cross-linking of lipid- and/or cutin-type and suberin-type materials. Results of CPMAS ^{13}C–NMR studies have shown that forest soil organic matter contains from 20 to 30 percent aromatic C.[35]

^{1}H-NMR SPECTROSCOPY

Based on both natural isotopic abundance and gyromagnetic ratio (see Eq. [1]), the proton (i.e., ^{1}H) is the most easily observed nucleus. However, as applied to humic substances,[48-52] resolution of resonance groups has been generally poor. A critical review of ^{1}H–NMR spectroscopy of humic substances has been given by Wershaw.[6]

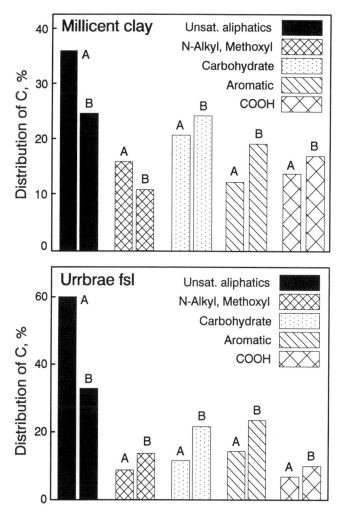

Fig. 11.11 Composition of organic C in the fine clay (A) and coarse clay (B) fractions of two Australian soils as determined by CPMAS ^{13}C–NMR. Adapted from Baldock et al.[26]

Solution ^{1}H–NMR is the preferred method, but a major problem exists because of interference from proton signals due to water and hydroxide. To circumvent these difficulties, dimethyl sulfoxide (DMSO) or sodium deuteroxide (NaOD) has been used to dissolve the humic material, in which case proton resonances due to the sample are observed as small signals in the presence of a dominant peak due to the DMSO or HOD (formed by exchange of acid protons in the humic material and from proton impurities in the deuterated solvent). A representative ^{1}H–NMR spectrum of a soil fulvic acid using DMSO as the solvent is given in Fig. 11.12.

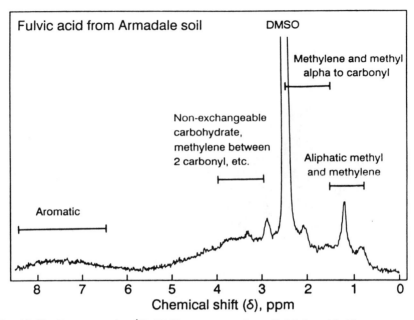

Fig. 11.12 Representative ^{1}H–NMR spectrum of a soil fulvic acid. The pronounced peak near 2.5 ppm is due to the solvent (DMSO). Adapted from Wershaw,[6] reproduced by permission of John Wiley & Sons, Inc.

In a recent study, Wilson et al.[52] found that a considerable percentage of the protons in a soil humic acid was associated with carbohydrates (35 percent); proton aromaticity (excluding exchangeable protons) was 17 percent. Other work, reviewed by Wershaw,[6] has shown that the percentage of aromatic protons to total protons in humic substances is of the order of from 19 to 35 percent. Resonance due to aromatic structures in humic substances invariably appears as a slight rise of the baseline on the NMR spectra (see Fig. 11.12); accordingly, estimates for proton aromaticities are subject to considerable error. It should also be noted that proton aromaticities would be expected to be lower than aromaticities based on ^{13}C–NMR spectroscopy due to extensive substitution of the aromatic ring, such as with OH groups and aliphatic side chains.

The technique of ^{1}H–NMR spectroscopy shows considerable promise for the determination of functional groups in humic substances through formation of derivatives, such as by acetylation of hydroxyl groups.[46]

^{15}N–NMR SPECTROSCOPY

Nitrogen contains two isotopes (^{14}N and ^{15}N) that are suitable for NMR spectroscopy. With the most abundant ^{14}N isotope (99.63 percent abundance), lines are broad and little resolution is observed. The ^{15}N isotope is much more suitable but is only 0.37 percent in natural abundance and is thus difficult to observe. However, with the very sensitive high-field spectrometers now available, ^{15}N–NMR spectroscopy of soil organic matter appears feasible. Thus far, success using the technique has been restricted to ^{15}N tracer studies, including ^{15}N-enriched soil,[42] ^{15}N-labeled melanoidins,[53] and a synthetic humic acid prepared from *p*-benzoquinone and NH_4Cl.[54] In the study with ^{15}N-labeled melanoidins,[53] secondary (and possibly tertiary) amines were found as well as evidence for the presence of pyrrole-type N.

^{31}P–NMR SPECTROSCOPY

Although present in low concentrations in soil organic matter, ^{31}P is a sensitive nucleus and great potential exists for studies on soil organic P. A brief discussion of this work is given in Chapter 5.

ANALYTICAL PYROLYSIS

Two types of degradative methods have been applied to soil organic matter: chemical (hydrolysis, reduction, and oxidation) and thermal (pyrolysis). The latter is discussed in this section. Chemical degradation approaches are covered in Chapter 10.

Pyrolysis, when used in tandem with gas–liquid chromatography (Py–GC), mass spectroscopy (Py–MS), or a combination of the two (Py–GC–MS) has emerged as a valuable approach for characterization of soil organic matter. Advantages of Py–MS, in which the pyrolysis products are passed directly into the mass spectrometer, include better reproducibility, detection of highly polar species, fast analysis time, and amenability to multi-variate or other chemometric methods of pattern analysis. However, Py–Ms does not distinguish between compounds of similar molecular weights and thereby fails to provide positive identification of all pyrolysis products. This can best be done by an initial separation of degradation products by gas chromatography, in which case the column effluent is allowed to flow directly into the mass spectrometer (i.e., Py–GC–MS). Positive identifications of pyrolysis products are then made from the mass spectrometric data in combination with a knowledge of retention times for standard compounds on the gas chromatograph.

The literature on the application of pyrolysis techniques to soil organic matter is extensive and has been covered in several reviews.[6,55,56] Pyrolysis methods have been used in structural studies of isolated components of soil organic matter,[57-62] to distinguish between the organic matter in soils of different origins,[63-65] to characterize organic matter in size and density fractions of the soil,[26] to delineate the raw humus layers of forest soils,[66] to follow the humification process in peat,[67,68] to examine litter decomposition and humification in forest soils,[38,68] and to establish horizon differentiation in Histosols.[56] The method has been found useful for detecting "biomarkers" for terrestrial plant input into sediments.[69]

In analytical pyrolysis, a pulse of thermal energy is applied to the sample, thereby causing fracture of weaker linkages of associated macromolecules, with release of products characteristic of their structures. Two techniques have been used for pyrolysis: 1) quasi-instantaneous heating, or Curie-point pyrolysis, and 2) controlled temperature programming. A criticism of 2) is that secondary reactions during pyrolysis can lead to the formation of compounds unrelated to the material being pyrolyzed. Curie-point methods are highly reproducible, and they can be applied to small sample sizes. Ratios obtained for pyrolysis products derived from polysaccharides, polypeptides, lignin, and so on, can serve as a semiquantitative index of soil humus type and "degree of humification."

The pyrolysis–mass spectrometer system used by MacCarthy et al.[58] for analysis of humic substances is shown in Fig. 11.13. The basic system consists of a Curie-point pyrolyzer, an expansion chamber, an ion source, a quadrupole, and an electron multiplier. Essentially, a mass spectrometer bombards the substance under investigation (i.e., pyrolysis products) with an electron beam and records the results as a spectrum of positive ion fragments. Mass spectra are scanned continuously during passage of the pulse of products through the source and are "averaged" to obtain a representative spectrum (i.e., plot of relative abundances versus mass of the ion fragments divided by charge, m/z). A computer is generally used for instrument control and data collection. A "pattern recognition" program was used by MacCarthy et al.[58] to detect differences between humic and fulvic acids from diverse sources.

Figure 11.14 shows the mass spectrum for pyrolysis products from a humic acid from a Udic Boroll in Germany.[56] The spectrum shows high relative

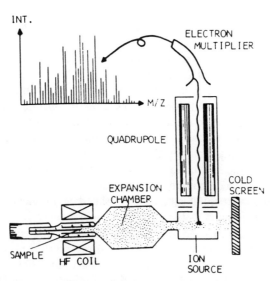

Fig. 11.13 Schematic diagram of a pyrolysis–mass spectrometer system. From MacCarthy et al.,[58] reproduced by permission of Pergamon Press, Ltd.

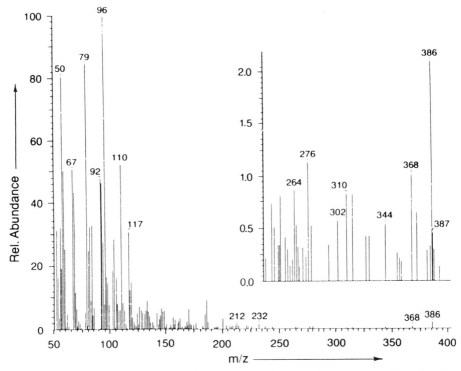

Fig. 11.14 Mass spectrum of pyrolysis products from a soil humic acid. From Schulten.[56] Reproduced by permission of Elsevier Science Publishers, Amsterdam

abundances of signals up to m/z values of 117; only weak signals are detected in the higher mass range (up to m/z of 386). A complete interpretation of the spectrum is beyond the scope of this chapter, and the reader is referred to Schulten[56] and the review of Schnitzer[4] for additional information. Briefly, signals at m/z values of 60, 82, 96, 110, and 126 have tentatively been assigned to polysaccharides; the presence of lignin-like components is indicated by signals at m/z values of 212, 302, and 244. Nitrogen containing ions are evident at m/z values of 79 (pyridine), 92 (methyl pyridine), and 117 (indole).

In a comparison study of some well-characterized humic and fulvic acids, MacCarthy et al.[58] found that carbohydrate and phenolic components were more pronounced in pyrolysis products from the fulvic acids; saturated and unsaturated hydrocarbons were more pronounced in pyrolysis products of the humic acids. Evidence was also obtained indicating that fundamental differences exist between aquatic and soil humic substances.

One problem with pyrolysis techniques is that the data obtained are selective for the more volatile components, such as carbohydrates and N-containing constituents. The method is particularly useful for detecting lipid-like constituents (e.g., esters and polyesters of long-chain alcohols and fatty acids), which are sometimes associated with humic substances. With humic and fulvic acids, less than one-half of the C is usually recovered as volatile products; accordingly, the data cannot be regarded as being representative of the entire sample under examination.

It should also be noted that, unlike ^{13}C–NMR spectroscopy, pyrolysis does not provide quantitative data for structural units or components of humic substances. Another limitation is that some of the products may be artifacts of the degradation process. The production of artifacts can be minimized by carrying out pyrolysis under vacuum or in a stream of inert gas.

Some specific points relative to Curie-point pyrolysis are:

1. Pyrolysates of humic substances contain a rich mixture of products that can be related to their constituent biopolymers and building blocks. Spectra of pyrolyzed humic acids have generally shown an abundance of signals due to polysaccharides, as well as pyridine and indole units. The contribution of lignin to the aromatic character of humic acids is indicated by signals from lignin monomers (i.e., guaiacyl and syringyl units).
2. Pyrolysis–mass spectroscopy of humic substances generates a vast amount of data that must be reduced to a few principal component variables, which requires modern computer capability.[63] Selection of mass ions is somewhat arbitrary and has been based on fragmentation patterns of known biopolymers.
3. Curie-point pyrolysis methods have the potential for differentiating between the different forms of organic matter (i.e., lignin and lignin-like products, polysaccharides, proteinaceous constituents, lipids, etc.) in the "intact" soil. In this respect, the approach has the potential for determining the "quality" of soil organic matter and for serving as a "fingerprint" technique for comparative purposes.
4. With whole soils, peat, and the litter layer of forest soils, the method provides an indication of the extent of humification of plant remains. Results of pyrolysis studies indicate that, during humification, complex polysaccharides of crop residues (e.g., cellulose and xylans) are lost, and lignin is transformed through loss of methoxyl groups and oxidation of C-3 side chains. As decomposition proceeds, a new set of products is produced through synthesis by microorganisms, including secondary or pseudo-polysaccharides. A portion of the aromatic rings of lignin remains behind and contributes to the substituted benzenes and phenols obtained on pyrolysis.

Finally, results from pyrolysis compliment those obtained by ^{13}C–NMR spectroscopy and other analytical procedures. Several investigators have combined the two techniques of Py–Ms pyrolysis and ^{13}C–NMR spectroscopy for characterization of soil organic matter.[11,26] Pyrolysis has the potential for providing valuable information when testing molecular models for the structures of humic substances.[55]

SUMMARY

The techniques described in this Chapter, notably ^{13}C–NMR and analytical pyrolysis, offer exception potential for characterization of soil organic matter components, including humic acids and "generic" fulvic acids. The two approaches are unique in that they also have the potential for distinguishing the main classes and functional groups of organic matter in the intact soil and its size fractions without the need for extraction. They complement, but do not eliminate, approaches described in earlier chapters, which for the most part were based on extraction and fractionation to yield major components (e.g., humic acids, fulvic acids, polysaccharides) or hydrolysis to liberate individual biochemical compounds (e.g., amino acids, amino sugars, monosaccharides).

The technique of ^{15}N–NMR spectroscopy shows considerable promise in studies where soil organic matter has been enriched with the stable isotope ^{15}N. Further use needs to be made of ^{31}P–NMR spectroscopy.

REFERENCES

1   R. L. Malcolm, "Application of Solid-state ^{13}C NMR Spectroscopy to Geochemical Studies of Humic Substances," in M. H. B. Hayes, P. MacCarthy, R. L. Malcolm, and R. S. Swift, Eds., *Humic Substances II: In Search of Structure*, Wiley, New York, 1989, pp. 339–372.

2   R. L. Malcolm, "Variations Between Humic Substances Isolated from Soils, Stream Waters, and Groundwaters as Revealed by ^{13}C–NMR Spectroscopy," in P. MacCarthy, C. E. Clapp, R. L. Malcolm, and P. R. Bloom, Eds., *Humic Substances in Soil and Crop Sciences: Selected Readings*, American Society of Agronomy, Madison, 1990, pp. 13–35.

3   R. L. Malcolm and P. MacCarthy, "The Individuality of Humic Substances in Diverse Environments," in M. S. Wilson, Eds., *Advances in Soil Organic Matter Research: The Impact of Agriculture and the Environment*, Redwood Press Ltd., Wiltshire, England, 1991, pp. 23–34.

4   M. Schnitzer, "Selected Methods for the Characterization of Soil Humic Substances," in P. MacCarthy, C. E. Clapp, R. L. Malcolm, and P. R. Bloom, Eds., *Humic Substances in Soil and Crop Sciences: Selected Readings*, American Society of Agronomy, Madison, 1990, pp. 65–89.

5   C. Steelink, R. L. Wershaw, K. A. Thorn, and M. A. Wilson, "Application of Liquid-State NMR Spectroscopy to Humic Substances," in M. H. B. Hayes, P. MacCarthy, R. L. Malcolm, and R. S. Swift, Eds., *Humic Substances II: In Search of Structure*, Wiley, New York, 1989, pp. 281–308.

6   R. L. Wershaw, "Application of Nuclear Magnetic Resonance Spectroscopy for Determining Functionality in Humic Substances," in G. R. Aiken, D. M. McKnight, R. L. Wershaw, and P. MacCarthy, Eds., *Humic Substances in Soil, Sediment, and Water: Geochemistry, Isolation, and Characterization*, Wiley, New York, 1985, pp. 561–582.

7   M. A. Wilson, "Solid-State Nuclear Magnetic Resonance Spectroscopy of Humic Substances: Basic Concepts and Techniques," in M. H. B. Hayes, P. MacCarthy, R. L. Malcolm, and R. S. Swift, Eds., *Humic Substances II: In Search of Structure*, Wiley, New York, 1989, pp. 309-338.

8   M. A. Wilson, "Application of Nuclear Magnetic Resonance Spectroscopy to Organic Matter in Whole Soils," in P. MacCarthy, C. E. Clapp, R. L. Malcolm, and P. R. Bloom, Eds., *Humic Substances in Soil and Crop Sciences: Selected Readings*, American Society of Agronomy, Madison, 1990, pp. 221-260.

9   R. M. Silverstein, G. C. Bassler, and T. C. Morrill, Spectrometric Identification of Organic Compounds, Wiley, New York, 1981.

10 B. Dehorter, C. Y. Kontchou, and R. Blondeau, *Soil Biol. Biochem.*, **24**, 667 (1992).

11 I. Kögel-Knabner, P. G. Hatcher, and W. Zech., *Soil Sci. Soc. Amer. J.*, **55**, 241-247 (1991).

12 M. Krosshavn, J. O. Bjørgum, J. Krane, and E. Steinnes, *J. Soil Sci.*, **41**, 371 (1990).

13 J. C. Lobartini and K. H. Tan, *Soil Sci. Soc. Amer. J.*, **52**, 125 (1988).

14 R. H. Newman and K. R. Tate, *J. Soil Sci.*, **42**, 39 (1991).

15 J. M. Novak and M. E. Smeck, *Soil Sci. Soc. Amer. J.*, **55**, 96 (1991).

16 A. Piccolo, L. Campanella, and B. M. Petronio, *Soil Sci. Soc. Amer. J.*, **54**, 750 (1990).

17 C. M. Preston and B. A. Blackwell, *Soil Sci.*, **139**, 88 (1985).

18 C. M. Preston and M. Schnitzer, *Soil Sci. Soc. Amer. J.*, **48**, 305 (1984).

19 C. M. Preston and M. Schnitzer, *J. Soil Sci.*, **38**, 667 (1987).

20 M. Schnitzer and C. M. Preston, *Soil Sci. Soc. Amer. J.*, **50**, 326 (1986).

21 M. Schnitzer and C. M. Preston, *Soil Sci. Soc. Amer. J.*, **51**, 639 (1987).

22 M. Schnitzer, K. Kodama, and J. A. Ripmeester, *Soil Sci. Soc. Amer. J.*, **55**, 745 (1991).

23 P. G. Hatcher, I. R. Breger, G. E. Maciel, and N. M. Szeverenyi, "Geochemistry of Humin," in G. R. Aiken, D. M. McKnight, R. L. Wershaw, and P. MacCarthy, Eds., *Humic Substances in Soil, Sediment, and Water: Geochemistry, Isolation, and Characterization*, Wiley, New York, 1985, pp. 275-302.

24 C. M. Preston, M. Schnitzer and J. A. Ripmeester, *Soil Sci. Soc. Amer. J.*, **53**, 1442 (1989).

25 M. A. Arshad, J. A. Ripmeester, and M. Schnitzer, *Can. J. Soil Sci.*, **68**, 593 (1988).

26 J. A. Baldock, G. J. Currie, and J. M. Oades, "Organic Matter as Seen by Solid State ^{13}C NMR and Pyrolysis Tandem Mass Spectrometry," in M. S. Wilson, Ed., *Advances in Soil Organic Matter Research: The Impact of Agriculture and the Environment*, Redwood Press Ltd., Wiltshire, England, 1991, pp. 45-60.

27 J. M. Oades, A. M. Vassallo, A. G. Waters, and M. A. Wilson, *Aust. J. Soil Res.*, **25**, 71 (1987).

28 J. O. Skjemstad, R. C. Dalal, and P. F. Barron, *Soil Sci. Soc. Amer. J.*, **50**, 354 (1986).

29 M. A. Wilson, P. F. Barron, and K. M. Goh, *J. Soil Sci.*, **32**, 419 (1981).

30 M. A. Wilson, P. F. Barron, and K. M. Goh, *Geoderma*, **26**, 323 (1981).
31 W. Zech, L. Haumaier, and R. Hempfling, "Ecological Aspects of Soil Organic Matter in Tropical Land Use," in P. MacCarthy, C. E. Clapp, R. L. Malcolm, and P. R. Bloom, Eds, *Humic Substances in Soil and Crop Sciences: Selected Readings*, American Society of Agronomy, Madison, 1990, pp. 187-202.
32 C. M. Preston, R. L. Dudley, C. A. Fyfe, and S. P. Mathur, *Geoderma*, **33**, 245 (1984).
33 C. M. Preston and J. A. Ripmeester, *Can. J. Spectros.*, **27**, 99 (1982).
34 C. M. Preston, S.-E. Shipitalo, R. L. Dudley, C. A. Fyfe, S. P. Mathur, and M. Lévesque, *Can. J. Soil Sci.*, **67**, 187 (1986).
35 I. Kögel, R. Hempfling, W. Zech, P. G. Hatcher, and H.-R. Schulten, *Soil Sci.*, **146**, 124 (1988).
36 G. Ogner, *Geoderma*, **35**, 343 (1985).
37 M. A. Wilson, S. Heng, K. M. Goh, R. J. Pugmire, and D. M. Grant, *J. Soil Sci.*, **34**, 83 (1983).
38 I. Kögel-Knabner, W. Zech, P. G. Hatcher, and J. W. de Leeuw, "Fate of Plant Components during Biodegradation and Humification in Forest Soils: Evidence from Structural Characterization of Individual Biomacromolecules," in M. S. Wilson, Eds., *Advances in Soil Organic Matter Research: The Impact of Agriculture and the Environment*, Redwood Press Ltd., Wiltshire, England, 1991, pp. 61-70.
39 S. A. Boyd and L. E. Sommers, "Humic and Fulvic Acid Fractions from Sewage Sludges and Sludge-Amended Soils," in P. MacCarthy, C. E. Clapp, R. L. Malcolm, and P. R. Bloom, Eds., *Humic Substances in Soil and Crop Sciences: Select Readings*, American Society of Agronomy, Madison, 1990, pp. 203-220.
40 J. A. Baldock, J. M. Oades, A. M. Vassallo, and M. A. Wilson, *Aust. J. Soil Res.*, **28**, 193 (1989).
41 C. M. Preston and J. A. Ripmeester, *Can. J. Soil Sci.*, **63**, 495 (1983).
42 L. Benzing-Purdie, W. V. Cheshire, B. L. Williams, C. I. Ratcliffe, J. A. Ripmeester, and B. A. Goodman, *J. Soil Sci.*, **43**, 113 (1992).
43 L. Benzing-Purdie and J. A. Ripmeester, *Soil Sci. Soc. Amer. J.*, **47**, 56 (1983).
44 M. Schnitzer and Y. K. Chan, *Soil Sci. Soc. Amer. J.*, **50**, 67 (1986).
45 Y. Inbar, Y. Chen, and Y. Hadar, *Soil Sci. Soc. Amer. J.*, **53**, 1695 (1989).
46 J. A. Leenheer and T. I. Noyes, "Derivatization of Humic Substances for Structural Studies," in M. H. B. Hayes, P. MacCarthy, R. L. Malcolm, and R. S. Swift, Eds., *Humic Substances II: In Search of Structure*, Wiley, New York, 1989, pp. 257-280.
47 M. A. Mikita, C. Steelink, and R. L. Wershaw, *Anal. Chem.*, **53**, 1715 (1981).
48 P. Ruggiero, F. S. Interesse, and O. Sciacovelli, *Soil Biol. Biochem.*, **12**, 297 (1980).
49 P. Ruggiero, F. S. Interesse, L. Cassedei, and O. Sciacovelli, *Soil Biol. Biochem.*, **13**, 361 (1981).
50 P. Ruggiero, O. Sciacovelli, C. Testini, and F. S. Interesse, *Geochim. Cosmochim. Acta*, **42**, 411 (1978).
51 P. Ruggiero, F. S. Interesse, and O. Sciacovelli, *Geochim. Cosmochim. Acta*, **44**, 603 (1980).

52. M. A. Wilson, P. J. Collin, and K. R. Tate, *J. Soil Sci.*, **34**, 297 (1983).
53. L. Benzing-Purdie, J. A. Ripmeester, and C. M. Preston, *J. Agric. Food Chem.*, **31**, 913 (1983).
54. C. M. Preston, B. S. Rauthan, C. Rodger, and J. A. Ripmeester, *Soil Sci.*, **134**, 277 (1982).
55. J. M. Bracewell, K. Haider, S. R. Larter, and H.-R. Schulten, "Thermal Degradation Relevant to Structural Studies of Humic Substances," in H. M. B. Hayes, P. MacCarthy, R. L. Malcolm, and R. S. Swift, Eds., *Humic Substances II: In Search of Structure*, Wiley, New York, 1989, pp. 181–222.
56. H.-R. Schulten, *J. Anal. Appl. Pyrol.*, **12**, 149 (1987).
57. R. W. L. Kimber and P. L. Searle, *Geoderma*, **4**, 47, 57 (1970).
58. P. MacCarthy, S. J. Deluca, K. J. Voorhees, R. L. Malcolm, and E. M. Thurman, *Geochim. Comochim. Acta*, **49**, 2091 (1985).
59. C. Saiz-Jimenez and J. W. de Leeuw, *Organic Geochem.*, **6**, 287 (1984).
60. C. Saiz-Jimenez and J. W. de Leeuw, *J. Anal. Appl. Pyrol.*, **9**, 99 (1986).
61. C. Saiz-Jimenez and J. W. de Leeuw, *J. Anal. Appl. Pyrol.*, **11**, 367 (1987).
62. M. Schnitzer and H.-R. Schulten, *Soil Sci. Soc. Amer. J.*, **56**, 1811 (1992).
63. J. M. Bracewell and G. W. Robertson, *J. Soil Sci.*, **35**, 549 (1984).
64. J. M. Bracewell and G. W. Robertson, *Geoderma*, **40**, 333 (1987).
65. G. Halma, M. A. Posthumus, R. Miedema, R. van de Westeringh, and H. L. C. Meuzelaar, *Agrochimica*, **22**, 372 (1978).
66. J. M. Bracewell and G. W. Robertson, *J. Soil Sci.*, **38**, 191 (1987).
67. J. M. Bracewell, G. W. Robertson, and B. R. Williams, *J. Anal. Appl. Pyrol.*, **2**, 53 (1980).
68. R. J. Hempfling, F. Ziegler, W. Zech, and H.-R. Schulten, *Z. Pflanzenernähr. Bodenk.*, **150**, 179 (1987).
69. C. Saiz-Jimenez and J. W. de Leeuw, *Organic Geochem.*, **10**, 869 (1986).

12

STRUCTURAL BASIS OF HUMIC SUBSTANCES

The study of the molecular structures of humic substances is one of the most intriguing research areas of soil science, and the most difficult. From the very beginning (see review of chapter 2), scientists have speculated on structures responsible for the color, reactivity, and physical/chemical properties of soil humic substances.

A knowledge of the basic structures of humic and fulvic acids is required for a full understanding of the role and function of these constituents in soil. However, because of the multiplicity of component molecules of which they are composed, together with the numerous types of linkages that bind them together, accurate structural formulas are unattainable. Each fraction (humic acid, fulvic acid, etc.) must be regarded as consisting of a series of molecules of different sizes, few having the same structural configuration or array of reactive functional groups.

Although precise formulas for humic substances cannot be made, information provided in previous chapters allows for predictions to be made regarding typical or average structures. Type structures recorded in the literature are reproduced herein as a benchmark for future studies.

Also included in this chapter is the introduction of a simplified structural unit (a dimer) that is useful for undergraduate classroom presentations.[1] The dimer is used to illustrate many of the reactions for which humic substances are famous, including contribution to the cation-exchange-capacity of the soil (see Chapter 13), complexation of micronutrient cations (Chapters 16 and 17), formation of clay–organic complexes (Chapter 18), and binding of pesticides (Chapter 19).

GENERAL CONSIDERATIONS

Structural models have arisen from the need to represent the known chemistry of humic substances in terms of structures which, although hypothetical, are based on knowledge derived from studies on elemental composition, functional groups, degradation products, spectroscopic data, and physical chemical properties as discussed earlier in this book.

The main uses of hypothetical structural models are: 1) as a means of representing the average properties of humic and fulvic acids, 2) to help in the formulation of new hypotheses regarding their structures and the development of new experimental schemes for their investigation,[2] and 3) for illustrating mechanisms for the binding of metal ions and xenobiotics.

As noted by Bracewell et al.,[2] several dangers are encountered in devising hypothetical structures for humic substances. One danger is that features may be incorporated in addition to what can strictly be deduced from experimental data, and these features, although experimentally ungrounded, may be propagated by others in formulating structural models. A second danger is the natural tendency to focus on those aspects of the structures that are thought to be known, whereas new research should continually seek out and identify unknown aspects of the structures.

In discussing molecular structures, the following points should be kept in mind.

1. Aromatic nuclei are important in the molecular structures of humic substances, particularly humic acids.
2. Aliphatic structures are of greater importance than previously thought. Frequently, two-thirds or more of the component molecules of fulvic acids can be accounted for in aliphatic linkages. In general, aquatic humic substances have more of an aliphatic character than those from soil.
3. Humic and fulvic acids share some structural features, but significant differences exist between them. In contrast to humic acids, the low-molecular-weight fulvic acids have higher oxygen but lower C contents, and they contain considerably more acidic functional groups, particularly COOH. Another important difference is that, while the oxygen in fulvic acids can be accounted for largely in known functional groups (COOH, OH, C=O), a high portion of the oxygen in humic acids occurs as a structural component of the nucleus (as ether or ester linkages, etc.). The reader is referred to Chapter 11 for a discussion of compositional variations between humic and fulvic acids from different environmental sources.
4. Fatty acids and long-chain hydrocarbon structures appear to be components of some, but not all, humic and fulvic acids, where they may contribute to the hydrophobic characteristics of these substances (Chapter 10). The fatty acids may exist, in part, as esters formed with phenols or other structures containing hydroxyl groups. In some humic acids, al-

TYPE STRUCTURES FOR HUMIC AND FULVIC ACIDS

kanes may be present as isolated sidechains, or as links between aromatic structures.

5. Condensed ring structures, although present under some circumstances, are not a prominent feature of soil humic or fulvic acids.

TYPE STRUCTURES FOR HUMIC AND FULVIC ACIDS

Of historical significance is the structural formula for humic acid as proposed by the German scientist Fuchs (see Chapter 1). His structure (Fig. 12.1) was based largely on results with coal humic acids and consisted of a condensed ring system containing attached COOH and OH groups. As noted above, condensed rings would not appear to be a prominent feature of soil humic substances.

Results obtained for reactive functional groups (Chapter 9) and chemical degradation products (Chapter 10) suggest that the "core" structure of soil humic acids consist, at least in part, of aromatic rings of the di- or trihydroxyphenol-type bridged by $-O-$, $-(CH_2)_n-$, $-NH-$, $-N-$, and other groups and containing both COOH and OH groups, as well as quinone-type linkages. The typical dark color of humic acids, and their ability to be reduced to leucohumic acids with Na-amalgam, is consistent with this concept. In the natural state, the molecule may contain proteinaceous and carbohydrate residues, some of which may be linked to the "core" through covalent linkages. Some of the aromatic moieties have their origin in lignins; some are of microbial origin.

An early structural representation by Dragunov[3] (Fig. 12.2) closely conforms to this concept of structural components in soil humic acids. However, a fully acceptable model would need to account for the occurrence of significant

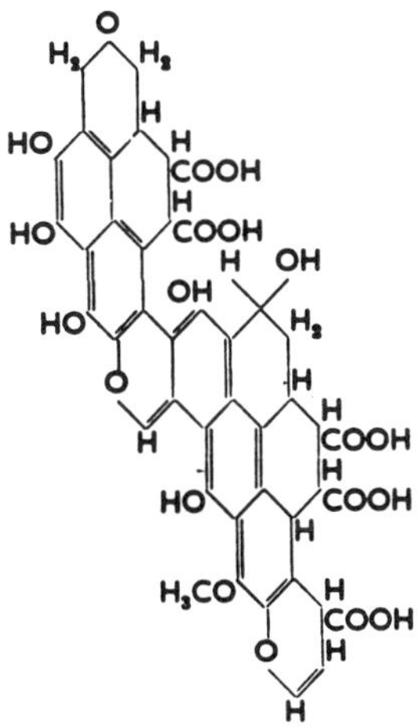

Fig. 12.1 Structure of humic acid according to Fuchs

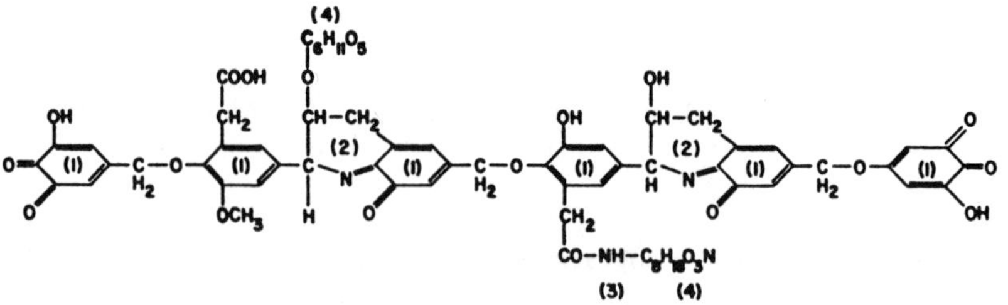

Fig. 12.2 Dragunov's structure of humic acid as recorded by Kononova[2] shows: 1) aromatic rings of the di- and trihydroxybenzene type, part of which has the double linkage of a quinone group, 2) nitrogen in cyclic forms, 3) nitrogen in peripheral chains, and 4) carbohydrate residue

amounts of aromatic COOH groups, some arranged in positions such that cyclic anhydrides can be formed by various chemical treatments (see Chapter 11). Flaig's[4] model (Fig. 12.3) contains abundant quantities of phenolic OH and quinones but is generally lacking in COOH groups.

The hypothetical structure devised by the author for humic acid (Fig. 12.4) contains many of the requirements for a "typical" soil humic acid, including free and bound phenolic OH groups, quinone structures, N and oxygen as bridge units, and COOH groups variously placed on aromatic rings. This model was devised by adding an additional structural unit (i.e., a condensed aromatic ring presumed to be of lignin origin) to the hypothetical structure shown by Fig. 8.16 of Chapter 8.

It should be noted that the models shown by Figs. 12.2 to 12.4 contain lower amounts of aliphatic substances (other than saccharide and protein constituents) than would be expected based on ^{13}C–NMR studies (see item 2).

Orlov's[5] hypothetical structure of the humic acid of a Mollisol (Fig. 12.5) contains both aromatic rings and aliphatic constituents. However, the structure is void of COOH groups and contains few phenolic OH groups. As was the case for the structures shown by Figs. 12.2 and 12.4, N is shown to be an integral part of the molecule.

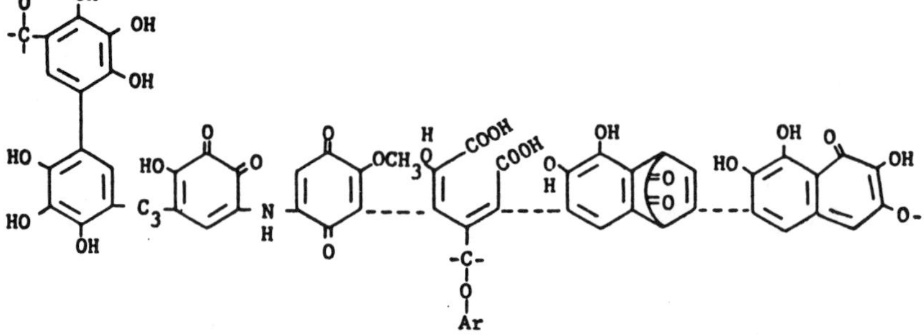

Fig. 12.3 Hypothetical structure of humic acid according to Flaig[4]

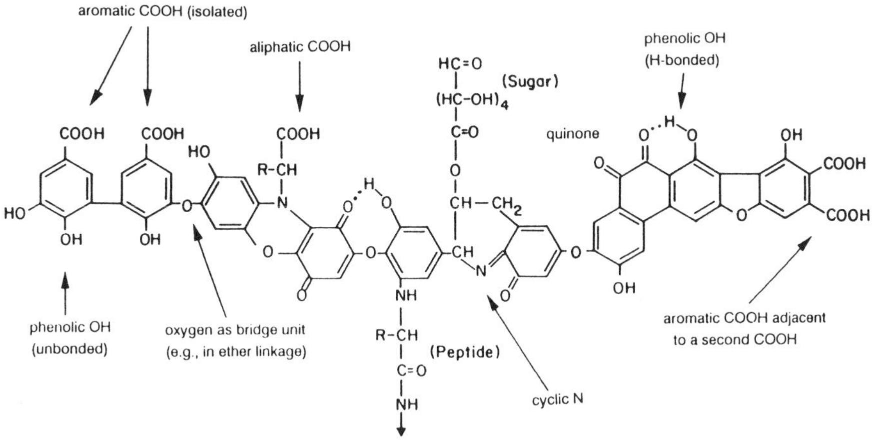

Fig. 12.4 Hypothetical structure of humic acid showing free and bound phenolic OH groups, quinone structures, nitrogen and oxygen as bridge units, and COOH groups variously placed on aromatic rings

On the basis of elemental data for humic acids and content of COOH and phenolic OH groups, Steelink[6] has proposed a tetramer structure for humic acid (Fig. 12.6). His structure has a much more aliphatic flavor than the models shown earlier. Unlike Fig. 12.4, the COOH groups are shown to be associated with the aliphatic portion of the molecule.

Schulten and Schnitzer[7] developed a schematic structure for humic acids based on ^{13}C-NMR, analytical pyrolysis, and oxidative degradation data. The structure (Fig. 12.7) consists of aromatic rings linked by long-chain alkyl structures so as to form a flexible network containing voids that trap and bind other organic components. Both COOH and OH groups are present in abundance,

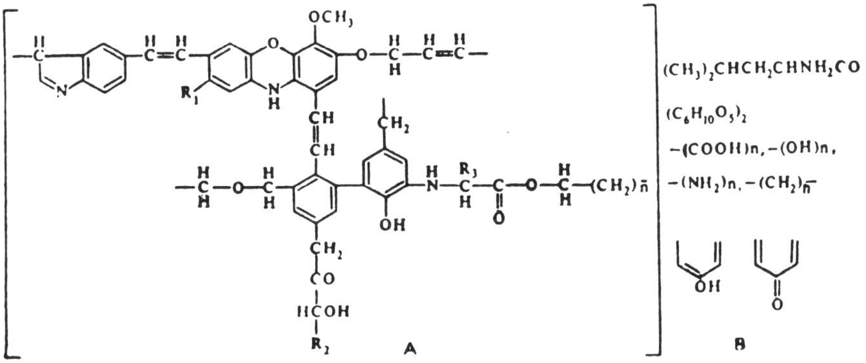

Fig. 12.5 Scheme of structural cell of humic acid from a Mollisol soil. A: nucleus; B: peripheral part. From Orlov[5]

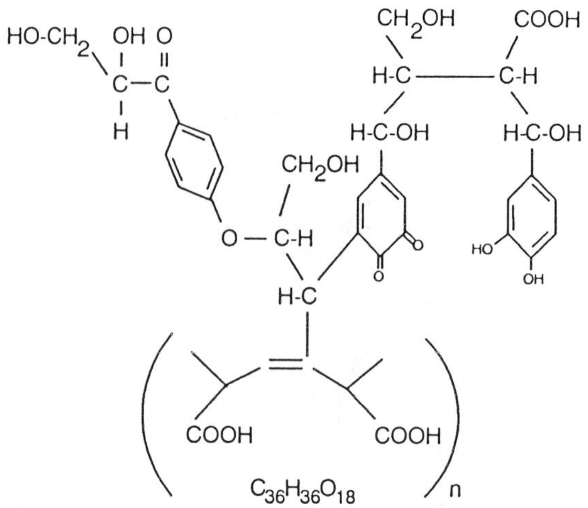

Fig. 12.6 Proposed tetramer structure for humic acid. From Steelnik[6]

and they occupy positions on both the aromatic ring and aliphatic side chains. Like the structure shown in Fig. 12.4, some of the COOH groups occupy adjacent positions on the aromatic ring.

Fig. 12.8 shows the relationship of structures in freshwater humic acids as deduced from data for alkaline permanganate degradation reactions.[8] A unique feature of this model is the occurrence of aliphatic (C_2-C_x) bridges between aromatic units. Whereas the model is consistent with criteria based on degradations with alkaline $KMnO_4$ (detection of aliphatic and benzenepolycarboxylic acids in the digests), the structure contains no N or quinone structural units.

Bergmann's hypothetical structure for the humic material from activated sewage sludge (Fig. 12.9), as recorded by Steinberg and Muenster,[9] shows a preponderance of aliphatic structures enriched with COOH groups. Humic substances from sludges (as well as aquatic environments) have been found to be more aliphatic in nature than those of agricultural soils.

Schnitzer[10] concluded that fulvic acids, or a portion thereof, consists of phenolic and benzenecarboxylic acids held together through H-bonds to form a polymeric structure of considerable stability (Fig. 12.10). A distinguishing characteristic of the structure is that the molecule is punctured by voids or holes of different dimensions that can trap low-molecular-weight organic and inorganic compounds, such as pesticides and metal ions. As applied to generic fulvic acids in general, the structure is untenable in that H-bonded structures would be expected to disperse in alkaline media, in which case the component molecules would become "detached" and the structure would disintegrate.[11]

Buffle's[12] model structure of fulvic acid (Fig. 12.11) contains both aromatic and aliphatic structures, both extensively substituted with oxygen-containing functional groups. This structure is consistent with results of ^{13}C–NMR studies indicated the prominence of aliphatic structures in fulvic acids.

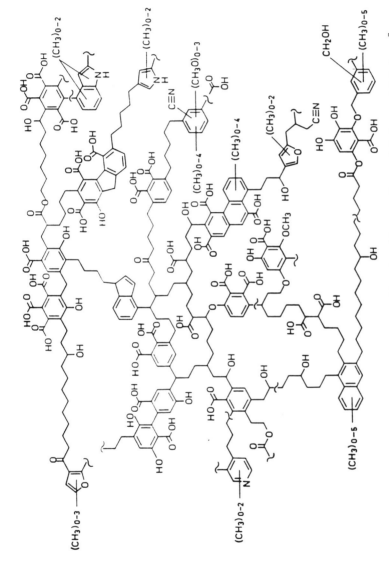

Fig. 12.7 Chemical network structure of humic acids according to Schulten and Schnitzer.[7] Reproduced by permission of Springer-Verlag.

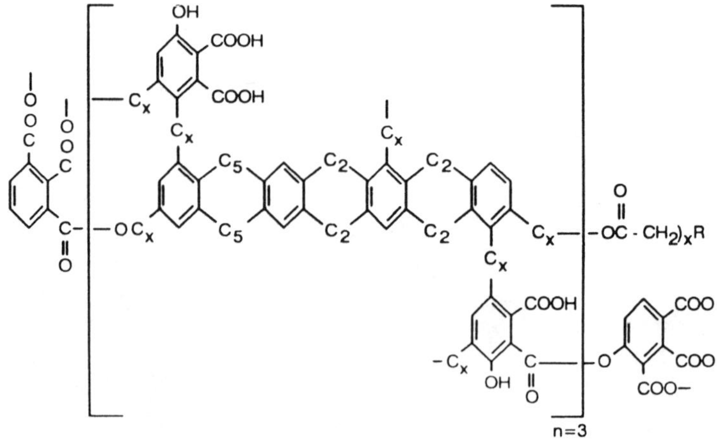

Fig. 12.8 Hypothetical relationship of structures in freshwater humic acids as deduced from data for alkaline $KMnO_4$ degradations. From Christman et al.[8]

One of the most extensively studied fulvic acid is from the Suwannee river in southeastern Georgia.[13] This river is unique in that it contains highly colored water along its entire length, extending from southeastern Georgia to its estuary in the Gulf of Mexico. On the basis of exhaustive studies, several models have been proposed for the average structure of the Suwannee river fulvic acid.[13] The various models differ only in minor detail and the one shown in Fig. 12.12

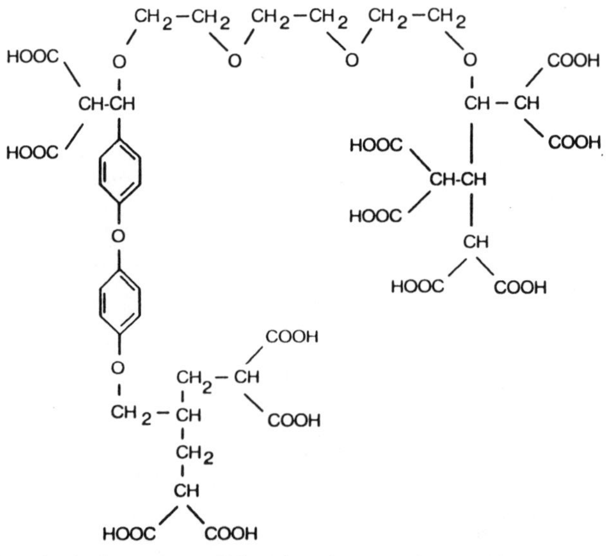

Fig. 12.9 Hypothetical structure of humic substance from activated sludge system as recorded by Steinberg and Muenster[9]

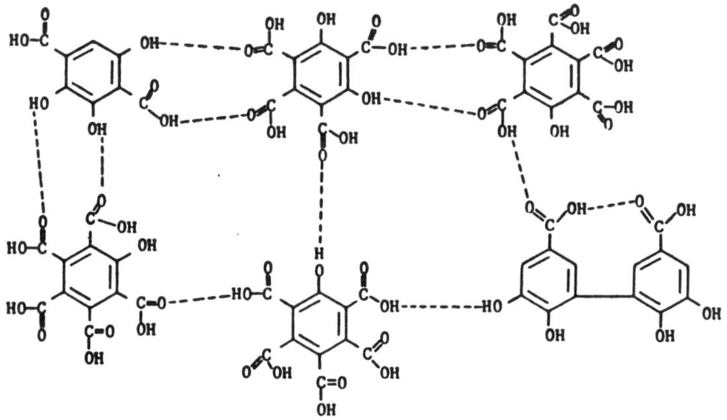

Fig. 12.10 Structure of soil fulvic acid as proposed by Schnitzer[10]

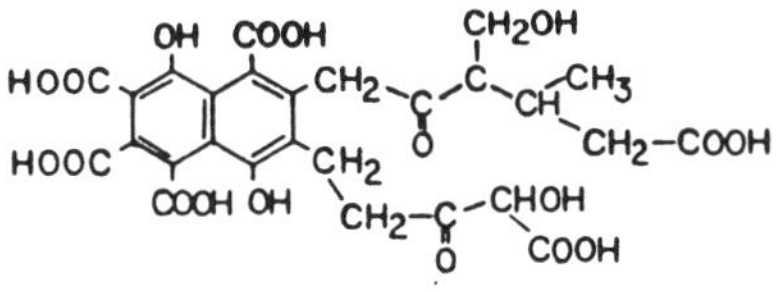

Fig. 12.11 Model structure of fulvic acid according to Buffle[12]

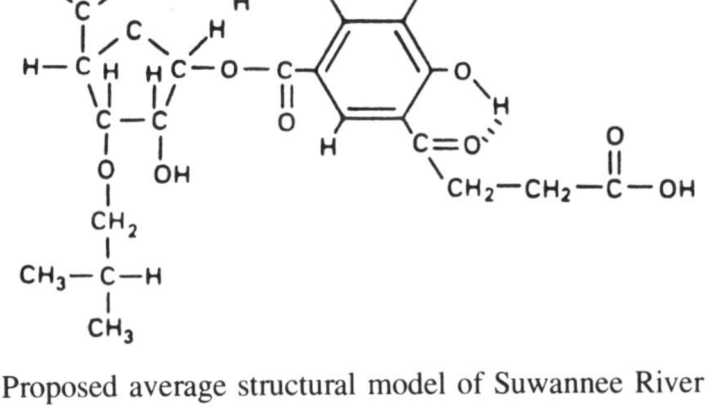

Fig. 12.12 Proposed average structural model of Suwannee River fulvic acid[13]

is representative. An interesting feature of the structure is the occurrence of condensed ring structures to which alkyl units are attached. Most COOH groups are shown to be associated with aliphatic side chains.

A generalized model for humic substances has been proposed by Wershaw[14] as a means of understanding their interactions with hydrophobic xenobiotics (see Chapter 19). Briefly, humic substances are regarding as consisting of partially decomposed plant degradation products that are stabilized by incorporation into humic aggregates. Functional groups in components of the aggregates are bound together by weak bonding mechanisms (e.g., hydrogen bonding, hydrophobic interactions) so as to form "membrane-like" structures with hydrophilic exterior surfaces and hydrophobic interiors into which hydrophobic xenobiotics will partition.

UNIQUE FEATURE OF HUMIC SUBSTANCES IN SEAWATER

Major differences exists between humic substances of seawater and those in soils and sediments, both in the pathways of their formation and in chemical structures.[15] Investigators in this field have concluded that: 1) other than metal complexation and redox functions, the only resemblance between humic substances from the open ocean (marine) and terrestrial environments is that they are both colored organic acids soluble in water, and 2) marine humic substances are formed *in situ* and only in the coastal zone is there an admixture of terrestrially derived humic substances from rivers.

A popular concept at the present time is that marine humic and fulvic acids are formed from the free radical autoxidative cross-linking of unsaturated lipids released into the water by plankton, as illustrated in Fig. 12.13. As the structures suggest, marine humic substances are much more aliphatic in nature than their terrestrial counterparts. The basis for the structures of marine humic substances are discussed in detail by Harvey and Boran[15] and Harvey et al.[16]

A SIMPLIFIED STRUCTURAL REPRESENTATION OF SOIL HUMIC SUBSTANCES

In this section, a comprehensible structural representation of humic substances is presented.[1] The simplified model, which consists of a dimer formed from two individual structural units (building blocks), is used to illustrate the important roles played by soil organic matter. In using the model, some precision is sacrificed for clarity, simplicity, and usefulness. The treatment is oriented more toward the undergraduate student than to graduate students or researchers.

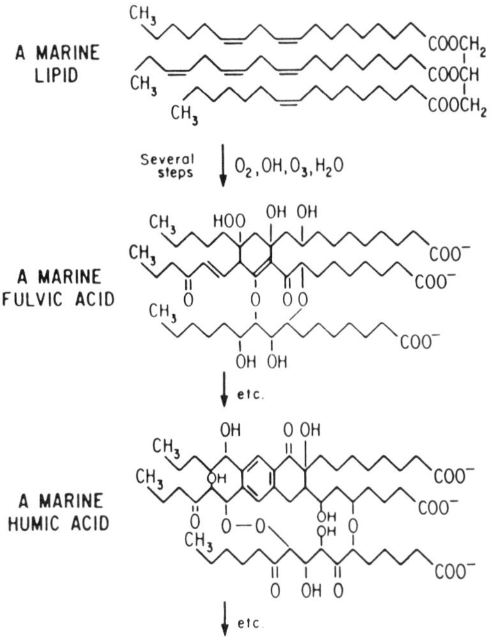

SIMPLIFIED REPRESENTATION OF SOIL HUMIC SUBSTANCES

Fig. 12.13 Representative structures in the dynamic continuum of fulvic acid and humic acid formation in sea water[15, 16]

Basic Considerations

The inconvenience and awkwardness of writing structural formulas and reactions for humic substances is readily apparent to those involved in teaching. In some textbooks on soil chemistry, one or more "type" structures are given for humic and/or fulvic acids; in other cases, empirical formulas are provided (e.g., about $C_{10}H_{12}O_5N$ for humic acids[4]). *More frequently, humic substances are merely described as being exceedingly complex and of unknown composition.* With respect to both teaching and learning, this state of affairs is unfortunate. In order to explain the properties and reactions of humic substances to students, it would be advantageous to express their basic chemical structure in terms of a simple structural unit or building block.

As indicated in previous chapters, humic substances in soil are high-molecular-weight compounds (macromolecules) that have been chemically structured, in part, through the linking together of individual building blocks or monomers. A key basic unit is the aromatic ring substituted with COOH and OH functional groups and to which various aliphatic substances are attached.

To some extent, chemical reactions of the macromolecules are similar to those of the monomers from which they were formed, primarily because they both contain similar types of reactive functional groups. As noted in previous chapters, and in structures depicted above, humic substances may contain the following functional groups:

Carboxyl (COOH)
  a. On aromatic rings
  b. On side chains

Hydroxyl (OH)
  a. Aliphatic
  b. Phenolic

Amine
  a. Primary ($-NH_2$)
  b. Secondary ($-NH-$)
  c. Tertiary ($-N=$)

Carbonyl (C=O)
  a. Ketonic
  b. Quinone

One popular concept is that, for the most part, humic substances in soil are formed from polyphenols of lignin origin or synthesized by microorganisms (see Chapter 8). A key step is the transformation of the polyphenols to quinones. Phenols and quinones are highly reactive and readily combine one with another, or with other biochemicals (e.g., amino acids), to form polymeric molecules of increasing molecular weight.

The Dimer Concept

As noted in Chapter 8, synthesis of humic substances can be thought of in terms of card playing, where each structural unit (e.g., a polyphenol or quinone) represents a separate card in the deck and a given "hand" represents the combinations in which the structural units combine to form a humic molecule. Because the types of monomers are numerous, and the ways in which they combine is great, structures of the resulting macromolecules are exceedingly variable and complex. The odds of getting the same hand (structure) twice are extremely small indeed.

To some extent, the relevant chemistry of humic substances can be learned or taught by regarding the "central core" of the macromolecules as consisting of the dimer represented by structure **I**, where the dashes indicate the location of H and functional groups and the letter R indicates an additional monomer(s). Structure **II** shows attachment of two important oxygen-containing functional groups found in humic substances (COOH and phenolic OH), as well as an amine group ($-NH_2$). Whereas dimer **II** is but one of many that would exist in humic substances (most would not contain an amine group), it is appropriate for instructional purposes at the undergraduate level.[1]

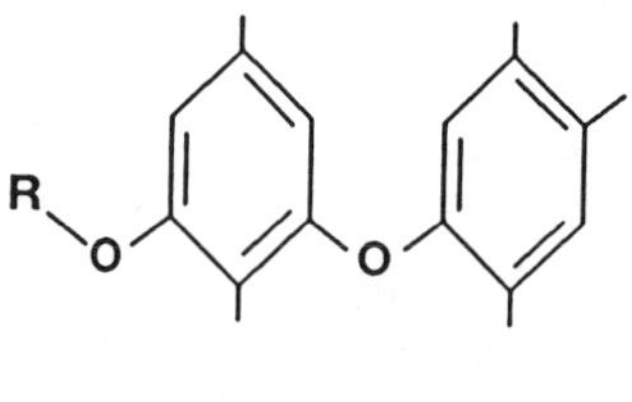

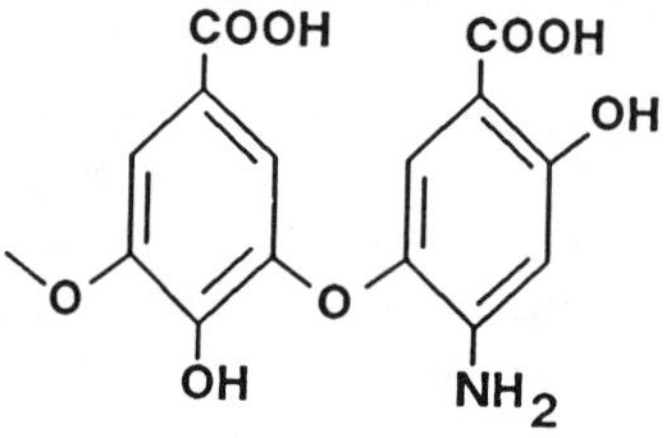

The dimer concept is more readily visualized by the student than the more complex macromolecular structures (e.g., see Fig. 12.7) and can easily be remembered and drawn. Structure **II** has a molecular weight of approximately 250 and it has a C/N ratio 14/1, which is of the order of that found for soil humus.

Application of the Dimer Concept to an Understanding of the Role of Organic Matter to Soil Processes

The dimers (or components thereof) have been used to illustrate the various chemical reactions that organic substances undergo in soil.[1] A brief résumé is given below but it should be noted that more extensive details can be found in other chapters.

1. By consideration of the acidic properties of COOH and phenolic-OH groups, the dimer (or component monomers) facilitates discussion of the buffering capacity of soil organic matter, the lime requirement of acid soils, and related topics. The reaction of added base (MOH) with COOH groups in the H^+-form, for example, can be written as

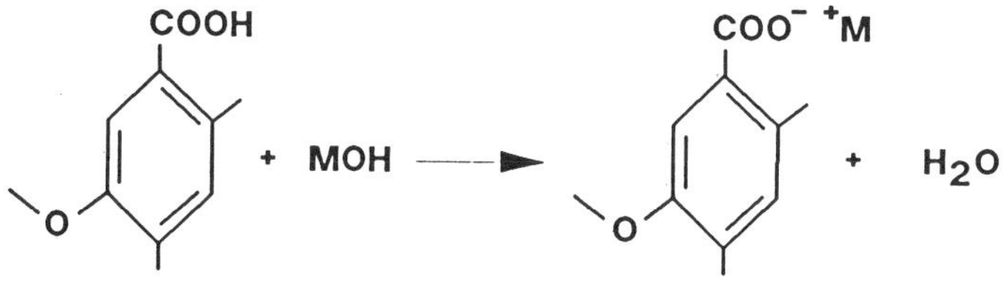

    [Reaction diagram: aromatic dimer with COOH + MOH → aromatic dimer with $COO^- \, {}^+M$ + H_2O]

2. The dimers can be used to illustrate the known influence of pH on the cation exchange capacity (CEC) of the soil, namely, as the pH of the soil increases so does the CEC, due to an increase in negative charges on the soil organic colloids. The following diagram shows the ionization reactions of acidic functional groups in humic substances. The initial reaction shows the formulation of the anion COO^-, which occurs at slightly acidic to neutral pHs; the second reaction illustrates the formation of the phenolate ion ($-O^-$), which occurs in the alkaline pH range.

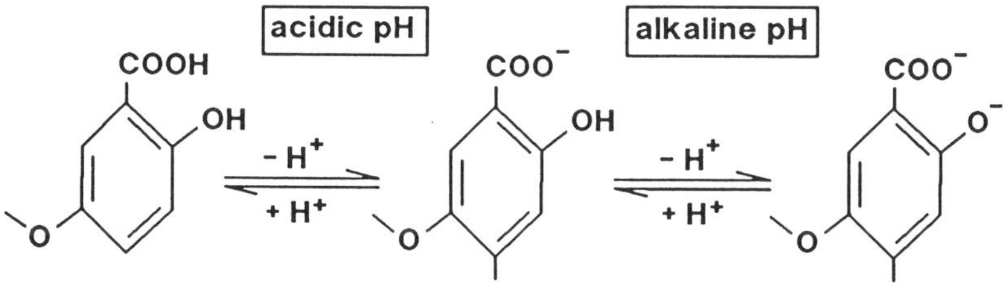

    [Reaction diagram showing ionization: COOH, OH form ⇌ ($-H^+$/$+H^+$, acidic pH) COO^-, OH form ⇌ ($-H^+$/$+H^+$, alkaline pH) COO^-, O^- form]

    The diagram can be used to show the influence of mass action in shifting the reaction either to the protonated or fully dissociated form, depending on pH.

3. Reference can be made to the dimers as being representative of stored forms of organic N that have different availabilities to plants. Waksman's[17] lignin–protein theory of humus formation was at one time widely accepted in scientific circles and is still extensively quoted in textbooks on soils. Whereas the theory is no longer regarded as acceptable in its original form, the idea that proteinaceous substances are attached to the central core of the humic molecule is probably valid. The linkage of protein (or peptide) with the dimeric structure is illustrated below.

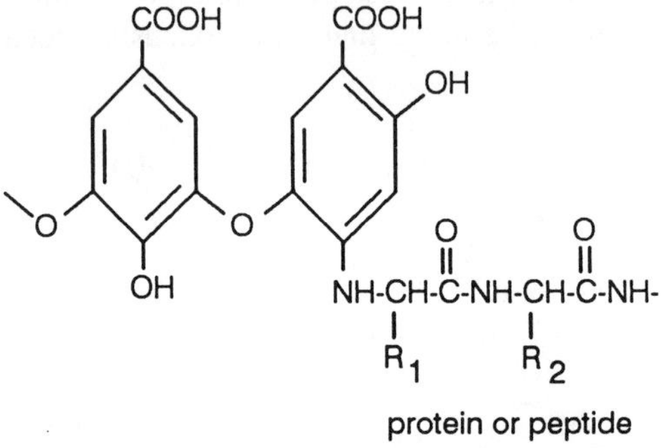

$$\text{NH-CH(R}_1\text{)-C(=O)-NH-CH(R}_2\text{)-C(=O)-NH-}$$

protein or peptide

III

The attached protein (or peptide), while partially stabilized, would be subjected to microbial attack with release of N to available mineral forms. In the process, the terminal $-NH_2$ group may remain attached to the central core (structure II), thus accounting for some of the unknown N in humic substances. Some of the recalcitrant organic N in soils may exist as a linkage between monomeric units, as shown below.

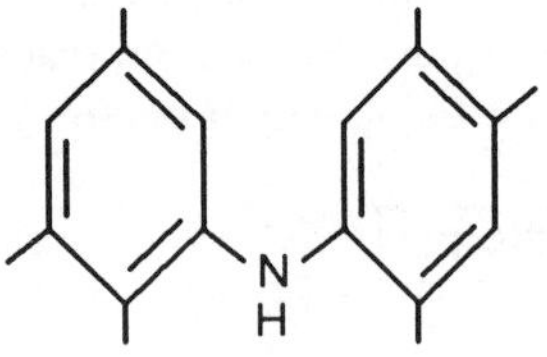

IV

4. The tendency of humus to be bonded to clay and other silicate surfaces can easily be illustrated, namely, through linkages with polyvalent cations (V) and by H-bonding (VI).

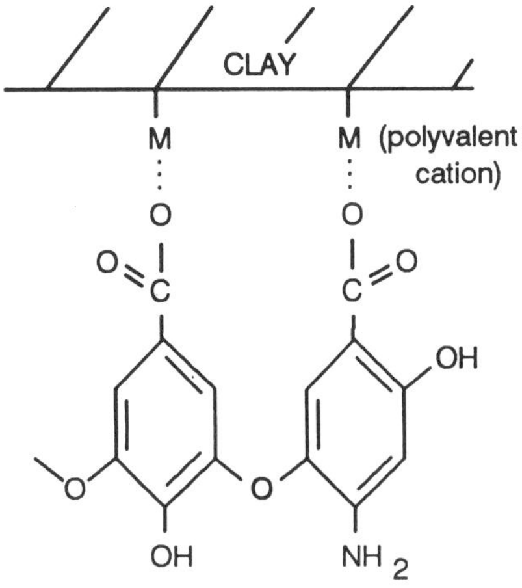

SIMPLIFIED REPRESENTATION OF SOIL HUMIC SUBST...

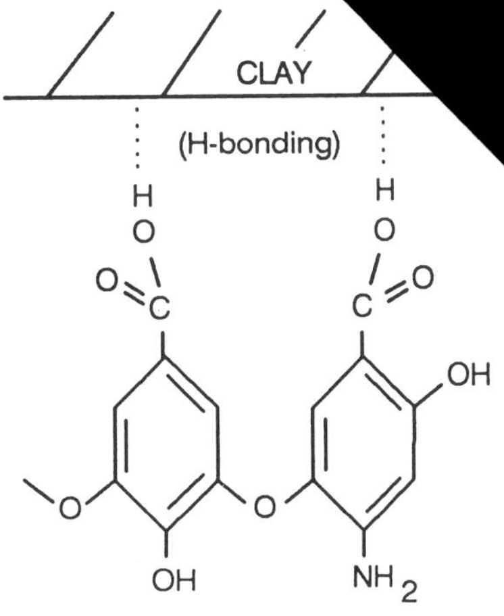

V — clay–metal–organic complex (M = polyvalent cation)

VI — clay–organic complex (H-bonding)

The ability of humus to resist decomposition in soil can be accounted for in two ways: chemical resistance as noted later in item 7, and physical protection through interactions with clay minerals to form clay–metal–organic (**V**) and clay–organic (**VI**) complexes.

5. The dimer can be used to illustrate the ability of humus to form stable complexes with trace elements and heavy metals (M). Chelation can occur when the oxygen-containing functional groups are in adjacent positions on the aromatic ring, as follows.

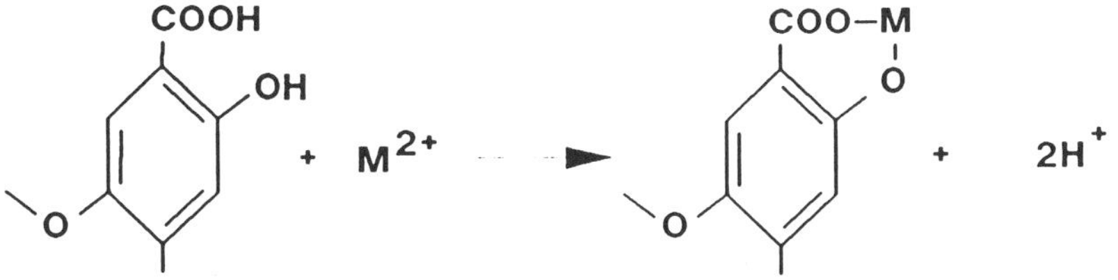

[Reaction: aromatic dimer with COOH and OH groups + M^{2+} → chelated COO–M complex + $2H^+$]

6. The dimer can be used to illustrate the bonding between soil organic matter and pesticides. This interaction is of interest to agronomists because application rates for pesticides must often be adjusted upward on soil rich in organic matter due to adsorption by the organic fraction (Chapter 19). The reaction is also of interest to environmentalists who are concerned about contamination of ground and surface waters. A representative structure is shown below.

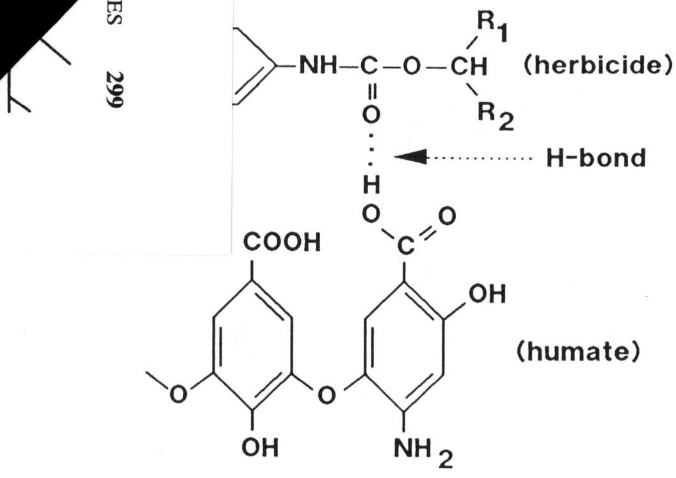

SIS OF HUMIC SUBSTANCES

VII

Humic and fulvic acids from whatever source invariably contain some carbohydrate material (possibly as polysaccharides); these constituents may be bound to the central core through H-bonding in much the same manner as noted in **VII** for pesticides.

7. The simplified representation can be used to illustrate the ways in which individual monomers can be linked together to form chain-like matrixes that are relatively resistant to microbial degradation. When considered along with item 4, the model provides an understanding of the ability of humus to resist decomposition in soil.

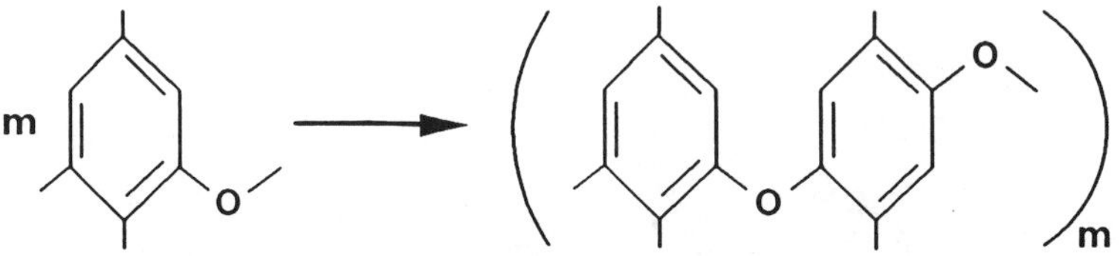

resistant core of humic substances

8. Use of the dimer serves to explain why soil organic matter exhibits anion exchange capacity at low pH (acid soils) but little or no capacity in near neutral or alkaline soils. The following reaction shows the formation of a positively charged site on soil organic matter.

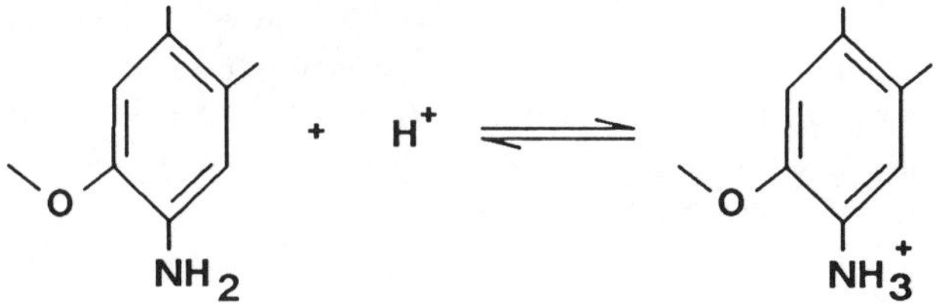

9. The dimer concept can be used to explain why organic matter that is saturated with Na^+ is highly charged (negatively), highly dissociated, and dispersed, such as in black alkali soils. Conversely, when saturated with Ca^{2+} or other di- or trivalent cations, cross-linking occurs by a mechanism similar to that shown for the clay–metal–organic complex (see structure **V**), the cations themselves are largely unionized, and the colloids are in a more flocculated state (i.e., Ca-humates are insoluble).

The above examples are by no means all-inclusive and other examples can be found in other chapters. However, they serve to illustrate the usefulness of the dimer concept for learning and/or teaching humus chemistry at the undergraduate level. The dimer concept also bridges the crucial gap between the precursors of humus (lignin decomposition products) and formation of the humus matrix.

SUMMARY

In this and previous chapters, emphasis has been placed on the polydispersity in the composition of humic substances. Accordingly, there is no single "repeating unit" and no "regularity" in the structures of the macromolecules. The structural formulas reproduced herein are extremely variable, both with regard to the ratio of aromatic to aliphatic components but in the kind and arrangement of reactive functional groups. Taken as a whole, the formulas are useful in that they provide an indication of: 1) the types of structural units that make up the core or backbone of humic and fulvic acids, 2) the possible arrangement of reactive functional groups, 3) the extreme heterogeneity of humic substances from different sources, and 4) the occurrence of aliphatic constituents as side chains or as linkages between aromatic rings, particularly in the humic substances of sediments and natural waters.

In addition to the more complex structures, a simple and useful model is presented for the basic chemical structure of humic substances. The model, which is based on a simple dimer, is of particular value for teaching at the undergraduate level. By focusing attention upon the dimer, the relative inertness of the humus matrix to microbial attack is made understandable; furthermore, simple reactions can be used to illustrate the important roles played by organic matter in soil, including binding of metal ions and pesticides and formation of complexes with clay minerals, details of which are given in subsequent chapters of this book.

REFERENCES

1 F. J. Stevenson and R. A. Olson, *J. Agron. Ed.* **18,** 84 (1989).
2 J. M. Bracewell et al., "The Characterization and Validity of Structural Hypotheses Group Report," in F. H. Frimmel and R. F. Christman, Eds., *Humic*

Substances and their Role in the Environment, Wiley, New York, 1988, pp. 151–164.
3. M. Kononova, *Soil Organic Matter,* Pergamon, London, 1966.
4. W. Flaig, *Suomen Kem.,* **A33,** 229 (1960).
5. D. S. Orlov, *Humus Acids of Soils* (English translation), Oxonian Press, New Delhi, 1985.
6. C. Steelink, "Implications of Elemental Characteristics of Humic Substances," in G. R. Aiken, D. M. McKnight, R. L. Wershaw, and P. MacCarthy, Eds., *Humic Substances in Soil, Sediment, and Water,* Wiley, New York, 1985, pp. 457–476.
7. H.-R. Schulten and M. Schnitzer, *Naturwissenschaften,* **80,** 29 (1993).
8. R. F. Christman, D. L. Norwood, Y. Seo, and F. R. Frimmel, "Oxidative Degradation of Humic Substances from Freshwater Environments," in M. H. B. Hayes, P. MacCarthy, R. L. Malcolm, and R. S. Swift, Eds., *Humic Substances II,* Wiley, New York, 1989, pp. 33–67.
9. C. Steinberg and U. Muenster, "Geochemistry and Ecological Role of Humic Substances in Lakewater," in G. R. Aiken, D. M. McKnight, R. L. Wershaw, and P. MacCarthy, Eds., *Humic Substances in Soil, Sediment, and Water,* Wiley, New York, 1985, pp. 105–145.
10. M. Schnitzer, "Humic Substances: Chemistry and Reactions," in M. Schnitzer and S. U. Kahn, Eds., *Soil Organic Matter,* Elsevier, New York, 1978, pp. 1–64.
11. S. M. Griffith and M. Schnitzer, "Oxidative Degradation of Soil Humic Substances," in M. H. B. Hayes, P. MacCarthy, R. L. Malcolm, and R. S. Swift, Eds., *Humic Substances II,* Wiley, New York, 1989, pp. 69–98.
12. J. A. E. Buffle, "Les Substances Humiques et leurs Interactions avec les Ions Mineraux," in *Conference Proceeding de la Commission d'Hydrologie Appliquee de l'A.G.H.T.M.,* l'Universite d'Orsay, 1977, pp. 3–10.
13. U. S. Geological Survey Staff, "Humic Substances in the Suwannee River, Georgia; Interactions, Properties, and Proposed Structures," Open File Report 87-557, U. S. Geological Survey, Denver, Colorado, 1989.
14. R. L. Wershaw, *J. Contam. Hydrol.* **1,** 29 (1986).
15. G. R. Harvey and D. A. Boran, "Geochemistry of Humic Substances in Seawater," in G. R. Aiken, D. W. McKnight, R. L. Wershaw, and P. MacCarthy, Eds., *Humic Substances in Soil, Sediment, and Water,* Wiley, New York, 1985, pp. 233–247.
16. G. R. Harvey, D. A. Boran, L. A. Chesal, and J. M. Tokar, *Mar. Chem.* **12,** 119 (1983).
17. Waksman, S. A. 1936. *Humus,* Williams & Wilkins, Baltimore, MD.

13

SPECTROSCOPIC APPROACHES

Spectroscopic measurements in the different regions of the electromagnetic spectrum have wide application to the study of soil organic matter, especially humic substances. Attractive features of most of these methods are: 1) they are nondestructive, 2) only small sample weights are required, and 3) most are experimentally simple and do not require special manipulative skills.

The spectroscopic techniques that have been used, or have potential use, in studies of soil organic matter and its subfractions, together with the transitions involved, are as follows:

Ultraviolet-visible absorption spectroscopy: Electronic transitions of bonding electrons. The ratio of absorbance at 465 nm to that at 665 nm (i.e., the E_4/E_6 ratio) has been used for diagnostic purposes; absorption in the visible region is useful for quantitative analyses.

Fluorescence spectroscopy: Also electronic transitions of bonding electrons. The method is uniquely applicable to studies on metal ion binding (see Chapter 16 and 17).

Infrared (IR) spectroscopy: Molecular vibrations. The technique is of diagnostic value for reactive functional groups (COOH, phenolic-, enolic-, and alcoholic OH, $C=O$, $-NH_2$, etc.), as well as for identification of molecular structures (i.e., aliphatic and aromatic components, peptide linkages, others).

Raman spectroscopy: Also molecular vibrations.

Electron spin resonance spectroscopy (ESR): Used for detection and characterization of free-radical species (i.e., structures containing an unpaired electron).

Nuclear magnetic resonance (NMR) spectroscopy: A technique that provides quantitative data for structural components.

Mössbauer spectroscopy: Nuclear absorption of gamma rays. This method is not directly applicable to structural studies of organic substances but has been used to study the binding of Fe and other paramagnetic metal ions by humic substances (see Chapter 16).

Emphasis will be given in this chapter to UV-visible, IR, and ESR spectroscopy. Raman spectroscopy is subject to interference from fluorescence and has not yet been successfully applied to soil organic matter. Of the various approaches, NMR spectroscopy has provided the greatest amount of information on the chemical composition of soil organic matter; this subject is discussed separately in Chapter 11.

For detailed information on the application of spectroscopic approaches (other than NMR) in studies of soil organic matter, the reader is referred to several reviews on the subject.[1-5]

ULTRAVIOLET (UV) AND VISIBLE REGIONS

Absorption in the UV (200–400 nm) and visible (400–800 nm) regions is caused by atomic and electrometric vibrations, and involves elevation of electrons in σ-, π-, and η-orbitals from the ground state to higher energy levels. Constituents containing unbonded electrons on oxygen and sulfur atoms are capable of showing absorption, as well as systems containing conjugated C=C double bonds.

Color-Producing Groups

The part of an organic molecule responsible for producing color is called a chromophore, or chromophoric group. The groupings responsible for the dark color of humic substances have not been firmly established but a combination of several structural types are suspect. Some of the more common chromophoric groups are

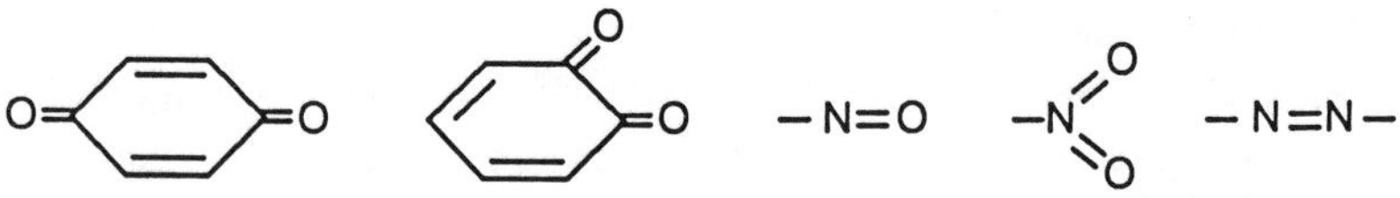

```
    I              II            III        IV         V
  O=⟨⟩=O      ⟨⟩=O          -N=O      -NO₂      -N=N-
```

Other unsaturated structures which are chromophoric include

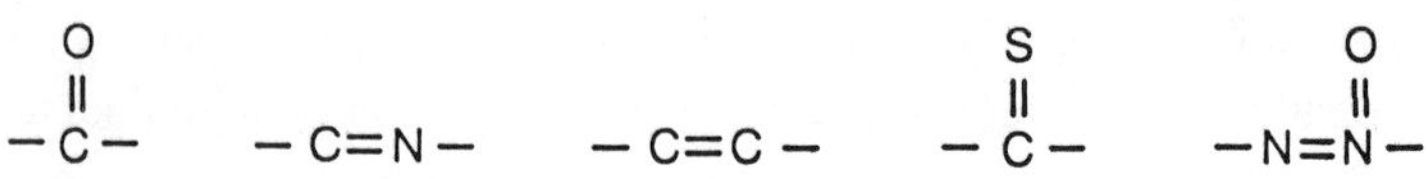

```
   VI         VII        VIII        IX         X
   O                                  S          O
   ‖                                  ‖          ‖
  -C-        -C=N-      -C=C-        -C-       -N=N-
```

A single double bond is usually insufficient to produce color, but if several are present in conjugations, intense color can develop. Many scientists are of the opinion that the dark color of humic substances is due primarily to quinone-like structures (**I** and **II**) and ketonic C=O in conjugation, as follows:

$$-CH_2-\underset{\underset{O}{\|}}{C}-CH_2-\underset{\underset{O}{\|}}{C}- \quad \rightleftharpoons \quad -CH_2-\underset{\underset{OH}{|}}{C}=CH-\underset{\underset{O}{\|}}{C}-$$

keto form   enol form
XI   XII

Spectra of Humic and Fulvic Acids

Absorption spectra of humic and fulvic acids in both the UV (200–400 nm) and visible (400–800 nm) regions are somewhat featureless in that well-defined maxima and minima are absent (Figs. 13.1 and 13.2). However, a slight maximum is often indicated in the 260–300 nm region of the ultraviolet.

Spectra of humic acids of diverse origins vary somewhat and may be related to differences in degree of humification. Tsutsuki and Kuwatsuka[6] found that absorbance at any given wavelength decreased stepwise with decreasing pH, a result that was attributed to changes in structural properties, such as degree of dissociation of COOH and phenolic OH groups. In a recent study on solvent and pH effects on the UV–visible spectra of fulvic acids, Baes and Bloom[7] concluded that absorbances at variable wavelengths were not consistent with the behavior of simple quinones but might be due to chromophores with extended conjugations, possibly polyaromatic structures.

Absorption near 465 nm has been used for quantitative analysis, but, as noted below, considerable variation exists in the optical density of humic substances from various sources.

Quantitative Analysis

The relationship between absorption and concentration is given by the Beer–Lambert law:

$$\log (I_0/I) = kcd$$

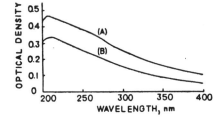

Fig. 13.1 Ultraviolet spectra of humic acid (A) and fulvic acid (B) from a Spodosol. From Schnitzer,[1] reproduced by permission of Marcel Dekker, Inc.

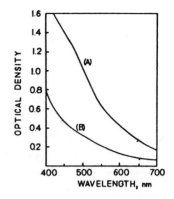

Fig. 13.2 Visible spectra of fulvic acid (A) and humic acid (B) from a Spodosol. E_4/E_6 ratios were 7.5 and 5.7, respectively. From Schnitzer,[1] reproduced by permission of Marcel Dekker, Inc.

where I_0 is the intensity of the incident light, I is the intensity of the transmitted light, k is the extinction coefficient, c is the concentration, and d is the path length of the cell. Essentially, the Beer–Lambert law states that the amount of light absorbed is proportional to the number of molecules of absorbing substances through which the light passes.

The extinction coefficient, k, is equal to optical density or absorbance (log I_0/I) when the cell length is 1 cm and the concentration of the sample is 1 mole/L. For humic and fulvic acids, where molecular weight is variable or unknown, the extinction coefficient is expressed in terms of a given quantity of C (0.1 to 0.2 mg/mL is often used).

On the basis of equal concentrations, the extinction coefficient of humic compounds has been observed to increase with: 1) an increase in molecular weight, 2) C percentage, 3) degree of condensation, and 4) ratio of C aromatic rings to C in aliphatic structures.[2] Fulvic acids have a relatively low capacity for light absorption and the extinction coefficient is somewhat similar for fulvic acids from various sources.

From the above, it can be seen that the Beer–Lambert law is valid only for humic substances of similar origins. For quantitative analysis, an extinction coefficient must be determined for each type of material under study. Once the coefficient is known, the concentration of humic material can be calculated. The estimates are best carried out at a near neutral pH, a concentration of humate C of the order of 0.14 mg/mL, and a wavelength of 465 nm. An appropriate solvent for most work is $0.05N$ $NaHCO_3$.

Colorimetric methods are useful for: 1) monitoring the results of fractionation studies, such as gel filtration, 2) determining the degree (extent) of humification in compost heaps and peats, and 3) estimating the humic acid content of lignites. As will be noted in Chapter 20, colorimetric analysis for humates in pyrophosphate or alkali extracts has been used as a means of discriminating between humus types on the forest floor, as well as for the classification and recognition of Spodosol B horizons (see Chapter 20).

The E_4/E_6 Ratio

The ratio of absorbances at 465 and 665 nm, referred to as the E_4/E_6 ratio, has been widely used for characterization purposes. Ratios for humic acids are usually < 5.0; those for fulvic acids range from 6.0 to 8.5.[1] According to Chen et al.[8] the best procedure for determining the E_4/E_6 ratio is to dissolve from 2 to 4 mg of the humic or fulvic acid in 10 mL of $0.05N$ $NaHCO_3$, which gives an optimum pH for absorbance measurements.

The E_4/E_6 ratio decreases with increasing molecular weight and condensation and is believed to serve as an index of humification. Thus, a low ratio may be indicative of a relatively high degree of condensation of aromatic constituents; a high ratio infers the presence of relatively more aliphatic structures. Chen et al.[8] concluded that the E_4/E_6 ratios of humic and fulvic acids were governed primarily by particle sizes and weights (due in part to scattering of light) but this was not confirmed by Baes and Bloom.[7] An inverse relationship has been observed between the E_4/E_6 ratio and the mean residence time of humic material; specifically, humic substances with the highest ratios had the lowest mean residence time. From this result, the conclusion was reached that the older material was more highly condensed and aromatic in nature.[9]

The E_4/E_6 ratios for humic acids extracted from soils belonging to several great soil groups are given in Table 13.1.

INFRARED (IR) SPECTROSCOPY

In contrast to the relatively few absorption bands observed in the visible and ultraviolet regions, the IR spectra of humic substances and their derivatives

Table 13.1 E_4/E_6 Ratios of Humic and Fulvic Acids[a]

Acids	E_4/E_6 Ratio
Humic acids	
Podzol	±5.0
Gray forest	±3.5
Chernozem	3.0–3.5
Chestnut	3.8–4.0
Serozem	4.0–4.5
Krasnozem	±5.0
Fulvic acids	6.0–8.5

[a]From Schnitzer.[1] In the Comprehensive Soil Classification system, the first two soils are a Spodosol and an Alfisol, respectively. The Chernozem and Chestnut soils are Mollisols; the Sierozem soil is an Aridisol.

contain a variety of bands that are diagnostic of specific molecular structures.[1,10,11] Infrared spectroscopy is of considerable value in humus research for the following reasons: 1) key information is provided regarding the nature, reactivity, and structural arrangement of oxygen-containing functional groups; 2) the occurrence of protein and carbohydrate constituents can be established; 3) the presence or absence of inorganic impurities (metal ions, clay) in isolated humic fractions can be demonstrated; and 4) the technique is suitable for quantitative analysis. Other applications of IR spectroscopy have included studies on metal–organic and pesticide–organic matter interactions (see Chapters 16 and 19, respectively).

Theory and Methods

Absorption in the IR region is due to rotational and vibrational movements of molecular groups and chemical bonds of a molecule. Essentially, there are two fundamental vibrations; stretching, where the atoms remain in the same bond axis but the distance between atoms increases or decreases, and bending (deformation), where the positions of the atoms change relative to the original bond axis. When infrared light of the same frequency as any given stretching or bending vibration is incident on the sample, energy is absorbed and the amplitude of that vibration is increased. Due to energy absorption at the resonance frequency, the detector of the IR spectrometer records an absorption peak for that wavelength.

Some typical vibrations for a group of atoms are illustrated below, where the + and − signs signify vibrations perpendicular to the plane of the paper.

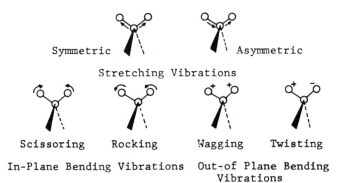

Bending vibrations generally require less energy and are found at lower frequencies than stretching vibrations. Stretching due to the triple bond (2300–2000 cm^{-1}) is somewhat stronger than for the double bond (1900–1500 cm^{-1}), which in turn is stronger than for such single bonds as C—C, C—O, and C—N (1300–800 cm^{-1}). Stretching vibrations for single bonds involving the very small proton (e.g., C—H, O—H, and N—H) occur at frequencies between 3700 and 2650 cm^{-1}. The band for O—H stretching occurs at a higher frequency (3700–3200 cm^{-1}) than for C—H stretching (3050–2850 cm^{-1}). For detailed information on chemical applications of IR spectroscopy, the early works of Bellamy[12] and Rao[13] are recommended.

To obtain an IR spectrum, the experimental sample is illuminated with IR radiation of successive wavelengths from 2.5 to 25 μm (4000–400 cm^{-1}) and the amount of light transmitted by the sample is measured by a recording spectrometer, which computes the percentage of light transmitted at each wavelength and produces in a relatively short time (5–20 min.) a curve of transmittance versus wavelength or frequency.

The spectrum of a solid sample is usually best determined as an alkali halide pellet. About 1 mg of the sample and from 100 to 200 mg KBr are ground together and pressed at an elevated pressure (at least 25,000 psi) into a small disc about 10 mm in diameter and 1 to 2 mm thick. Since KBr does not absorb light in the IR region, a complete spectrum of the sample is obtained.

The spectrum can also be determined as a mull, in which case about 5 mg of material is ground to a very fine dispersion with a drop of a suitable mulling agent, which is placed between two NaCl plates. A common mulling agent is Nujol, which consists of a mixture of high-molecular-weight paraffinic hydrocarbons. Interfering absorption bands are presented in the C—H stretching region between 3030 and 2860 cm^{-1}.

A major problem in the application of IR spectroscopy for analysis of humic substances is interference due to adsorbed moisture, which produces bands in the 3300–3000 cm^{-1} and 1720–1500 cm^{-1} regions. Interference is particularly serious when pressed discs of alkali halides (e.g., KBr) are used. In the technique used by Theng and Posner,[14] adsorbed moisture is eliminated by evacuating the die at 75°C for 30 min. before forming the pellet under pressure. Stevenson and Goh[15] recommended heating preformed KBr discs under vacuum at 100°C for 2 h., but prolonged heating should be avoided. More widespread use of a nonhygroscopic pelleting matrix (e.g., thallous bromide) is recommended.

A technique that shows promise for eliminating interference due to water absorption is diffuse reflectance Fourier-transform IR.[16,17] By this method, bands due to water are digitally subtracted out. Another approach for eliminating OH absorption bands due to water is deuteration.[18]

Spectra of Humic and Fulvic Acids

Typical IR spectra of humic and fulvic acids are shown in Figs. 13.3 and 13.4, respectively.[19] Assignments generally given to the various absorption bands are tabulated in Table 13.2. Main absorption bands are in the regions of 3300 cm^{-1} (H-bonded OH groups), 2900 cm^{-1} (aliphatic C—H stretching), 1720 cm^{-1} (C=O stretching of COOH and ketones), 1610 cm^{-1} (aromatic C=C and H-bonded C=O), and 1250 cm^{-1} (C—O stretching and OH deformation of COOH). In addition, small bands are often evident at about 1500 cm^{-1} (aro-

310 SPECTROSCOPIC APPROACHES

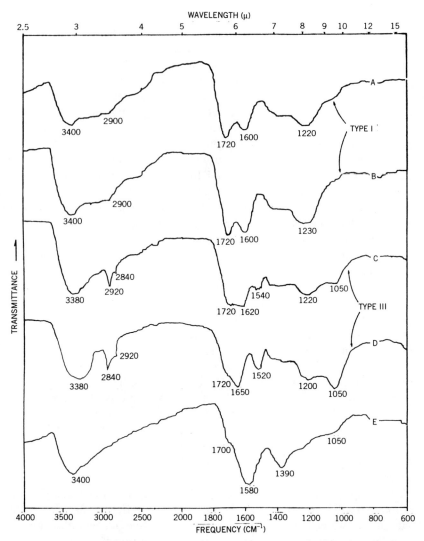

Fig. 13.3 Infrared spectra of humic acids from several sources as recorded by Stevenson and Goh[19]: (A) Mollisol, (B) North Dakota lignite (Leonardite), (C) B horizon of a Spodosol, (D) Mud Lake, Florida, and (E) sodium salt of A

matic C=C), 1460 cm^{-1} (C—H deformation of CH_2 or CH_3 groups), and 1390 cm^{-1} (O—H deformation, CH_3 bending, or C—O stretching).

Humic and fulvic acids have somewhat similar spectra, the main difference being that the intensity of the 1720 cm^{-1} band is considerably stronger in the fulvic acids because of the occurrence of more COOH groups (see previous chapter). Also, the 1600 cm^{-1} band is centered at a higher frequency (near 1640 cm^{-1}).

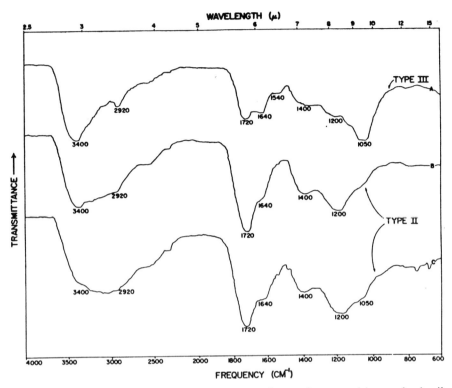

Fig. 13.4 Infrared spectra of fulvic acid-type constituents from a prairie grassland soil as recorded by Stevenson and Goh[19]; (A) Colloid (nondialyzable) preparation, (B) dialyzable component I, (C) dialyzable component II

Stevenson and Goh[19] classified the IR spectra of humic substances into three general types, as follows:

Type I: Spectra of this type are typical of those generally shown for humic acids. Strong bands are evident near 3400, 2900, 1720, 1600, and 1200 cm^{-1}. The 1600 cm^{-1} band is about equal in intensity to the one at 1720 cm^{-1}.

Type II: These spectra, shown by low-molecular-weight fulvic acids, are characterized by very strong absorption near 1720 cm^{-1}. A second prominent feature is that absorption in the 1600 cm^{-1} region is weak and centered near 1640 cm^{-1}.

Type III: In addition to the major absorption bands shown by Types I and II, relatively strong bands are evident neat 1540 cm^{-1}. Absorption near 2900 cm^{-1} (aliphatic C—H absorption) is also more pronounced. A unique feature of these spectra is the presence of bands indicative of proteins and carbohydrates.

Table 13.2 Main IR Absorption Bands of Humic Substances

Frequency (cm^{-1})	Assignment
3400–3300	O—H stretching, N—H stretching (trace)
2940–2900	Aliphatic C—H stretching
1725–1720	C=O stretching of COOH and ketones (trace)
1660–1630	C=O stretching of amide groups (amide I band), quinone C=O and/or C=O of H-bonded conjugated ketones
1620–1600	Aromatic C=C, strongly H-bonded C=O of conjugated ketones?
1590–1517	COO$^-$ symmetric stretching, N—H deformation + C=N stretching (amide II band)
1460–1450	Aliphatic C—H
1400–1390	OH deformation and C—O stretching of phenolic OH, C—H deformation of CH$_2$ and CH$_3$ groups, COO$^-$ antisymmetric stretching
1280–1200	C—O stretching and OH deformation of COOH, C—O stretching of aryl ethers
1170–950	C—O stretching of polysaccharide or polysaccharide-like substances, Si—O of silicate impurities.

The humic substances of any given soil show a range of spectra intermediate between Types I and II. Thus, the 1720 cm^{-1} band becomes progressively weaker with an increase in color intensity, indicating a progressive decrease in COOH content with increasing molecular weight. Systematic changes also occur in the intensities and positions of bands in the 1660–1600 cm^{-1} region.[15] Spectra of Type III are typical of lake humic acids.

A more elaborate classification of humic acids according to IR characteristics was made by Kumada,[20] who classified humic acids into four major types (A, B, R_p, and P) based on the relative intensities of specific absorption bands. Type A was further subdivided into subtypes A_1 and A_2. The scheme is as follows:

Type A: Humic acids of this type have a fairly strong aromatic C—H stretching band near 3075 cm^{-1} and the aliphatic C—H stretching band near 2940 cm^{-1} is smaller than for other types. The intensity of the band near 2630 cm^{-1}, assigned to OH stretching of COOH, is strong. Two sharp absorption bands exist near 1710 and 1615 cm^{-1} for C=O and C=C stretching vibrations, respectively. The 1640 and 1515 cm^{-1} bands, which are obvious in Types B and R_p, are absent in A_1 while the latter is a shoulder in A_2. Two very small bands occur in the region of 1470–1370 cm^{-1} and a very broad and strong band occurs in the 1280–1200 cm^{-1} region.

Type B: The intensity of the 2940 cm^{-1} is very strong and sharp while the 3050 and 2630 cm^{-1} bands are weak. Also, the band near 1690 cm^{-1} is weak. Besides a strong 1615 cm^{-1} band, a pronounced one is present near 1515 cm^{-1} and a band is found as a shoulder near 1640 cm^{-1}. Four fairly sharp bands are present near 1450, 1480, 1380, and 1330 cm^{-1} and a strong band is present in the 1280–1205 region.

Type R: Spectra of these humic acids are similar to those of Type *B* except that the band near 2630 cm^{-1} is weak and a small shoulder is found near 1540 cm^{-1}. Other minor differences were also observed.

Type P: These spectra are similar to Type *A* but the intensities of the 3075, 2940, and 2630 cm^{-1} bands are relatively weak and the intensity of the 1695 band relative to the 1615 cm^{-1} band is smaller. Differences were also observed at lower frequencies.

Applications of the scheme for characterizing humic substances from diverse sources can be found in the book by Kumada.[20]

Spectra of Methylated and Acetylated Derivatives

While IR analyses of humic and fulvic acids have proved useful for characterization purposes, substantially more information has been obtained from spectra of derivatives produced by various chemical treatments, such as through methylation and acetylation as discussed previously. Some of the changes are noted in Table 13.3. For example, methylation with methanolic–HCl (Fig. 13.5) results in an increase in the intensity of the bands caused by C—H stretching (2900 cm^{-1}), C=O stretching (1720 cm^{-1}), and C—H deformation of CH$_3$

Table 13.3 Assignments of Absorption Bands for Methylated and Acetylated Humic Acids

Sample and frequency (cm^{-1})	Assignment
Methylated humic acid	
2920	C—H stretching of added methyl groups
1720	C=O of methyl esters
1440	H-bending of added CH$_3$
1220	C—O of methyl esters
Acetylated humic acid	
2920	C—H stretching of added acetyl group
1840	C=O of 5-membered ring anhydrides
1815	C=O of mixed anhydrides
1775	C=O of 5-membered ring anhydrides plus alcoholic and phenolic acetates
1740	C=O of mixed anhydrides
1440	H-bending of added CH$_3$ groups
1365	CH$_3$—C deformation of added acetyl groups
1250–1150	C—O of anhydrides plus phenolic and alcoholic acetates

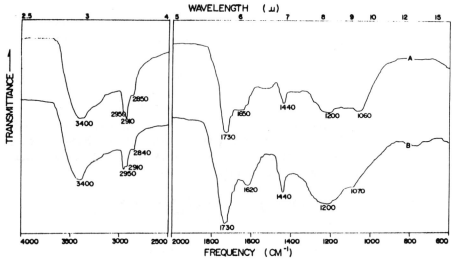

Fig. 13.5 Infrared spectra of a soil humic and fulvic acid methylated with methanolic-HCl as recorded by Stevenson and Goh:[15] (A) Humic acid (B) fulvic acid

groups (1460 cm^{-1}). In addition, the 1250 cm^{-1} band is sharpened considerably. The significance of these and other changes brought about by chemical modification is discussed in the following subsections.

The 3300–3400 cm^{-1} Region Humic substances show broad absorption in the 3300–3400 cm^{-1} region and this is usually attributed to O—H stretching. In the undissociated state, OH groups absorb near 3600 cm^{-1}; when associations occur, such as through H-bonding, the frequency is reduced somewhat. Accordingly, the broad band in the 3300–3400 cm^{-1} region is assigned to H-bonded OH groups.

Spectra of methylated or acetylated derivatives invariably show prominent residual absorption in the 3300–3400 cm^{-1} region (see Fig. 13.5). This may be due, in part, to N—H stretching or to OH groups resistant to methylation or acetylation. However, it should be noted that absorption in the OH region must be interpreted with caution because of moisture contamination, especially when KBr discs are used.[14, 15]

The 2900 cm^{-1} Band Absorption due to aliphatic C—H stretching varies considerably, being strong in some preparations but not in others (see Fig. 13.3). Absorption in this region is enhanced by methylation (see Fig. 13.5) or acetylation, due to introduction of CH$_3$ and —CH$_2$—CH$_3$ groups.

Aromatic C—H absorption generally occurs at frequencies slightly higher than 3000 cm^{-1} and the absence of a pronounced band in this region may be due to extensive substitution of the aromatic ring or masking from the broad band due to OH stretching.

The 1720 cm^{-1} Band This pronounced band largely disappears when humic substances are reduced with diborane or are converted to salts (see Fig. 13.3), thereby providing conclusive evidence that the major part of the absorbance in this region is due to C=O stretching of COOH groups. In the latter case, two new bands are introduced near 1550 and 1400 cm^{-1} for adsorption due to the carboxylate (COO$^-$) ion. The residual absorption may be due to saturated open-chain ketones or aldehydes, which also absorb at about 1720 cm^{-1}, The absence of strong absorption bands between 1785 and 1735 cm^{-1} is reasonable proof that few, if any, C=O groups occur as normal saturated esters.

The 1720 cm^{-1} band often shifts to a slightly higher frequency (by about 5 cm^{-1}) when humic substances are methylated, which can be attributed to the fact that the C=O of esters generally absorb at higher frequencies than the C=O of the corresponding acids. The frequency of absorption for both acid and ester forms in humic substances are lower than normal which can be explained by the occurrence of conjugated C=O groups and/or aromatic acids.

Wood et al.[21] found that when a lignite humic acid was treated with acetic anhydride or heated (> 100°C) in sulfolane, bands (1850 and 1785 cm^{-1}) typical of five-membered and seven-membered ring anhydrides were formed. Nearly 80 percent of the COOH groups (350 cmole out of 430 cmole/kg) were either paired on mutually adjacent sites, such as on aromatic rings, or located on attached ring structures of the diphenic acid type.

XIII   XIV

In a similar study,[22] about one-third of the COOH groups (128 cmole out of 389 cmole/kg) in a soil humic acid was found to occur in sites such that cyclic anhydrides could be formed by heating with acetic anhydride. Bands typical of anhydrides have also been observed by heating preformed KBr pellets of humic substances at elevated temperatures (Fig. 13.6). The possible presence of anhydrides is indicated by introduction of new bands at 1850 and 1780 cm^{-1} in spectra of the heated samples; the sharp band at 2315 cm^{-1} is due to CO_2 trapped in the KBr matrix, apparently from decarboxylation of COOH groups.

The 1610 cm^{-1} Band Absorption at about 1610 cm^{-1} is usually attributed to C=C vibrations of aromatic structures. However, several other reasons can be given for absorption in this region; for example, non-aromatic double bonds and rings formed by H-bonding of OH to C=O of quinones (see structures

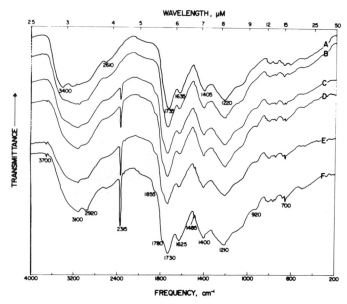

Fig. 13.6 Infrared spectra of a fulvic acid showing the effects of heating preformed pellets at 50°C for up to 27 days: (A) Vacuum dried at 25°C, (B) heated for 2 h, (C) heated for 2 days, (D) heated for 7 days, (E) heated for 14 days, (F) heated for 27 days. Unpublished results of author

I–III). Quinones normally absorb between 1690 and 1635 cm^{-1}, but the C=O band shifts to lower frequencies when H-bonding occurs.

If the 1610 cm^{-1} band is due primarily to C=C vibrations of aromatic rings, one might expect to find a more pronounced band at about 3030 cm^{-1} for aromatic C—H stretching. As noted earlier, the absence of a distinct band at this frequency (see Fig. 13.3) can be explained by extensive substitution of the ring. Also, the band may be masked by strong absorption due to H-bonded OH groups, which can extend from about 3660 to 2900 cm^{-1}.

Simple aromatic compounds invariably produce a band at 1500 cm^{-1} that is more intense than the band at 1610 cm^{-1}. The reversal of the frequency for humic substances strongly suggests that structures other than aromatic C=C contribute to absorption in the 1610 cm^{-1} region.

Particular attention has been given to the contribution of C=O stretching to the 1610 cm^{-1} band of humic acids.[10] One possibility is that H-bonded C=O of quinones absorb in this region. The suggestion has been made that the release of H-bonding by methylation with CH_2N_2 should shift the C=O band to that of the parent quinone but attempts to demonstrate a quinone band in humic acids by this approach have been unsuccessful. An explanation given for this result is that H-bonded OH groups may not be methylated with CH_2N_2, and, consequently, absorption in this region remains unchanged. Moschopedis,[23] using humic acids from lignites, attempted to solve this problem by removing H-bonding through acetylation; when this was done, a band at 1660 cm^{-1} appeared in the IR spectrum. This appears to have been the first indication for quinone structures in humic acids. Wagner and Stevenson[22] were unable to detect a quinone band in the spectra of acetylated humic acids from soil. However, these investigators used the KBr disc technique and difficulty was encountered in eliminating interference due to moisture.

In other work, Schnitzer et al.[24] failed to detect quinones in fulvic acids by reductive acetylation while a 1660 cm^{-1} band (possibly due to quinones) has been observed in methylated (CH$_2$N$_2$) humic acids after prolonged heating.[15] The contribution of H-bonded conjugated C=O groups to the band near 1610 cm^{-1} has been stressed by Theng and Posner.[25]

When humic substances are methylated with CH$_2$N$_2$, the intensity of the 1720 cm^{-1} band is greatly increased, whereas that at 1610 cm^{-1} appears to be reduced. Several explanations have been given for this result, including; 1) release of H-bonding due to methylation of OH groups, with a shift of C=O absorption from 1610 cm^{-1} to 1720 cm^{-1}, 2) increased intensity of the 1720 cm^{-1} band due to ester formation, and 3) elimination of the stretching mode of COO$^-$ ions. The first hypothesis is unlikely because a shift of such magnitude (over 100 cm^{-1}) is extremely rare. As suggested by Farmer and Morrison,[26] C=O absorption of COOH groups must extend over a broad range due to varying degrees of H-bonding, and that the esters formed by methylation should absorb within a much narrower range, thereby giving a more intense peak.

Other Absorption Bands Humic and fulvic acids also show broad absorption in the 1540–1470 cm^{-1} region (aromatic C=C and/or the amide II band of proteins), near 1400 cm^{-1} (OH deformation, C—O stretching of phenolic OH, and CH deformation of CH$_2$ and CH$_3$ groups), at 1200 cm^{-1} (C—O stretching and OH deformation of COOH), and in the region near 1040 cm^{-1} (C—O stretching of polysaccharides, Si—O of silicate impurity). Absorption near 1575 and 1390 cm^{-1} may also be due to the COO$^-$ ion.

ELECTRON SPIN RESONANCE (ESR) SPECTROMETRY

Humic substances are known to contain relatively high concentration of stable free radicals, which are indigenous to their structures and that are of special significance because they may influence many of the reactions in which humic and fulvic acids are known to participate, such as complexing of metal ions, adsorption and chemical modification of pesticides, and physiological activity.[27-35] Free radicals are undoubtedly involved in humus synthesis and degradation.

The first discovery of stable free radicals in humic substances was by Rex.[30] Since then, numerous studies have been conducted in attempts to determine their nature and properties. This work has been reviewed elsewhere.[4,5,35]

Theory and Methods

An atom or group of atoms that contain an odd (unpaired) electron is called a free radical. Substances that contain a free radical are attracted to a magnetic field, and it is this property that permits their detection by ESR spectrometry.

The electrons surrounding an atom are not static but exert a spinning action. When two electrons are present on an atom they have opposite spins that cancel each other out. In as much as a free radical is not paired, the spin of the single electron is not canceled and the radical exhibits a net magnetic moment, which is the cause of its paramagnetism.

Essentially, a spinning electron acts like a small magnet, a gyromagnetic top. When a molecule containing one or more unpaired electrons is held in a magnetic field, and electromagnetic radiation is applied whose wavelengths are in the microwave radio frequency range, energy is absorbed and an absorption peak is registered. From known values of magnetic field intensity and frequency required to induce resonance, a g-factor can be calculated that provides insight into the orbital state of the unpaired electron and thereby the kind of chemical bond in the system. For a free electron, the g-factor is 2.0023; deviation from this value is a measure of the interaction of the electron with its environment.

Under a given set of conditions, the intensity of ESR absorption is proportional to the number of unpaired spins in the sample. However, an absolute determination of spin content is subject to considerable error.

ESR Spectra of Humic Substances

A typical ESR spectrum of humic acid is shown in Fig. 13.7. A characteristic feature of most ESR spectra of humic and fulvic acids is the lack of detailed hyperfine structures, due undoubtedly to the heterogeneous nature of humic substances. Spectra showing resolved hyperfine structures characteristic of simple organic compounds are rare.[4,5,35]

For a discrete organic compound, four types of information are provided from an ESR spectrum:

1. The number of unpaired electrons, usually given in spins/g. This parameter is calculated from the double integration of the spectrum, which is then compared to a spectrum of a standard compound of known spins/g
2. The width of the absorption line (Gauss), which is influenced by such factors as free radical concentration, temperature, and state of aggregation of the sample
3. A g-value (discussed earlier)
4. The hyperfine structure, which gives the number and types of nuclei interacting with the free electron and describes the chemical structure of the free radical

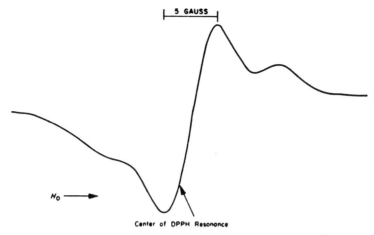

Fig. 13.7 ESR spectrum of humic acid. From Steelink and Tollin,[35] reproduced by permission of Marcel Dekker, Inc.

Due to the lack of detailed hyperfine structures in most spectra of humic substances (see Fig. 13.7), only the first three ESR values are generally given: free radical concentration (spins/g), line width, and g-value. A selection of values for free radical concentrations in humic and fulvic acids is given in Table 13.4; values for additional parameters are given in Table 13.5.

Values given in the tables for free radical concentrations (spins/g) are within the range usually reported for humic and fulvic acids. In general, spin content

Table 13.4 Spin Concentrations of Various Humic and Lignin Preparations[a]

Sample	Spins (g)	Spins (mole)
Fulvic acids		
Wisconsin Spodosol	3×10^{17}	—
Humic acids		
Wisconsin Spodosol	0.8×10^{18}	—
Muck soil	0.3×10^{18}	—
California Spodosol	2×10^{18}	4×10^{22}
English Spodosol	1.4×10^{18}	3×10^{22}
Arizona brown forest soil	0.8×10^{18}	—
Lignins		
Brauns native lignin	5×10^{16}	5×10^{19}
Norway spruce (H_2SO_4)	3×19^{16}	1.5×10^{20}
Yellow pine (kraft)	3×10^{17}	2.1×10^{21}

[a]From Steelink and Tollin.[35]

Table 13.5 Electron Spin Resonance Parameters for the Humic and Fulvic Acids From Six Soils of the Mediterranean Region[a]

Sample	Free-Radical Concentration (Spins $g^{-1} \cdot 10^{-17}$)	Line Width (Gauss)	Spectroscopic Splitting Factor (G-value)
Humic acids	5–10	4.8–5.2	2.0032–2.0047
Fulvic acids	1–2	4.5–7.5	2.0037–2.0050

[a]From Chen et al.[29]

follows the order: stream fulvic acid > soil fulvic acid > stream humic acid > soil humic acid.[5] Riffaldi and Schnitzer[31] found that the free radical content of humic substances from soils of widely different geographical regions followed the order: fulvic acids > humic acids > humins. High spin contents are usually associated with high O/C atomic ratios and low H/C atomic ratios.[31]

On a per mole basis, the absolute amount of unpaired electrons in humic and fulvic acids is rather small, but nevertheless important to their chemical reactivity. A concentration of from 10^{17} and 10^{18} spins/g in a fulvic acid having a molecular weight of 10,000 daltons represents about one radical per from 60 to 600 molecules. In the case of the higher molecular weight humic acids (100,000 daltons), the value increases to one radical for from 6 to 60 molecules.[4,5]

The g-value for most humic substances (see Table 13.4) is of the order of 2.0040, which is consistent with semiquinone radical units possibly conjugated to aromatic rings.[4] The rather broad line widths (often > 4 Gauss) can be ascribed, at least in part, to superimposed resonances from different structural units.

Results of the extensive study by Atherton et al.[27] on the ESR spectra of humic acids are unique in that some spectra showed evidence for hyperfine structures, notably for humic acids from acidic environments. Their results can be summarized as follows:

1. The ESR spectra could be divided into two classes, depending upon the pH of the soil or sediment from which the humic acid was derived. Humic acids from acidic environments (pH 2.8 to 4.3), such as mor humus layers, bog peats, and humus-containing horizons of Spodosols, showed four-lined spectra. In contrast, humic acids from more basic environments (pH 4.3 to 7.2), such as mull humus layers, fen peats, and most soils, gave ill-defined, structureless spectra without distinct peaks
2. The signals were eliminated by reduction with sodium dithionite and recovered by exposure to air, thereby implicating semiquinone ion radicles
3. The spectra were dependent upon the overlying vegetation only in so far as this corresponded to the pH of the soil (see item 1). The nature of the signal was also independent of depth or age of the soil

Factors affecting the ESR properties of humic substances include pH, oxidation and reduction, UV-visible radiation, hydrolysis, and method of extraction.[27-34] An accounting of these effects is provided in the reviews of Senesi[4] and Senesi and Steelink.[5]

Nature of Stable Free Radicals in Humic Substances

Sources of stable free radicals in humic acids can be classified as follows.[35]

Hydroxyquinones: Chemical species, such as those shown in the following structures, may exist in humic acids through an interaction between OH and quinoid mono- and bi-radicals as follows:

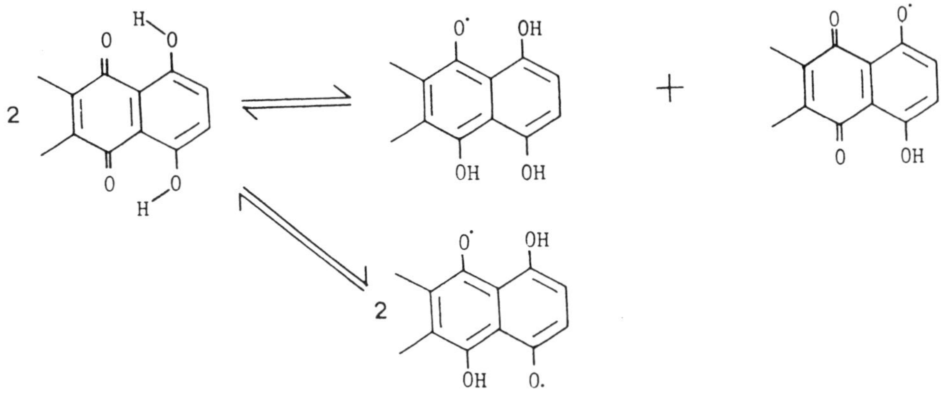

Semiquinone polymers: Polyradicals of the type shown below were considered possible because humic acids may contain quinone groups capable of being reduced by soil constituents, or reactive groups within the polymer.

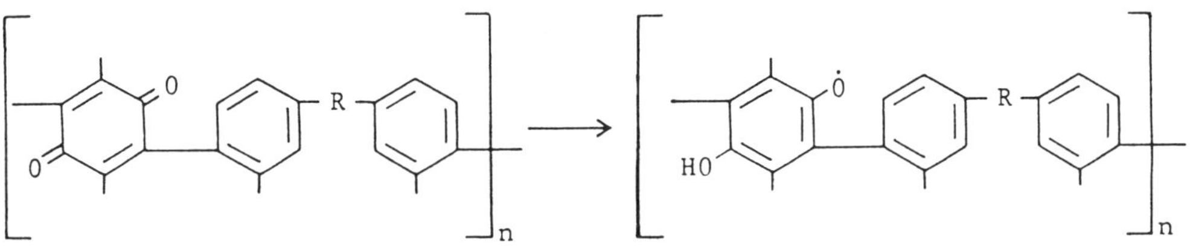

Adsorbent complexes: Humic acids, as normally prepared, contain considerable amounts of mineral matter (1–10 percent). The possibility that a mineral–organic complex contributes appreciably to the free radical content of humic substances has not been sufficiently investigated.

Trapped radicals: Monomeric free radicals may be trapped in the polymeric matrix of the humic acid. It should be noted that repeated solution of humic acid in base and reprecipitation with acid does not lead to a loss in free radical content.

Polynuclear hydrocarbons: Polymeric materials with high C contents give ESR signals but their presence in humic acids (as impurities) has yet to be established.

The likelihood that quinone and quinone-like substances are primarily responsible for the free-radical properties of humic substances has been substantiated by recent research on the subject.[4,5]

FLUORESCENCE SPECTROSCOPY

Electrons excited by absorption of ultraviolet or visible radiation return to the ground state, and, in doing so, a portion of the scattered radiation may have a different wavelength than the incident beam. Fluorescence occurs when the wavelength of the emitted radiation has a longer wavelength than the incident beam and when the excited state has a rather short lifetime, that is, when there is a short delay between the time of excitation and emission. Like UV-visible spectroscopy, the intensity of fluorescence in humic substances is affected by pH and the presence of electrolytes. Fluorescence is quenched by bound metal ions and this property has been used to differentiate between free and bound sites in metal ion binding studies (see Chapters 16 and 17).

Fluorescence is a useful technique for detecting chromophoric structures with extend π electrons, such as aromatic compounds and semiquinones. In general, fluorescence bands of humic substances are broad and as yet uncharacterized. In an examination of 50 humic and fulvic acids from various soils and related materials (e.g., peat, paleosols, composted organic matter, sewage sludges), Senesi et al.[36] concluded that the fluorescence properties of humic substances provided an adequate criteria for their differentiation and classification. They further found that fluorescence peaks for the high-molecular-weight humic acids of soil, peat, paleosol, and lignite could be mainly ascribed to: 1) the presence of linearly condensed aromatic rings, and 2) other unsaturated bond systems capable of a high degree of condensation and conjugation and bearing electron-withdrawing substituents such as COOH and C=O groups.

Results of spectroscopic studies, as applied to humic substances, have been reviewed by Bloom and Leenheer,[2] from which additional references can be found.

SUMMARY

Spectroscopic methods, particularly when combined with chemical approaches, can provide valuable information regarding the chemical structures of humic substances. Quantitative analysis in the visible region is suitable for specific humic and fulvic acids but cannot be applied to humic substances of unknown origin or composition.

Spectra in the ultraviolet and visible regions are somewhat featureless but absorbances at 465 and 665 nm (the E_4/E_6 ratio) are widely used as an index of "degree of condensation" of aromatic constituents. Considerably more information has been obtained by IR spectroscopy. Main absorption bands occur in the regions of 3300 cm^{-1} (O—H and N—H stretching), 2900 cm^{-1} (aliphatic C—H stretching), 1720 cm^{-1} (C=O stretching of COOH and ketonic groups), 1610 cm^{-1} (aromatic C=C and/or H-bonded C=O), 1400 cm^{-1} (OH deformation and C—O stretching of phenolic OH and C—H deformation of CH$_2$ and CH$_3$, and 1250 cm^{-1} (C—O stretching and OH deformation of COOH). Considerable controversy exists regarding the contribution of quinones to absorption in the 1660–1600 cm^{-1} region.

Electron spin resonance (ESR) spectra of humic substances have revealed the occurrence of stable free radicals. The origin of the free radicals is unknown but quinone groups of various types are suspect.

REFERENCES

1. M. Schnitzer, "Characterization of Humic Constituents by Spectroscopy," in A. D. McLaren and J. Skujins, Eds., *Soil Biochemistry,* vol. 2, Marcel Dekker, New York, 1971, pp. 60–95.
2. P. R. Bloom and J. A. Leenheer, "Vibrational, Electronic, and High-Energy Spectroscopic Methods for Characterizing Humic Substances," in M. H. B. Hayes, P. MacCarthy, R. L. Malcolm, and R. S. Swift, Eds., *Humic Substances II: In Search of Structure,* Wiley, New York, 1989, pp. 409–446.
3. P. MacCarthy and J. A. Rice, "Spectroscopic Methods (Other than NMR) for Determining Functionality in Humic Substances," in G. R. Aiken, D. McKnight, R. L. Wershaw, and P. MacCarthy, Eds., *Humic Substances in Soil, Sediment, and Water,* Wiley, New York, 1985, pp. 527–559.
4. N. Senesi, *Adv. Soil Sci.,* **14,** 77 (1990).
5. N. Senesi and C. Steelink, "Application of ESR Spectroscopy to the Study of Humic Substances," in M. H. B. Hayes, P. MacCarthy, R. L. Malcolm, and R. S. Swift, Eds., *Humic Substances II: In Search of Structure,* Wiley, New York, 1989, pp. 373–408.
6. Tsutsuki and S. Kuwatsuka, *Soil Sci. Plant Nutr.,* **25,** 365, 373 (1979).
7. A. U. Baes and P. R. Bloom, *Soil Sci. Soc. Amer. J.,* **54,** 1248 (1990).
8. Y. Chen, N. Senesi, and M. Schnitzer, *Soil Sci. Soc. Amer. J.,* **41,** 352 (1977).
9. C. A. Campbell, E. A. Paul, D. A. Rennie and K. J. McCallum, *Soil Sci.,* **104,** 152 (1958).
10. F. J. Stevenson and J. H. A. Butler, "Chemistry of Humic Acids and Related Pigments," in G. Eglinton and M. T. J. Murphy, Eds., *Organic Geochemistry,* Springer-Verlag, New York, 1969, pp. 534–557.
11. F. J. Stevenson and K. M. Goh, *Soil Sci.,* **113,** 34 (1972).
12. L. J. Bellamy, *The Infrared Spectra of Complex Molecules,* Chapman and Hall, London, 1975.

13. C. N. R. Rao, *Chemical Applications of Infrared Spectroscopy,* Academic Press, New York, 1963.
14. B. K. G. Theng and A. M. Posner, *Soil Sci.,* **104,** 191 (1967).
15. F. J. Stevenson and K. M. Goh, *Soil Sci.,* **117,** 34 (1974).
16. A. U. Baes and P. R. Bloom, *Soil Sci. Soc. Amer. J.,* **53,** 695 (1989).
17. J. Niemeyer, Y. Chen, and J.-M. Bollag, *Soil Sci. Soc. Amer. J.,* **56,** 135 (1992).
18. P. MacCarthy and H. B. Mack, *Soil Sci. Soc. Amer. Proc.,* **39,** 663 (1975).
19. F. J. Stevenson and K. M. Goh, *Geochim. Cosmochim. Acta,* **35,** 417 (1971).
20. K. Kumada, *Chemistry of Soil Organic Matter* (English translation), Japan Scientific Societies Press, Tokyo/Elsevier, 1987.
21. J. C. Wood, S. E. Moschopedis, and W. den Hertog, *Fuel,* **40,** 491 (1961).
22. G. H. Wagner and F. J. Stevenson, *Soil Sci. Amer. Proc.,* **29,** 43 (1965).
23. S. E. Moschopedis, *Fuel,* **41,** 425 (1962).
24. M. Schnitzer, D. A. Shearer, and J. R. Wright, *Soil Sci.,* **87,** 252, (1959).
25. B. K. G. Theng and A. M. Posner, *Soil Sci.,* **104,** 191 (1967).
26. V. C. Farmer and R. I. Morrison, *Sci. Proc. Roy. Dublin Soc., Ser. A,* **1,** 85 (1960).
27. N. M. Atherton, P. A. Cranwell, A. J. Floyd and R. D. Haworth, *Tetrahedron,* **23,** 1653 (1967).
28. K. Ghosh and M. Schnitzer, *Soil Sci. Soc, Amer. J.,* **44,** 975 (1980).
29. Y. Chen, N. Senesi, and M. Schnitzer, *Geoderma,* **20,** 87 (1978).
30. R. W. Rex, *Nature,* **188,** 1185 (1960).
31. R. Riffaldi and M. Schnitzer, *Soil Sci. Soc. Amer. Proc.,* **36,** 301 (1972).
32. N. Senesi, Y. Chen, and M. Schnitzer, *Soil Biol. Biochem.,* **9,** 371 (1977).
33. N. Senesi and M. Schnitzer, *Soil. Sci.,* **123,** 224 (1977).
34. D. Slawinska, J. Slawinska, and T. Sarna, *J. Soil Sci.,* **26,** 93 (1975).
35. C. Steelink and G. Tollin, "Free Radicals in Soil," in A. D. McLaren and G. H. Petersen, Eds., *Soil Biochemistry,* Marcel Dekker, New York, 1967, pp. 147–169.
36. N. Senesi, T. M. Miano, M. R. Provenzano, and G. Brunetti, *Soil Sci.,* **152,** 259 (1991).

14

COLLOIDAL PROPERTIES OF HUMIC SUBSTANCES

The size, shape, and molecular weight characteristics constitute basic physical properties of humic substances that have a profound effect on their interactions with other soil components (clay, metal ions, biota) and xenobiotics. Topics discussed in this chapter include the colloidal state, molecular weights and particle sizes, morphological features as deduced from electrical microscopy, shape factors, and separations by gel filtration.

The literature on the colloidal properties of humic substances have been covered in several reviews,[1-3] as well as in Chapters 15-20 of the volume edited by Hayes et al.[4]

THE COLLOIDAL STATE

The colloidal state represents a condition intermediate between true solutions, where the particles are of ionic or molecular dimensions, and particulate suspensions, where the particles are sufficiently large to settle under the influence of gravity. The colloidal range is frequently regarded as extending from 0.001 to 1.0 μm in diameter, which corresponds to particles ranging from 10^{-6} to 10^{-3} mm, or from 10 to 10,000 Ångstrom units (Å). As will be noted later, electron microscope studies show that humic acids fall within the colloidal size range. However, some of the lower-molecular-weight fulvic acids are too small to be classified as "colloids," although for convenience they will be referred to as such. The term "macromolecule" more adequately describes humic substances as a whole. Many of the techniques developed for characterizing biocolloids have application to soil organic colloids.

Chemical and physical reactions of various types are greatly enhanced in

Table 14.1 Relative Sizes of Particles[a]

Particle	Approximate Diameter
Atoms	0.1–0.6 mμ
Molecules	0.2–5.0 mμ
Molecular groups	0.5–10 mμ
Colloidal particles	0.001–10 mμ
Microscopically visible particles	>250 μm
Colloidal clay	<0.2 μm
Clay	2 μm
Silt	0.05–0.002 mm
Very fine sand	0.10–0.05 mm
Fine sand	0.25–0.10 mm

[a] 1 Å (Ångstrom) = 10^{-7} mm = 10^{-8} cm; 1 mμ (millimicron) = 10^{-6} mm = 10^{-7} cm; 1 μm (micrometer) = 10^{-3} mm = 10^{-6} m.

colloidal systems due to the relatively large surface areas associated with small particles. Table 14.1 indicates surface areas for particles of different sizes.

In the realm of colloidal chemistry, two general types of colloidal systems can be differentiated, hydrophobic (hatred toward water) and hydrophilic (love of or attraction for water). Humic colloids fall into the latter category in that they have high affinity for water and are solvated in aqueous solution. However, as noted later in Chapter 19, humic substances may contain hydrophobic surfaces to which nonionic xenobiotics can adsorb.

The high stability of many polyelectrolytes has been attributed to the presence of a "double layer" of charges of opposite sign, the inner one being immediately adjacent and fixed to the colloidal surface and the outer one, of opposite sign, being movable and thus extending into the supporting medium. Humic substances are sometimes regarded in this light, namely, they are assumed to consist of rigid, spherical particles, which in aqueous solution contain an electric double layer plus a mantle of hydration water. However, this concept does not account for all of the properties attributed to soil organic colloids. One popular viewpoint is that they are rather long-chain molecules, or two- or three-dimensional cross-linked macromolecules, that can adopt the configuration of a flexible coil in solution. As noted in Chapters 11 and 15, electric charges on the molecule originate largely from ionization of acidic groups variously distributed on the particles. The presence of charged sites ($R-$COOH $\rightarrow R-$COOH$^-$ + H$^+$) results in mutual repulsion of acidic groups and causes maximum expansion of the particles under neutral and near-neutral pH conditions.

MOLECULAR WEIGHTS AND PARTICLE SIZES

One of the first questions asked about an organic compound is its molecular weight. A pure compound has a single molecular weight that can accurately be determined. Macromolecules are of two types: 1) those that are homogeneous and where all particles have the same molecular weight, and 2) those that are polydispersed and thereby exhibit a range of molecular weights. Humic and fulvic acids fall into the second category. As noted in Chapter 2 (see Fig. 2.4), molecular weights of humic substances may vary from as low as a few hundred for fulvic acids to as much as several hundred thousand for humic acids.

Approaches used for determining molecular weights and sizes of humic substances include ultracentrifugation, viscometry, freezing-point depression (cryoscopy), vapor pressure osmometry, light scattering, small-angle X-ray scattering, and gel filtration; other methods are X-ray diffraction and electron microscopy. These will be discussed briefly in the following sections.

Because of the polydispersed nature of most humic preparations, the best method of determining molecular weights would be one from which molecular weight distribution patterns could be obtained. This has rarely been done and most recorded values represent average molecular weights. Orlov et al.[3] concluded that there may be cases where: 1) molecules are identical except for the length of the chain (size), 2) molecules have the same or approximately the same size but have variable composition, 3) the size and composition of molecules change simultaneously, and 4) homogeneous or nonhomogeneous molecules interact with one another owing to H-bonds, or mineral bridges. The last item characterizes the formation of molecular aggregates and can lead to high estimates for molecular weight. The degree to which molecular interactions occur depends upon concentration, presence or absence of mono- or polyvalent cations, pH, and ionic strength. These factors have not always been taken into account in determining molecular weights of humic substances.

The "average" molecular weight of polydispersed systems can be expressed in several ways, depending on the method of determination. Some of the more common techniques, such as freezing-point depression and osmotic pressure measurements, involve estimates for the number of molecules in solution, in which case molecular weight is obtained by dividing the "weight" of the colloidal material in the experimental solution by the "number of molecules" "present." This procedure yields the so-called number-average molecular weight, $\overline{M}_n$. Thus, if the number of molecules of molecular weight M_i is given by n_i, the total weight, W, of the sample will be $n_i M_i$ and the "number-average" molecular weight will be given as

$$\overline{M}_n = \frac{\Sigma n_i M_i}{\Sigma n_i} \qquad (1)$$

Other types of measurements give a somewhat different average molecular weight. Light scattering and sedimentation in an ultracentrifuge, for example, give a "weight-average" molecular weight, $\overline{M}_w$, defined as

$$\overline{M}_w = \frac{\Sigma w_i M_i}{\Sigma w_i} = \frac{\Sigma n_i M_i^2}{\Sigma n_i M_i} \qquad (2)$$

where w_i is the weight of the ith molecule in the mixture ($w_i = n_i M_i$).

On occasion, a third average molecular weight is obtained, namely, the "Z-average" molecular weight, $\overline{M}_z$, defined as

$$\overline{M}_z = \frac{\Sigma w_i M_i^3}{\Sigma w_i M_i^2} \tag{3}$$

For homogeneous materials $\overline{M}_n = \overline{M}_w = \overline{M}_z$; for heterogeneous systems $\overline{M}_z > \overline{M}_w > \overline{M}_n$. Orlov et al.[3] pointed out that $\overline{M}_w$ values would be expected to correlate better with known properties of humic substances and should be given high priority; at the same time, $\overline{M}_n$ values are needed for computing the degree of polydispersity (in this case the $\overline{M}_w/\overline{M}_n$ ratio).

Before proceeding with a discussion of individual methods, an overview of the findings is appropriate. As can be seen from Table 14.2, there is very little agreement between molecular weights as determined by the various methods. In accordance with results of other polydispersed systems, molecular weights based on changes in colligative properties (numbers of particles) are lower than those obtained by most other methods. Few measurements have been made on extensively fractionated samples, the main exception being the work of Cameron et al.[5] As noted by Wershaw and Pinckney,[6] comparisons of molecular weights from one laboratory to another are complicated by the possibility that molecular associations may have occurred in some instances, the extent being dependent on pH, ionic strength, and concentration.

Table 14.2 Molecular Weights Recorded for Humic Substances[a]

	Type of Material		
Method	Humic Acid	Fulvic Acid	Unspecified
Ultracentrifugation			
Sedimentation–diffusion	53,000–100,000[b]		
Sedimentation–viscosity	22,000–28,000		
Equilibrium sedimentation	24,000–230,000		
Viscosimetric properties	36,000		
Vapor pressure lowering			
Freezing-point depression	25,000	640–1,000	670–1,680
Osmotic pressure measurements		951[c]	47,000–53,800
Light scattering	65,000–66,000		
Electron microscopy	<20,000		
X-ray diffraction	1,390		
Small-angle X-ray scattering	200,000–1,000,000		

[a] For specific references see text.
[b] Range of 2,000–1,360,000 for highly fractionated samples.
[c] Range of 275–2,110 following fractionation by gel filtration.

Ultracentrifuge Measurements

A popular method for determining molecular weights, sizes, and shapes of high-molecular-weight biocolloids is by sedimentation in an analytical ultracentrifuge. Several approaches can be used, including sedimentation velocity and diffusion, sedimentation velocity and viscosity, and equilibrium ultracentrifugation.

Sedimentation Velocity and Diffusion Under a centrifugal force of sufficient magnitude, colloidal particles in solution are thrown outward from the center of rotation, thereby forming a distinct boundary at the colloid–solvent interface. The rotor of the analytical ultracentrifuge contains a window for observing the sedimentation process and an optical system is used to follow the boundary automatically by producing a plot of the change in refractive index versus distance of migration. With opaque materials, such as humic acids, the boundary can be followed photographically.

The equation of motion in a centrifugal field can be written as

$$dx/dt = s\omega^2 x \tag{4}$$

where x is the distance from the center of rotation, t is the time, s is the sedimentation coefficient, and ω is the angular velocity.

By integration of the above equation and solving for the sedimentation coefficient, s, the following is obtained.

$$s = \frac{2.3 \log (x_2/x_1)}{\omega^2 (t_2 - t_1)} \tag{5}$$

In the sedimentation velocity method, the molecular weight of the colloid is obtained by application of the Svedberg equation:

$$M = \frac{RTs}{D(1 - \bar{v}\rho)} \tag{6}$$

where R is the gas constant, T is the absolute temperature, D is the diffusion coefficient, $\bar{v}$ is the partial specific volume of the colloid, and ρ is the density of the solvent, which is usually known. The partial specific volume of the colloid is determined by density measurements of the solvent in the absence and presence of variable amount of colloidal material.

The diffusion coefficient, D, which by definition is the quantity of solute (colloid) that diffuses through a unit cross sectional area in unit time at a concentration gradient of unity, is determined in separate experiments by application of Fick's first law of diffusion.

$$\frac{dS}{dt} = -AD\frac{dc}{dx} \qquad (7)$$

where S is the quantity of solute, t is the time, A is the cross-sectional area of diffusion cell, and ds/dx is the concentration gradient.

For colloidal systems where there is a gradient in particle sizes, a curve can be obtained giving the distribution of S values as a function of position in the cell, from which molecular weight distribution patterns can be obtained. Weight-average molecular weights ($\overline{M}_w$) of the order of 20,000 to 100,000 daltons have been obtained for humic acids by the sedimentation velocity and diffusion methods.[7-9] Extreme polydispersity was indicated in these studies; thus, particles having much lower and much higher molecular weights were undoubtedly present. Orlov et al.[3] obtained molecular weights of from 19,000 to 71,000 daltons for the humic acids from a series of Russian soils (method of calculation not given).

The polydisperse nature of humic acids was investigated in some detail by Cameron et al.,[5] who determined the molecular weights of humic acids after extensive fractionation by means of gel filtration. In this way, a molecular weight range, rather than a mean value, was obtained. As noted from Table 14.3, molecular weights extended over an extremely wide range (from 2,600 to 1,360,000 daltons). The upper value is one of the highest yet recorded for humic acids and may reflect molecular interactions to form aggregates. The most abundant portion of the molecular weight distribution was believed to be of the order of 100,000; the high value was attributed to an extended high molecular weight tail.[5] The wide variation in $\overline{M}_z$ values for humic acids is also

Table 14.3 Molecular Weights of Humic Acids Following Extensive Fractionation by Gel Filtration[a]

Extractant	Molecular Weight	f/f_0
Sodium pyrophosphate		
A_1	2,600	1.14
A_2	4,400	1.28
NaOH at 20°C		
B_1	12,800	1.41
B_2	20,400	1.46
B_3	23,800	1.52
B_4	83,000	1.96
B_5	127,000	2.18
B_6	199,000	2.35
B_7	412,000	2.12
NaOH at 60°C		
C_1	408,000	2.11
C_2	1,360,000	2.41

[a]From Cameron et al.[3]

shown in the study of Ritchie and Posner,[10] where a range of from 20,040 to 26,400 daltons was obtained for a low-molecular-weight fraction of a muck humic acid; the range for a high-molecular-weight fraction was 125,900 to 168,800 daltons, depending on pH. In the latter study, $\overline{M}_z$ values for both the low and high molecular weight fractions decreased as the pH became more alkaline (Table 14.4). On the other hand, f/f_0 ratios and shapes of the molecules (regarded as approximately spherical) were unaffected by pH.

From these and other data obtained with the ultracentrifuge (see subsequent section), the conclusion can be made that the upper weight-average molecular weight of humic acids lies above 200,000 daltons and the lower limit of the order of a few thousand. The average will be in the 50,000 to 70,000 range.

Data obtained with the analytical ultracentrifuge can also be used to estimate particle shapes. This is done by calculating a frictional ratio, f/f_0, where f is the real frictional coefficient and f_0 is the frictional coefficient of an unsolvated sphere of the same mass. These values are obtained from the equations:

$$f = \frac{RTs}{M} \qquad (8)$$

and

$$f_0 = 6\pi\eta (3M\nu/4\pi N)^{1/3} \qquad (9)$$

where N is Avogadros number and η is the viscosity. Other symbols were identified earlier.

The ratio f/f_0 will exceed unity when the true shape differs from that of a sphere (or if the molecule undergoes interation with the solvent). Equations are available for interpreting f/f_0 in terms of rods, ellipsoids, and so on.

Frictional ratios for the humic acids examined by Cameron et al.[5] are in-

Table 14.4 Molecular Parameters for a Low- (LMW) and a High-Molecular-Weight (HMW) Fraction of a Soil Humic Acid at Various pH Values. From Ritchie and Posner[11]

Sample	pH	$\overline{M}_z$	f/f_0
LMW	5	26,400	1.37
	7	23,900	1.36
	9	21,700	1.38
	11	20,400	1.39
HMW	5	168,800	1.45
	7	146,200	1.44
	9	125,900	1.44
	11	108,300	1.42

cluded in Table 14.3. All ratios exceeded unity, with a range of 1.14 for the lower molecular weight materials to 2.41 for those of high molecular weight (Fig. 14.1). These finding are in agreement with results obtained by Ritchie and Posner,[10] where f/f_0 ratios were lower for a low-molecular-weight fraction of humic acid than for a high-molecular-weight component, although to a lesser extent (see Table 14.4).

Frictional coefficients, f/f_0, obtained by Flaig and Beutelspacher[7] were much closer to unity than those recorded in Table 14.3, from which it was concluded that humic acids were approximately spherical or globular in shape. In contrast, f/f_0 ratios calculated by Piret et al.[11] and Stevenson et al.[8] were within the range of from 1.14 to 2.41.

After consideration of several possible structural models, Cameron et al.[5] selected a "random-coil" model with the possibility of branching, particularly at the higher molecular weights. In solution, the humic acid molecule was visualized as a series of charged, occasionally branched strands, which coil and wind randomly with respect to space and time. With branching, coil density was postulated to increase within the molecule, thereby giving rise to more compact spheres than for linear molecules of equivalent weights. The structure was pictured as being perfused with solvent molecules; at higher molecular weights, the solvent, while flowing freely through the periphery of the sphere, would be trapped in the central regions.

Sedimentation Velocity and Viscosity Molecular weights can also be estimated from the sedimentation constant (Eq. [5]) and the frictional coefficients

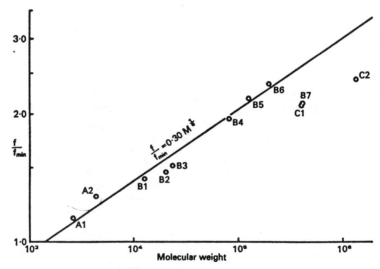

Fig. 14.1 Experimental relationship between the frictional ratio and molecular weight for a series of humic acid preparations. From Cameron et al.,[5] reproduced by permission of Oxford University Press, Oxford

f_0 and f (Eqs. [8] and [9]), where f_0 is obtained from viscosity data. By using this approach, Piret et al.[11] obtained a molecular weight of from 22,000 to 28,000 daltons for a humic acid preparation.

Equilibrium Ultracentrifugation In an attempt to avoid some of the problems involved in determining molecular weights of humic acids by sedimentation velocity and diffusion, Posner and Creeth[12] applied the technique of equilibrium ultracentrifugation. In this method, the sample is centrifuged until equilibrium is established between the centrifugal force pulling the material to the bottom of the cell and the concentration gradient tending to disperse the material. From the data obtained, estimates can be made for weight-average ($\overline{M}_w$), number-average ($\overline{M}_n$), and Z-average ($\overline{M}_z$) molecular weights by a single run. As noted earlier, all three values are equal when the material is monodispersed; for polydispersed systems, ratios of these molecular weights provide a measure of polydispersity. A major disadvantage of the equilibrium method is that a long time is required to attain equilibrium, of the order of 16 h.

Results obtained by Posner and Creeth[12] for six humic acid preparations are given in Table 14.5. Polydispersity was clearly indicated by the differences in $\overline{M}_w$ and $\overline{M}_z$ values. Values for $\overline{M}_w$ ranged from 24,000 to over 230,000 daltons.

Viscosimetric Properties

Important information about the size and shapes of hydrophobic colloids can be obtained by viscosimetric measurements. An Oswald-type viscosimeter is frequently used, in which case the time required for the solution to flow from a mark above the bulb of the viscosimeter to a mark below the bulb is recorded. If a given volume of solvent of known viscosity η_0 and density d_0 take t_0 seconds

Table 14.5 Molecular Weights ($\overline{M}_n$, $\overline{M}_w$ and $\overline{M}_z$) for Several Humic Acids as Determined by Equilibrium Ultracentrifugation[a]

Source of Humic Acid	$\overline{M}_n$	$\overline{M}_w$	$\overline{M}_z$
Red-Brown earth			
1	6,300	24,000	126,000
2	19,000	74,000	161,000
3	>150,000–<1,000,000	>230,000	>450,000
Lateritic Podzol	>120,000–<2,070,000	>230,000	>400,000
Bog Peat			
1	>17,000–<47,000	>48,000	>156,000
2[b]	>58,000–<246,000	>83,000	>192,000

[a]From Posner and Creeth.[12]
[b]Acid-boiled sample.

to flow through the viscosimeter and the same volume of the experimental sample of density d takes t seconds, the viscosity is given as

$$\eta = \frac{td}{t_0 d_0} \eta_0 \qquad (10)$$

The relative viscosity η_R is

$$\eta_R = \frac{\eta}{\eta_0} = \frac{td}{t_0 d_0} \qquad (11)$$

The ratio of the change is viscosity, $\eta - \eta_0$, to that of the solvent is referred to as the specific viscosity, η_{sp}, given as

$$\eta_{sp} = \frac{\eta - \eta_0}{\eta} = \frac{\eta}{\eta_0} - 1 \qquad (12)$$

Essentially, η_{sp} is a measure of the increase in viscosity due to the colloid.

The value of η_{sp}/c at any given concentration, c, is sometimes referred to as the viscosity number, η_z.

$$\eta_z = \eta_{sp}/c \qquad (13)$$

where c is given as grams per 100 mL of solution.

For spherical particles, η_z generally ranges between 0.02 and 0.05; for threadlike particles, the range is 0.5 to 5 or higher. The reason for this is that the viscosity of linear colloids is of the order of 20 to 250 times greater than for spherical colloids.

An equally significant value is the intrinsic viscosity, $[\eta]$, defined as the limit of η_{sp} when plotted against colloid concentration, c, and extrapolated to zero concentration.

$$[\eta] = \lim_{c \to 0} \frac{\eta_{sp}}{c} \qquad (14)$$

For macromolecules, values for $[\eta]$ and η_{sp} can also be related to particle shapes. The theoretical basis for these calculations has been given by Flaig et al.,[2] who also cite specific examples where attempts have been made to estimate shapes of humic acids from viscosimetric data.

For synthetic polyelectrolytes, the intrinsic viscosity $[\eta]$ has been related to molecular weight, M, by the equation:

$$[\eta] = kM^a \qquad (15)$$

where k and a are constants. Using constants obtained for polystyrene, the molecular weight of a soil humic acid has been estimated at 36,000 daltons.[13]

Visser[14] calculated values of a and k (Eq. [15]) from $[\eta]$ values for the molecular weight fractions of a humic acid as obtained by ultrafiltration. For measurements carried out in a $0.1M$ NaCl solution, values for a varied from 0.34 to 0.52; values for k varied from 7.226×10^{-5} to 7.776×10^{-4}. The k values were somewhat lower than reported by Chen and Schnitzer[17] for a humic acid and a fulvic acid in salt-free solutions (7.8×10^{-4} and 3.06×10^{-4}, respectively). Molecular weights calculated by Chen and Schnitzer[17] ($a = 0.67$) were found to increase with a decrease in pH (i.e., pH 10 → pH 1), with a range of from 2625 to 3145 daltons for the humic acid and from 1742 to 6366 daltons for the fulvic acid.

In addition to temperature and electrolyte effects, the viscosimetric properties of humic acids vary with nitrogen content and pH.[2, 15-17] With regard to the latter, the viscosity of some humic acids has been shown to decrease to a minimum as the pH approached 5 and then to increase again in the neutral and alkaline pH ranges, as shown in Fig. 14.2. Viscosity is reduced in the presence of electrolyte, due undoubtedly to a decrease in degree of dissociation of ionizing groups and associated contraction of the particles.

Research conducted by Ghosh and Schnitzer[16] indicate that humic and fulvic acids behave like rigid ''spherocolloids'' at high sample concentrations, low pH, and high amounts of neutral electrolyte. In contrast, at low sample con-

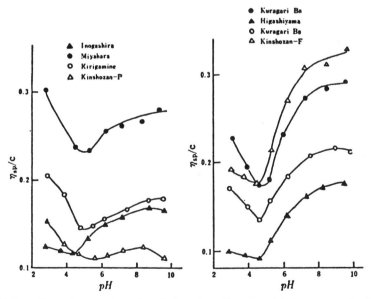

Fig. 14.2 Effect of pH on the reduced viscosity of humic acids from several soil types. From Kumada and Kawamura[15]

centrations, neutral pH, or low ionic strengths they behaved like "flexible linear colloids." Their postulated macromolecular structures are illustrated in Table 14.6.

A diagrammatic representation of changes in molecular contraction of a macromolecule as influenced by added electrolyte is shown below: A = random coil, B = expanded sphere, C = partially collapsed, D = precipitated or solid-state.

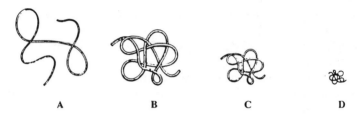

A discussion of parameters affecting shapes and sizes of the humic macromolecule has been given by Hayes et al.[4] (pp. 455–465), from which the above sequence was derived.

Number-Average Molecular Weight ($\overline{M}_n$)

Colloidal substances function as nonvolatile solutes and thereby lower the partial pressure of the solvent in which they are dissolved. Reduction of the vapor pressure, in turn, leads to lowering of the freezing point and elevation of the boiling point; in addition, the process of osmosis is affected. These properties,

Table 14.6 Model Macromolecular Structures of Humic and Fulvic Acids as Influenced by pH and Concentration of Neutral Salt (NaCl)[a]

Sample conc.	Electrolyte conc., M				pH		
	0.001	0.005	0.010	0.050	2.0	3.5	6.5
Fulvic acid							
Low	～	～	⌇	⊚	⊚⊚⊚⊚	～	～
High	⊚	⊚	⊚	⊚	⊚⊚⊚⊚	⊚	⊚
Humic acid							
Low	～	～	⌇	⊚			～
High	⊚	⊚	⊚	⊚			⊚

[a] From Ghosh and Schnitzer,[6] used by permission of Williams and Wilkins Co.

which are all closely related, serve as a basis for the determination of molecular weights. Since it is the number of particles or separate units, rather than their nature, which determines the extent of vapor pressure lowering, values obtained for polydispersed systems represent number-average molecular weights ($\overline{M}_n$).

Methods for determining molecular weights based on the number of molecules in solution are beset with difficulties when applied to colloidal systems. The magnitude of the changes are not only exceedingly small but miscellaneous ions are difficult to remove and each ion is as effective as a colloidal particle in lowering the freezing point or vapor pressure. It should be noted that 6×10^{23} molecules dissolved in 1 L of water will lower the freezing point 1.86°, and if one thousand of these units are combined into a single colloidal particle, freezing point lowering will only be 0.00186°. Methods based on changes in vapor pressure are usually not accurate for molecular weights of large macromolecules ($M > 100,000$ daltons).

Freezing-Point Depression (Cryoscopic Measurements) The molecular weight of small molecules can conveniently be determined by measuring the lowering of the freezing point of a solvent in the presence of solute. The change in freezing point is given by

$$\Delta T_f = K_f \frac{w_2 M_1}{w_1 M_2} \qquad (17)$$

where K_f is the cryoscopic constant of the solvent, w_1 and M_1 are the respective weight and molecular weight of the solvent, and w_2 and M_2 are the corresponding values for the solute. For an aqueous solution, K_f equals 1.86, the extent of freezing-point lowering of a molar solution.

To calculate molecular weight, apparent values are obtained at several concentrations of solute and plotted versus concentration, from which the intercept is obtained by extrapolation to infinite dilution.

As noted earlier, considerable difficulty is encountered in determining molecular weights of humic substances by this approach. Changes in freezing point are small, especially in water where the cryoscopic constant is low ($K_f = 1.86$). Considerable improvement has been obtained using sulfolane (tetramethylene sulfane) as the solvent ($K_f = 65–66$). However, some humic materials do not dissolve readily in sulfolane.

Schnitzer and Desjardins,[18] using sulfolane as solvent, obtained molecular weights ($\overline{M}_n$) of 1,684 and 669 daltons for organic matter from the Ao and Bh horizons of a Spodosol, respectively. A lignite humic acid examined by Wood et al.[19] was found to have a molecular weight of the order of 10,000 daltons. Russian work (summarized by Orlov et al.[3]) show values between 300 to 900 daltons for fulvic acids to from 9,000 to 10,000 daltons and from 68,000 to 74,000 daltons for some peat humic acids.

Vapor Pressure Measurements (Osmometry) Closely allied to freezing-point lowering is the change is osmotic (vapor) pressure resulting from the passage of a solvent through a membrane from a dilute solution into a more concentrated one. The pressure of the dissolved material lowers the escaping tendency of the solvent, and the greater the concentration the greater is the lowering. The basic equation relating molecular weight to osmotic pressure is as follows:

$$\text{OP} = \frac{g}{M} \frac{RT}{v} \tag{18}$$

where OP is the osmotic pressure, R is the gas constant, T is the temperature, v is the volume of the solution, and g is the weight of solute having molecular weight M. As with other approaches, several measurements are made using different amounts of the colloid and M_n is obtained by extrapolation to zero concentration; the equation is

$$\overline{M}_n = \frac{RT}{\underset{c \to 0}{\text{Limit}} \text{ OP}/c} \tag{19}$$

where c is the concentration of the solute (grams/100 mL of solvent).

For this determination, several experimental values of OP/c are plotted against c and the limiting value is inserted in the above equation for calculating $\overline{M}_n$. Molecular weights of from 47,000 to 53,800 daltons were obtained for the nondialyzable organic matter from a Spodosol.[20]

Hansen and Schnitzer[21] pointed out that $\overline{M}_n$ values obtained for humic substances by this method may be in error due to dissociation of acidic functional groups (COOH → COO$^-$ + H$^+$). A correction for H$^+$ concentration, as influenced by degree of dissociation, was used by Hansen and Schnitzer[21] to obtain $\overline{M}_n$ for an unfractionated fulvic acid from the Bh horizon of a Spodosol. An $\overline{M}_n$ value of 951 daltons was obtained. The sample was then separated into six different size fractions by gel filtration, in which case individual $\overline{M}_n$ values ranged from 275 to 2110 daltons. These data are recorded in graphical form in Fig. 14.3, where $\overline{M}_n$ values are given above the bars.

Isometric Distillation Still another technique for determining molecular weights based on reduction of the vapor pressure is isothermal distillation, which essentially reflects the influence of the solute on raising the boiling point of the solvent (see earlier discussion). For humic acid, a molecular weight of about 1000 daltons has been obtained using this technique. (See Flaig et al.[2]).

Molecular Weight by Light Scattering

Another technique for determining the molecular weight of high-molecular-weight polymeric substances is by light scattering. The basis for the measurement is that a direct proportionality has been found to exist between the intensity

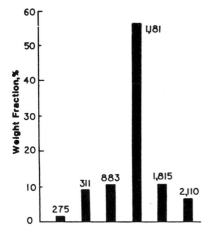

Fig. 14.3 Distribution of number-average molecular weights ($\overline{M}_n$) for fractions obtained by gel filtration of a fulvic acid. Numbers above the bars indicate molecular weight. Adapted from data of Hansen and Schnitzer[21]

of scattered light at any given concentration and molecular weight. The equation is

$$\frac{kC}{R_0} = \frac{1}{M} + \frac{2B}{RT} c \qquad (20)$$

where c is the concentration of the colloid particles in solution, M is the molecular weight, R_0 is the intensity of the scattered light (dependent on angle of observation and wavelength), B and C are virial coefficients of osmotic pressure, and k is the solvent-dependent constant of unpolarized primary light. The latter is given by the equation

$$k = \frac{4\eta^2 n_1^2}{N_L \gamma^4} \left(\frac{dn}{dc}\right)^2 \qquad (21)$$

where n_1 is the refractive index of the solvent, dn/dc is the change is refractive index of the solution with concentration, γ is the wavelength of primary light, and N_L is the Loschmidt number.

For equation [20], a plot of kC/R_0 versus c results in a line of slope $2B/RT$ and an intercept value of $1/M$, from which the molecular weight is obtained. Application of the method by Russian workers (see Orlov[3]) have given molecular weights of 65,300 and 66,200 daltons for two humic acids.

MORPHOLOGICAL FEATURES BY ELECTRON MICROSCOPY

The electron microscope provides a direct method for observing the sizes and shapes of colloidal particles. Difficulties arise in the examination of humic substances by this means, because the particles are very transparent to electron rays and cannot easily be distinguished from the supporting film. The measurements are made on samples obtained by drying of dilute solutions and considerable care is required in order to obtain single particles rather than aggregates. Samples must also be free of inorganic contaminants, such as clay. Maintenance of the original molecular shapes, as pertains to the solution phase, is highly desired, although probably unattainable.

Flaig and his associates have made extensive studies on the morphological features of humic acids by electron microscopy. This work has been reviewed by Flaig et al.[2] Peptized humic acids were added as droplets on supporting films and freeze-dried to remove the solvent. For better contrast, the samples were shadow-casted with platinum. Single particles were obtained from humates in alkaline medium, indicating complete peptization. On the other hand, clusters were formed at low pH values, apparently due to particles being held together through H-bonding.

The results of electron microscopic investigations, as summarized by Flaig et al.,[2] are as follows:

1. A remarkable number of particles, with diameters up to 100 Å, occur beyond the upper limit of the electron microscope (20 Å)
2. Single particles occur in the form of globular spheres, as shown by oval shadows on the micrographs
3. Humic acids, even after purification by repeated precipitations with acid, often show varying amounts of platey particles, pointing to admixtures with clay

Other workers (e.g., Tan[22]) have produced electron micrographs of humic substances showing the occurrence of tissue- and fiber-like structures; micrographs of Khan[23] for humic acids in the H^+-form indicate a loose spongy structure with a large number of internal spaces. Schnitzer and Kodama[24] made an electron microscopic examination of a fulvic acid and found that crystallinity, shapes, dimensions, and extent of aggregation varied with pH. The most outstanding features exhibited under the conditions prevailing in highly acidic soils (e.g., Spodosols) were: 1) a sponge-like polymeric structure perforated by voids or holes of relatively large dimensions; 2) the occurrence in the polymeric structure, apparently in free form, of very small spheroids 20 to 30 Å in diameter and of aggregates of spheroids; and 3) a structure sensitive to relatively small changes in pH, aggregating when the pH is lowered and dispersing when raised. Chen and Schnitzer[25] found that a fulvic acid changed from elongated fibers or bundles of fibers at low pH (<3) to a finely woven network at intermediate pH (4 to 7), giving a sponge-like structure. With a further increase in pH, distinct change in structure occurred until, at pH 10, fine homogeneous grains were visible.

For humic acids, spherical particles with diameters of from 60 to 100 Å have been recorded.[2, 26, 27] From these dimensions, molecular weights greater than 20,000 are indicated.[1] In the work reported by Orlov et al.,[3] large oval particles with molecular weights of the order of millions were observed to lie on an almost continuous background of minute particles with diameters of approximately 30 Å. This observation is in disagreement with other studies (mentioned above) indicating more of an ellipsoidal or elongated configuration. The formation of large oval particles was attributed by Orlov et al.[3] to deformation and association of molecules upon transition from the aqueous or gel form to the dry state. In aqueous solution, the molecule would be fully expanded, contraction occurring upon drying.

More than most other experimental techniques, electron microscopy is affected by the method of sample preparation (e.g., solute concentration, pH, ionic strength), as well as method of freezing.[22] At high solute concentrations, low pH, and high ionic strength, sheet-like structures can be formed due to coagulation. For optimum results, extremely dilute solutions must be used and quick-freezing or freeze-drying techniques must be used for water removal. A critical examination of preparation methods is provided in Chapter 22 of the volume edited by Hayes et al.[4]

STRUCTURAL EVALUATION BY X-RAY DIFFRACTION

A particularly useful technique for examining crystalline substances is by X-ray diffraction. The basis for these measurements is that a substance which contains atoms or molecular groups oriented in definite planes so as to form a three-dimensional space lattice will reflect X-rays in an orderly fashion from the repeating parallel units and thus can be differentiated from amorphous materials, which reflect X-rays in a disordered fashion.

Considerable discrepancy exists regarding the occurrence of crystalline structures in humic and fulvic acids and evidence can be cited for both their presence and absence. Much of the early work is difficult to evaluate because of the possibility of clay contamination. In view of the contradictory evidence, van Dijk[1] concluded that humic acids having both types of structures (crystalline and amorphous) might be present in nature, one type consisting of conversion products of lignin and condensates of microbial metabolites and the second to the "weathering" of carbonized organic matter. Diffraction patterns of humic acids often show a broad band near 3.5 Å and this has been attributed to condensed aromatic rings.[28] Kodama and Schnitzer[29] examined in some detail the X-ray diffraction properties of a Spodosol fulvic acid. A diffuse band was observed at 4.1 Å, accompanied by a few minor humps. The conclusion was reached that the carbon skeleton of the fulvic acid examined consisted of a broken network of condensed aromatic rings, with appreciable numbers of disoriented aliphatic or alicyclic chains around the edges. In conjunction with molecular weight and density measurements, a particle volume of 690 Å^3 was calculated.

Visser and Mendel[30] deduced the molecular weight of a fungal humic acid from X-ray data. From the dimensions of the unit cell, molecular weight was calculated from the formula

$$M = dNV \qquad (22)$$

where d is the density of the material (1.3 g/cm), N is Avogadro's number, and V is the volume of the unit cell. A value of 1392 daltons was obtained, which is somewhat lower than the molecular weights of soil humic acids as estimated from ultracentrifuge measurements (see Table 14.2)

SMALL-ANGLE X-RAY SCATTERING

A potentially valuable tool for determining shape factors, molecular weights, and molecular volumes of soil humic colloids is by the technique of small-angle X-ray scattering. These parameters are calculated from measurements of the intensity of the scattered X-ray beam, at low angles, as a function of the scattering angle 2θ. In practice, the distance of the detector above the primary bean, m, is measured, which is related to the scattering angle by the equation

$$2\theta = m/a \qquad (23)$$

where a is the distance of the sample from the detector.

For randomly oriented, identical scattering particles, scattered intensity $I(h)$ follows the relationship:

$$\ln I(h) = \frac{-R^2}{3} h^2 + C \qquad (24)$$

where R is the radius of gyration of the particle, C is a constant, and $h = 2\pi \sin 2\theta/\gamma$, where γ is the wavelength of the impinging radiation.

The radius of gyration (R), which can be defined as the root mean square distance of the electrons in a particle from the center of charge, is a useful parameter of comparative molecular or particle size and is calculated from the slope of the line obtained by plotting $\ln I$ versus h^2, the so-called Guinier plot. In systems where all particles are of equal size, the Guinier plot will be a straight line.

Guinier plots obtained by Wershaw et al.[31] for a Na-humate (Fig. 14.4) consisted to two intersecting tangents (A and B), indicating the presence of two different particle sizes. Assuming a mean specific gravity of 1.6 for the particles, molecular weights of 200,000 and 1,000,000 daltons were calculated. The higher value was probably due to molecular associations rather than molecules, as suggested by the studies mentioned below. The radius of gyration was 28 Å for the smaller particles and 100 Å for the larger ones. The ratio of the radius of gyration to that expected for a spherical particle of equal volume

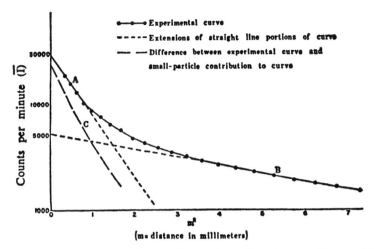

Fig. 14.4 Guinier plot by small-angle X-ray of a sodium humate. Slopes of the two straight-line portions are related to the radii of gyration of the scattered particles. From Wershaw et al.,[31] reproduced by permission of the American Association of Advancement of Science, Washington, D.C.

(the shape factor) was 1.3 and 2.5 for the smaller and larger particles, respectively, which indicated that the particles were ellipsoidal, with the smaller ones being more spherical. This observation relative to the effect of molecular weight on particle shape agrees with findings obtained with the analytical ultracentrifuge (see previous section).

In subsequent work, Wershaw and Pinckney[6, 32] found that the Guinier plot for a Na-humate was a straight line at pH 7 but was concave upward at higher and lower pH values. Three different types of aggregation were observed in humic acid systems: 1) disaggregation up to pH 7 and then reaggregation at high pHs; 2) increased aggregation below pH 3.5 with little disaggregation about this pH; and 3) a continual decrease in aggregation with increasing pH. Lindqvist[33] found that a sodium humate that contained different sized particles in solution yielded a monodispersed system by acid hydrolysis, with radii of gyration between 17 and 19 Å.

SIZE EVALUATION BY GEL FILTRATION

Gel filtration is a method of separation based on size by elution through beds of porous beads. One popular product is Sephadex, a modified dextran obtained by cross-linking the linear macromolecule (polysaccharide) to form a three-dimensional network whose pores are determined by the degree of cross-linking of the polymer. Seven grades are available, each with a different pore size (Table 14.7). Glass beads having variable pore size are also available. A major

Table 14.7 Sephadex Types and Fractionation Range

Type	Approximate Limit for Complete Exclusion (MW)	Fractionation Range (MW)
G-10	700	0–700
G-15	1,500	0–1,500
G-25	5,000	100–5,000
G-50	10,000	500–10,000
G-75	50,000	1,000–50,000
G-100	100,000	5,000–100,000
G-150	150,000	5,000–150,000
G-200	200,000	5,000–200,000

problem with gel filtration as applied to soil humic substances is adsorption onto gel surfaces. However, these gel–solute interactions may be avoided by proper choice of operating conditions, especially with respect to gel and buffer types.[34, 35]

Separation by gel filtration is accomplished by a type of molecular sieving. A sample of the material to be examined is passed on top of the gel column and eluted with an appropriate aqueous solvent. Small particles move with the elutant both within and without the gel particle while large particles move outside the gel particles only. Thus, the large particles are eluted first, followed in order by the smaller particles. This effect is shown schematically in Fig. 14.5.

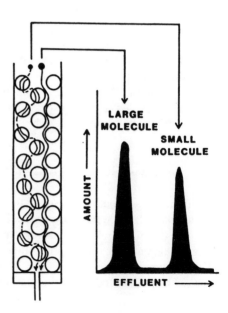

Fig. 14.5 Separation of particles of different sizes by gel filtration

By selecting a series of gels of varying pore sizes, it has been possible to separate humic and fulvic acids into more homogeneous molecular weight fractions.[34-47] For example, the fulvic acid fraction has been separated into six components (Fig. 14.6), all with different C, hydrogen, and oxygen contents (Table 14.8) Also, total acidity increased slightly with decreasing molecular size. The values for M_n (obtained by osmometry) are those shown earlier in graphical form (see Fig. 14.3). Fractionations of humic acids by gel filtration have indicated a mixture of macromolecules whose elemental composition and optical extinction coefficients varies with molecular weight.[36,44]

Attempts have been made from time to time to obtain absolute molecular weights for humic and fulvic acids from separations obtained by gel filtration. Most estimates have been based on molecular weight exclusion values quoted by the gel manufacturers, which cannot be applied directly to molecules having markedly different properties from the calibrating species. It should be noted that the separations achieved by gel filtration are related to hydrodynamic sizes of the particles rather than molecular weight *per se* and will depend on such properties as charge, shape, and hydration. Humic and fulvic acids of known molecular weights (i.e., from ultracentrifuge and osmometry measurements), as well as phenolic acids, benzene–carboxylic acids, and polyphenolic compounds, have been used for calibration.[41,42,48] This work has shown that the behavior of humic substances on gels cannot be predicted from calibration curves obtained for proteins or dextrans but that primary humic and fulvic acid standards would be required. By comparing M_n values of Table 14.8 with exclusion limits shown in Fig. 14.6 it can be seen that the relationship between the two is fortuitous.

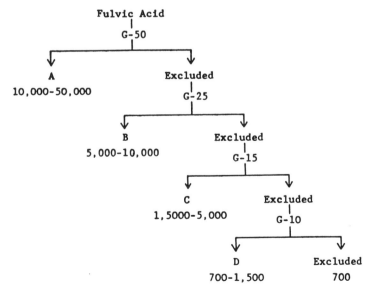

Fig. 14.6 Fractionation of fulvic acid by gel filtration, From Schnitzer and Skinner[40]

Table 14.8 Analytical Characteristics of Fulvic Acid and Fractions Separated From It by Gel Filtration[a]

Fraction[b]	Yield (%)	C (%)	H (%)	N(%)	O (%)	$\bar{M}_n$
Original	1.00	49.5	4.5	0.8	45.0	951
A	0.16	53.9	6.2	1.7	33.9	2110
B	0.23	55.4	6.1	0.7	37.6	1815
C	0.14	53.1	5.0	0.7	41.0	1181
D	0.27	54.2	4.9	0.8	39.9	883
D_1	0.06	49.8	6.3	0.5	43.2	311
D_2	0.02	—	—	—	—	275

[a] From Schnitzer and Skinner.[40]
[b] S content in the original fulvic acid = 0.3%; the fractions contained <0.2% S. The letters A through D_2 refer to the fractions shown in Fig. 12.6.

Variable success has been obtained by gel filtration of humic and fulvic acids on single Sephadex columns.[43–47] Results for three fulvic acid preparations are shown in Fig. 14.7. A feature of most published chromatograms is that a portion of the material is recovered at the void volume (V_0), representing the molecular fraction excluded without retention. This is followed by a broad peak

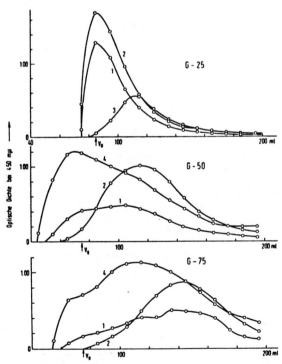

Fig. 14.7 Gel filtration of three fulvic acid preparations on Sephadex gels G25, G50, G75. From Dubach et al.,[44] reproduced by permission of Pergamon Press, Inc.

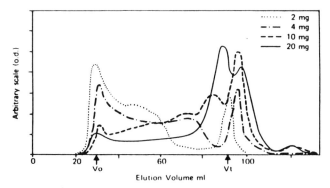

Fig. 14.8 Fractionation of humic acid on Sephadex G-100 at various sample concentrations. From De Nobili et al.[35a] as prepared from data published by Swift and Posner[35]

at higher elution volumes, indicating humic material that has entered and passed through the gel pores.

Two polydispersed fractions have been obtained by gel filtration of humic acids.[34,36,45,46] Lindqvist[46] concluded that a significant portion of the humic acid that is retained by the gel, and recovered in the low-molecular-weight fraction, results from adsorption of humic acid to the gel. Dubach et al.[44] observed slight adsorption of brown humic acids on Sephadex gels but extensive adsorption of gray humic acids.

In other work, Swift and Posner[35] found that elution patterns of humic acids were not independent of sample concentration, as required by theory. From Fig. 14.8, it can be seen that, with increasing dilution, a greater percentage of the sample appeared in the excluded and near-excluded regions. One explanation for this effect is that dilution allowed for greater extension of the molecules, thereby preventing entry "into" pores of the gel.

From the above it can be seen that considerable difficulty is encountered in interpreting published results on gel filtration of humic and fulvic acids. Adsorption problems seem to have been encountered by some investigators but not by others. The technique seems to be of value for separating organic compounds from natural waters,[38] and for preliminary fractionation of organic matter in purified soil extracts.[47] For additional information and applications, the reader is referred to Chapter 20 of the volume edited by Hayes et al.[4]

SUMMARY

The humic fraction of the soil consists of a complex system of molecules that have a wide range in molecular weights. The average range for humic acids is of the order of 50,000 to 100,000 daltons with few molecules having molecular weights exceeding 250,000 daltons. A "typical" fulvic acid will have a molecular weight in the 500 to 2000 dalton range.

The modern view of the colloidal nature of humic acids is that they are rather long-chain molecules, or two- or three-dimensional cross-linked macromolecules, that can adopt the configuration of a flexible coil in solution. The shape of the molecule in solution, however, is strongly influenced by pH, the presence of neutral salts, and possibly concentration. Under neutral or slightly alkaline conditions, the molecules are in an expanded state due to mutual repulsion of charged acidic groups (e.g., COO^-); at low pHs, and in the presence of a neutral electrolyte, contraction occurs due to charge reduction and the molecule becomes more spherical shape in shape. At very low pHs, and at high concentrations, molecular aggregates are formed.

The polymeric nature of humic substances has recently been questioned.[49] Rather, humic substances are visualized as consisting of membrane-like microaggregates containing hydrophilic exterior surfaces and hydrophobic interior surfaces to which nonionic organics can be bound through partition (see Chapter 19). The validity of this theory requires confirmation.

REFERENCES

1 H. van Dijk, "Colloid Chemical Properties of Humic Matter," in A. D. McLaren and G. H. Peterson, Eds., *Soil Biochemistry: Vol. 2*, Marcel Dekker, New York, 1972, pp. 16–34.
2 W. Flaig, H. Beutelspacher, and E. Rietz, "Chemical Composition and Physical Properties of Humic Substances," in J. E. Gieseking, Ed., *Soil Components. Vol. 1*, Springer-Verlag, New York, 1975, pp. 1–211.
3 D. S. Orlov, Ya. M. Ammosova, and G. I. Glebova, *Geoderma*, **13**, 211 (1975).
4 M. H. B. Hayes, P. MacCarthy, R. L. Malcolm, and R. S. Swift, Eds., *Humic Substances II: In Search of Structure*, Wiley, New York, 1989.
5 R. S. Cameron, B. K. Thornton, R. S. Swift, and A. M. Posner, *J. Soil Sci.*, **23**, 394 (1972).
6 R. L. Wershaw and D. J. Pinckney, *J. Res. U. S. Geol. Survey*, **1**, 701 (1973).
7 W. Flaig and H. Beutelspacher, "Investigations of Humic Acids with the Analytical Ultracentrifuge," in *Use of Isotopes in Soil Organic Matter Studies*, International Atomic Energy Agency, Vienna, 1968, pp. 23–30.
8 F. J. Stevenson, Q. van Winkle, and W. P. Martin, *Soil Sci. Soc. Amer. Proc.*, **17**, 31 (1953).
9 W. Flaig and H. Beutelspacher, *Landbauwk. Tijdschr.*, **66**, 306 (1954).
10 G. S. Ritchie and A. M. Posner, *J. Soil Sci.*, **33**, 233 (1982).
11 E. L. Piret, R. G. White, H. C. Walter, and A. J. Madden, *Proc. Roy. Dublin Soc.*, **A1**, 69 (1960).
12 A. M. Posner and J. M. Creeth, *J. Soil Sci.*, **23**, 333 (1972).
13 S. A. Visser, *J. Soil Sci.*, **15**, 202 (1964).
14 S. A. Visser, *Plant Soil*, **87**, 209 (1985).
15 K. Kumada and Y. Kawamura, *Soil Sci. Plant Nutr.*, **14**, 190 (1968).
16 K. Ghosh and M. Schnitzer, *Soil Sci.*, **129**, 266 (1980).
17 Y. Chen and M. Schnitzer, *Soil Sci. Soc. Amer. J.*, **40**, 866 (1976).

18. M. Schnitzer and J. G. Desjardins, *Soil Sci. Soc. Amer. Proc.*, **26**, 362 (1962).
19. J. C. Wood, S. E. Moschopedis, and R. M. Elofson, *Fuel*, **40**, 193 (1961).
20. J. R. Wright, M. Schnitzer, and R. Levick, *Can. J. Soil Sci.*, **38**, 14 (1958).
21. E. H. Hansen and M. Schnitzer, *Anal. Chem. Acta*, **46**, 247 (1969).
22. K. H. Tan, *Soil Sci. Soc. Amer. J.*, **49**, 1185 (1985).
23. S. U. Khan, *Soil Sci.*, **112**, 401 (1971).
24. M. Schnitzer and H. Kodama, *Geoderma*, **13**, 279 (1975).
25. Y. Chen and M. Schnitzer, *Soil Sci. Soc. Amer. J.*, **40**, 682 (1976).
26. S. A. Visser, *Soil Sci.*, **96**, 353 (1963).
27. W. Wiesemuller, *Albrecht Thaer Archiv.*, **9**, 419 (1965).
28. S. S. Pollack, H. Lentz, and W. Ziechmann, *Soil Sci.*, **112**, 318 (1971).
29. H. Kodama and M. Schnitzer, *Fuel*, **46**, 87 (1967).
30. S. A. Visser and H. Mendel, *Soil Biol. Biochem.*, **3**, 259 (1971).
31. R. L. Wershaw, P. G. Burcar, C. L. Sutula, and B. J. Wiginton, *Science*, **157**, 1429 (1967).
32. R. L. Wershaw and D. J. Pinckney, *Geol. Survey Prof. Paper*, **750D**, 216 (1971).
33. I. Lindqvist, *Acta Chem.*, **24**, 3068 (1970).
34. A. M. Posner, *Nature*, **198**, 1161 (1963).
35. R. S. Swift and A. M. Posner, *J. Soil Sci.*, **22**, 237 (1971).
35a. M. De Nobili, E. Gjessing, and P. Sequi, "Sizes and Shapes of Humic Substances by Gel Chromatography," in M. H. B. Hayes, P. MacCarthy, R. L. Malcolm, and R. S. Swift, Eds., *Humic Substances II: In Search of Structure*, Wiley, New York, 1989, pp. 561–591.
36. K. H. Tan and J. E. Giddens, *Geoderma*, **8**, 221 (1972).
37. E. T. Gjessing, *Physical and Chemical Characteristics of Aquatic Humus*, Ann Arbor Science Press, Ann Arbor, 1976.
38. E. Gjessing and G. F. Lee, *Environ. Sci. Tech.*, **1**, 631 (1967).
39. S. U. Khan and M. Schnitzer, *Soil Sci.*, **112**, 231 (1971).
40. M. Schnitzer and S. I. M. Skinner, "Gel Filtration of Fulvic Acid, a Soil Humic Compound," in *Isotopes and Radiation in Soil Organic Matter Studies*, International Atomic Energy Agency, Vienna 1968, pp. 41–55.
41. R. S. Swift, B. K. Thornton, and A. M. Posner, *Soil Sci.*, **110**, 93 (1970).
42. R. S. Cameron, R. S. Swift, B. K. Thornton, and A. M. Posner, *J. Soil Sci.*, **23**, 342 (1972).
43. A. Piccolo and J. S. C. Mbagwu, *Plant Soil*, **123**, 27 (1990).
44. P. Duback, N. C. Mehta, J. Jakab, F. Martin, and N. Roulet, *Geochim. Cosmochim. Acta*, **28**, 1567 (1964).
45. J. N. Ladd, *Soil Sci.*, **107**, 303 (1969).
46. I. Lindqvist, *Acta Chem. Scand.*, **21**, 2564 (1967).
47. K. M. Goh and M. R. Reid, *J. Soil Sci.*, **26**, 207 (1975).
48. H. J. Dawson, B. F. Hrutfiord, R. J. Zasoski, and F. C. Ugolini, *Soil Sci.*, **132**, 191 (1981).
49. R. L. Wershaw, *J. Contam. Hydrol.*, **1**, 29 (1986).

15

ELECTROCHEMICAL AND ION-EXCHANGE PROPERTIES OF HUMIC SUBSTANCES

Humic and fulvic acids behave like weak-acid polyelectrolytes and are amenable to examination by techniques based on the ionization of acidic functional groups. In addition to COOH groups, negative charges may arise from the presence of phenolic OH, enolic OH, imide (= NH), and possibly other groups. The occurrence of charged sites (e.g., COO^-) accounts for the ability of soil organic matter to retain cations in nonleachable forms.

The binding and exchange of cations by the soil organic fraction is of importance in soil fertility because the supply of K^+, Ca^{2+}, Mg^{2+}, and certain micronutrients (Cu^{2+}, Mn^{2+}, Zn^{2+}, Fe^{3+}, others) to plants is strongly dependent on ion exchange and up to 80 percent of the cation-exchange capacity (CEC) of the soil may be due to organic matter. Coulombic or electrostatic forces are primarily involved (COO^-K^+) but the bonds may be partly covalent, particularly for the micronutrient cations. The topic of metal–organic matter (chelate) complexes is discussed in Chapter 16 and will not be considered herein. In terms of quantity, the main exchangeable cations in soil are Na^+, K^+, H^+, Ca^{2+}, and Mg^{2+}.

A unique feature of humic substances is that they exhibit buffering over a wide pH range. This buffering capacity is of considerable practical significance in that most plants grow best within a rather narrow pH range. The exact contribution of humic material to the total buffering capacity of the soil is unknown but it is believed to be appreciable. In general, soils that are rich in humus are well buffered.

Since humic substances carry net negative charges, they exhibit mobility in the presence of an applied electrical field. Electronegative groups may also be present that can participate in oxidation–reduction reactions.

ACIDIC NATURE OF HUMIC AND FULVIC ACIDS

In considering the acidic properties of humic substances, a brief review of the nature of acids and bases is appropriate.[1] Arrhenius defined an acid as a substance that ionizes to produce a proton (H^+) and a base is a substance that ionizes to provide a hydroxyl ion (OH^-). This definition was later modified by Lowry and Brønsted to include the role of the solvent in acid–base equilibrium. These investigators defined an acid as a substance that gives up a proton and a base as a substance that accepts a proton. A particularly valuable feature of this concept is that a compound can only function as an acid (or base) in the presence of a conjugate base (or acid). In aqueous solution, water function as a base by accepting a proton from the acid, as follows:

$$HA + H_2O \rightleftharpoons H_3O^+ + A^- \qquad (1)$$

In the above example, the hydrated proton (H_3O^+) is written to emphasize the role of the solvent. Hereafter, the unhydrated form will be abbreviated in the usual ways as H^+. For aqueous titrations, H_2O is present in large excess and is thereby not shown when illustrating ionization reactions.

Base titration, because it is economical and easy to perform, is an excellent technique for determining the acidic properties of humic and fulvic acids. The only apparatus required is a pH meter, of which several excellent types are available. Essentially, the potentiometric method provides a measure of the quantity of H^+ ions not bound to acidic functional groups and is calculated from the well-known relationship:

$$pH = -\log H^+ \qquad (2)$$

A more general concept of acids and bases has been given by G. N. Lewis, who defined an acid as a substance that can take up an electron pair to form a covalent bond and a base as a substance that can furnish an electron pair. Thus, an acid is an electron-pair acceptor and to be acidic the molecule must be electron-deficient. One might expect that humic acids would function as Lewis acids due to the occurrence of such electron deficient groups as OH, NH, and SH.

For acid–base titrations (see next section), the Lowery–Brønsted definition applies (molecule must contain a proton). The degree of acidity or acid strength of the colloid will depend on the nature of the reactive group involved and of associated structures on the molecule. In general, the OH group of carboxylic acids ($R-COOH$) dissociates more readily than aromatic or aliphatic alcohols. Phenolic compounds are stronger acids than water or alcohols but weaker than most carboxylic acids. The acidic nature of humic substances is usually attributed to ionization of COOH and phenolic OH groups, although other structures may be involved (e.g., keto–enol tautometers).

As noted later, the acidic nature of polyelectrolytes, as opposed to monomeric acids, is complicated by electrostatic charge accumulation on the polymer as neutralization proceeds, which causes the remaining acidic groups to become weaker in acidity. An increase in the concentration of supporting electrolyte generally promotes ionization of weak-acid polyelectrolytes, due to contraction of the polymer molecule. A major effect of dilution is to raise the pH and to reduce the dissociation of COOH groups.

Potentiometric Titration in Aqueous Solution

Data from potentiometric titrations are important as a basis for: 1) comparing acidic properties of humic substances from various sources, 2) determining total- and exchange acidity (i.e., $\approx$ COOH) contents, 3) evaluating results of chemical fractionations and chromatographic separations, 4) determining charge densities and radii, and 5) determining metal-ion binding properties, including stability constant measurements (see Chapter 17).

A number of factors can affect the association–dissociation of proton-carrying functional groups in humic substances, as follows:[2]

1. Electrostatic attraction or repulsion due to charges located on the molecule
2. Inter- and intra-molecular H-bonding
3. Burial of acidic groups in hydrophobic environments. The hydrophilic/hydrophobic properties of humic substances are discussed in Chapters 12 and 19

Several procedures have been used for potentiometric acid–base titrations of humic substances. They include discontinuous titration, in which case pH is measured on a series of samples containing increasing amounts of titrant, continuous titration using an automatic titrator, and normal acid–base titrations with a conventional pH meter. In the latter case, titration is conveniently carried out in tall-form beakers of 50- to 200-mL sizes. The beaker is fitted with a stopper containing five openings, two of which are for the electrodes; one is for a N_2 gas inlet, one for a thermometer, and one for the burette or automatic syringe. Good temperature control is essential for meaningful results and CO_2-free alkali must be used for base titrations.

Oden[3] is usually credited as being the first to carry out a comprehensive study of the electrochemical properties of humic substances. The literature since then is extensive and only select references can be listed herein.[6-18] The early literature is covered in Chapter 1 and the review of Flaig et al;[4] for recent research, the review of Perdue[5] is recommended.

In the absence of secondary side reactions, acid–base reactions occur rapidly. The results of Borggaard[7] show that equilibrium during base titration of humic acid is very rapid and that optimum results can be obtained by sequential additions of base to a single sample over a short period of time, as contrasted with discontinuous titration over prolonged periods. At high pH values, alkali is consumed through autoxidation of organic matter in the presence of even trace amounts of O_2, with a corresponding drop in pH upon standing.[8,15] Base consumption from this mechanism is particularly serious at pHs above 8, where phenolic OH groups dissociate.

Typical titration curves of humic acids are shown in Fig. 15.1. The gradual rise in pH with added base attests to the high buffering capacity of humic substances and is consistent with the concept that they behave as weak-acid polyelectrolytes. Buffering is shown over a wide pH range. Because of electrostatic effects, and the likelihood that configurational changes occur with increasing pH, titration curves of humic (and fulvic) acids would be expected to have a more "smeared out" appearance than those of monomeric acids. Also, being heterogenous in nature, their potentiometric titration properties would deviate somewhat from those expected for a typical weak-acid polyelectrolyte containing identical repeating units. As discussed in other chapters, COOH of humic substances cannot be considered to be regularly spaced on the molecule; furthermore, their acidic properties may be influenced by other reactive groups in close proximity (notably OH).

The titration curves shown in Fig. 15.1 can be broken down into three zones (I, II, and III). The first zone is the lower acid region where COOH groups dissociate; zone III represents dissociation of phenolic OH and other very weak acid types. The zone marked II is an intermediate area where ionization of

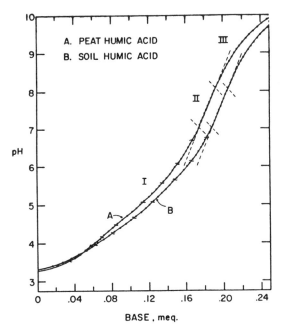

Fig. 15.1 Titration curves of a soil and peat humic acid. The small wavy lines on the curves indicate end points for ionization of weak-acid groups having different, but overlapping ionization constants

weak (COOH) and very weak acid groups (i.e., phenolic OH) overlap.[15] The discussion of this chapter will be limited to potentiometric titration as applied to zone I (i.e., COOH groups).

Several difficulties are encountered in the preparation and interpretation of titration curves of humic substances. A factor of some importance is the method of sample preparation. In many cases, proper account has not been taken of inorganic impurities, including excess mineral acid.

Humic acids, as normally prepared, are not readily soluble in water, but must first be converted to the salt form by addition of base (NaOH or KOH). Following solution of the humic acid, some investigators have lowered the pH by addition of an equivalent amount of mineral acid (usually HCl); others have arbitrarily lowered the pH to a predetermined set value. These practices are open to criticism, because of the difficulty of protonating all acidic functional groups without addition of excess mineral acid, which would be titrated along with weakly acidic groups of the humic acid. Furthermore, addition of excess mineral acid can lead to partial flocculation of humic acid, especially for titrations carried out at high ionic strengths (I). Secondary reactions may also occur, such as electrophillic attack on unsaturated carbonyl structures[17] and counter-ion condensation.[18]

In a study on the potentiometric titration of humic acids by dialysis and treatment on cation- and anion-exchange resins, the author[17] concluded that the acidic properties of humic (and fulvic) acids cannot be explained entirely by simple ionization of acidic functional groups (e.g., COOH), but that secondary chemical reactions are involved at both low and high pHs. Two major types of reactions were considered responsible: 1) electrophillic and nucleophilic attack on double bonds of unsaturated ketonic structures, and 2) acid- and base-catalyzed keto–enol transformations. Reactions for electrophillic attack under acidic conditions, with binding of the anion (i.e., CL^-), are illustrated in Fig. 15.2. In other work, Sposito et al.[18] suggested that the titration properties of humic acids at low pH may be affected by counter-ion condensation (i.e., retention of the cation used for titration).

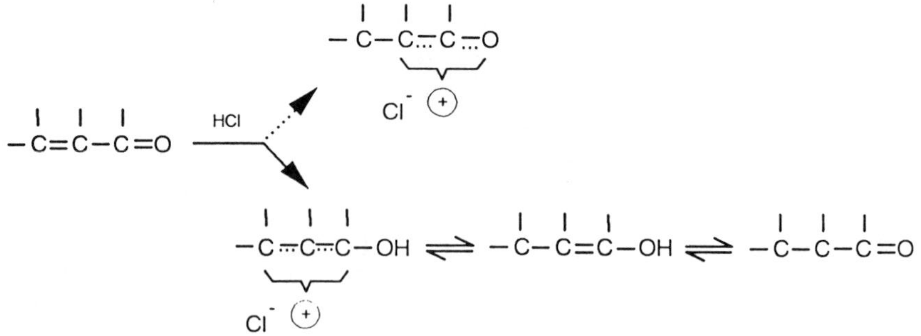

Fig. 15.2 Formation of carbonium ions by electrophilic attack of an acid (HCl) on C=C bonds of unsaturated carbonyl structures. A similar reaction may take placed during acid–base titrations of humic substances (see text)

A convenient method for removal of Na⁺ or (K⁺) from alkaline extracts of humic acids is by passage through a cation-exchange resin in the H⁺-form, as done by Posner.[15] The titration curves shown in Fig. 15.1 were prepared in this manner.

Another problem in the interpretation of titration curves of humic substances arises from selection of the end point for ionization of COOH groups. Posner[15] used the point where the rate of change of pH with added alkali became maximum (initial portion of region II in Fig. 15.1). This value was found to occur at pHs between 7.0 to 7.6, depending on the ionic strength of the solution. Other workers, have arbitrarily selected a given pH (a value of 8.0 has sometimes been used).

In more recent times, graphical methods have been applied. End points (EP) for COOH groups are established through first- or second-derivative plots, as follows:

$$d\text{pH}/dV \text{ versus titratant volume } V \quad (3)$$

or

$$d^2\text{pH}/dV^2 \text{ versus titratant volume } V \quad (4)$$

For the determination of equivalence points for strong (mineral) acid and base, Gran[19,20] plots have been used. For strong acid (before EP), a plot is made of

$$10^{-\text{pH}}(V_0 + V) \text{ versus } V \quad (5)$$

where V_0 is the initial volume of the experimental solution.

For strong base (after EP), the plot is

$$10^{\text{pH}}(V_0 + V) \text{ versus } V \quad (6)$$

Gran plots using Eq. [5] have been used to estimate the content of strong mineral acids in natural waters.[21-23]

Application of the above-mentioned plotting techniques can be illustrated from the titration data of Barak and Chen,[6] in which a strong acid (HCl) was added to the original samples to lower the pH to between 2.5 and 3.0. Gran plots for the family of curves shown in Fig. 15.3 are given in Fig. 15.4. Thus, the end point for the weak acid ($\approx$ COOH) was estimated from a second differential plot (Eq. [4]) and the amounts of strong acid and base (excess titrant) from the relationships shown in Eqs. [5] and [6], respectively. Analysis of the titration data gave values of 269 cmole/kg for weakly acidic groups ($\approx$ COOH), 104 cmole/kg for very weakly acidic groups (phenolic OH or equivalent), and 373 cmole/kg for total acidic groups.

It should be noted that estimates for strong mineral acid from Gran plots

356 ELECTROCHEMICAL PROPERTIES OF HUMIC SUBSTANCES

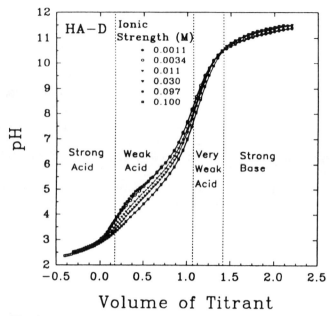

Fig. 15.3 Titration curves for a soil humic acid at six different ionic strengths. Curves have been offset for presentation of the equivalent amount of strong acid added to all but the 0.001M titration. Vertical dashed lines mark equivalence points as determined by Gran plots and inflection analysis (see Fig. 15.4). From Barak and Chen,[6] reproduced by permission of Williams and Wilkins Co.

(see Eq. [5]) only applies when the mineral acid is present in large excess. At low concentration of mineral acid, curvilinear plots are obtained, and attributed to the occurrence of COOH groups in humic substances that ionize in the lower pH range (< pH 4.0). With natural waters, the sample is often spiked with a known amount of mineral acid, in which case the content of mineral acid in the original sample is estimated as the difference between the plotted value and the amount added.

Problems encountered with the addition of a large excess of mineral acid to humic acids (notably, formation of precipitates at high I) can be avoided using a plotting technique devised by McCallum and Midgley.[24] The equation is

$$F_{mix} = (V_0 + V)\left([H] - \frac{K_w}{[H]}\right)\left(1 + \frac{1}{K_A[H]}\right)$$
$$+ \frac{m(V - V_e)}{K_A[H]} = m(V_s - V) \quad (7)$$

where $[H]$ is the molar concentration of H$^+$, K_w is the ionization constant for H$_2$O, K_A is the association constant (reciprocal of the ionization constant) of

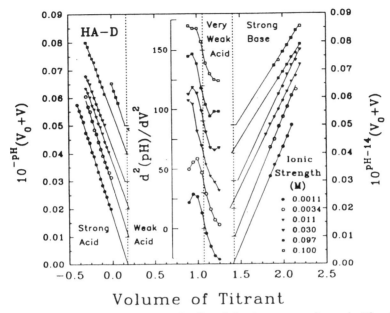

Fig. 15.4 Equivalence points for the family of titration curves shown in Fig. 15.3: Gran plot for strong acid (left), inflection point (d^2(pH)$/dV^2 = 0$) for weak acid (center), and Gran plot for strong base (right). From Barak and Chen,[6] reproduced by permission of Williams and Wilkins Co.

the weak acid group (taken as $1/pK_a$), m is the base molarity, and V_s is the titrant volume assigned to the strong acid. A plot of F_{mix} (composite values in the brackets) versus V yields V_s as the intercept.

It should be noted that for very weak organic acids with high pK_as, the value for K_A can be ignored and Eq. [7] reduces to that of the Gran plot (Eq. [5]).

Interpretation of Titration Curves Special difficulties are encountered in interpreting titration curves of soil humic substances. As noted earlier, selection of the appropriate end point is rather arbitrary. In addition, macromolecules that contain a large number of ionizing groups, such as humic and fulvic acids, tend to expand in aqueous solution until all charges are as far removed from one another as possible. When titrant is added, the molecule may contract in response to an increase in I. The net result is a shift in acid–base equilibrium and a concomitant change in pK values. To minimize this effect, a swamping concentration of electrolyte (usually KCl) is added to keep the molecule in contracted form. Many titration curves published in the literature for humic substances have been prepared in the absence of a supporting electrolyte. The shape of the titration curve, and average pK values, are also strongly influenced by di- and trivalent metal ions (see Chapter 17).

Another problem in the base titration of humic substances is that several acidic groups may be present on the macromolecule but the respective end points may not be evident because the titration of one type begins before that of the other is completed. Most workers have reported a single inflection point whereas the titration curves shown in Fig. 15.1 indicate several inflections. Gamble[9] postulated the occurrence of two general types of COOH groups in fulvic acid, one being ortho to a phenolic OH group. As noted in Chapter 13, results of IR studies indicate that some COOH groups of humic substances may occupy adjacent positions on the aromatic ring. In addition, some COOH groups may exist on aliphatic side chains (see Chapter 12).

Calculation of Ionization Constants Several attempts have been made to analyze titration curves of humic acids using approaches known to be successful for weak-acid polyelectrolytes. A brief outline of the theory is given below.

The dissociation of an acidic group (i.e., COOH) proceeds as follows:

$$HA \rightleftharpoons A^- + H^+ \tag{8}$$

for which the ionization constant is given by

$$K = \frac{(A^-)(H^+)}{(HA)} \tag{9}$$

where the values in parentheses indicate concentration.

The titration data provide estimates for (A^-) and (HA) at any given pH, from which the ionization constant can be determined. In making these calculations, corrections must be made for H^+ and OH^- should either become excessively large. To obtain the "molar concentration" of H^+, the measured value (activity) is divided by the activity coefficient of H^+ for the appropriate ionic strength. The relationship between the activity of H^+ and molar concentration H_m^+ is given by

$$H^+ = \gamma H_m^+ \tag{10}$$

where γ is the activity coefficient. Corrections of measured pH values have not always been made in base titrations studies of humic substances.

For any given point on the titration curve, the following relationship is valid:

$$(A^-) + (OH^-) = (K^+) + (H_m^+) \tag{11}$$

The quantity (K^+) equals the concentration of titrant (KOH) after allowance for dilution. Hence,

$$(A^-) = (KOH) + (H_m^+) - (OH^-) \tag{12}$$

Since the ionization constant is normally a very small number, the negative log is often recorded, the *pK* of the acid group. From Eq. [9], the following can be obtained:

$$pK = -\log \frac{(A^-)(H^+)}{(HA)} = pH - \log \frac{(A^-)}{(HA)} \quad (13)$$

The ionization constant of an acid is of value because it reveals the proportion of the material which occurs in ionized and unionized forms at any chosen pH. At half-neutralization (i.e., where $HA = A^-$), the *pK* is equal to pH and this is commonly referred to as the pK_a of the acid or acidic group.

The Extended Henderson–Hasselbalch Equation Attempts to apply the simple mass-action law to weak polyacids has led to variable dissociation constants which change markedly with concentration of polyelectrolyte and addition of neutral salt. For any given point on the titration curve, the apparent ionization constant (K_{app}) decreases with an increase in degree of dissociation (α), a result that has been attributed to greater stretching of the macromolecule as repulsive forces are enhanced through ionization (generation of COO^- from COOH as base is added). For a randomly "kinked" molecule, the reaction can be pictured as follows:

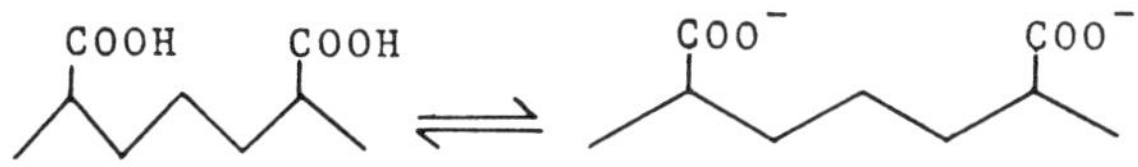

Due to this stretching effect, the remaining protons become increasingly difficult to remove from the macromolecule, hence a continuous decrease in K_{app}. In practice, the humic molecule may be a three-dimensional cross-linked molecule of considerable complexity with opportunity for stretching and contraction among the branched chains.

A variety of equilibrium functions have been developed to characterize the acid–base behavior of synthetic and naturally occurring polyelectrolytes, the most common one being the so-called "extended" Henderson–Hasselbalch equation.[25,26]

$$pH = pK_a - n \log \frac{(HA)}{(A^-)} \quad (14)$$

where *n* is a constant that depends on such factors as concentration and *I*. The equation is applicable for only a limited range of α around $\alpha = 0.5$ and cannot be applied to the entire titration curve. (*note:* α is the degree of dissociation.)

It should be noted that the pK_a (equivalent to the pH at half-neutralization, $\alpha = 0.5$), is an intrinsic value that, on a theoretical basis, should apply to all COOH groups on the macromolecule; the constant *n* reflects the extent to which individual pK_{app} values are modified by electrostatic effects.

From Eq. [14], it can be seen that a plot of pH versus log $[(HA)/A^-]$ should yield a straight line with slope n and an intercept corresponding to the pK_a. Henderson–Hasselbalch plots for the family of curves shown earlier in Fig. 15.3 are illustrated in Fig. 15.5, where log $[\alpha/(1-\alpha)]$ is equal to $-\log (HA/A^-)$]. At all value for I, the lines are not linear but are concave upward in the lower pH range. Values for pK_a declined from approximately 5.6 to 4.9 as I increased from near 0.001 to 0.1M. Values for n were of the order of 1.95. The increase in pK_a with decreasing I is in accord with results expected for a weak acid polyelectrolyte.

The question has arisen as to whether titration curves of humic substances can be described by the extended Henderson–Hasselbalch equation. In the literature, both linear and nonlinear plots have been obtained. There is also disagreement in recorded pK_a values but this is not surprising in view of the different conditions under which the titrations have been performed. As noted earlier, the pK_as of weak-acid polyelectrolytes are strongly affected by I and not all workers have taken this into account in reporting their data.

In addition to indicating the pH at which half-neutralization occurs, the pK_a provides a convenient way of comparing the strength of COOH groups, both between humic substances from various sources and for any given humic or fulvic acid as affected by neutral electrolytes. The lower the pK_a the stronger is the acidic group. The pK_a also provides a rough guide to the expected degree of ionization at higher and lower pHs. At one pH unit above and below the pK_a, the acidic group will be about 90 and 10 percent ionized, respectively; at two pH units above and below this value, the acidic group will be 99 and 1 percent ionized, respectively. Humic substances undoubtedly deviate from this general rule but perhaps only slightly so. Information on pK_a values for humic substances are useful for predicting the sorption of such herbicides as the s-triazines by soil organic matter (see Chapter 19).

For humic substances, slopes of the curves obtained by application of Eq. [14] have generally been independent of I (see Fig. 15.5). For known poly-

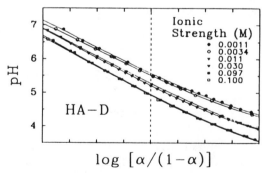

Fig. 15.5 Henderson–Hasselbalch plots for family of titration curves shown in Fig. 15.3 (x-axis units are ±0.2). Adapted from Barak and Chen,[6] reproduced by permission of Williams and Wilkins Co.

electrolytes where repeating COOH groups are regularly distributed on the chain, the constant n has been shown to increase with increasing concentration of electrolyte. Posner[15] concluded that the slope of the titration curve for humic acids was not due to interactions of "similar" groups on the same molecule but rather to a distribution of pK values within and between molecules. In another study, Posner[16] attempted to classify humic acids according to the magnitude of their pK values.

From the above, it can be seen that the pH range of maximum buffering for humic acids (and fulvic acids as well) depends upon salt concentration. A second "intrinsic" dissociation constant, pK_0, has been derived by obtaining pK_a values for several levels of neutral salt and extrapolating to infinite dilution. Results obtained for a soil humic acid are shown in Fig. 15.6, where pK_a values are plotted against the square root of the ionic strength. The limiting value for pK_0 was 5.5.

Application of the Equation of Katchalsky A related equation was introduced by Katchalsky and his coworkers,[25,26] in which account is taken of electrostatic free energy changes accompanying the accumulation of negative charges on a flexible linear polyelectrolyte. The general form of the equation is

$$\text{pH} = pK_{\text{int}} + \log \frac{\alpha}{1 - \alpha} - 0.868\omega n\alpha \tag{15}$$

where pK_{int} is an intrinsic constant equivalent to pK_{app} at zero concentration, n is the average number of COOH groups per macromolecule, and ω is a composite term that depends on I.

By rearrangement and combination with Eq. [13], the following is obtained:

$$pK_{\text{app}} = \text{pH} - \log \frac{\alpha}{1 - \alpha} = pK_{\text{int}} - 0.868\omega n\alpha \tag{16}$$

From Eq. [16], it can be seen that a plot of pK_{app} versus α should yield a straight line with a slope of $0.868\omega n$ and with pK_{int} as the intercept. Posner[15]

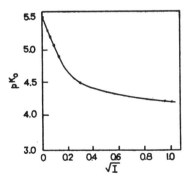

Fig. 15.6 Plot of pK_a versus $\sqrt{(\text{KCl})}$ ionic strength. The limiting value represents the intrinsic dissociation constant (pK_0). From Posner[15]

found that titration data for humic acids gave a better straight line when plotted in this way than by using the modified Henderson–Hasselbalch equation. In disagreement with theory, the slope of the line did not decrease with increasing electrolyte concentration but remained constant. For ideal polyelectrolytes, a decrease in slope is expected because of the effect of electrolyte on the ionic atmosphere around the molecule. In other work, Young et al.[13] obtained curvilinear plots of pK_{app} versus α for some soil fulvic acids.

Marinsky and Ephraim[12] pointed out that Eqs. [14] and [16] do not necessarily apply to cross-linked resins and gels. They propose a model in which the existence of a "gel phase" for humic substances is assumed; a method of correcting for Donnan potential terms in such systems was presented.

As noted earlier, the slope of the line obtained by plotting pK_{app} versus α yields $0.868\omega n$ as the slope. The factor ω includes a number of terms (e.g., electronic charge, dielectric and Boltzmann constants, radii of the molecule) from which estimates can be made for molecular dimensions of the macromolecules.[25,26] Thus far, the approach has not been exploited in studies with humic substances. However, based on potentiometric titration data, Barak and Chen[6] determined the equivalent radii of a soil humic acid; for an assumed infinitely long cylinder, the radii ranged from about 1.1 nm at 10 percent ionization to 0.24 nm at 90 percent ionization.

Gaussian Distribution Model Perdue et al.[14] have presented a model based on the assumption that COOH groups in humic substances can be described in terms of a mixture of low molecular weight organic acids in which the dissociation constants are normally distributed and can thereby by described by a symmetrical Gaussian distribution function.

The Gaussian distribution function for proton binding is

$$\frac{C_i}{C_L} = \frac{1}{\sigma\sqrt{2\pi}} e^{-1/2 \left(\frac{\mu - \log K_{app}}{\sigma}\right)^2} d\log K_{app} \qquad (17)$$

where C_i/C_L is the mole fraction of COOH groups in the interval dpK_{app}, and σ is the standard deviation for the distribution of pK_{app} values about the mean (μ).

In terms of degree of ionization, α, for a mixed normal distribution of COOH groups,

$$\alpha_1 = (\sigma\sqrt{2\pi})^{-1} \int_a^b \frac{K_{app}}{K_{app} + [H^+]} \exp\left[-\frac{1}{2}\left[\frac{\mu - pK_{app}}{\sigma}\right]^2\right] dpK_{app} \qquad (18)$$

A similar equation can be written for OH groups. Application of the Gaussian function to proton binding by humic substances has been discussed by Perdue.[5] Manunza et al.[11] found that the model provided a good description of titration curve of soil humic acids; mean pK_{ave} values for COOH groups (I of $0.1M$) were of the order of 4.6.

The author has applied Eq. [17] to titration data for humic acids as affected by neutral electrolytes. As can be seen from Fig. 15.7, distribution curves for experimentally determined pK_{app} values did exhibit the typical bell-shaped Gaussian pattern, although some resemblance of symmetry can be noted. For the sample at I of $0.001M$, the pattern shows evidence of being bimodal.

Continuous Distribution Models Most of the approaches described earlier can be regarded as intrinsic binding site models in that extrapolations are used to obtain an intrinsic constant (i.e., pK_a, pK_0, or pK_{int}) characteristic of the acidic group. A limitation of these methods arises from the assumption that all COOH groups are of the same type and that differences arise because of coulombic interactions between charges on the molecule as ionization progresses.

In recent years, more general modeling approaches have evolved. They include the continuous distribution model of Gamble[9] and the affinity spectrum models of Shuman,[27] as applied to metal complexes. A full discussion of these models is beyond the scope of the present chapter and the reader is referred to the review of Perdue[5] for additional information.

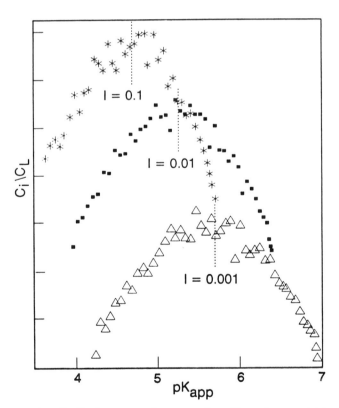

Fig. 15.7 Gaussian distribution curves for successive pK_{app} values as obtained by base titration of a soil humic acid at three ionic strengths. Unpublished data of author

Equivalent Weights of Humic Substances

Equivalent weight refers to the quantity of material (in grams) having the combining power of 1 g of H^+. The lower the equivalent weight the greater the exchange acidity per unit weight. Fulvic acids have lower equivalent weights than humic acids, which in turn have lower equivalent weights than clay minerals.

Reported values for equivalent weights of humic acids range from 160 to 540, corresponding to exchange acidities of 625 to 185 cmole/kg. The equivalent weight of montmorillonite is of the order of 1000; illite and kaolonite have much higher equivalent weights.

Conductometric Titration

When base is added to an aqueous solution of an acidic substance, the highly conducting H^+ ion is replaced by a metal ion having a lower conductance. Thus, the conductance of the solution decreases. When neutralization of the acid is complete, further addition of alkali causes the conductance to rise. Thus, like potentiometric titrations, end points for acidic groups and values for total acidity can be obtained by conductometric titrations.

Variable results have been obtained using conductometric titration to examine humic acids. The review of Flaig et al.[4] shows that from as few as one to as many as four breaks occur in conductometric titration curves of humic substances. Conductometric titration curves obtained by Schnitzer and Skinner[28] for the organic matter from the Bh horizon of a Spodosol showed one minimum by titration with NaOH but a total of four with $Ca(OH)_2$. Gamble[9] obtained two end points by conductometric titration of fulvic acid (Fig. 15.8). The first end point was believed to be due to COOH groups ortho to OH groups on aromatic rings and the second to total COOH groups.

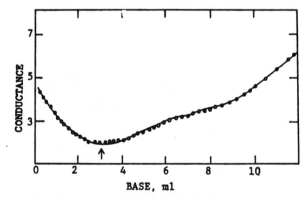

Fig. 15.8 Conductometric titration of fulvic acid showing the possible presence of two types of acidic groups. Conductance is given in mho × 10^3. Only the first end point is indicated. From Gamble,[9] published with permission of the National Research Council of Canada

Titration in Nonaqueous Solvents

The technique of nonaqueous titration offers a simple and convenient means for the quantitative determination of acids and bases. The procedure, which utilizes conventional apparatus, has the merit of providing end points for very weakly acidic substances that normally do not ionize to any extent in water.

The electrolytic dissociation theory of Arrhenius (discussed earlier) is inadequate for explaining titrations in nonaqueous media since such solvents contain neither OH ions or water. The concept of Brønsted and Lowry does apply, however. According to Brønsted and Lowry, an acid is a substance that tends to dissociate to yield a proton and a base is a substance that combines with a proton. In the nonaqueous titration of acids, a basic solvent substitutes for water as a proton acceptor.

$$\underset{\text{Acid}}{\text{HA}} + \underset{\substack{\text{Basic} \\ \text{solvent}}}{\text{S}} \rightleftharpoons \underset{\substack{\text{Solvated} \\ \text{proton}}}{\text{SH}^+} + \underset{\substack{\text{Conjugate} \\ \text{base}}}{\text{A}^-} \qquad (20)$$

By appropriate selection of solvent, the acidic strength of a weak acid can be enhanced, thereby shifting the equilibrium position to the right. Solvents which have been used in soil organic matter studies include pyridine, dimethylformamide [$(CH_3)_2N-CHO$)] and ethylenediamine ($NH_2-CH_2-CH_2-NH_2$). Special titrants are required, the most common ones being KOH in isopropyl alcohol or tetrabutylammonium hydroxide in isopropyl alcohol.

Wright and Schnitzer[29] titrated some humic and fulvic acid preparations from the Ao and Bh horizons of a Spodosol with sodium aminoethoxide (pyridine, dimethylformamide, and ethylenediamine were the solvents) and obtained a single inflection that corresponded to the total COOH and phenolic OH contents. In other work, van Dijk[30] failed to distinguish between different kinds of acid groups in humic acids by potentiometric and conductometric titrations with tetrabutylammonium hydroxide in nonaqueous solvents. However, subsequent studies by van Dijk[31] indicated that, under suitable conditions, conductometric and high frequency titrations might be suitable for determining COOH groups in humic acids; maxima corresponding to weak acid (COOH) groups were shown on titration curves using sodium isopropylate as titrant and dimethylformamide as solvent.

In the study of Thompson and Chester,[32] nonaqueous titrations of humic acids were carried out using pyridine as the solvent, potassium methoxide as the titrant, and an electrode assembly consisting of a platinum electrode in combination with a sleeve-type calomel reference electrode in which the saturated aqueous KCl solution was replaced with a saturated methanol solution of KCl. Several distinct breaks were evident on the titration curves (Fig. 15.9). No attempt was made to assign the inflections to particular acidic groups on the humic acid molecule. Similar studies with plant lignins indicate higher acidic-H contents than generally recognized.[32]

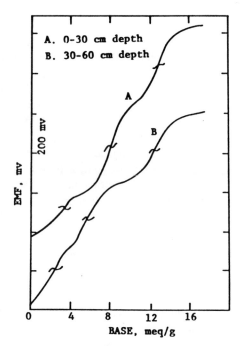

Fig. 15.9 Nonaqueous titration curves of humic acids from an uncultivated sandy loam soil. The samples were titrated in dimethylformamide using CH_3OK in benzene–methanol as titrant and a Pt–calomel electrode combination. From Thompson and Chesters,[32] reproduced by permission of Oxford University Press, Oxford

High-Frequency Titrations

A disadvantage of potentiometric titration is the need to insert electrodes into the titration liquid which, unless care is taken, can be contaminated with reagents from the electrodes. This drawback does not occur with high-frequency titration in which the electrodes are placed outside of the titration vessel. The approach is suitable for both aqueous and nonaqueous titrations.

The measurements obtained by high-frequency titration are a composite function of the dielectric constants and conductances of both the titrating solution and the wall of the titration vessel plus any air space between. Since the properties of the titration solution vary during the course of a titration, a high-frequency end point is obtained in much the same way as conventional titration.

The most extensive high-frequency titrations of humic substances appear to be those of van Dijk.[31] The titration curves obtained corresponded to a high degree with those obtained by conductometric titration but positions of the maximum and minimum were more pronounced.

Thermometric Titrations

The technique of thermometric titration consists in monitoring the temperature change of the solution as a function of the amount of titrant added, or as a function of time when the titrant is added at a constant rate. The technique was first applied to soil organic matter by Ragland,[33] and more recently by Khalaf et al.[34] and Perdue.[35] Thermograms obtained for humic acids show a continuously varying slope, suggesting the occurrence of many classes of ionizable groups.

SELECTIVITY OF HUMIC SUBSTANCES FOR EXCHANGE CATIONS

When a cation exchanger is placed in a solution containing a counter-ion (cation different from the one on the exchange site), a redistribution of cations takes place. For two monovalent cations, the reaction is

$$AM_x + M_y^+ \rightleftharpoons AM_y + M_x^+ \tag{21}$$

As a general rule, the exchanger selects one species in preference to the other, with the result that the concentration ratio of the two on the exchanger is different from that in the solution phase. Polyvalent cations are usually held preferentially to monovalent cations. For ions of equal valency, those that are the least hydrated have the highest energy of adsorption. The hydration order for monovalent cations is: $Li^+ > Na^+ > K^+ > Rb^+ = Cs^+$. For divalent cations the order is $Ba^{2+} > Sr^{2+} > Ca^{2+} > Mg^{2+}$.

The exchange capacity of humates for the alkaline earth metal ions (Ca^{2+} and Mg^{2+}) has been found to be substantially higher than for the alkali ions (Na^+ and K^+), possibly due to some chelation of the former. For monovalent cations, replacement follows the order of decreasing radius of the hydrated ion as noted above. Zadmard[36] has made an extensive study of the selectivity of humic acids for mono- and divalent cations.

CONTRIBUTION OF ORGANIC MATTER TO THE CATION-EXCHANGE CAPACITY (CEC) OF THE SOIL

According to the Glossary of Soil Science Terms of the Soil Science Society of America,[37] the CEC of the soil is

> The sum total of exchangeable cations that a soil, soil constituent, or other material can absorb at a specific pH. It is usually expressed in centimoles of charge per kilogram of exchanger (cmole$_c$/kg).

Both clay and organic matter contribute to the CEC of the soil. Because cations are positively charged, they are attracted to surfaces that are negatively charged. The charge on the clay fractions arises from isomorphous replacement, the ionization of OH groups at the edges of clay plates, and pH-dependent charges associated with Al oxides. For the organic fraction, the charge arises largely from ionization of COOH groups, although some contribution from phenolic OH is suspect.

From 25 to 90 percent of the total CEC of the top layer of mineral soils is due to organic matter. For many soils, a direct relationship exists between CEC and organic matter content, as illustrated in Fig. 15.10 for 18 coastal plain soils.[38] Since organic matter content generally increases with clay content, the increase in CEC cannot be attributed entirely to the organic matter.

As would be expected, practically all of the CEC of highly organic soils (peats), as well as the humus layer of forest soils, is due to organic matter. For these special cases, the greater the degree of humification the higher the CEC. The small amount of organic matter normally found in sandy soils is extremely important in retaining cations against leaching.

Unlike clay minerals, soil organic matter does not have a fixed capacity for binding exchangeable cations. Not only does the exchange acidity of humic substances vary widely but the exchange capacity increases markedly with increasing pH. The latter has been attributed to increased ionization of acidic groups, mostly COOH, at the higher pH values.

The contribution of organic matter to the CEC of the soil has been determined in several ways, as follows:

1. Measurement of the CEC of the whole soil before and after destruction of organic matter with H_2O_2 or by ignition. The loss in CEC is attributed to organic matter. This method may result in incomplete oxidation of organic matter, alteration and partial dehydration of clay minerals, and destruction of exchange sites due to the organic matter–clay complex
2. Determination of CEC following destruction of clay with HF. New exchange sites may be exposed and some of the organic matter may be solubilized by the HF
3. Determination of the CEC of organic matter after extraction. This approach is inadequate because the amount of organic matter recovered is

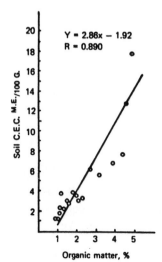

Fig. 15.10 Relationship between soil organic matter and the CEC of Coastal Plain soils. From Kamprath and Welch[38]

incomplete and depends on the extraction method. Furthermore, new exchange sites may be exposed, thereby giving high results

4. Measurement of total CEC for a range of soils having variable clay and organic matter contents followed by regression analysis of the accumulated data

The latter procedure has been adopted in most studies.[39-44] The CEC of both clay and organic matter is obtained directly from the regression equation and the average relative contribution of each to the CEC is expressed by the partial regression coefficients. The assumption is made that the compositions of the clay and organic matter are identical from one sample to another and that the soils vary only in the amounts of the components present. For this reason, regression equations can only be used for predicting the contribution of organic matter to the CEC within soils of a confined geographical and climatic zone.

Results of these studies have shown that organic matter makes a significant contribution to the CEC of the soil but that the actual contribution depends upon soil pH. The results of Helling et al.[40] show that for each unit change in pH, the change in CEC of organic matter was severalfold greater than for clay. At an adjusted pH of 2.5, only 19 percent of the CEC of several grassland and forest soils was due to organic matter, but this value rose to 45 percent at pH 8.0. From Fig. 15.11, it can be seen that the CEC of organic matter is much more strongly influenced by soil pH than is the CEC of clay.

Regression equations obtained by Hallsworth and Wilkinson[39] for five major soil types are given in Table 15.1, where the overall regression equation is given by

$$Y = Y_0 + b_1 x_1 + b_2 x_2 \tag{22}$$

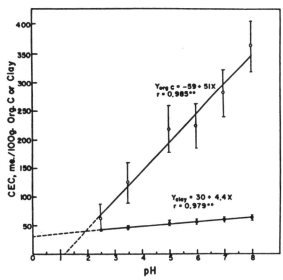

Fig. 15.11 Influence of pH on the CEC of humus and clay. From Helling et al.[40]

Table 15.1 Regression Equations Relating CEC of Some Australian Soils to Organic C (x_2) and Clay (x_1)[a]

Soil type[b]	Description	Number of Soils	Mean pH	Regression Equation
Chernozemic and Sierozemic	Medium to heavy textured soils of low rainfall area	88	7.20	$Y = 4.12 + 0.82x_1 + 5.12x_2$
Stony downs	Medium to heavy textured soils of low rainfall area. Carbonate concentrations in subsoil and often in surface soil	24	7.00	$Y = 4.37 + 0.44x_1 + 2.25x_2$
Euchrozems	Derived from bauxite Laterites. Slight acid and well-supplied with organic matter	19	6.65	$Y = 11.48 + 0.01x_1 + 5.01x_2$
Miscellaneous acid soils	Selected from the same climatic zone as the Chernozemic and Sierozemic group	65	5.58	$Y = 5.13 + 0.23x_1 + 2.27x_2$
Alpine humus	A group on acid alpine soils	15	4.82	$Y = 3.60 + 0.18x_1 + 1.12x_2$

[a] From Hallsworth and Wilkinson.[39]
[b] For equivalents in the comprehensive soil classification system, a beginning textbook on soils can be consulted.

where Y is the exchange capacity and x_1 and x_2 represent percent clay and organic C, respectively.

Coefficients obtained for the clay, the b_1 values, provide an estimate for the average CEC for the clay. These values ranged from 0.01 and 0.82 cmole$_c$/g and were considered to be in accord with clay mineralogy data for the soils used. The contribution of organic C to the CEC, indicated by the b_2 coefficients, ranged from 1.12 to 5.12 cmole$_c$/g. In agreement with results shown in Fig. 15.11, the CEC of organic matter increased markedly with increasing pH. On the assumption that organic C constituted 58 percent of the organic matter, a CEC of 297 cmole$_c$/kg was estimated for organic matter of the "Chernozem" and "Sierozem" soils; for the miscellaneous acid soils from the same area, a CEC of 134 cmole$_c$/kg organic matter was estimated.

The CEC of the soil is determined not only by humus content but by the kind and amount of clay that is present. The CEC of different clays is of the order of 3–5 cmole$_c$/kg for kaolinite, 30–40 cmole$_c$/kg for illite, 80–150 cmole$_c$/kg for montmorillonite, and 100–150 cmole$_c$/kg for vermiculite. As noted in Chapter 9, total acidities of humic acids usually range from 485 to 870 cmole$_c$/kg; for fulvic acids, values up to 1400 cmole$_c$/kg have been recorded. These comparisons explain why humus can make a significant contribution to the CEC of the soil even though the amounts present may be quite low relative to clay. Organic matter will be particularly important as a cation exchanger in soils where kaolinitic-type clays predominate, such as in soils of southern United States.

In the natural soil, the CEC of organic matter and clay cannot be considered additive, for the reason that some sites are lost through associations between the two. Also, many organic sites may be tied up as complexes with polyvalent cations. Thus, CEC values for organic matter *in situ* will be somewhat less than for the isolated components. Kamprath and Welch[38] estimated a range of from 62 to 279 cmole$_c$/kg for the apparent CEC of organic matter in 18 Coastal Plain soils; at pH 7, an average of 55 percent of the total CEC was due to organic matter. They quote research data for some soils of Puerto Rico where the values ranged from 100 to 150 cmole$_c$/kg and accounted for about 25 percent of the total CEC.

Schnitzer[45] suggested that two types of CEC for organic matter should be considered: 1) "measured" CEC as determined by exchange with NH_4^+ or any other appropriate cation, and 2) "potential" CEC, the sum of the above and CEC due to blocked sites. The "blocked" sites, which may exceed the "measured" ones, are exposed when organic matter is extracted from soil.

The studies of McLean and his associates[42,46] indicate that "blocked" exchange sites of organic matter are released when soils are limed. Several factors may be responsible, including increased ionization of acidic groups and release of exchange sites previously inactivated with Al.

The contribution of organic matter to the CEC of the soil nearly always decreases with depth in the profile, for the reason that the proportion of organic matter to clay decreases.

COAGULATION OF HUMIC SUBSTANCES BY POLYELECTROLYTES

Humic and fulvic acids, like other organic colloids, can be precipitated from solution by electrolytes. The extent to which flocculation occurs depends upon such factors as pH and the nature of the electrolyte. Colloid concentration is also of importance, particularly for monovalent cations. Under proper pH conditions, trivalent cations, and to some extent divalent ones, are effective in precipitating humic substances from very dilute solutions; monovalent cations are generally effective only at relatively high particle concentrations. Because of their high molecular weights, humic acids are more easily coagulated than fulvic acids, and, for the former, gray humic acids flocculate at lower electrolyte concentrations than brown humic acids. Attempts have been made to subdivide humic acids on the basis of molecular weight by fractional precipitation with ammonium sulfate.[47]

In agreement with the well-known Schultz–Hardy rule, trivalent cations are more effective in coagulating humic matter than divalent cations, which in turn are more effective than monovalent ones. For a variety of tri-, di-, and monovalent cations, flocculation of humic substances follows the order:

$$Fe^{3+} \approx Al^{3+} > H^+ > Ca^{2+} > Sr^{2+} > Mg^{2+} > K^+ > Na^+ > Li^+$$

The H^+ ion behaves more like a divalent cation whereas Mg^{2+} acts like a monovalent one. With the former, an increase in H^+ concentration is accompanied by a corresponding decrease in pH and a shift in ionization equilibrium ($R-COO^- + H^+ \rightarrow R-COOH$).

When a trivalent ion (Fe^{3+}, Al^{3+}) is added to humic acid, precipitation occurs at rather low concentrations but the precipitate dissolves as the pH rises above 7.0 during titration. This has been attributed to the formation of soluble hydroxy complexes. For both mono- and divalent cations, coagulating power increases with decreasing radii of the hydrated ion, the order of increasing effectiveness being $K^+ > Na^+ > Li^+$ and $Ca^{2+} > Mg^{2+}$. Data obtained by Ong and Bisque[48] for the critical concentration of electrolyte required to produce precipitation of humic material from a peat soil are shown in Table 15.2. It should be emphasized that complete precipitation with salts of monovalent cations is seldom achieved. For the anions of K^+, coagulation follows the order: $SO_4^{2-} > NO_3^- > Cl^-$.

Various attempts have been made to explain the unusually high coagulating power of $CaCl_2$ relative to the monovalent salt (KCl or NaCl), the former being far more than twice as effective (of the order of 100 time or more). According to one theory of coagulation of colloids, as reviewed by Ong and Bisque,[48] the critical concentration of ions required to initiate flocculation is inversely proportional to the sixth power of their valency. Thus,

$$\text{monovalent}:\text{divalent}:\text{trivalent} = (1/1)^6: (1/2)^6: (1/3)^6$$
$$= 1:0.016:0.0014$$

Table 15.2 Critical Concentration of Electrolyte Required to Initiate Coagulation of the Soluble Humic Material from a Peat Soil[a]

Electrolyte	Radius of cation (A°)		Moles electrolyte (liter)	Mean value	Ratio[b]
	Ionic	Hydrated			
Monovalent					
LiCl	0.60	3.40	826		
NaCl	0.95	2.76	598	596	1
KCl	1.33	2.32	335		
Divalent					
$MgCl_2$	0.65		30	18.6	0.0317
$CaCl_2$	0.99		7.2		
Trivalent					
$AlCl_3$	0.50		1.10	1.31	0.0022
$FeCl_3$	0.64		1.52		

[a] From Ong and Bisque.[48]
[b] Inverse valency of cation to the sixth power (calculated values).

The results obtained for soil organic colloids follow this relationship rather closely. However, as Ong and Bisque[48] pointed out, this coagulation theory is only valid for hydrophobic colloids. To account for agreement of the theory with humus colloids, they postulated that, in the presence of electrolyte, the humus macromolecule changes from a hydrophilic to a hydrophobic-type colloid. This was believed to occur in the following way. In aqueous solution, acidic functional groups of the humic colloid are more or less dissociated and the polymer assumes a stretched configuration due to mutual repulsion of negatively charged functional groups (e.g., COO^-). When electrolyte is added, the cation is attracted to negative groups, thus causing a reduction in intermolecular coulombic repulsion in the polymer chain. This in turn favors coiling of the chain and causes a reduction in the amount of hydration water held by the colloid.

A limitation of the above-mentioned explanation is that it fails to take into account the possible formation of mixed complexes with polyvalent cations, as described in Chapter 16.

ELECTROPHORESIS

Humic and fulvic acids, by virtue of negative charges on their surfaces, move to the positive electrode under the influence of an applied electric field. This movement relative to that of the suspending medium is called electrophoresis.

Electrophoresis has had wide application for the separation and purification of biocolloids from complex mixtures. The approach was initially applied to soil organic colloids with considerable enthusiasm because it was felt that the technique offered a way of obtaining homogeneous preparations in which all molecules would be rather similar. Unfortunately, these high expectations failed to materialize and the technique has been used sparingly in recent years.

Initial electrophoresis studies with soil organic colloids were carried out using the so-called moving boundary method.[49] In this approach, the colloidal material is dissolved in an appropriate buffer and placed into a U-shaped tube in such a way that a sharp boundary exists between the colloid solution and the supporting electrolyte (buffer). An electric current is then applied and the negatively charged particles migrate to the positive electrode; positively charged particles migrate to the negative electrode. Colloids of dissimilar charge move at different rates and can thus be separated from one another. With soil humic colloids, most of the material moves as a single broad band, indicating one main component very heterogeneous with respect to surface charge. A second component has been found to be present in small amounts.

The techniques of paper and gel electrophoresis have also been applied (see reviews of Flaig et al.[4] and Duxbury[50]). Although two and sometimes three migrating components have been observed, a clear separation of colloidal components has seldom been achieved. With paper electrophoresis, one migrating component produces a white fluorescence under ultraviolet light.

More recently, the technique of isoelectric focusing (also called electrofocusing) has been applied, with somewhat better success.[51-54] This technique can be defined as electrophoresis in a pH gradient and is carried out in a supporting gel. This work, reviewed by Duxbury,[50] has shown that as many as 30 distinct bands have been obtained by analysis of NaOH and $Na_4P_2O_7$ extracts of soil. In some studies, electrophoretic separations have been carried out on subfractions obtained by gel filtration.[52]

In evaluating the results of electrophoresis studies, Duxbury[50] concluded that much more work needs to be done before firm conclusions can be drawn. A pressing problem in the interpretation of results, including those of electrofocusing, is how to deal with interactions between humic substances and inorganic materials and of the possible existence of intermolecular aggregates.

OXIDATION–REDUCTION POTENTIAL

Oxidation–reduction potentials can be defined as the difference in voltage between a measuring electrode (e.g., platinum) and a reference electrode (e.g., saturated calomel electrode). The absolute value is given in reference to the standard hydrogen electrode with an assigned potential of zero at all temperatures. The oxidation–reduction potential of a system is related to pH and is a measure of the tendency of ions or molecules to gain (reduction) or lose (oxidation) electrons under given conditions. The measured difference is thus a characteristic of the oxidized or reduced state of the system.

The redox potential, E, is given by the Nernst equation.

$$E = E_0 + \frac{RT}{nF} \ln \frac{\text{(oxidized state)}}{\text{(reduced state)}} \qquad (18)$$

where E_0 is the standard potential (all substances at unit activity against the normal hydrogen electrode), R is the gas constant, T is the temperature, n is the number of electrons participating in the system, and F is the Faraday constant.

Relatively few studies have been conducted on the oxidation–reduction properties of humic substances. The review of Flaig et al.[4] indicates that humic acids have reductive properties. In the study of Visser,[55] redox potentials of humic acids from a tropical Sphagnum peat deposit were obtained with a platinum and hydrogen gas electrode upon titration with potassium ferriccyanide. Depending on the depth at which the humic acids were recovered, redox potentials ranged from +0.38 to +0.32.

SUMMARY

Soil humic and fulvic acids have properties similar to weak-acid polyelectrolytes in that charges on the molecules are strongly influenced by pH and the presence of neutral salts. A variety of ionizable acidic groups are present but they cannot easily be discerned by acid–base titrations because the titration of one type begins before that of the other is ionized. In their protonated forms, the molecules are in a more or less contracted state; when completely dissociated, the molecules assume a stretched configuration due to mutual repulsion of charges groups (e.g., COO^-). Trivalent cations, and to some extent divalent ones, are effective in precipitating humic substances from very dilute solutions; monovalent cations are relatively ineffective and then only at high particle concentrations.

Organic matter makes a significant contribution to the CEC of many soils, the absolute value behind dependent upon such factors as organic matter content, kind and amount of clay, and pH. Organic colloids exhibit buffering over a wide pH range. Being negatively charged, humic and fulvic acids can be examined by electrophoretic techniques, although results obtained thus far have been somewhat disappointing.

REFERENCES

1 A. H. Beckett and E. H. Tinley, *Titration in Non-Aqueous Solvents*, British Drug Houses, Ltd., Poole, England, 1957.
2 P. A. Arp, *Can. J.Chem.*, **61,** 1671 (1983).
3 S. Oden, *Kolloidchem. Beihefte,* **11,** 75 (1919).
4 W. Flaig, H. Beautelspacher, and E. Rietz, "Chemical Composition and Physical Properties of Humic Substances," in J. E. Gieseking, Ed., *Soil Components, Vol. I*, Springer-Verlag, New York, 1975, pp. 1–211.
5 E. M. Perdue, "Acidic Functional Groups of Humic Substances," in G. R. Aiken, D. M. McKnight, R. L. Wershaw, and P. MacCarthy, Eds., *Humic Substances in Soil, Sediment, and Water,* Wiley, New York, 1985, pp. 493–526.

6   P. Barak and Y. Chen, *Soil Sci.*, **154**, 184 (1992).
7   O. K. Borggaard, *J. Soil Sci.*, **25**, 189 (1974).
8   H. Davis and C. J. B. Mott, *J. Soil Sci.*, **32**, 379, 393 (1982).
9   D. S. Gamble, *Can. J. Chem.*, **48**, 2662 (1970).
10 J. E. Gregor and M. K. J. Powell, *J. Soil Sci.*, **39**, 243 (1988).
11 B. Manunza, C. Gessa, S. Deiana, and R. Rausa, *J. Soil Sci.*, **43**, 127 (1992).
12 J. A. Marinsky and J. Ephraim, *Environ. Sci. Tech.*, **20**, 349, 354, (1986).
13 S. D. Young, B. W. Bache, D. Welch, and H. A. Anderson, *J. Soil Sci.*, **32**, 592 (1981).
14 E. M. Perdue, J. H. Reuter, and R. S. Parrish, *Geochim. Cosmochim. Acta*, **48**, 1257 (1984).
15 A. M. Posner, *Trans. 8th Intern. Congr. Soil Sci.*, **3**, 161 (1964).
16 A. M. Posner, *J. Soil Sci.*, **17**, 65 (1966).
17 F. J. Stevenson, *Trans. 11th Intern. Cong. Soil Sci.*, **1**, 55 (1978).
18 G. Sposito, K. M. Holtzclaw, and D. A. Keech, *Soil Sci. Soc. Amer. Proc.*, **41**, 1119 (1977).
19 G. Gran, *Acta Chem. Scand.*, **4**, 559 (1950).
20 G. Gran, *Analyst*, **77**, 661, (1952).
21 A. Hendriksen and H. M. Seip, *Water Res.*, **14**, 809 (1980).
22 Y.-H. Lee and C. Brosset, *Water, Air, Soil Pollut.*, **10**, 457 (1978).
23 K. Molværsymr and W. Lund, *Water Res.*, **17**, 303 (1983).
24 C. McCallum and D. Midgley, *Anal. Chem. Acta*, **78**, 171 (1975).
25 A. Katchalsky and P. Spitnik, *J. Polymer. Sci.*, **2**, 432 (1947).
26 A. Katchalsky, H. Schavit, and H. Eisenberg, *J. Polymer. Sci.*, **13**, 69 (1954).
27 M. S. Schuman, G. J. Collins, P. J. Fitzgerald, and D. L. Olson, "Distribution of Stability Constants and Dissociation Rate Constants Among Binding Sites on Estaurine Copper-Organic Complexes: Rotated Disk Electrode Studies and an Affinity Spectrum Analysis of Ion-Selective Electrode and Photometric Data," in F. R. Christman and E. T. Gjessing, Eds., *Aquatic and Terrestrial Humic Materials*, Ann Arbor Science Press, Ann Arbor, 1983, pp. 349-370.
28 M. Schnitzer and S. I. M. Skinner, *Soil Sci.*, **96**, 86 (1963).
29 J. R. Wright and M. Schnitzer, *Trans. 7th Intern. Congr. Soil Sci.*, **2**, 120 (1960).
30 H. van Dijk, *Z. Pfl. Ernähr. Dung, Bodenk.*, **84**, 150 (1959).
31 H. van Dijk, *Sci. Proc. Roy. Dub. Sci.*, **A1**, 163 (1960).
32 S. O. Thompson and G. Chesters, *J. Soil Sci.*, **20**, 346 (1969).
33 J. L. Ragland, *Soil Sci. Soc. Amer. Proc.*, **26**, 133 (1962).
34 K. Y. Khalaf, P. MacCarthy, and T. W. Gilbert, *Geoderma*, **14**, 319, 331 (1975).
35 E. M. Perdue, *Geochim. Cosmochim. Acta*, **42**, 1351 (1978).
36 H. Zadmard, *Kolloid-Beihefte*, **49**, 315 (1939).
37 Anonymous, *Glossary of Soil Science Terms*, Soil Science Society of America, Madison, 1987.
38 E. J. Kamprath and C. D. Welch, *Soil Sci. Soc. Amer. Proc.*, **26**, 263 (1962).
39 E. G. Hallsworth and G. K. Wilkinson, *J. Agr. Sci.*, **51**, 1 (1958).

40 C. S. Helling, G. Chesters, and R. B. Corey, *Soil Sci. Soc. Amer. Proc.*, **28,** 517 (1964).
41 Y. A. Martel, C. R. DeKimpe, and M. R. Lavordiere, *Soil Sci. Soc. Amer. J.*, **42,** 764, (1978).
42 E. O. McLean and E. J. Owen, *Soil Sci. Soc. Amer. Proc.*, **33,** 855 (1969).
43 E. H. Drake and H. L. Motto, *Soil Sci.*, **133,** 281 (1982).
44 W. R. Wright and J. E. Foss, *Soil Sci. Soc. Amer. Proc.*, **36,** 115 (1972).
45 M. Schnitzer, *Nature*, **207,** 667 (1965).
46 E. O. McLean, D. C. Reicosky, and C. Lakshmanan, *Soil Sci. Soc. Amer. Proc.*, **29,** 374 (1965).
47 B. K. G. Theng, J. R. H. Wake, and A. M. Posner, *Plant Soil*, **29,** 305 (1968).
48 H. L. Ong and R. E. Bisque, *Soil Sci.*, **106,** 220 (1968).
49 F. J. Stevenson, J. D. Marks, J. E. Varner, and W. P. Martin, *Soil Sci. Soc. Amer. Proc.*, **16,** 69 (1952).
50 J. M. Duxbury, "Studies of the Molecular Size and Charge of Humic Substances by Electrophoresis," in M. H. B. Hayes, P. MacCarthy, R. L. Malcolm and R. S. Swift, Eds., *Humic Substances II: In Search of Structure*, Wiley, New York, 1989, pp. 593–620.
51 M. De Nobili, *J. Soil Sci.*, **39,** 437, (1988).
52 N. R. Curvetto and G. A. Orioli, *Plant Soil*, **66,** 205 (1982).
53 G. Govi, O. Francioso, C. Ciavatta, and P. Sequi, *Soil Sci.*, **154,** 8 (1992).
54 M. M. de Gonzales, M. Castagnola, and D. Rossetti, *J. Chromatogr.*, **209,** 421 (1981).
55 S. A. Visser, *Nature*, **204,** 581 (1964).

16

ORGANIC MATTER REACTIONS INVOLVING METAL IONS IN SOIL

Practically every aspect of trace element chemistry and reactions in soil is related to the formation of stable complexes with organic substances. Whereas monovalent cations (Na^+, K^+, etc.) are held primarily by simple cation exchange through formulation of salts with carboxyl groups ($R-COO^-Na^+$, $R-COO^-K^+$), multivalent cations (Cu^{2+}, Zn^{2+}, Mn^{2+}, Fe^{3+}, and others) have the potential for forming coordinate linkages with organic molecules.

A schematic diagram of organic matter reactions involving trace elements in soil is given in Fig. 16.1. The trace elements present in the soil solution as soluble organic complexes (MCh_e), and as charge inorganic species, are shown to be influenced by the activities of microorganisms and higher plants, both of which serve as sources of water soluble ligands for complex formation. Some metals are held in insoluble organic and inorganic complexes and are relatively unavailable to plants. Organic matter reactions involving trace elements have been discussed in several reviews.[1-8]

The quantity of any given trace element in the soil solution at any one time (often in the parts per billion range) is normally trivial in comparison to the total amount held by clay and humus colloids, or as precipitates. However, from the standpoint of plant nutrition, the soluble cations are of greatest importance.

With regard to complex formation and plant nutrition, the metals can be placed in the following groups:

1. Those which are essential to plants but that are not bound in coordinate compounds. Included are all of the monovalent cations, such as K^+, and the divalent cations Ca^{2+} and Mg^{2+}
2. Those metals which are essential to plants and that form coordinate link-

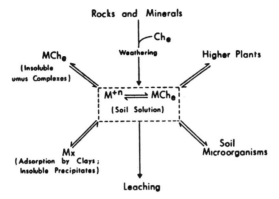

Fig. 16.1 Schematic diagram of organic matter reactions involving metal ions in soil. Ch_e refers to a chelating agent, but complexes are also formed in which the metal ion does not form an internal ring structure with the ligand (see text). Adapted from Stevenson and Ardakani[6]

ages with organic ligands. They include nearly all of the metals in the first transitions series, including Cu^{2+}, Zn^{2+}, and Mn^{2+}, as well as Mo of the second transition series

3. Those without a known function in plants but that are essential for animals, the most notable example being Co^{2+}
4. Those without a known biochemical function in plants or animals but which accumulate in the environment. Included with this group are Cd^{2+}, Pb^{2+}, and Hg^{2+}, which are introduced into soil as contaminants. Renewed interest in organic matter–metal complexes in soils, sediments, and natural waters has been generated by the nocuous introduction of toxic heavy metals into the environment

Emphasis is given in this chapter to the significance of complexation reactions involving soil organic matter, properties of organic matter–metal complexes, nature of naturally occurring organic ligands, organically bound forms of trace elements in soil, and environmental significance. Stability constants of metal complexes with humic substances are outlined in Chapter 17; pedological aspects are covered in Chapter 20.

PROPERTIES OF METAL–ORGANIC MATTER COMPLEXES

A metal ion in aqueous solution contains attached water molecules oriented in such a way that the negative (oxygen) end of the water dipole is directed towards the positively charged metal ion. A complex arises when water molecules surrounding the metal ion are replaced by other molecules or ions, with the formation of a coordination compound. The organic molecule that combines with the metal ion is commonly referred to as the ligand.

A covalent bond consists of a pair of electrons shared by two atoms, and occupying two orbitals, one of each atom. Essentially, a coordinate complex arises because the outer electron shell of the metal ion is not completely filled and can accept additional pairs of electrons from atoms that have a pair of electrons available for sharing. Examples of groupings in organic compounds that have unshared pairs of electrons, and that can form coordinate linkages with metal ions, are shown by structures **I** to **IV**.

$$\underset{\textbf{I}}{R-\overset{H}{\underset{|}{C}}=\overset{..}{\underset{..}{O}}:} \qquad \underset{\textbf{II}}{R-\overset{H}{\underset{|}{\overset{..}{\underset{..}{S}}}}:} \qquad \underset{\textbf{III}}{R-\overset{H}{\underset{|}{N}}H} \qquad \underset{\textbf{IV}}{R-\overset{H}{\underset{|}{\overset{..}{\underset{..}{O}}}}:}$$

The order of decreasing affinity of organic groupings for metal ions is as follows[9]:

$$\underset{\text{enolate}}{-O-} > \underset{\text{amine}}{-NH_2} > \underset{\text{azo}}{-N=N-} > \underset{\text{ring N}}{N} > \underset{\text{carboxylate}}{-COO^-} > \underset{\text{ether}}{-O-} > \underset{\text{carbonyl}}{C=O}$$

Most metal ions can accept more than one pair of electrons and a corresponding number of donor atoms can be coordinated simultaneously. The number of donor atoms held is called the coordination number of the metal ion.

A chelate complex is formed when two or more coordinate positions about the metal ion are occupied by donor groups of a single ligand to form an internal ring structure. The word chelate is derived from the Greek chele, meaning a crab's claw, and refers to the pincer-like manner in which the metal is bound. If the chelating agent forms two bonds with the metal ion it is said to be bidentate; similarly, there are terdentate, tetradentate, and pentadentate complexes. The formation of more than one bond between the metal and the organic molecule usually imparts high stability to the complex.

The reaction between an amino acid and Cu^{2+} to first form a 1:1 chelate complex and then a 2:1 chelate complex is illustrated in Fig. 16.2.

The stability of a metal–chelate complex is determined by such factors as the number of atoms that form a bond with the metal ion, the number of rings that are formed, the nature and concentration of metal ions, and pH. The stability sequence for some select divalent cations is as follows[10]:

$$Cu^{2+} > Ni^{2+} > Co^{2+} > Zn^{2+} > Fe^{2+} > Mn^{2+}$$

Metal ions can be classified into two main classes based on their ability to form a coordinate linkage with specific atoms of the ligand.[11] Class A metal ions are those that form complexes with ligands that contain oxygen as a donor atom; class B metal ions are those that coordinate preferentially with ligands containing N, P, and S donor atoms. The Cu^{2+} ion fits both categories and will thus coordinate with all active groups expected to be present in humic and fulvic acids. The Zn^{2+} ion is an example of a class B metal ion and therefore

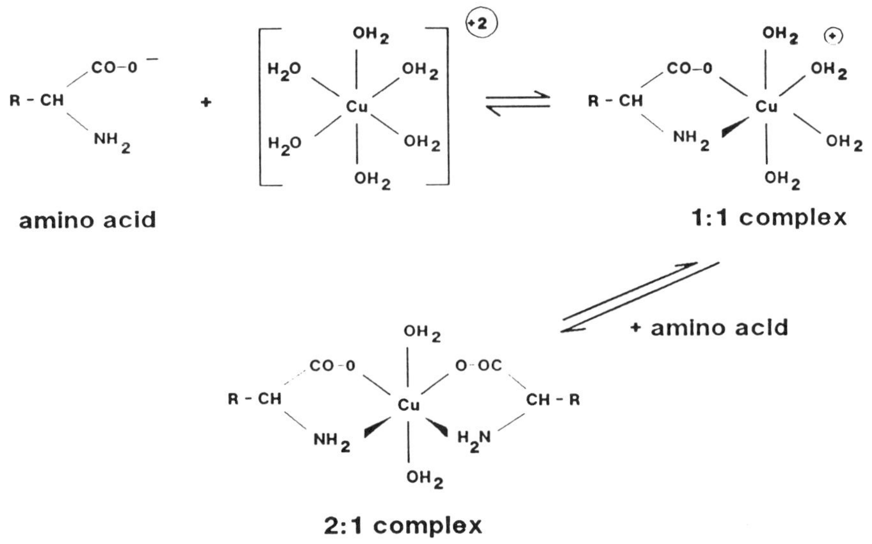

Fig. 16.2 Formation of 1:1 and 2:1 Cu^{2+} complexes with an amino acid

should form high-energy bonds with any N or S donor groups that might be present. Insufficient attention has been given thus far to the selective nature of divalent cations for specific reactive sites in humic and fulvic acids.

SIGNIFICANCE OF CHELATION REACTIONS IN SOIL

The formation of metal–organic complexes can have the following effects in soil.

1. Organic substances are involved in the weathering of rocks and minerals, and they serve as agents for the transport of sesquioxides in leached terrestrial soils, notably Spodosols. Furthermore, they are involved in the neogenesis of certain minerals (see Chapter 20).
2. The cycling of trace elements in soils is profoundly influenced by complexation reactions with organic substances. A unique feature of the cycle is the enrichment of the upper portion of the solum (and the organic matter contained therein) with trace elements due to long-term upward translocation by plant roots and subsequent incorporation into the surface layer of the soil through plant litter decay. Overall aspects of micronutrient cycling have been discussed elsewhere.[3,4]
3. Organic complexing agents act as carriers of trace elements in the soil solution, thereby enhancing the availabilities of trace elements to higher plants, as well as to soil microorganisms. Low-molecular-weight organic acids and other biochemicals form water-soluble complexes with trace

elements. Complexes of trace elements with fulvic acids are also water soluble.

4. Organic substances can enhance the availabilities of insoluble mineral phosphates through complexation of Fe and Al in acid soils and Ca in calcareous soils.
5. Under certain conditions, metal-ion concentrations in the soil solution may be reduced to a nontoxic level through complexation with soil organics. This is particularly true when the metal-organic complex has low solubility, such as in the case of complexes with humic acids and other high-molecular-weight components of organic matter. The interaction of Al^{3+} with organic substances is of considerable importance in controlling soil solution levels of the highly toxic Al^{3+} ion in acid soils.[12,13]
6. Polyvalent cations serve as linkages between humic substances and clay minerals, thereby affecting the physical properties of soils.
7. Natural complexing agents are of importance in the transport of trace elements to other ecosystems, such as lakes and streams. Research on acid deposition has implicated organic substances in the solubilization and transport of Al from terrestrial environments to natural waters.
8. A sludge-like deposit, consisting of a mixture of Fe-oxides and insoluble organic matter (bacterial cell bodies and waste products) is often formed in tile drainage systems, such as are used in Florida citrus groves. The deposit prevents the tile from functioning properly by blocking the passage way.
9. Organic matter acts as a "buffer" in ameliorating the adverse effects of toxic heavy metals in soil, such as Pb and Cd, which are introduced into soil by atmospheric deposition, notably near smelters and in the proximity of highways.

Individual effects, as itemized above, are difficult to quantify and will vary from soil to soil, depending on organic matter content, pH, kind and amount of clay minerals, and soil management practice. The role of organic substances in the weathering of rocks and minerals (Item 1) and the solubilization of mineral phosphates (Item 4) are discussed in Chapters 20 and 5, respectively.

Influence of Organic Matter in Ameliorating Al Toxicities

Considerable interest has recently been directed to Al-organic matter complexes because of toxicity of Al to plants grown on acidic soils ($<$pH 5.5).[12,13] Numerous studies have shown that organically complexed forms of Al in the soil solution are less toxic to plants (and aquatic life) than Al^{3+} or its hydrated monomers ($Al(OH)^{2+}$, $Al(OH)_2^+$). Research by Ares[14] and others indicate that organic ligands play a dynamic and important role in defining the speciation of Al in the aqueous phase of forest soils.

Toxicities due to Al^{3+} have been noted in several regions of the eastern

United States, Canada, and the tropics, where acid soils are found. However, acid soils rich in native organic matter, or amended with large quantities of organic residues, give low Al^{3+} concentrations in the soil solution and permit good growth of crops under conditions where toxicities would otherwise occur. Studies conducted at various pHs, Al concentrations, and quantities of organic amendment added have shown that better plant growth is achieved with an increase in the amount of organic matter added.[12,13]

Borate Complexes with Organic Matter

Boron and Mo are unique among the micronutrient elements in that they normally occur in anionic forms ($H_2BO_3^-$) and (MoO_4^{2-}). Thus, they are subject to losses through leaching. However, the main form of boron in soil may be as borate complexes (**V**) with compounds that contain the *cis*-hydroxyl group, such as saccharides.

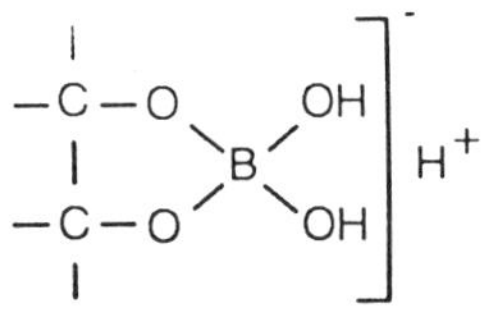

$$\left[\begin{array}{c} -\underset{|}{C}-O \\ -\underset{|}{C}-O \end{array} \diagdown B \diagup \begin{array}{c} OH \\ OH \end{array}\right]^{-} H^+$$

V

Yermiyaho et al.[15] found that the sorption capacity of composted organic matter for boron (on a weight basis) was at least four times greater than for clay and soils. This was attributed to chemical association between boron and the organic matter.

Chelation Reactions in the Rhizosphere

The rhizosphere (soil in the intermediate vicinity of plant roots), contains microorganisms that synthesize high quantities of organic acids and other biochemical chelating agents.[5-7] By inference, it is usually assumed that a beneficial effect has accrued through enhanced availability of micronutrients to the plant. Plant roots are also known to exude a wide variety of organic chelates. Differences in the susceptibilities of plant species to trace element deficiencies have often been attributed to variations in the plant's ability to synthesize and excrete metal complexing organics.

Micronutrient-Enriched Organic Products as Soil Amendments

Considerable interest has been shown to the fertilizer value of micronutrient-enriched organic products.[2] Most work has focused on Fe–organo complexes as sources of Fe for sensitive crops growing on deficient soils; a few investigations have been concerned with Mn and Zn. Among the Fe-enriched products that have shown promise for increasing Fe uptake are composts of plant refuse, forest by-products (lignosulfates and polyflavonoids), peat, lignites, and animal manures. Their effectiveness in improving the Fe nutrition of the test crop is usually attributed to their similarity to soil organic matter, and in particular to the humic substances they contain.

In general, it would appear that enriched products can improve the uptake of trace elements by plants growing on deficient soils, although less efficiently than for synthetic chelates (i.e., EDTA, EDDHA, etc.). On a unit cost basis, the natural products have often compared favorably with the synthetic chelates because larger quantities can be applied for a lower cost, and possible longer duration of effect.[2]

Farmyard manures are rich sources of trace elements. The application of 5000 kg farmyard manure/ha (2.23 tn./a.) results in the addition of the following approximate quantities of micronutrients (in kg): boron, 0.1; Mn, 1.0; Co, 0.005; Cu, 0.08; Zn, 0.48; Mo, 0.01. The rate at which these micronutrients are released will depend upon conditions affecting microbial activity and will be highest in warm, moist, well-aerated soils that have a near-neutral reaction.

Complexation of Toxic Heavy Metals in Soil

As noted earlier, organic matter serves as a buffer in ameliorating the adverse effects of heavy metals that are introduced into soil as contaminants. Lead and Cd, for example, are highly toxic contaminants that have been added in large quantities to agricultural and forested ecosystems worldwide through atmospheric deposition. In forested ecosystems, much of the Pb and Cd has been immobilized through complexation with humified organic matter of the forest floor, with enrichment of the organic layer with these heavy metals.[16] There is fear that the Pb and Cd will eventually be mobilized and transported into lakes and streams as soluble organic complexes.

FORMS OF TRANSITION METAL IONS IN SOIL

Fractionation Schemes

The trace elements in soil can be partitioned into several pools by sequential extraction, as outlined below[17-19]:

Form	Method of Determination
1. Water soluble, as the free cation and as organic and inorganic complexes	Extraction of field-moist soil by miscible displacement, by centrifugation with and without an immiscible liquid, or by use of ceramic or plastic filters.
2. Exchangeable	Extraction of soil with $1M$ $MgCl_2$ at pH 7.0 or $1M$ NaOAc at pH 8.2 and correction for water-soluble forms.

3. a. Bound to carbonates
   b. Specifically adsorbed

Leaching of residue from above with $1M$ NH$_4$OAc (pH 5.0), or extraction from above with 2.5 percent (v/v) CH$_3$COOH.

4. Bound to Fe and Mn oxides

Extraction of residue from above with $0.3M$ NH$_2$OH:HCl in 25 percent (v/v) CH$_3$COOH.

5. Organically complexed

Extraction of residue from above with a complexing agent (e.g., pyrophosphate at $0.1M$ concentration), or release to exchangeable forms by oxidation of organic matter with sodium hypochlorite.[20]

6. Held in primary minerals

Digestion of final soil residue with an HF:HClO$_4$ mixture.

The percentage distribution of any given trace element in the above pools is highly variable, depending on type of metal cation and such factors as pH, kind and amount of clay minerals, and organic matter content. In general, the quantities that exist in water-soluble and exchangeable forms (pools most readily available to plants) are generally low (<2 percent of the total amount present in the soil). For some micronutrients (notably Cu), a significant fraction of the cation in the water-soluble pool may occur in chelated form.

The organically complexed pool (5) represents a major "storehouse" form of micronutrient cations in many soils. McLaren and Crawford[17] concluded that the amount of Cu available to plants (soluble and exchangeable Cu) was controlled by equilibria involving specifically adsorbed forms (Cu extracted with 2.5 percent acetic acid) and the organically bound pool. The suggested relationship between the three forms was as follows:

$$\text{Exchangeable and soluble Cu} \rightleftharpoons \text{Specifically adsorbed Cu} \rightleftharpoons \text{Organically bound Cu}$$

Results obtained for the distribution of Cu in 24 contrasting soil types are given in Table 16.1.[17] For the soils examined, from one-fifth to one-half of the Cu occurred in organically bound forms. Oxide occluded forms are im-

Table 16.1 Distribution of Cu in 24 Contrasting Soil Types. From McLaren and Crawford[17]

Form	Percentage of Total Cu
Soluble + exchangeable	0.1–0.2
Specifically adsorbed by clay	0.2–2.7
Organically bound	16.2–46.9
Oxide (and organic matter) occluded	0.0–35.9
Mineral lattice	33.6–77.2

portant only in those particularly soils that contain appreciable amounts of Fe or Mn oxides. Essentially all of the trace elements in the uppermost organic layer of forest soils (Aflisols, Spodosols) occur in organically bound forms.

Shuman[21] obtained the following values for the percentages of three trace elements that occurred in organically bound forms in 10 representative soils of the southeastern section of the United States: Cu, 1.9–68.6 percent; Mn, 9.5–82.0 percent; Zn, 0.2–14.3 percent. A somewhat similar range for Zn (0.1 to 7.4%) has been obtained for soils of the Appalachian, Coastal Plain, and Piedmont regions of Virginia.[22]

Speciation of Trace Elements in the Soil Solution

Considerable interest has been shown in recent years to water soluble forms of metal ions in soil, both because of the importance of the soluble fraction to plant nutrition but because the cationic forms of some metal ions are highly toxic to plants. Quantitative data on speciation of trace elements in the soil solution are limited and attributed to the following:[7]

1. The concentration of trace elements in the soil solution are normally very low, often of the order of 10^{-9} to 10^{-8} M, thereby creating severe analytical problems in their determination
2. The trace element may exist in a large number of different chemical forms
3. Extracts typical of the soil solution are not easily obtained
4. The amounts and chemical forms of any given micronutrient in the soil solution varies with time and may be affected by the method of sample preparation, including drying and storage of the soil

A common method of determining the concentrations of trace elements in aqueous solution is by atomic absorption spectrometry. However, this technique does not differentiate between free (M^{2+}) and bound forms of the trace element.

Organically bound forms have been determined in the following ways:

1. Addition of a chelating agent to the soil extract that forms a complex with the free form of the trace element and that is subsequently removed with an immiscible solvent. Organically bound forms of the trace element are taken as the difference between the total amount in solution and that recovered as the free cation. A limitation of this method is that an account is not given for any inorganically bound species that might be present (e.g., chloro complexes)
2. Separation of charged species on a cation-exchange resin column; organically complexed forms of the trace element are not retained. Inorganic complexes also pass through the resin but their amounts can be calculated from thermodynamic data

3. Recovery of the metal–organic complex through dialysis. This method is only applicable when the organic material is of high molecular weight
4. Direct analysis of the free cation with an ion-selective electrode (ISE) or by anodic stripping voltammetry (ASV)

Increasing use has been made of ISE and ASV for determining the speciation of trace elements in the soil solution, as well as natural waters (approach 4). A major limitation of ISE is its low sensitivity; furthermore, only a few divalent cations can be measured (Cu^{2+}, Pb^{2+}, Cd^{2+}, Ca^{2+}). The technique has the greatest potential where high concentrations of the cation of interest would be expected in the soil solution, such as in sludge-amended soils. For both ISE and ASV, electrode response is affected by pH, ionic strength, and sorption of organics on the electrode surface.[23-25]

By calibration of the Cu(II) ISE with a known chelating agent (malonic acid), Sanders[26] was able to determine rather low concentrations of Cu^{2+} ($<10^{-6}$ M) in displaced soil solutions. In contrast to results obtained for Mn and Zn, where the ion-exchange equilibrium method was used,[27] essentially all of the soluble Cu occurred in organically bound forms. In other work, a method involving reduction in F^- content through complexation with Al^{3+} (as determined with a F^- ISE) has been used for the determination of organically bound Al in the soil solution.[28]

Advantages of ASV for determining the concentration of free metal cations are:

1. The method is highly sensitive and is capable of measuring free forms of most metal ions at the low levels normally found in soil extracts and natural waters. Small volumes of solution can be analyzed (i.e., ~2 mL)
2. Only free and labile species of the metal are measured. Some information about chemical forms may be obtained from potential shifts and analysis of the voltammetric waveshape obtained

Oxidation of natural waters by ultraviolet irradiation in the presence of H_2O_2 generally increases the concentration of metal ions detectable by ASV and the amount released has been designated organically bound.[29-31] Application of this approach has shown that substantial amounts of the Cu, and lesser amounts of the Cd and Zn, occur in river and reservoir water in organically bound forms. Little use has been made of the method for analysis of soil solutions.

Speciation of metal ions in the soil solution has been predicted on the basis of analytical data for cations, anions, and soluble organic matter, for which computer models (e.g., GEOCHEM, MINEQ, etc.) have been applied.[32,33] The approach is only applicable to systems well characterized with regard to cationic (e.g., K^+, Na^+, NH_4^+, H^+, Ca^{2+}, Mg^{2+}, Cu^{2+}, etc.) and anionic (Cl^-, HCO_3^-, NO_3^-, SO_4^{2-}, etc.) species, and, ideally, for the kinds and amounts of organic ligands.

Since the condition of electrochemical neutrality must be maintained in the soil solution, the total quantity of cations in ionic forms must equal total anionic content (inorganic + dissociated acidic functional groups of humic substances). Cronan et al.[34] estimated organically bound forms of trace elements in the leachates of some New Hampshire subalpine forests from the deficit between total cations and inorganic anions.

Results using approach 1 (removal of the free cation with a chelating agent) have given the following values for the percentage of three trace elements in displaced soil solutions that occurred in organically complexed forms: Cu, 98–99 percent, Mn, 84–99 percent, Zn, greater than 75 percent.[35,36] These values may be high in that account was not given to any inorganic complexes that might have been present.

Values obtained by Sanders[26] for complexed forms of Co, Mn, and Zn in aqueous extracts of some English soils (an ion-exchange equilibrium technique was used) were considerably lower than those noted above. Camerynck and Kiekens[37] used a combination of cation- and anion-exchange resins to determine speciation of select metals in the water-soluble fraction of a sandy soil. Copper and Fe were largely present as stable complexes, Mn was largely in the free ionic form, and Zn was evenly distributed between the two. Other work[38] has shown that the affinity of ligands in aqueous solutions of sludge amended soils was approximately $10^{5.5}$ times greater for Zn and Cd than for Ca. Research on dissolved Al in forest soils indicate that Al–organic forms dominate in the forest floor but decline in importance with increasing soil depth.[39]

The content of dissolved organic matter in soils varies both in content and composition; thus, the percentage of trace elements in organically bound forms are highly variable. Other factors affecting speciation include pH and types of competing inorganic ligands. As was the case for river and reservoir water, the work done thus far indicates that those trace elements capable of forming strong chelate complexes with organic matter (e.g., Cu^{2+}) occur largely in organically complexed forms; those that form weak complexes (e.g., Zn^{2+}) occur mostly as free ionic or inorganic complexed forms.

BIOCHEMICAL COMPOUNDS AS CHELATING AGENTS

The organic compounds in soil that form complexes with trace elements and other polyvalent cations are of two major types, as follows:

1. Defined biochemicals synthesized by living organisms, such as simple aliphatic acids, amino acids, hydroxamate siderophores, phenols and phenolic acids, and complex polymeric phenols
2. A series of acidic, yellow- to black-colored polyelectrolytes formed by secondary synthesis reactions and referred to as humic and fulvic acids (see Chapters 8–10)

In general, biochemical compounds form complexes that are water soluble; complexes with humic substances (particularly humic acids) are, for the most part, insoluble. Metal complexes with humic substances are discussed in the next section.

Biochemicals having the ability to chelate metal ions are produced periodically in soil through microbial activity; others are found in root excretions, in the rhizosphere, and in leachates of decomposing plant residues, including leaf litter of the forest floor. Key references for this work can be found in several reviews.[4-8] An incomplete list of biochemical chelating compounds, together with a description of the environments under which they occur, is presented in Table 16.2.

A major source of soluble organics in forest soils (Alfisols, Spodosols) are residues that reach the soil in the form of leaves, branches, and other organic debris. During the course of decay by microorganisms, a wide array of biochemical chelating agents are synthesized and washed into the mineral layer of the soil (A horizon) in percolating waters. Organic chelating substances are also found in canopy drippings and stem flow.

The concentrations of individual biochemical species in the soil solution (20 percent moisture level) are approximately as follows.

Simple organic acids	$1 \times 10^{-3} - 4 \times 10^{-3}$ M
Amino acids	$8 \times 10^{-5} - 6 \times 10^{-4}$ M
Phenolic acids	$5 \times 10^{-5} - 3 \times 10^{-4}$ M
Hydroxamate siderophores	$1 \times 10^{-8} - 1 \times 10^{-7}$ M

Table 16.2 Key Biochemical Compounds that Form Complexes with Micronutrient Cations

Compound	Occurrence
Citric, tartaric, lactic, and malic acids	Produced by bacteria in the rhizosphere and during decay of plant remains. Identified in root exudates, aqueous extracts of forest litter, and canopy drippings
Oxalic acid	Produced by fungi in forest soils, including mycorrhizal fungi. Particularly abundant in acid soils
Hydroxamate siderophores	Produced in the rhizosphere and by extomycorrhizal fungi. Greater amounts may be produced when organisms are under Fe stress
Phenolic acids	Formed through decay of plan residues (lignin). Abundant in canopy drippings and leachates of forest litter. Involved in the mobilization and transport of Fe in acid soils
Polymeric phenols	Present in high amounts in leachates of forest litter. Produced by lichens growing on rock surfaces
2-Ketogluconic acid	Synthesized by bacteria living on rock surfaces and in the rhizosphere. Particularly abundant in habitats rich in decaying organic matter

In most cultivated soils, binding will be carried out by a relatively large number of ligands present in small amounts rather than by a few dominate species present at high concentrations. Thus, while the concentration of any given biochemical or class of biochemicals in the soil solution may be slight, the accumulating effect of all complexing species may be appreciable. In many soils, the combined total of potential chelate formers in the aqueous phase is probably sufficient to account for the minute quantities of metal ions normally present, often in the $10^{-9} - 10^{-8}$ M range.

Factors affecting the production of biochemical compounds include the moisture status of the soil, plant type and stage of growth, cultural practice, and climate (temperature and rainfall). Relatively high amounts would be expected in early spring as decomposition of plant residues commences, to decrease during the hot summer months due to reduced microbial activity, and to increase once again in early fall when plant growth ceases and crop residues start to decay. Soils amended with manures and other organic wastes would be expected to be relatively rich in metal-binding biochemicals.

As noted earlier, binding is influenced by the nature of the organic ligand. Organic compounds having the greatest potential for binding Fe^{3+} are those that contain oxygen (i.e., COOH and phenolic-, enolic-, and aliphatic-OH groups); nitrogen-containing substances (amino acids, porphyrins) have a high affinity for Cu^{2+} and Ni^{2+}.[11]

Typical Fe^{3+} complexes with biochemical compounds are shown below:

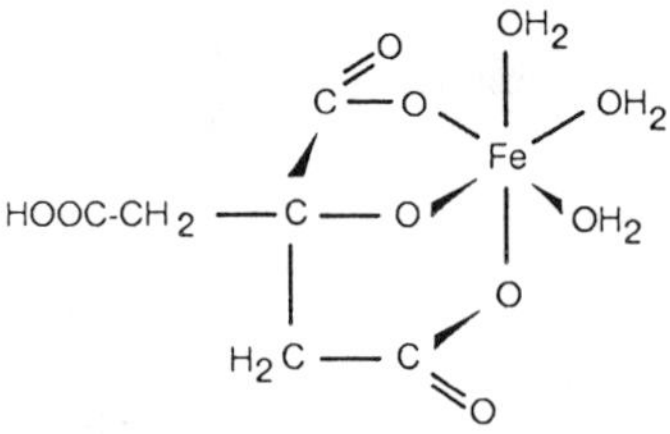

Citrate
VI

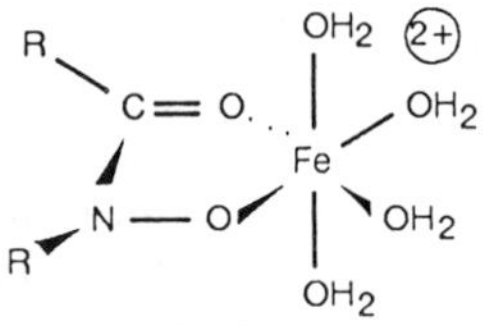

Hydroxamate
VII

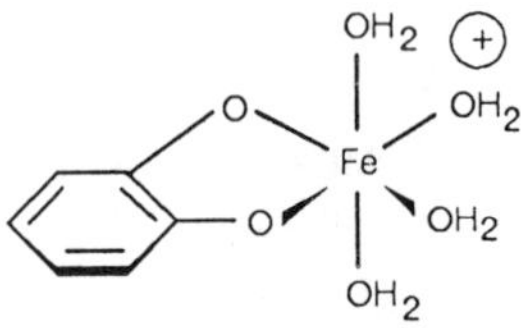

Catecholate
VIII

Although discussed separately, the relative importance of any given biochemical compound or ligand type in metal-ion binding will vary with time and environmental conditions, including the nature of the organic ligand being synthesized and the kind and amount of competing cations.

Organic Acids

Organic acids are of special interest in chelation because of their ubiquitous nature and because hydroxy derivatives, such as citric acid, are effective solubilizers of mineral matter. Data provided in Chapter 7 show that a variety of simple organic acids are secreted from plant roots and that soil is a particularly favorable habitat for organic acid-producing bacteria.

The concentration of organic acids in the soil solution is normally low (1 × 10^{-3} to 4 × 10^{-3} M) but higher amounts occur in the rhizosphere and in leachates from the forest canopy and organic layer of forest soils.

Considerable emphasis has recently been given to the importance of oxalic acid as a chelator of Fe^{3+} (as well as Al^{3+} and Ca^{2+}) in forest soils.[40] Many fungi are prolific producers of oxalic acid, including the vesicular arbuscular mycorrhizal fungi, where calcium oxalate crystals can form at the soil–hyphae interface.[41,42]

Amino Acids

A wide array of free amino acids have been reported in soils (see Chapter 3). Under optimum conditions for microbial activity, their concentration in the soil solution is of the order of 8 × 10^{-5} to 6 × 10^{-4} M. *A priori*, it would appear that amino acids play a role secondary to organic acids and other biochemical compounds as chelating agents in soil.

Hydroxamate Siderophores

Although relatively abundant in soil (to 6 percent), Fe is often unavailable to plants because of its low solubility under near neutral and alkaline conditions. The availability of Fe in calcareous soils is thought to be highly dependent on organic chelating agents, such as those secreted from plant roots or produced in the rhizosphere by microorganisms.

Current evidence indicates that hydroxamate siderophores play an important role in the Fe nutrition of plants growing on calcareous soils. These substances, which contain the reactive group —CO—NOH—, represent a class of microbially synthesized Fe^{3+} transport moieties with exceptionally high stability constants ($\sim 10^{32}$). Greater amounts appear to be produced when the organism is under severe Fe stress. A chelate structure of Fe^{3+} with hydroxamate was shown earlier in structure (**VII**).

Biologically significant levels of hydroxamate siderophores have been observed in soil (10^{-8} to 10^{-7} M). The amounts contained in the rhizosphere of plants appear to be 10 to 50 percent higher than in the bulk soil.[43] Hydroxamate siderophores are also synthesized by ectomycorrhizal fungi, which live in intimate association with plant roots.

Phenols and Phenolic Acids

Phenols and phenolic acids are formed during decay of lignin, and they are synthesized by such microscopic fungi as *Stachybotrys atra*, *S. chartarum*, and *Epicoccum nigrum* when grown on nonlignin carbon sources (see Chapter 8). Phenolic constituents are abundant in forest canopy leachates, in decomposing plant and animal remains, and in root exudates. The concentration of phenolic acids in the solution phase of agricultural soils has been estimated at 5 × 10^{-5} to 3 × 10^{-4} M.

Phenols and phenolic acids are believed to be of considerable importance in the complexation and translocation of Fe and Al in forest soils. High concentrations of phenolics have been observed in aqueous extracts of decomposing organic matter on the forest floor (O_1 and O_2 horizons), as well as in canopy leachates. The phenolic composition of forest soils depends partly on tree species.[44]

Of the phenolic acids, those containing two (or more) OH groups on adjacent positions of the aromatic ring are particularly effective in forming complexes with metal ions. Typical examples include protocatechuic **(IX)**, gallic **(X)**, and caffeic **(XI)** acids.

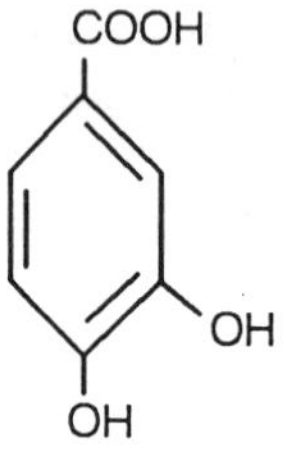

Protocatechuic
IX

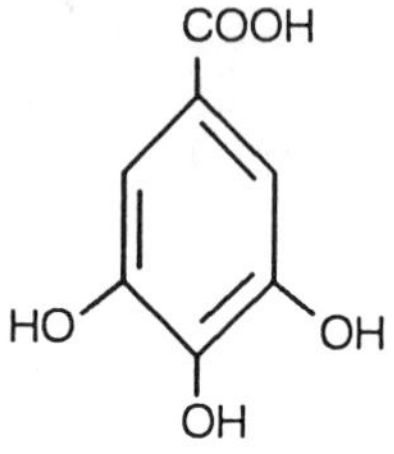

Gallic
X

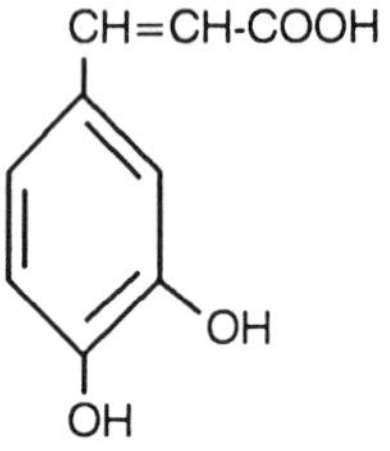

Caffeic
XI

Polymeric Phenols

Polymeric phenols refer to those substances that contain more than one aromatic ring and the phenolic OH group. They include the flavonoids, which comprise one of the largest and most widespread groups of secondary plant products. Structures for some common flavonoids are as follows:

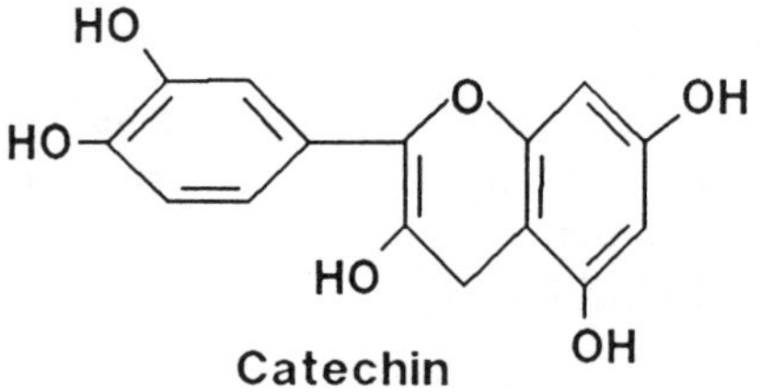

Catechin
XII

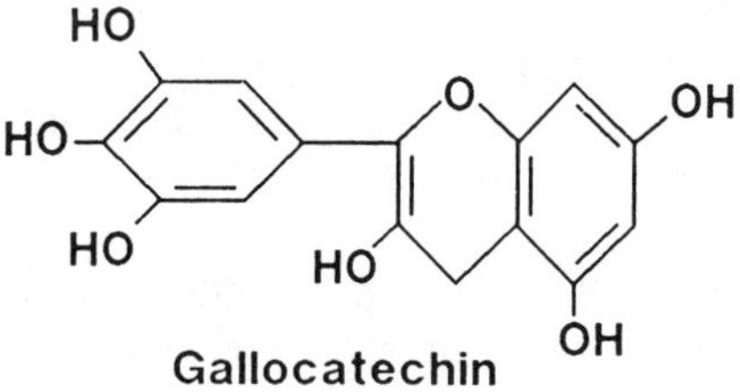

Gallocatechin
XIII

Flavonoids of the types shown above have been found in aqueous extracts of the leaves and needles of forest trees.

Another group of potential chelate formers, particularly in forest soils, are the tannins. They represent an ill-defined group of substances with molecular weights between 500 to 3000 and that contains at least one or two phenolic OH groups per 100 molecular weight. They are of two main types: hydrolyzable and condensed. Hydrolyzable tannins consist of gallic and/or hexahydroxydiphenic acids (digallic acid) bound to a sugar moiety through a glycosidic linkage. They have numerous phenolic OH groups to which Fe^{3+} and other polyvalent cations can be bound. Condensed tannins have a flavonoid nature in that they contain catechin and gallocatechin as primary constituents.

The ability of lichens to dissolve mineral substances during the weathering of rocks and minerals is well known. These organisms synthesize a variety of complex polymeric phenols (often referred to as "lichen acids"), that form highly stable complexes with metal ions. Typical examples are shown by structures **XIV** and **XV**.

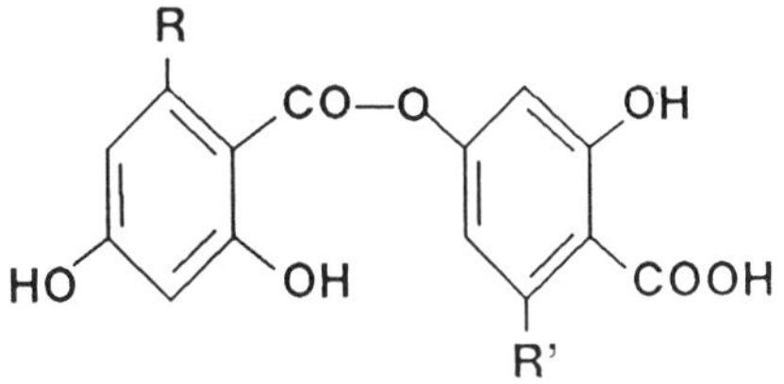

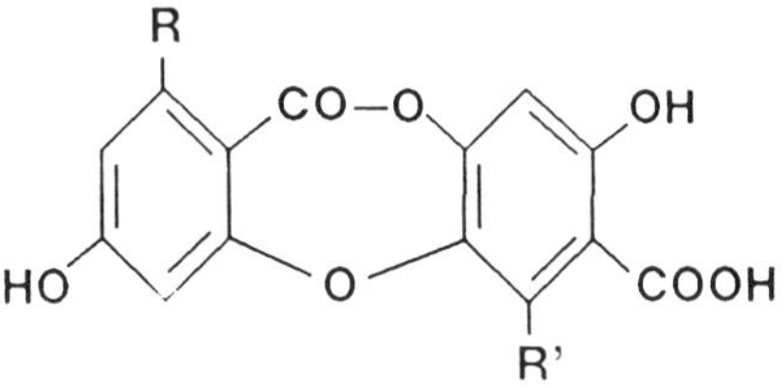

Geographically, lichens are widely distributed in nature. They are often the initial colonizers of virgin landscapes, where their activity leads to rock weathering and mobilization of micronutrient cations.

Sugar Acids

Sugar acids may also be important natural chelators in soils. Gluconic, glucuronic, and galacturonic acids are common metabolites of microorganisms. Habitats rich in organic matter have been found to contain large numbers of microorganisms that synthesize 2-ketogluconic acid; this compound was shown to make up over 25 percent of the organic acids in the rhizosphere. A high proportion of the bacteria in soil, as well those living on rock surfaces, produced 2-ketogluconic acid. For a discussion of this work, the reader is referred to references 5–7.

Miscellaneous Compounds

Other chelating compounds occurring naturally in soil, albeit in minute amounts, include organic phosphates, phytic acid, chlorophyll, chlorophyll-degradation products (porphyrins), simple sugars (formation of borate complexes), and auxins. The significance of these constituents in complexing trace elements in soil is unknown.

Proteinaceous substances and polysaccharides are also capable of forming complexes with trace elements. As much as 30 percent of the soil–organic matter occurs as saccharides, only a small portion of which can be accounted for as polysaccharides. Polysaccharides extracted from soil usually contain complexed Fe, Al, and Si. Evidence for complexing of Al by soil polysaccharides has been obtained by Saini.[45]

The bonding of polysaccharides to clay minerals through an Al linkage has been suggested as a mechanism of the formation of stable aggregates in soil (see Chapter 16).

TRACE METAL INTERACTIONS WITH HUMIC SUBSTANCES

As noted in Chapter 2, the bulk of the organic matter in most soils consists of a complex mixture of substances referred to by such names as humic and fulvic acids. Their abilities to form complexes with trace elements is due to their unusually high content of oxygen-containing functional groups, which include COOH, phenolic-, enolic-, and alcoholic OH, and C=O. Amino and imino groups may also be involved.

The humate structures of Chapter 10 present a variety of potential sites for complexation, as depicted by structures XVI to XXI.

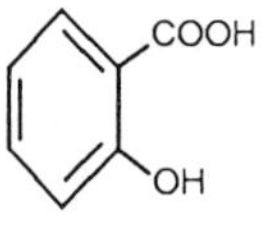

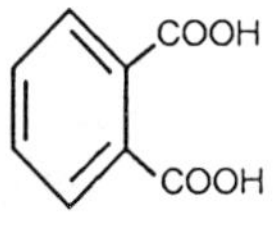

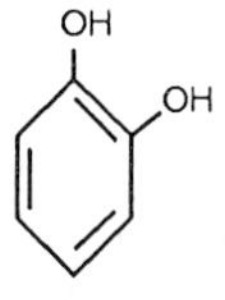

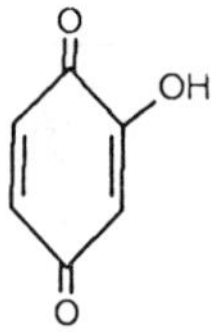

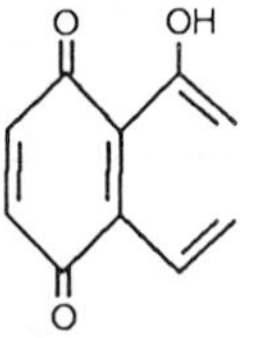

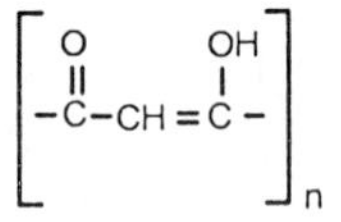

Due to the heterogeneous nature of humic substances, complexation of trace elements can be regarded as occurring at a large number (continuum) of reactive sites with binding affinities that range from weak forces of attraction (i.e., ionic) to formation of highly stable coordinate linkages. Binding of Cu^{2+}, for example, could occur through: (1) a water bridge (XXII), (2) electrostatic attraction to a charged COO^- group (XXIII), (3) formation of a coordinate linkages with a single donor group (XXIV), and (4) formation of a chelate (ring) structure, such as with a COO^--phenolic OH site combination (XXV).

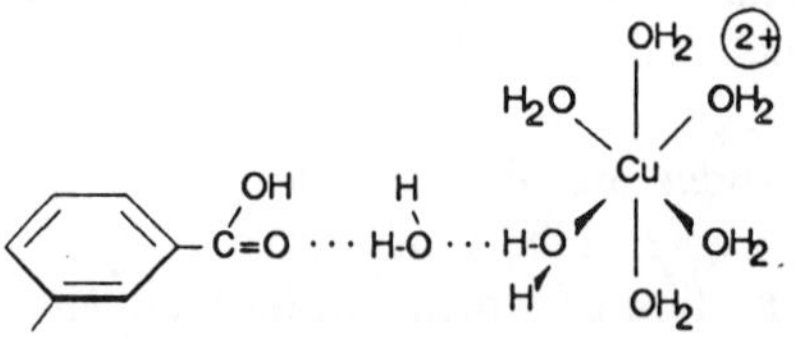

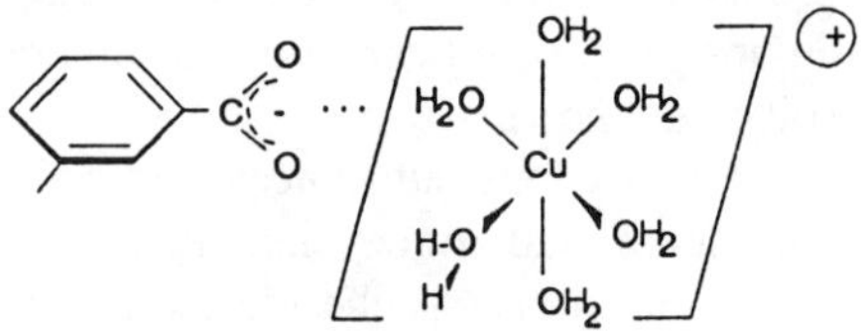

XXIV   XXV

Binding of a trace element would occur first at those sites that form the strongest complexes (i.e., formation of coordinate linkages and ring structures). Thus, structures of types **XXIV** and **XXV** represent the predominant forms of complexed trace elements when humic substances are present in abundance. Binding at the weaker sites (**XXII** and **XXIII**) become increasingly important as the stronger sites become saturated. Many investigators have emphasized the formation of chelate rings (see **XXV**), but, as shown later, they cannot be considered to be the sole, or even most prominent, structural unit of complexation. Indirect evidence for the formation of highly stable complexes has come from experimental difficulties in obtaining metal-free humic acids from soil.

Methods of Study

Diverse methods have been used to examine the ways in which trace elements interact with humic substances, some of which are outlined below.

Potentiometric Titration The formation of complexes between polyvalent cations and humic substances leads to release of protons (H^+) from COOH and other acidic groups that participate in complexation. Titration methods have been used primarily in studies designed for determining stability constants and are discussed in detail in Chapter 17.

Infrared Spectroscopy (IR) Results of IR studies have confirmed that COOH groups play a prominent role in the complexation of di- and trivalent cations by humic substances. The basis for the analysis is that the C=O absorption band for COOH at 1720 cm^{-1} disappears upon reaction with metal ions (see Chapter 13) and new bands for asymmetric and symmetric stretching vibrations of the carboxylic ion (COO$^-$) appear near 1600 and 1380 cm^{-1}, respectively.

The position of the asymmetric COO^{-1} stretching band near 1600 cm^{-1} provides an indication as to whether the linkages are ionic or covalent. For ionic bonds, absorption occurs in the 1630–1575 cm^{-1} region; when coordinate linkages are formed, absorption shifts to higher frequencies (i.e., to between 1650 and 1620 cm^{-1}). Results obtained thus far for metal complexes of humic substances have been inconclusive in that frequency shifts have been variable and slight, a result that may be due to the formation of mixed complexes. For Cu^{2+}, strong covalent linkages are apparently formed at low levels of metal-ion additions, but bonding becomes increasingly ionic as the humate becomes saturated with the metal.[46]

Some IR evidence indicates that OH, C=O, and NH groups are also involved in the complexing of metal ions by humic substances.[46-48] Conjugated ketonic structures may participate in metal-ion binding, according to the following reaction.[46]

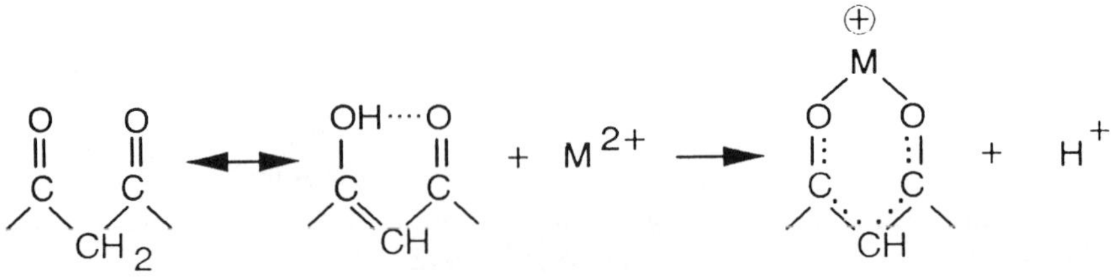

$$\underset{CH_2}{\overset{O}{\underset{\|}{C}}\overset{O}{\underset{\|}{C}}} \longleftrightarrow \underset{CH}{\overset{OH\cdots O}{\underset{|}{C}}\overset{}{\underset{\|}{C}}} + M^{2+} \longrightarrow \underset{CH}{\overset{\overset{\oplus}{M}}{\overset{O}{\underset{}{C}}\overset{O}{\underset{}{C}}}} + H^+$$

Electron Spin Resonance Spectroscopy (ESR) Substances that contain an odd (unpaired) electron are attracted to a magnetic field, and it is this property of certain transition metal ions, such as Cu^{2+}, that permit examination of their complexes with humic substances. A brief outline of ESR theory and methods is given in Chapter 15.

As was the case for IR, characterization of the binding of Cu^{2+} by humic substances by ESR has given somewhat variable results. Lakatos et al.[49] reported that binding of Cu^{2+} by humic acid occurred through a nitrogen donor atom and two carboxylates (COO^-). On the other hand, McBride[50] concluded that only oxygen donors (i.e., COO^-) were involved; furthermore, a single bond was observed. In contrast, spectral data obtained by Boyd et al.[51,52] were consistent with the formation of Cu-chelate complexes with two oxygen donor groups (see **XXV**). McBride[53] concluded that only a small fraction of the acidic groups of fulvic acid were involved in the formation of "inner-sphere" complexes with Mn^{2+}.

Electron spin resonance spectra of improved resolution were obtained by Senesi et al.[54] using a procedure that involved removal of excess Cu^{2+} by a cation-exchange resin; the spectra gave evidence for three different coordination environments for bound Cu^{2+}. Additional ESR evidence suggests that small quantities of Cu^{2+} can be bound to humic acids through a porphyrin-type linkage.[55,56]

Fluorescence Spectroscopy Fluorescence of organic ligands is often quenched when complexation occurs; thus, differentiation between free from bound sites of the ligand can be made. Because of its high sensitivity, fluorescence spectroscopy provides data for binding under conditions typical of the natural environment (e.g., low organic matter concentrations in the soil solution). The method has been used to separate solution-phase complexation from solid-phase adsorption.[57]

Ultraviolet Spectroscopy (UV) Limited application has been made of UV spectroscopy in studies of metal-ion binding by humic substances, due undoubtedly to the fact that UV spectra of humic and fulvic acid are broad and rather featureless (see Chapter 13). The technique was used by Alberts and Dickson[58] to study Al binding by a humic acid, tannic acid, and a lignosulfonic acid; only the spectrum of the tannic acid showed prominent shifts that could be attributed to Al binding. However, a small plateau near 270 nm was introduced in the spectrum of the humic acid.

Thermogravimetric (TG) Analysis In TG analysis, the mass of the substance being analyzed is monitored as the temperature is increased. Two approaches are used, namely, differential thermogravimetry (DTG), in which the rate of weight loss is determined, and differential thermal analysis (DTA), where the difference in temperature between a sample and reference is measured. Both approaches have been applied to metal-ion complexes of humic substances.[59-62]

For soil fulvic acids, the main decomposition reaction determined by DTG occurs at 420°C.[59] When molar 1:1 Al–fulvic acid complexes (number-average molecular weight of the fulvic acid was 670 daltons) were analyzed, DTG curves were similar to the original fulvic acid. However, upon increasing the molar ratio to 3:1 and 6:1, rather broad peaks appeared at 350 to 450°C. A weaker band at 50 to 100°C increased in size when the amount of Al in the complexes increased. Since the pH was maintained at 4.0 to minimize the formation of Al-hydroxide, this last peak may be due to increased water retention. In a study using DTA, Tan[62] found that the thermal stability of humic acid was increased through the binding of Al. Contrary to this, trivalent cations have been reported to cause lower thermal stability of humic and fulvic acids due to increased strain within the molecule when complexed.[60,61]

Differential Pulse Polarography Analysis of a soil humic acid and its Al complex by differential pulse polarography (DPP) showed that, whereas humic acid produced a reduction peak at pH 3.4 or less, the complex was electroinactive.[63] Stable complexes were formed at pH 4.0, with the amount of free Al in equilibrium with the complex being rather low (<0.1 percent of the total Al). These results confirm that humic acids have a high affinity for Al, and that they serve as carriers of Al in the soil solution, as well as natural waters.

Molecular Weight Distribution Characterization of metal–humate complexes based on molecular weight distribution has provided information on major binding components. By Sephadex gel chromatography, Kribek et al.[64] observed that a reduction in the low-molecular-weight components was accompanied by an increase in high-molecular-weight components upon Al complexation. Davis and Gloor[65] used gel chromatography to separate the humic substances of a Swiss lake into three molecular-weight-size classes. Humic material in the intermediate range (1000–3000 molecular weight) was the most effective in reacting with suspended Al particles. In the study of Kribek et al.,[64] acid hydrolysis of the high-molecular-weight Al complexes failed to produce any low-molecular-weight material upon further analysis by gel chromatography. One explanation of these results is that the low-molecular-weight substances participated in Al complexation to a greater extent than the high molecular weight substances, and that the Al complexes were stable towards acid degradation.

Solubility Characteristics of Metal–Humate Complexes

Humic substances form both soluble and insoluble complexes with trace elements, depending on pH, presence of salt (ionic strength effect), and degree of saturation of binding sites. In the natural soil, the complexes are largely insoluble, due in part to interactions with clay minerals. Because of their lower molecular weights and higher contents of acidic functional groups, metal complexes of fulvic acid are less susceptible to precipitation than humic acid.

When humic and fulvic acids are dissolved in water, dissociation of acidic functional groups occurs ($R-COOH \rightleftharpoons R-COO^- + H^+$) and the molecule assumes a stretched configuration due to repulsion of charged groups. Upon addition of metal ions, the charge is reduced through salt formation and the molecule collapses, thereby reducing solubility. Polyvalent cations also have the potential for linking individual molecules together to produce chainlike structures. Stevenson[66] concluded that metal complexes of humic acid are soluble at low metal–humic acid ratios (few combined molecules in the chain), but precipitation occurs as the chain-like structure grows and the isolated COOH groups become neutralized through salt bridges. The point at which visible precipitation occurs will be influenced by such factors as ionic strength, pH, humic acid concentration, and type of metal cation.

Mechanisms that affect the solubility of humic substances in the presence of polyvalent cations are as follows:

1. Precipitation due to protonation, with reduction of charge on the humic polymer (i.e., molecule becomes more hydrophobic)
2. Formation of hydroxy complexes of the metal ion at high pH values
3. Formation of chain-like structures through metal-ion bridges, as shown below by structure **XXVI**

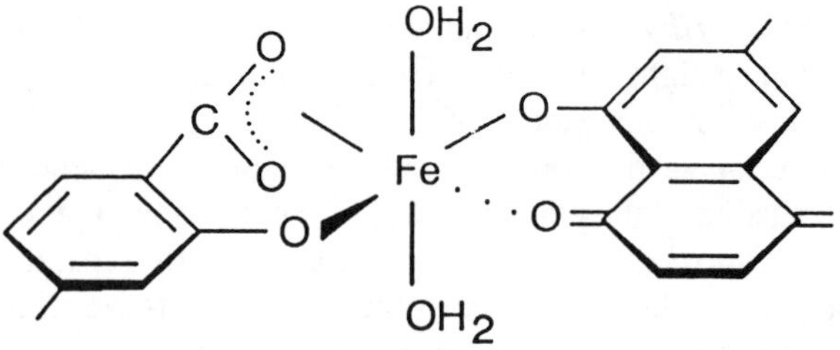

XXVI

4. Attachment to clay particles and oxide surfaces, such as through metal-ion linkages. The bulk of the organic matter in most soils is bound to clay minerals, probably through linkages with Fe, Al, and other polyvalent cations (see structure **VI** of Chapter 12).

Reduction of the net charge on the humate molecule occurs through reaction of the COO^- group with polyvalent cations. This, in turn, causes the "stretched" configuration of the humic molecule to collapse, thereby reducing solubility. As chain-like structures are formed, and as oxygen-containing functional groups become neutralized, precipitation increases.

Immobilization of trace elements by interaction with humic substances can occur through either the formation of insoluble complexes or through solid-phase complexation to humates present as a coating on clay surfaces. Sorption is possible through direct exchange at the clay–organic interface or through the formation of soluble complexes which subsequently become associated with mineral surfaces through adsorption. Some cations link humic complexes to clay surfaces; others occupy peripheral sites and are available for exchange with ligands of the soil solution.

Metal-Ion Binding Capacities

Approaches used to determine the binding capacities of humic substances for metal ions include proton release,[66-68] metal-ion retention as determined by competition with a cation-exchange resin,[69] dialysis,[70] and ASV or ISE measurements.[24,25,71] Other approaches involve the determination of binding sites that are occupied using the techniques of ultraviolet (UV) and fluorescence spectroscopy.[72,73]

The maximum binding capacity of humic substances for any given metal ion is approximately equal to the content of acidic functional groups, primarily COOH. An exchange acidity 500 $cmol_c$/kg humic acid corresponds to a retention of about 160 mg of Cu^{2+}/g. Assuming a C content of 56 percent for humic acid, one Cu^{2+} atom is bound per 60 C atoms in the saturated complex.

Factors influencing the quantity of metal ions bound by humic substances include pH, kind and amount of acidic functional groups, ionic strength, and molecular weight. For any given pH and ionic strength, trivalent cations are bound in greater amounts than divalent ones; for the latter, those forming the strongest coordination complexes (e.g., Cu^{2+}) are bound to a greater extent (and at the stronger binding sites) than those that form weak complexes (e.g., Mn^{2+} and Zn^{2+}).

Effect of pH on Metal-Ion Complexation

The effect of pH on metal-ion complexation results from: 1) changes in extent of ionization of COOH groups, and 2) hydrolysis reactions involving the formation of monomeric species and polymers of the metal ions. For most trace elements, an increase in pH above about 4.5 leads to hydrolysis of the metal ion, with formation of oxide hydrates.

Humic and fulvic acids act as weak-acid polyelectrolytes in which ionization of COOH groups is controlled by pH, thereby affecting their ability to bind metal ions. Binding can be influenced in other ways, such as through conformational changes in the macromolecule.

Influence of Electrolytes

The binding of metal ions by humic substances is influenced by electrolytes (i.e., salts) in two ways. First, activity coefficients of charged inorganic species are dependent on ionic composition of the solution. At the same ionic strength, the activities of trivalent cations are reduced more than with divalent cations, which, in turn, are reduced to a greater extent than monovalent cations. For solutions with ionic strengths between 0.001 and 0.1, the physical size of the ion must also be taken into consideration.

A second effect is due to competition of cations for binding sites on the ligand. For macromolecules, there exist the potential for a variety of configurational arrangements based on the type and nature of the interacting cation. As noted in Chapter 12, humic substances behave like rigid "spherocolloids" at high concentrations, at low pH, or at high concentrations of electrolyte. On the other hand, they behave like "flexible linear" colloids at low humic concentrations, neutral pH, or at low ionic strength.

The following effects must be taken into account when considering competing cations[74]:

1. Changes in activity coefficients of inorganic ions in the reaction mixture
2. Configurational changes in the macromolecule with changes in pH or ionic strength
3. Counterion condensation in the diffuse double layer of the macromolecule

Relative Importance of Organic Matter and Clay in Retention of Applied Trace Elements

Clay and organic colloids are major soil components involved in retention of applied trace elements. However, individual effects are not as easily ascertained as might be supposed, for the reason that, in most mineral soils, organic matter is intimately bound to the clay, probably as a clay–metal–organic matter complex. Accordingly, clay and organic matter function more as a unit than as separate entities and the relative contribution of organic and inorganic surfaces to adsorption will depend on the extent to which the clay is coated with organic substances. The amount of organic matter required to coat the clay varies from one soil to another and depends on both kind and amount of clay. For soils with similar clay and organic matter contents, the contribution of organic matter to the binding of trace elements is highest when the predominant clay mineral is kaolinite and lowest when montmorillonite is the main clay mineral.

Reduction Properties

Humic substances may catalyze the reduction of Fe^{3+} to Fe^{2+} and anionic MoO_4^{2-} to Mo^{5+}.[49,75,76] Reduction of ionic species is of considerable importance because the solubility characteristics of the metal ions are modified, and hence availabilities to plants and microorganisms. Electron spin resonance (ESR) spectroscopy, in conjunction with Mössbauer spectroscopy, has been used to obtain information on oxidation states and site symmetries of Fe bound by humic and fulvic acids.[77]

SUMMARY

Renewed attention has been given in recent years to the complexation of micronutrient cations (Cu^{2+}, Zn^{2+}, Mn^{2+}, Fe^{3+}, others) and toxic heavy metals (Pb^{2+}, Cd^{2+}) by organic constituents in soil, including: 1) biochemical compounds as chelating agents, 2) mechanisms of metal-ion binding by humic substances, 3) chemical speciation of trace elements in the soil solution, 4) chelation reactions in the rhizosphere, and 5) influence of organic matter in ameliorating Al^{3+} toxicities in acid soils. Natural complexing agents are of considerable importance in the weathering of rocks and minerals, and they are involved in the movement of sesquioxides into the subsoil.

Organic constituents form both soluble and insoluble complexes with metal ions and thereby play a dual role in soil. Low-molecular-weight compounds (biochemicals, fulvic acids) serve as carriers of trace elements in the soil solution. In contrast, high-molecular-weight compounds (e.g., humic acids) function as a "sink" for polyvalent cations. Although estimates vary, the results obtained thus far indicate that those trace elements in soil that form strong complexes (e.g., Cu^{2+}) occur mostly in organically complexed forms; those that form weak complexes (e.g., Zn^{2+}) occur mostly in free forms or as inorganic complexes.

REFERENCES

1. P. R. Bloom, "Metal–Organic Matter Interactions in Soil," in R. H. Dowdy et al., Eds., *Chemistry of the Soil Environment*, Special Publication 40, American Society of Agronomy, Madison, 1981, pp. 129–150.
2. Y. Chen and F. J. Stevenson, "Soil Organic Matter Interactions with Trace Elements," in Y. Chen and Y. Avnimelech, Eds., *The Role of Organic Matter in Modern Agriculture*, Martinus Nijhoff Publishers, Dordrecht, 1986, p. 73–116.
3. G. Sposito and A. L. Page, "Cycling of Metal Ions in the Soil Environment," in

H. Sigel, Ed., *Metal Ions in Biological Systems: Vol. 8. Circulation of Metals in the Environment*, Marcel Dekker, New York, 1984, pp. 287–322.

4. F. J. Stevenson, *Cycles of Soil: Carbon, Nitrogen, Phosphorus, Sulfur, Micronutrients*, Wiley, New York, 1986.

5. F. J. Stevenson, "Organic Matter–Micronutrient Reactions in Soil," in J. J. Mortvedt, F. R. Fox, L. M. Shuman, and R. M. Welch, Eds., *Micronutrients in Agriculture*, American Society of Agronomy, Madison, 1991, pp. 145–186.

6. F. J. Stevenson and M. S. Ardakani, "Organic Matter Reactions Involving Micronutrients in Soils," in J. J. Mortvedt, P. M. Giordano, and W. L. Lindsay, Eds., *Micronutrients in Agriculture*, American Society of Agronomy, Madison, 1972, pp. 79–114.

7. F. J. Stevenson and A. Fitch, "Chemistry of Complexation of Metal Ions with Soil Solution Organics," in P. M. Huang and M. Schnitzer, Eds., *Interactions of Soil Minerals with Natural Organics and Microbes*, Special Publication 17, American Society of Agronomy, Madison, 1986, pp. 29–58.

8. F. J. Stevenson and G. F. Vance, "Naturally Occurring Aluminum–Organic Complexes," in G. Sposito, Ed., *The Environmental Chemistry of Aluminum*, CRC Press, Boca Raton, Florida, 1991, pp. 117–145.

9. S. Chaberek and A. E. Martel, *Organic Sequestering Agents*, Wiley, New York, 1959.

10. H. M. N. H. Irving and R. J. P. Williams, *Nature*, **162**, 746, 1058.

11. S. Arland, J. Chatt, and N. R. Davies, *Chem. Soc. London. Quart. Rev.*, **12**, 265 (1958).

12. W. L. Hargrove and G. W. Thomas, "Effect of Organic Matter on Exchangeable Aluminum and Plant Growth in Acid Soils," in R. H. Dowdy, Ed., *Chemistry of the Soil Environment*, Special Publication 40, American Society of Agronomy, Madison, 1981, pp. 151–166.

13. P. B. Hoyte and R. C. Turner, *Soil Sci.*, **119**, 227 (1975).

14. J. Ares, *Soil Sci.*, **142**, 13 (1986).

15. U. Yermiyaho, R. Keren, and Y. Chen, *Soil Sci. Soc. Amer. J.*, **52**, 1309 (1988).

16. H. Hendrichs and R. Mayer, *J. Environ. Qual.*, **9**, 111 (1980).

17. R. G. McLaren and D. V. Crawford, *J. Soil Sci.*, **24**, 172 (1973).

18. L. M. Shuman, *Soil Sci.*, **140**, 11 (1985).

19. W. P. Miller, W. W. McFee, and J. M. Kelly, *J. Environ. Qual.*, **12**, 279 (1983).

20. L. M. Shuman, *Soil Sci. Soc. Amer. J.*, **47**, 656 (1983).

21. L. M. Shuman, *Soil Sci.*, **127**, 10 (1979).

22. S. S. Iyengar, D. C. Martens, and W. P. Miller, *Soil Sci. Soc. Amer. J.*, **45**, 735 (1981).

23. W. T. Bresnahan, C. L. Grant, and J. H. Weber, *Anal. Chem.*, **50**, 1675 (1978).

24. F.-L. Greter, J. Buffle, and W. Haerdi, *J. Electroanal. Chem.*, **101**, 211 (1979).

25. G. A. Bhat, R. A. Saar, R. B. Smart, and J. H. Weber, *Anal. Chem.*, **53**, 2275 (1981).

26. J. R. Sanders, *J. Soil Sci.*, **33**, 679 (1982).

27. J. R. Sanders, *J. Soil Sci.*, **34**, 315 (1983).

28. S. C. Hodges, *Soil Sci. Soc. Amer. J.*, **51**, 57 (1987).

29 H. Blutstein and J. D. Smith, *Water Res.*, **12**, 119 (1978).
30 G. E. Batley and D. Gardner, *Estuarine Coastal Marine Sci.*, **7**, 59 (1975).
31 J. Gardiner and M. J. Stiff, *Water Res.*, **8**, 517 (1975).
32 B. Lighthart, J. Baham, and V. V. Volk, *J. Environ. Qual.*, **12**, 543 (1983).
33 A. C. M. Bourg and J. C. Vedy, *Geoderma*, **38**, 279 (1986).
34 C. S. Cronan, W. A. Reiners, R. C. Reynolds, Jr., and G. E. Lang, *Science*, **200**, 309 (1978).
35 J. F. Hodgson, H. R. Geering, and W. A. Norvell, *Soil Sci. Soc. Amer. Proc.*, **29**, 665 (1965).
36 J. F. Hodgson, W. L. Lindsay, and J. F. Trierweiler, *Soil Sci. Soc. Amer. Proc.*, **30**, 723 (1966).
37 R. Camerynck and L. Kiekens, *Plant Soil*, **68**, 331 (1982).
38 L. L. Hendrickson and R. B. Corey, *Soil Sci. Soc. Amer. J.*, **47**, 467 (1983).
39 S. I. Nilsson and B. Bergkvist, *Water, Air, and Soil Pollution*, **20**, 311 (1983).
40 T. R. Fox and N. B. Camerford, *Soil Sci. Soc. Amer. Proc.*, **54**, 1139 (1990).
41 K. Cromack, P. Sollins, W. C. Graustein, K. Speidel, A. W. Todd, G. Spycher, C. Y. Li, and R. L. Todd, *Soil Biol. Biochem.*, **11**, 463 (1979).
42 N. Malajczuk and M. Cromack, Jr., *New Phytol.*, **92**, 527 (1982).
43 P. E. Powell, P. J. Szaniszlo, G. R. Cline, and C. P. P. Reid, *J. Plant Nutr.*, **5**, 653 (1982).
44 A. I. Kuiters and C. A. J. Denneman, *Soil Biol. Biochem.*, **19**, 765 (1997).
45 G. R. Saini, *Current Sci.*, **10**, 259 (1966).
46 A. Piccolo and F. J. Stevenson, *Geoderma*, **27**, 195 (1981).
47 S. A. Boyd, L. E. Sommers, and D. W. Nelson, *Soil Sci. Soc. Amer. J.*, **43**, 893 (1979).
48 P. Vinkler, B. Lakatos, and J. Meisel, *Geoderma*, **15**, 231 (1976).
49 B. Lakatos, T. Tibai, and J. Meisel, *Geoderma*, **19**, 319 (1977).
50 M. B. McBride, *Soil Sci.*, **126**, 200 (1978).
51 S. A. Boyd, L. E. Sommers, and D. W. Nelson, *Soil Sci. Soc. Amer. J.*, **45**, 1241 (1981).
52 S. A. Boyd, L. E. Sommers, D. W. Nelson, and D. X. West, *Soil Sci. Soc. Amer. J.*, **45**, 745 (1981).
53 M. B. McBride, *Soil Sci. Soc. Amer. J.*, **46**, 1137 (1982).
54 N. Senesi, D. F. Bocian, and G. Sposito, *Soil Sci. Soc. Amer. J.*, **49**, 114 (1985).
55 A. L. Abdul-Halim, J. C. Evans, C. C. Rowlands, and J. H. Thomas, *Geochim. Cosmochim. Acta*, **45**, 481 (1981).
56 M. V. Cheshire, M. I. Berrow, B. A. Goodman, and C. M. Mundie, *Geochim. Cosmochim. Acta*, **41**, 1131 (1977).
57 D. S. Gamble, C. H. Langford, and A. W. Underdown, *Org. Geochem.*, **8**, 35 (1985).
58 J. J. Alberts and T. J. Dickson, *Org. Geochem.*, **8**, 55 (1985).
59 M. Schnitzer and S. U. Khan, *Humic Substances in the Environment*, Marcel Dekker, New York, 1972.
60 T. Dupuis, *J. Thermal. Anal.*, **3**, 281 (1971).

61 M. Schnitzer and H. Kodama, *Geoderma*, **7,** 93 (1972).
62 K. H. Tan, *Soil Biol. Biochem.*, **10,** 123 (1978).
63 G. S. P. Ritchie, A. M. Posner, and I. M. Ritchie, *J. Soil Sci.*, **33,** 671 (1982).
64 B. Kribek, J. Kaigl, and V. Oruzinsky, *Chem. Geol.*, **19,** 73 (1977).
65 J. A. Davis and R. Gloor, *Environ. Sci. Technol.*, **15,** 1223 (1981).
66 F. J. Stevenson, *Soil Sci.*, **123,** 10 (1977).
67 F. J. Stevenson, *Soil Sci. Soc. Amer. J.*, **40,** 665 (1976).
68 H. van Dijk, *Geoderma*, **5,** 53 (1971).
69 M. L. Crosser and H. E. Allen, *Soil Sci.*, **123,** 176 (1977).
70 H. Zunino and J. P. Martin, *Soil Sci.*, **123,** 188 (1977).
71 A. Fitch and F. J. Stevenson, *Soil Sci. Soc. Amer. J.*, **48,** 1044 (1984).
72 D. K. Ryan, C. P. Thompson, and J. H. Weber, *Can. J. Chem.*, **61,** 1505 (1983).
73 D. K. Ryan and J. H. Weber, *Anal. Chem.*, **54,** 986 (1982).
74 R. F. M. Cleven, *Heavy Metal/Polyacid Interactions: An Electrochemical Study of the Binding of Cd(II), Pb(II), and Zn(II) to Polycarboxylic and Humic Acids*, Thesis, Agricultural University, Wageningen, The Netherlands, 1984.
75 B. A. Goodman and M. V. Cheshire, *Nature*, **299,** 618 (1982).
76 R. K. Skogerboe and S. A. Wilson, *Anal. Chem.*, **53,** 228 (1981).
77 H. Kodama, M. Schnitzer, and E. Murad, *Soil Sci. Soc. Amer. J.*, **52,** 994 (1988).

17

STABILITY CONSTANTS OF METAL COMPLEXES WITH HUMIC SUBSTANCES

An important characteristic of a metal–organic complex is its stability constant, the value of which provides an index of the affinity of the cation for the ligand. Numerical values of stability constants for metal–humic complexes would be of considerable value in predicting the behavior of trace elements and toxic heavy metals in soils and sediments.

The utility of stability constant data can be illustrated by the following hypothetical example. Assume that the weight of an "acre-foot" of soil is 0.9×10^6 kg (2×10^6 lb.) and that 2.73 kg of a micronutrient (i.e., Zn) is introduced and mixed uniformly to plow depth. The total quantity of Zn in the top soil is increased by 3 μg/g and its concentration in the soil solution at field capacity (assumed moisture content of 20 percent) becomes 15×10^{-3} mg/mL, provided all of the added Zn is water soluble. On the assumption that most of the Zn is complexed by insoluble organic matter, and that 1×10^{-6} is dissociated, the concentration of Zn in the solution phase would only be 15×10^{-9} mg/mL.

One use that has been made of stability constant data is in computer models (e.g., GEOCHEM, MINEQ, etc.) designed to predict the speciation of metal ions in the soil solution. The subject of stability constants of metal complexes with humic substances has been covered in several reviews.[1-6]

GENERAL CONSIDERATIONS

Stability constants are classified according to the type of reaction they describe. The overall reaction is

$$nM + mL \rightleftharpoons M_nL_m \qquad [1]$$

where n and m are the number of moles of metal ion and ligand molecules in the complex, respectively. The terms M, L, and M_nL_m represent molar concentrations of the free metal ion, the free ligand, and the complex, respectively. The overall stability (equilibrium) constant is given by

$$K = \frac{(M_nL_m)}{(M)^n (L)^m} \qquad [2]$$

Several types of reactions can be visualized. The simplest case is 1:1 binding ($n = m = 1$). More complex, but mathematically solvable, are the formation of mononuclear complexes with two or more binding substrates. The central group may be either the macromolecule (formation of M_nL complexes) or the metal ion (formation of ML_m complexes). All three approaches have been used in metal binding studies of humic substances. A more difficult system to solve is one in which polynuclear complexes are formed (M_nL_m), of which greater attention needs to be given in the future.

Limitations of Stability Constant Measurements

Major difficulties are encountered in determining stability constants of metal complexes with humic substances. Most methods are those developed for well-defined, low-molecular-weight compounds and they apply in only a superficial way to complex macromolecules, such as humic and fulvic acids. Humic substances from different sources are heterogeneous with respect to molecular weight and content of reactive functional groups. Also, pH profoundly affects the ionization of acidic groups and thereby the number of sites available for binding. Several types of binding sites are undoubtedly present (see structures XVI to XXI of Chapter 16), and those site(s) forming the most stable complex(s) will react first. The possibility also exists that humic substances contain binding sites that are chemically identical but that interact with metal ions in such a way that binding at one site affects binding at subsequent sites. Another complication is that configurational changes in the macromolecule may accompany changes in pH or concentration of neutral salt. In most studies, stability constants have been reported as conditional constants that are a function of pH, ionic strength, and concentration of reactants.

In calculating stability constants, allowance has not always been made for side reactions involving the metal ion, the ligand, or both. Side reactions involving the ligand include competition of a proton for the reactive site ($R-COO^- + H^+ \rightarrow R-COOH$). To avoid problems with the formation of chloro complexes (MCl^+, MCl_2), nitrate or perchlorate can be used as the supporting electrolyte.

Determination of Binding Parameters

Irrespective of the approach used for calculating stability constants, some determination must be made for free and bound forms of the ligand or metal ion.

As noted in Table 17.1, both direct and indirect methods have been used to determine the desired parameter.[7-27] In most cases, the free metal ion (M) is the measured quantity, thereby leading to estimates for the amounts contained in the complex. Early work was done by competition of the metal ion with a cation-exchange resin. Increasing use is now being made of ion-selective electrodes (ISE) and anodic stripping voltammetry (ASV); other techniques include potentiometric base titrations, bioassay, spectrophotometric titration, spectrofluorometry, gel filtration, and dialysis. Advantages and limitations of ISE and ASV have been discussed by Bresnahan et al.[7] and Bhat et al.[16]

Choice of method for determining extent of binding is critical in two ways. First, humic substances may contain two or more specific reactive sites and the lower limit of detection will place a restriction on the highest stability constant that can be measured. A second aspect is in regard to the measured parameter. For example, with a divalent cation as the central group, the amount of ligand bound to the complex as determined by base titration is (ML^+ + $2ML_2$); for measurements of the free metal ion (e.g., by use of an ion-selective electrode), the quantity (ML^+ + ML_2) is determined. The two will only agree when 1:1 complexes are formed.

The concentrations of reactants and products ideally should be expressed in

Table 17.1 Methods Used to Measure Metal–Humic Acid Reaction Parameters

	Method	Value Measured[a]	Reference
One-step process: Analysis *in situ*	1 Ion-selective electrode (ISE)	M	Bresnahan et al.[7]; Buffle et al.[8]; Cheam and Gamble[9]; Fitch and Stevenson[10]; Fitch et al.[11]; Giesy et al.[12]; Saar and Weber[13]; Stevenson and Chen[14]; Stevenson et al.[15]
	2 Anodic stripping	M	Bhat et al.[16]; Greter et al.[17]
	3 Spectrofluorimetry	L_b or L	Saar and Weber[13]; Ryan and Weber[18]; Gamble et al.[19]
	4 Spectrophotometry	L	Langford and Khan[20]
	5 Potentiometric titration	L_b	Stevenson[21,22]; Takamatsu and Yoshida[23]
Two-step process: Separation and analysis[b]	1 Ion exchange/AA	M or M_b	Schnitzer and Hansen[24]
	2 Ion exchange/ASV	M or M_b	van den Berg and Kramer[25]
	3 Ion exchange/RA	M	
	4 Gel filtration/UV visible	M	Ardakani and Stevenson[26] Mantoura et al.[27]

[a] M refers to the free metal ion (i.e., M^{+n}).
[b] AA = atomic absorption; RA = radioactivity measurements (i.e., Zn^{65}); UV = ultraviolet.

molar units. Due to the heterogeneous nature and absence of molecular weight data for humic substances, this has seldom been possible. Most often, this parameter has been expressed in terms of the concentration of acidic groups (i.e., COOH), or metal-ion binding capacity (MBA). Methods for estimating MBA include potentiometric titration,[21-23] spectrophotometric titration,[20] fluorescence quenching,[13,18,19] and cation exchange with synthetic resins.[28]

MODELING APPROACHES

Formation of 1:1 Complexes

Early attempts to determine stability constants were done using Schubert's ion-exchange equilibrium method and the continuous variation method of Job. In most cases, the experimental data were interpreted in terms of the formation of 1:1 complexes ($n = m = 1$).

Ion-Exchange Equilibrium Method The ion-exchange equilibrium method is based on the competition between the ligand and a cation-exchange resin for the metal ion. The metal cation is distributed among the resin, M_R, the metal complex, M_c, and the aqueous phase as the free cation, M. The assumption is made that n of Eq. [1] is unity. A distribution coefficient, γ, is calculated for additions of various amounts of the ligand to a measured quantity of the metal ion:

$$\gamma = \frac{M_R}{(M + M_c)} = \frac{M_R}{(M + ML_m)} \quad [3]$$

A distribution coefficient is also obtained in the absence of the ligand:

$$\gamma_0 = \frac{M_R}{M} \quad [4]$$

By substituting M_R/γ_0 for M in Eq. [3], and solving for ML_m, the following is obtained:

$$ML_m = \frac{M_R}{\gamma} - \frac{M_R}{\gamma_0} \quad [5]$$

The concentration of unbound ligand (L) is obtained from the initial concentration of the ligand (L_t) minus the concentration of ligand in the complex (ML_m), or $L = L_t - (M_R/\gamma - M_R/\gamma_0)$. When the ligand is present in large excess, L can be considered as equal to L_t, and the various parameters of Eq. [1] can be expressed in terms of the quantities M_R/γ and $(M_R/\gamma - M_R/\gamma_0)$. By substitution and rearrangement of the resulting equation, and by conversion of the basic units in terms of γ or γ_0 (see Eqs. [3] and [4]), the following working equation is obtained.

$$\log [(\gamma_0/\gamma) - 1] = \log K + m \log (L_t) \qquad [6]$$

From an experimental standpoint, values for γ are obtained at several concentrations of ligand when added to a constant and known amount of the metal ion. A plot of log $[(\gamma_0/\gamma) - 1]$ versus log L_t is subsequently prepared, yielding i as the slope and log K as intercept. The results have usually been interpreted in terms of 1:1 complexes ($m = 1$).

Equation [6] applies only when m is an integral number, which has not always been the case in many studies using humic and fulvic acids. Other errors arise from the assumption that the concentration of metal ions in solution is negligible compared to the ligand. Modifications have been made in the ion-exchange equilibrium method in attempts to avoid the above problems.[26] Limitations of the ion-exchange equilibrium approach, as applied to humic substances, have been outlined elsewhere.[26,29,30]

Job's Method of Continuous Variation The method of continuous variation, or Job's method, is based on changes in the absorption characteristics of the metal ion or ligand when a complex is formed. Optical density measurements are made for a series of solutions containing variable ratios of ligand and metal ion while maintaining a constant reactant concentration. The point at which absorbance is maximum corresponds to the maximum number of ligands bound in the complex. A "Job" plot is subsequently prepared by plotting changes in optical density against the molar ratio of ligand to metal ion, from which the necessary information is obtained for calculating log K. Results obtained with humic substances can be affected by light scattering due to precipitation of the complexes, thereby leading to erroneous log K values.[19,31] In some studies, supplemental light scattering measurements have been made so as to detect the onset of aggregation.[19]

Log K Values for Assumed 1:1 Complexes An indication of the range of log K values that have been recorded for assumed 1:1 complexes of Co, Cu, Mn, and Zn with humic and fulvic acids are given in Table 17.2. Except for pH 8, the values fall within a rather narrow range. Stabilities of the complexes at pH 8 follow the approximate order of the well-known Irving-Williams stability series: Cu > Zn > Co > Mn.

Log K values for Zn^{2+} complexes with humic and fulvic acids as influenced by pH are shown in Fig. 17.1. Although log K values for any given pH are highly variable, a definite trend is evident. Matsuda and Ito[32] observed a range in log K of from 4.2 to 10.8 for 29 Zn–humic acid complexes (pH 7); the range for 32 Zn–fulvic acid complexes was 3.9–9.3. Adsorption strength of humic acid for Zn was believed to increase with an increase in "degree of humification."

Table 17.2 Conditional and Overall Stability Constants for the Complexes of Co, Cu, Mn, and Zn with Humic and Fulvic Acids from Various Sources[a]

Metal Ion and Source[b]	Supporting Electrolyte	pH 5	pH 6	pH 8	Reference
Co					
SFA	0.1 M KCl	4.10			Schnitzer and Hansen[24]
PFA	0.02 M Tris			4.51	Mantoura et al.[27]
Cu					
SFA	0.1 M KCl	4.00			Schnitzer and Hansen[24]
SFA	0.1 M NaClO$_4$	4.35			Cheam and Gamble[9]
SFA	0.1 M NaNO$_3$	4.00			Buffle et al.[8]
SFA	0.1 M KNO$_3$	4.68	5.03– 5.45		Ryan and Weber[18]
WFA	0.01 M KNO$_3$		7.80		van den Berg and Kramer[25]
Mn					
SFA	0.1 M KCl	3.70			Schnitzer and Hansen[24]
PFA	0.02 M Tris			4.17	Mantoura et al.[27]
Zn					
SHA(4)	0.1 M KCl		2.82– 4.93		Ardakani and Stevenson[26]
SHA(29)			4.20–10.83		Matsuda and Ito[32]
SFA(32)			3.88– 9.30		Matsuda and Ito[32]
SFA	0.1 M KCl	3.70			Schnitzer and Hansen[24]
PFA	0.02 M Tris			4.83	Mantoura et al.[27]

[a] Values at other pHs can be found in references cited.
[b] SHA = soil humic acid; SFA = soil fulvic acid; PFA = peat fulvic acid; WFA = water fulvic acid.

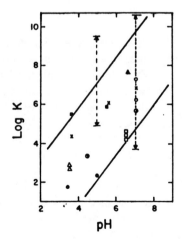

Fig. 17.1 Effect of pH on conditonal stability constants (log K) for Zn–organic matter complexes. The formation of 1:1 complexes is assumed. The vertical dashed line at pH 7 indicates the range of values reported by Matsuda and Ito[32] for 29 humic acids and 32 fulvic acids. Adapted from Stevenson and Ardakani[4]

Stability constants of the metal-fulvate complexes, as recorded in Table 17.2 appear to be generally lower than those for complexes with commercial chelating agents, such as EDTA. They are also lower than those of complexes with many, but not all, naturally occurring biochemical compounds, including certain amino acids and aliphatic hydroxy acids. These observations, which require confirmation, have significant practical implications regarding the transport and availability of micronutrients to plants. It should be noted that stability constants recorded in Table 17.2 for the Cu^{2+} complexes are generally lower than those obtained by other methods, as detailed below.

Macromolecule as the Central Group

In this approach, the macromolecule is assumed to contain two or more reactive sites to which the metal ion can be bound, as illustrated below:

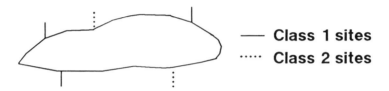

where the variable vertical lines represent different types of reactive sites.

The approach has been applied to humate and fulvate complexes with those divalent cations for which solid-state ion-selective electrodes (ISE) are commercially available (i.e., Cu^{2+}, Cd^{2+}, and Pb^{2+}). Ion-selective electrodes function in much the same way as the pH electrode in that it is the activity of the free cation in solution that is measured (e.g., Cu^{2+} in the case of the cupric ISE).

A typical experiment is carried out as follows[10,11]:

A 50- to 60-ml solution containing a known amount of humic material (about 25 mg) is added to a 200 ml tall-form beaker containing openings for the electrodes (ISE and pH), an N_2 gas inlet, and an automatic syringe for addition of titrant (solution containing the metal ion). A measured amount of supporting electrolyte is added (e.g., dilute $KClO_4$ or KNO_3) to give the desired ionic strength, the pH is set to a constant value with dilute KOH, and the volume is adjusted to 100 ml. Response of the electrode is recorded after each addition of titrant with the pH kept constant by addition of dilute KOH. The concentration of the free form of the metal ion is then determined by application of the Nernst equation, which relates electrode response to the activity (concentration) of the metal ion. A standard curve is obtained and checked immediately before each experimental analysis. For Cu^{2+}, the Nernst equation at 25°C is of the order of: $mV = mV_0 + 29.58 \log (Cu^{2+})$, where mV_0 is the intercept of the standard curve.

A responsive curve for the binding of Cu^{2+} to a soil humic acid is shown in Fig. 17.2. The curve can be partitioned into three sections: (I) an initial lag

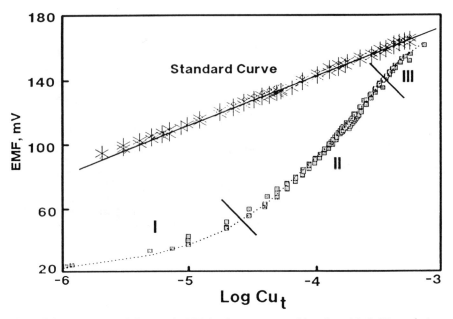

Fig. 17.2 Response of the cupric ISE in the presence of humic acid (0.25 mg/ml at pH 4; ionic strength of 0.005M). The response curve has been partitioned into three sections (see text). The data represent results for four replicates. From Stevenson et al.,[15] reproduced by permission of Williams and Wilkins Co.

phase at low Cu^{2+} additions, (II) a consistent increase in potential as more Cu^{2+} is added, and (III) a leveling off in potential at high additions. The leveling off in millivolt readings has been shown to coincide with flocculation-precipitation of humic acid or its complexes, with entrapment of Cu^{2+}.[11] Only data corresponding to section (II) are suitable for stability constant calculations.

With the macromolecule as the assumed central group, the formation of a series of $LM, LM_2, \cdots, LM_n$ complexes is assumed.

$$L + M \rightleftharpoons LM \qquad [7]$$

$$LM + M \rightleftharpoons LM_2 \qquad [8]$$

$$LM_{n-1} + M \rightleftharpoons LM_n \qquad [9]$$

The reactions can be described by n stability constants:

$$k_1 = \frac{(LM)}{(L)(M)}; \quad k_2 = \frac{(LM_2)}{(LM)(M)}; \quad k_n = \frac{(LM_n)}{(LM_{n-1})(M)} \qquad [10]$$

The extent of binding is expressed as a formation function, ν, defined as

$$\nu = \frac{\text{Molar concentration of bound metal ion}}{\text{Polymer concentration}} = \frac{M_b}{L_t} \qquad [11]$$

As noted above, the free metal ion (designated herein as M but more accurately the charged M^{2+} species) is the measured quantity. Values for the molar concentration of the metal ion bound to the complexes (M_b) is obtained from the difference between the total concentration (M_t) and the free metal ion (i.e., $M_b = M_t - M$). Since the molecular weights of humic substances are highly variable and usually unknown, concentration is generally expressed in terms of the molar concentration of COOH groups, as determined by base titration (see Chapter 13). Accordingly, ν indicates the fraction of potential sites occupied by the metal ion. Since values for the stability constants are obtained from slope values of the Scatchard plot (discussed below), they are unaffected by the unit of measurement chosen for L_t.

In terms of the molar concentration of the free ligand (L) and its complexes, Eq. [11] becomes

$$\nu = \frac{(LM) + 2(LM_2) + \cdots n(LM_n)}{(L) + (LM) + \cdots (LM_n)} \quad [12]$$

By substitution of the constants k_1, k_2, through k_n into Eq. [12] the following is obtained:

$$\nu = \frac{k_1(M) + 2k_1k_2(M) \cdots + nk_1k_2 \cdots k_n(M)}{1 + K_1(M) + k_1k_2(M) \cdots + k_1k_2 \cdots k_n(M)} \quad [13]$$

For binding at "identical and independent sites," ν is given by what is known as Adair's equation:

$$\nu = \frac{nK_0(M)}{1 + K_0(M)} \quad [14]$$

where K_0 is the microscopic binding constant and n is the number of binding sites per macromolecule.

For more than one class of sites, Eq. [14] becomes

$$\nu = \frac{K_1(M)}{1 + K_1(M)} + \frac{K_2(M)}{1 + K_2(M)} + \cdots \frac{K_n(M)}{1 + K_n(M)} \quad [15]$$

In some studies, the formation constant has been expressed in terms of metal binding capacity (MBA, or nL_t). Define:

$$\theta = \frac{M_b}{nL_t} = \frac{\text{Sites bound}}{\text{MBA}} \quad [16]$$

By appropriate substitution, Eq. [14] becomes

$$\theta = \frac{K_0(M)}{1 + K_0(M)} \quad [17]$$

Values for θ provide an indication of the fraction of the binding sites bound to the complexes. Estimates for MBA are obtained directly from the binding data, such as from the formation curve[14,15] or by application of the Langmuir equation.[10,11]

Equations [14] and [17] are cumbersome to use but can be rearranged in several ways (Fig. 17.3) so that the desired constants can be obtained by graphical means (i.e., from slope or intercept values of the plotting variables). As noted in Table 17.3, plotting can be done as a Scatchard plot (ν/M versus ν; θ/M versus θ), as a reciprocal plot (M/ν versus M), as a double reciprocal plot ($1/\nu$ versus $1/M$), and as a Hill plot (log $[\theta/(1 - \theta)]$ versus log M).

All of the plotting methods listed in Table 17.3 have been used for analysis of metal-ion binding data by humic substance, the most common being the Scatchard plot method. An approach used by Buffle et al.[8] to study the binding of Cu^{2+} to fulvic acid has the same form as the double reciprocal plot; the one of Zunino and Martin[33] is equivalent to a Hill plot. Fitch and Stevenson[10] pointed out that stability constants obtained by the Hill-plot method are unreliable when binding at one site decreases binding affinity at subsequent sites, which appears to be the case for metal complexes with humic substances, as noted below.

Scatchard Plot Method: Two-Component Model

The method of choice by most investigators has been the Scatchard plot approach.[7,8,10–15] Rearrangement of Eq. (14) gives (see also Table 17.3):

$$\frac{\nu}{M} = nK_0 - \nu K_0 \qquad [18]$$

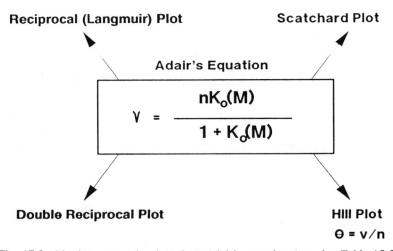

Fig. 17.3 Plotting approaches based on Adair's equation (see also Table 15.3)

MODELING APPROACHES 415

Table 17.3 Plotting Approaches for Analyzing Experimental Binding Data

Plot Title	Form of the Equation	Plot Y	X
Scatchard	$v/M = nK_0 - vK_0$	v/M	v
	or		
	$\theta/M = K_0 - \theta K_0$	θ/M	θ
Reciprocal (Langmuir)	$M/v = M/n + 1/nK_0$	M/v	M
	or		
	$M/M_b = M/nL_t + 1/nL_tK_0$	M/M_b	M
Double reciprocal	$1/v = 1/n + 1/nK_0M$	$1/v$	$1/M$
	or		
	$1/M_b = 1/nL_t + 1/nL_tK_0M$	$1/M_b$	$1/M$
Hill	$\log[\theta/(1-\theta)] = \log K^* + n \log M$	$\log[\theta/(1-\theta)]$	$\log M$

For independent binding at one class of sites, a plot of v/M versus v gives a straight line, yielding K_0 as the slope. However, as applied to metal-ion binding by humic substances, Scatchard plots (i.e., v/M versus v) are curvilinear, from which constants have been obtained for binding at two "classes" of sites (K_1 and K_2) by arbitrary segmentation of the plots into two straight line segments. A typical Scatchard plot is shown in Fig. 17.4.

The segmentation of Scatchard plots into two components is somewhat arbitrary and additional "sites" can be observed by assigning linear segments to various sections of the plots.[10,12] Accordingly, results obtained by the two-site

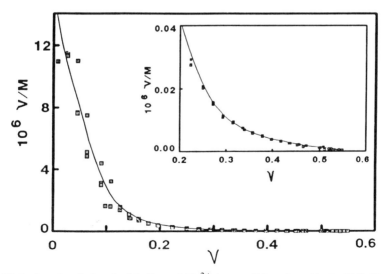

Fig. 17.4 Scatchard plot for binding of Cu^{2+} to a soil humic acid at pH 5. Results are for three replicates. An expanded scale for binding in the higher v range is given by the insert. Adapted from Fitch and Stevenson[10]

model are best expressed in terms of binding at the strongest measurable site (log K_1) and the weakest site (log K_n).

Stability constants (logs K_1 and K_n) for the complexes of Cu^{2+} with humic and fulvic acids from various sources are given in Table 17.4. As expected, log K values increase with increasing pH and decreasing ionic strength. It should be noted that log K_1 values are substantially higher than the conditional constants recorded in Table 17.2, where Schubert's ion-exchange equilibrium method was used.

Fitch et al.[11] pointed out that, with ISE, the magnitude of Log K_1 depends on the availability (and reliability) of data at low values of ν (i.e., low saturation of binding sites). Many estimates for log K_1 would appear to be low because of failure of ISE to measure the free form of the metal ion at low concentrations. By using a technique that permitted measurements to be made at Cu^{2+} concentrations $< 10^{-8}$ M, Fitch et al.[11] obtained a log K_1 of 8.2 for a Cu–humic acid at pH 4 (ionic strength of 0.005). This value is over an order of magnitude higher than the log K_1 (6.75) recorded in Table 17.4 for the same humic acid where the measurements were made in the conventional way by titration with Cu^{2+}.

A concentration effect has been observed when ISE is used in metal-ion binding studies of humic substances,[11,15] and attributed to inability of the ISE

Table 17.4 Comparison of Stability Constants for Binding of Cu^{2+} at the Strongest (Log K_1) and Weakest (Log K_n) Sites for the Humic and Fulvic Acids from Several Sources. Calculations were Made Using the Two-Site Scatchard Plot Model

Source	Supporting Electrolyte	Log K_1	Log K_n	Reference
		Fulvic Acids		
Soil				
pH 4	0.1 M KNO_3	5.60	3.95	Bresnahan et al.[7]
pH 5	0.1 M KNO_3	6.00	4.08	Bresnahan et al.[7]
pH 6	0.1 M KNO_3	6.30	3.07	Bresnahan et al.
Water				
pH 4	0.1 M KNO_3	5.48	4.00	Bresnahan et al.[7]
pH 5	0.1 M KNO_3	5.95	3.70	Bresnahan et al.[7]
pH 6	0.01 M KNO_3	6.11	3.85	Bresnahan et al.[7]
Water[a]				
pH 6.25	—	8.11	5.34	Tuschall and Brezonik[34]
pH 6.25	—	7.82	5.26	Tuschall and Brezonik[34]
		Humic Acids		
Soil				
pH 4	0.005 M $KCLO_4$	6.75	5.09	Fitch and Stevenson[10]

[a]Constants obtained for binding at three "sites" (log K_2 = 6.72–6.85)

technique to measure binding at the stronger sites when the humic material is present in low concentration. Since most work using the Scatchard plot method has been done on the basis of ISE measurements, the wide range of values reported for stability constants of metal–humate complexes may be a reflection of errors in determining binding at low saturation of binding sites.

In concluding this section, it can be said that the complexation of metal ions by humic substances cannot be adequately explained by the two-site model.[1,2,11-15] The curvilinearity of Scatchard plots may be due to one or more of the following:

1. The occurrence of a large number of different bindings sites, each with a slightly different affinity for the metal ion
2. Electrostatic effects, such that complexation at one site decreases the tendency of a neighboring functional group to complex another metal ion
3. Formation of ML_2 complexes (i.e., metal ion as the central group), as described later

Continuous Distribution Models

In view of the questionable validity of the two-component Scatchard plot approach continuous distribution models have been applied.[1,2,12,14,15] In one approach,[15] a series of "incremental conditional stability constants" (log K_is) was calculated from successive slope values of the Scatchard plot, and that decreased with an increase in saturation of binding sites. Individual K_is obtained in this way were considered to reflect variations in binding energies without regard to the manner in which the metal ion is bound. An intrinsic constant (log K_{int}) was obtained for binding at the strongest site by extrapolation to $v = 0$.

Typical plots for log K_i (as obtained from successive slope values of the Scatchard plot) versus the formation function (v) are shown in Fig. 17.5. Polynomial equations provided a satisfactory fit for the data, from which log K_{int} was obtained from the intercept value (i.e., $v = 0$). Results obtained for a soil–humic acid (Fig. 17.6) showed that values for log K_{int} increased with an increase in pH and decreased with an increase in the concentration of supporting electrolyte (I).

From an environmental standpoint, stability constants for binding at the stronger sites are of greatest significance.[12] This is the region of greatest difficulty in obtaining reliable binding data. To solve this problem, a ligand titration procedure has been used,[14] in which the humate is dispensed into a solution consisting of a 10^{-5} M Cu^{2+} solution. The rationale for the approach is that measurements at low saturation of binding sites would fall within the detection limits of the cupric ISE. Log K_{int} values for four humic acids and a fulvic acid at pH 4 and 5 are shown in Fig. 17.7. Affinity of the ligand for Cu^{2+} followed the order: soil humic acid > peat humic acid > lignite humic acid > soil fulvic acid = fungal melanin.

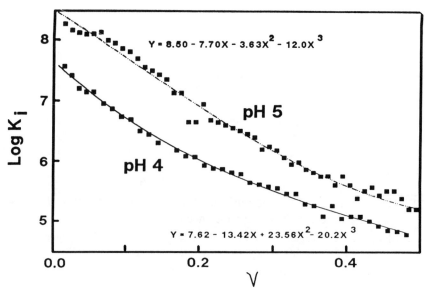

Fig. 17.5 Experimental plots relating log K_i to ν at 2 pH values. From Stevenson et al.,[15] reproduced by permission of Williams and Wilkins Co.

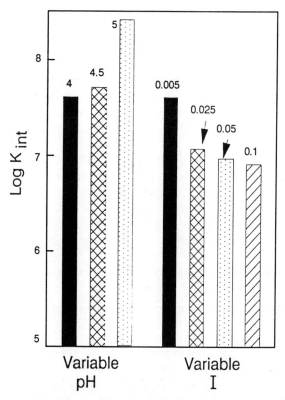

Fig. 17.6 Log K_{int} values for a soil humic acid as influenced by pH and ionic strength (I). Adapted from Stevenson et al.[10]

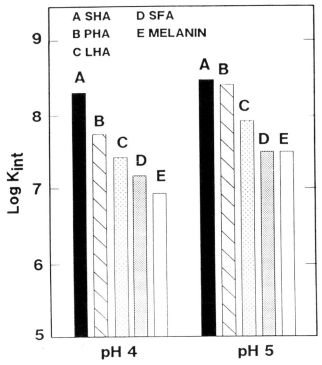

Fig. 17.7 Log K_{int} values for humic acids from soil (SHA), peat (PHA), and lignite (LHA), a soil fulvic acid (SFA), and a fungal melanin at pH 4 and 5, the general order being SHA > PHA > LHA > SFA = melanin. From Stevenson and Chen[14]

Two main concepts have evolved regarding metal ion binding by humic substances: 1) humic and fulvic acids, from whatever source, contain similar types of reactive functional groups; thus, log K values for their metal complexes will be similar, and 2) differences between humic substances occur due to structural variations and/or other attributes (e.g., aromaticity, degree of humification) and their binding affinities for metal ions will be affected accordingly. Results obtained in the above-mentioned study support the second hypothesis, namely, that differences exist in the binding affinity of humic substances for metal ions.

Data obtained for log K_i by the continuous distribution approach have also been examined using the Gaussian distribution function:

$$\frac{C_i}{C_L} = \frac{1}{\sigma\sqrt{2\pi}} e^{-1/2 \left(\frac{\mu - \log K_i}{\sigma}\right)^2}_{d\log K_i} \qquad [19]$$

where C_i/C_L is the mole fraction of binding sites in the interval $d\log K_i$, and σ is the standard deviation about the mean value (μ) of log K_i. Results obtained using this approach have shown that humic acids contain a higher number of weak binding sites, as compared to strong binding sites.[14] A modified Gaussian distribution function was used by Perdue and Lytle[1] to obtain stability constants for binding at high and low saturation of binding sites.

Conditional Average Stability Constants ($\overline{K}'$)

A modified version of Eq. [17] was used by Perdue and Lytle[1,2] to calculate "conditional average stability constants" ($\overline{K}'$) for a hypothetical ligand mixture designed to simulate binding of metal ions by humic substances. The approach was referred to as a "continuous multiligand model."

Rearrangement of Eq. [17] gives

$$\overline{K}' = \frac{1}{(M)} \left(\frac{\theta}{1 - \theta} \right) \qquad [20]$$

For the study reported earlier in Fig. 17.7, a sequencer of log $\overline{K}'$ values were obtained from the experimental data and compared to log K_i values for the soil, peat, and lignite humic acids.[14] From Fig. 17.8, it can be seen that the relationship between the two was nonlinear. Values for log $\overline{K}'$ were consistently lower than those for log K_i, particularly for binding at the stronger sites (i.e., high log K_is). It should be noted that values for $\overline{K}'$ will differ from K_i by the factor ν/n, and that the two will only be equal when all available sites enter into the reaction (i.e., when $\nu_{max} = 1$). Due to competition of H^+ for binding sites, this is not the case with metal ion binding by humic substances. Accordingly, the continuous distribution approaches based on log K_i must be considered to be more scientifically sound than the one for log $\overline{K}'$ (as well as the two-component Scatchard function).

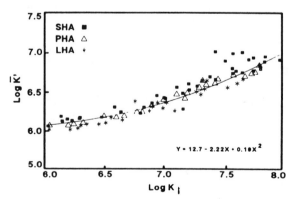

Fig. 17.8 Relationship between the logs of the conditional average stability constants ($\overline{K}'$) and the incremental stability constants (K_i) for humic acids from soil (SHA), peat (PHA), and lignite (LHA) at pH 5. From Stevenson and Chen[14]

Bjerrum Approach: Metal Ion as the Central Group

When the metal ion is the central group, which is usually the case with small molecules, a series of species of the type ML_n are obtained. For humic substances, the metal ion may: 1) combine with two separate molecules and thereby link the two together, and 2) form an internal ring structure by combining with two (or more) reactive sites on the same molecule.

The Bjerrum approach has been used to determine stability constants for the binding of the divalent cations Cd^{2+}, Cu^{2+}, Pb^{2+}, and Zn^{2+}, and the trivalent Al^{3+} ion, to humic and fulvic acids. Binding can be regarded as occurring at the dissociated form of the acidic functional group (i.e., COO^-), indicated herein as L^-, with release of H^+ from the protonated moiety (COOH) due to reestablishment of ionization equilibrium, as depicted below for Cu^{2+} (equation not balanced):

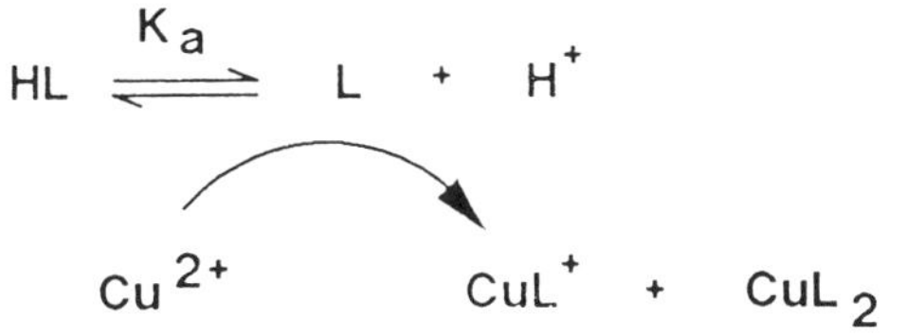

$$HL \underset{}{\overset{K_a}{\rightleftharpoons}} L + H^+$$

$$Cu^{2+} \longrightarrow CuL^+ + CuL_2 \quad [21]$$

where K_a is the dissociation constant for the acidic functional group. The reactions for divalent cations, using Cu^{2+} as an example, are

$$Cu^{2+} + L^- \rightleftharpoons CuL^+; \quad CuL^+ + L^- \rightleftharpoons CuL_2 \quad [22]$$

The overall reaction is

$$Cu^{2+} + 2L^- \rightleftharpoons CuL_2 \quad [23]$$

Two successive stability constants (k_1 and k_2), and the overall constant ($k_1 k_2$), are given by

$$k_1 = \frac{(CuL^+)}{(L^-)(Cu^{2+})}; \quad k_2 = \frac{(CuL_2)}{(L^-)(CuL^+)}; \quad k_1 k_2 = \frac{(CuL_2)}{(L^-)^2(Cu^{2+})} \quad [24]$$

A formation function, η, is determined and related to the stability constants as outlined below.

$$\eta = \frac{CuL^+ + 2CuL_2}{M_t} = \frac{L_t - L^- - HL}{M_t} \quad [25]$$

where L_t and M_t are the total concentrations of acidic (COOH) groups and the metal ion, respectively. The quantities HL and L^- are the protonated and dissociated forms of the acidic group, respectively.

A value of $\eta > 1$ indicates the formation of both 1:1 and 2:1 complexes (i.e., CuL^+ and CuL_2). In coordination chemistry, 1:1 and 2:1 complexes refer to the number of molecules of the ligand that combine with the metal ion. As used herein, η denotes the number of binding sites per metal ion.

In general terms, η is related to the stability constants through the relationship:

$$\Sigma(\eta - n)k_1 \cdots k_n(L^-) = 0 \qquad [26]$$

The desired constants, k_1 and k_2 (and hence k_1k_2), are normally obtained from a "formation curve" by plotting η versus pL^- and interpolating at η values of 0.5 and 1.5. However, formation curves for metal complexes with humic substances are atypical in that η values <0.7 are generally not obtained. This has been attributed to a pronounced tendency of humic substances to form 2:1 complexes with trace elements. Thus, $k_1 \approx k_2$, and an overall average for the system (k_{ave}^2), obtained by interpolation at $\eta = 1.0$, is the constant of greatest significance.[15,21,22,35]

Alternately, the desired constants can be calculated mathematically. Expansion of Eq. [26] for 2:1 complexes and rearrangement of the expanded equation gives

$$\frac{\eta}{(\eta - 1)(L^-)} = \frac{(2 - \eta)(L^-)}{(\eta - 1)} k_1k_2 - k_1 \qquad [27]$$

From Eq. [27], k_1 and the overall constant (k_1k_2) can be obtained graphically or by the method of least squares, the latter being the preferred method.

In the conventional Bjerrum approach, base titration curves are obtained for the ligand in the presence and absence of variable amount of the metal ion. A complication of using this procedure for metal complexes with humic substances is that the curves are horizontally displaced in the near neutral and alkaline pH ranges, as shown in Fig. 17.9. The horizontal displacement has been attributed, in part, to release of an otherwise nontitratable H^+ from the humate and/or protons from hydration water of the metal bound in 1:1 complexes.[21,22]

To avoid the problem of proton release from hydration water, a procedure has been used that involves sequential additions of the metal ion to a solution of humic material while maintaining the pH constant (pH adjusted after each addition).[22] At pH values below ~5.0, release of protons from hydration water of the metal ion is suppressed, thereby facilitating stability constant calculations.

Results obtained in a recent study[15] on the complexation of Cu^{2+} by a soil-humic acid at two pH values are illustrated in Fig. 17.10. Each addition of Cu^{2+} depressed the pH to a lesser extent than the previous addition, and less and less base was required to neutralize the liberated protons. Horizontal dashed lines in the diagrams indicate the content of "undissociated" COOH groups in the original sample. At pH 4 (upper diagram), the amount of base consumed

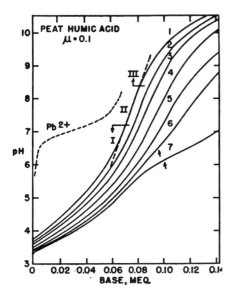

Fig. 17.9 Titration curves of a peat humic acid in the presence of variable amounts of Pb^{2+} at an ionic strength of 0.1: (1) No Pb^{2+}, (2) 0.01 meq, (3) 0.02 meq, (4) 0.04 meq, (5) 0.06 meq, (7) 0.08 meq. From Stevenson[21]

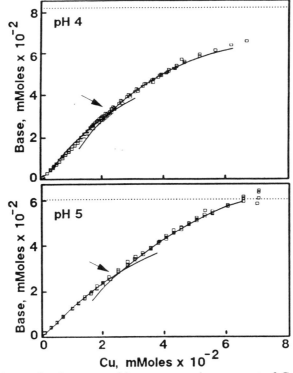

Fig. 17.10 Accumulated amount of base consumed vs amount of Cu^{2+} added to a soil humic acid at pH 4 (upper diagram) and pH 5 (lower diagram). The horizontal dashed lines indicate the content of COOH groups in the original sample; arrows indicate positions of inflection points. From Stevenson et al.,[15] reproduced by permission of Williams and Wilkins Co.

was somewhat less than that required for complete neutralization; at pH 5 (lower diagram), ionization was essentially complete. These results can be explained by greater competition of protons for reactive sites at the lower pH. Although not clearly defined, the curves gave evidence for the existence of an inflection point (noted by arrows in the diagram), indicating the formation of both 1:1 and 2:1 complexes.

In analyzing results of base titration studies (see Fig. 17.10), values for HL, the concentration of undissociated acidic groups, are first calculated for each data point from the amount of base required to return the pH to the initial starting position.

For any given additional of metal ion, the total amount of base (usually KOH) added to the system is

$$KOH = KOH_0 + KOH_t \quad [28]$$

where KOH_0 is the amount of base initially added to adjust the pH of the solution to the desired value and KOH_t is the amount required to return the pH to its initial value after addition of Cu^{2+}. The quantity HL is given by:

$$HL = L_t - KOH - H^+ + OH^- \quad [29]$$

where L_t is the concentration of COOH groups in the sample being analyzed; KOH is the concentration of added base after allowance for dilution.

Values for L^-, required for Eqs. [25] and [27], are then calculated from the pH-dependent ionization constants (K_a) of the acidic functional group (see Chapter 13).

$$L^- = \frac{(HL)(H^+)}{K_a} \quad [30]$$

In calculating H^+ from pH measurements, a correction is made using activity coefficients, as determined by the Debye-Huckel equation (see Chapter 13).

Several values for η are next obtained by substituting into Eq. [25] known values for L_t and M_t and the experimentally determined values of HL and L^-. Values for log k_{ave}^2 or the overall constant $(k_1 k_2)$ can then be obtained by interpolation from the formation curve or by use of Eq. [27], respectively.

Results obtained for the Cu^{2+} and Zn^{2+} complexes of a soil humic acid as influenced by ionic strength at two pH values are given in Fig. 17.11. A pronounced decrease in the overall constant (log K_2 = log $k_1 k_2$) is shown with an increase in ionic strength, which is to be expected. Log $k_1 k_2$ values obtained at low salt concentrations are more likely to represent conditions existing in the soil solution than those at high salt concentrations.

Overall stability constants, as obtained above, are not directly comparable with those reported in the sections on 1:1 complexes. However, estimated log k_1 values (i.e., [log $k_1 k_2$]/2) for the Cu^{2+} and Zn^{2+} complexes were of the

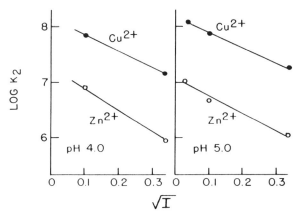

Fig. 17.11 Stability constants (log K_2 = log $k_1 k_2$) for metal complexes of a soil humic acid at two pH values as influenced by ionic strength (I) at pH 5. Adapted from Stevenson[22]

same order of magnitude as those reported in Table 15.2 for 1:1 complexes, where Schubert's ion-exchange equilibrium method was used. For Cu^{2+}, the overall constant (log $k_1 k_2$) is comparable to the log K_{int} value shown in Fig. 17.6 for the same humic acid.

Takamatsu and Yoshida[23] used a closely related approach to determine stability constant of a Cu^{2+}–humate complex. In this case, free Cu^{2+} was estimated by ISE measurements and the number of reactive sites (i.e., COOH) by base titration. Successive stability constants were obtained from the relationship:

$$\frac{M_t - (M^{2+})}{(M^{2+})(L^-)^2} = \frac{k_1}{(L^-)} + k_1 k_2 \qquad [31]$$

The left-hand side of Eq. [31] was plotted versus $1/L^-$, from which k_1 and $k_1 k_2$ were obtained from the slope and intercept, respectively. The total concentration of metal ions, M_t, was known and the concentration of unbound Cu^{2+} was obtained from the ISE measurements. Values for L^- were determined by differences between L_t (known) and HL plus bound sites. A limitation of Eq. [31] is that an inherent correlation exists between $1/(L^-)^2$ of the Y variable and $1/(L^-)$ of the X variable.

The equation relating log $k_1 k_2$ to pH for the Cu^{2+} complexes was

$$\log k_1 k_2 = 8.65 + 0.65 \, (\text{pH} - 5) \qquad [32]$$

Explanations given by Takamatsu and Yoshida[23] for the variation in log $k_1 k_2$ with pH were: 1) complexation at one site affects binding of the metal cation to an adjacent site, 2) in the higher pH region, metal ions are bound selectively to functional groups that form the strongest complexes but which are protonated in the lower pH region, and 3) stearic stabilization occurs through formation of chelate rings.

In some work, the constants have been expressed in terms of the reaction of the metal ion with the "protonated" form of the reactive site: for example, $M^{2+} + HL \rightleftharpoons ML^- + H^+$, for which the stability constant is: $b_1 = (ML^+)(H^+)/[(M^{2+})(HL)]$. The relationship between b_1 and k_1 is given by

$$b_1 = K_a k_1 \quad [33]$$

where K_a is the ionization constant for the acidic group.

This calculation method was used by Young et al.[35] to study the binding of Cu^{2+} to the soluble "polycarboxylates" from a peat and a mineral soil. Overall constants using both approaches were calculated for the following equilibria:

$$Cu^{2+} + 2L^- \rightleftharpoons CuL_2; \; \beta_{Cu} = (CuL_2)/(Cu^{2+})(L^-)^2 \quad [34]$$

$$Cu^{2+} + 2LH \rightleftharpoons CuL_2 + 2H^+; \; \beta_{Cu}^H = (CuL_2)(H^+)^2/(Cu^{2+})(LH)^2 \quad [35]$$

Log β_{Cu} values, which correspond to log k_1k as noted above, drastically increased with increasing dissociation of acidic groups, being of the order 0.5 to 2.8 for the completely protonated sample (low pH) and 7.6 to 9.3 for the fully dissociated sample (high pH). Log β_{Cu}^H values varied from -1.26 to -1.80 and were unaffected by degree of dissociation of acidic functional groups, or pH.

In conclusion, application of Bjerrum's base titration method indicates that polyvalent cations, such as Cu^{2+}, form both 1:1 and 2:1 complexes with humic substances. In view of structural considerations, 2:1 complexes would not be expected to be formed within the "same" molecule. Rather, the metal serves as a link between two macromolecules to form polynuclear complexes. Reactions leading to the formation of chain-like structures can be found in Chapter 16 (e.g., see structure **XXVI**).

The formation of polynuclear complexes provides an explanation for solubility characteristics of the complexes, namely, formation of soluble complexes at low saturation of binding sites (few molecules combined together) but formation of insoluble complexes as chain-like structures are formed through cross-linking.

Additional Approaches

In other work, Marinsky et al.[36] included a term for electrostatic interactions in their calculations for the stability constant of a Cu^{2+}–humic acid complex. Cheam and Gamble[9] used a differential equilibrium function to describe metal-ion binding by humic substances; MacCarthy and Smith[37] treated humic substances as mixtures of nonidentical molecules. Finally, Tipping and Hurley[38] described cation binding in terms of complexation at discrete sites, modified by electrostatic attraction and/or repulsion, and with account being taken of nonspecific binding due to counterion accumulation.

SUMMARY

Serious problems are encountered in the determination of stability constants of metal complexes with humic and fulvic acids. Several different approaches have been applied, and considerable progress has been made, but agreement has not yet been reached as to how the data can best be analyzed and interpreted. Comparisons of literature results cannot easily be made because of differences in approaches for modeling the experimental data. Continuous distribution models show considerable promise but additional research is required.

The wide range of stability constants that have been reported for metal complexes of humic substances poses a dilemma in using thermodynamic data to predict the speciation of trace elements in the soil solution and natural waters. For example, models based on the higher values may lead to the prediction that over 90 percent of a given trace element would exist in organically complexed forms; models based on the lower values may lead to the prediction that only 10 percent would be organically bound. Greater attention needs to be given to the formation of polynuclear complexes.

REFERENCES

1 E. M. Perdue and C. R. Lytle, "A Critical Examination of Metal–Ligand Complexation Models: Application to Defined Multiligand Mixtures," in R. F. Christman and E. T. Gjessing, Eds., *Aquatic and Terrestrial Humic Materials*, Ann Arbor Science Publishers, Ann Arbor, Michigan, 1983, pp. 295–313.

2 E. M. Perdue and C. R. Lytle, *Environ. Sci. Technol.*, **17**, 654 (1983).

3 F. J. Stevenson, "Organic Matter–Micronutrient Reactions in Soil," in J. J. Mortvedt et al., Eds., *Micronutrients in Agriculture, 2nd Edition*, American Society of Agronomy, Madison, 1991, pp. 145–186.

4 F. J. Stevenson and M. S. Ardakani, "Organic Matter Reactions Involving Micronutrients in Soils," in J. J. Mortvedt, P. M. Giordano, and W. L. Lindsay, Eds., *Micronutrients in Agriculture*, American Society of Agronomy, Madison, 1972, pp. 79–114.

5 F. J. Stevenson and A. Fitch, "Chemistry of Complexation of Metal Ions with Soil Solution Organics," in P. M. Huang and M. Schnitzer, Eds., *Interactions of Soil Minerals with Natural Organics and Microbes*, Special Publication 175, American Society of Agronomy, Madison, 1986, pp. 29–58.

6 F. J. Stevenson and G. F. Vance, "Naturally Occurring Aluminum–Organic Complexes," in G. Sposito, Ed., *The Environmental Chemistry of Aluminum*, CRC Press, Boca Raton, Florida, 1991, pp. 117–145.

7 W. T. Bresnahan, C. L. Grant, and J. H. Weber, *Anal. Chem.*, **50**, 1675 (1978).

8 J. Buffle, F.-L. Greter, and W. Haerdi, *Anal. Chem.*, **49**, 216 (1977).

9. V. Cheam and D. S. Gamble, *Can. J. Soil Sci.*, **54**, 413 (1974).
10. A. Fitch and F. J. Stevenson, *Soil Sci. Soc. Amer. J.*, **48**, 1044 (1984).
11. A. Fitch, F. J. Stevenson, and Y. Chen, *Org. Geochem.*, **9**, 109 (1986).
12. J. P. Giesy, J. J. Alberts, and D. W. Evans, *Environ. Toxicol. Chem.*, **5**, 139 (1986).
13. R. A. Saar and J. H. Weber, *Anal. Chem.*, **52**, 2095 (1980).
14. F. J. Stevenson and Y. Chen, *Soil Sci. Soc. Amer. J.*, **55**, 1586 (1986).
15. F. J. Stevenson, A. Fitch, and M. S. Brar, *Soil Sci.*, **155**, 77 (1993).
16. G. A. Bhat, R. A. Saar, R. B. Smart, and J. H. Weber, *Anal. Chem.*, **53**, 2275 (1981).
17. F.-L. Greter, J. Buffle, and W. Haerdi, *J. Electroanal. Chem.*, **101**, 211 (1979).
18. D. K. Ryan and J. H. Weber, *Anal. Chem.*, **54**, 986 (1982).
19. D. S. Gamble, C. H. Langford, and A. W. Underdown, *Org. Geochem.*, **8**, 35 (1985).
20. C. H. Langford and T. R. Khan, *Can. J. Chem.*, **53**, 2979 (1983).
21. F. J. Stevenson, *Soil Sci. Soc. Amer. J.*, **40**, 665 (1976).
22. F. J. Stevenson, *Soil Sci.*, **123**, 10 (1977).
23. T. Takamatsu and T. Yoshida, *Soil Sci.*, **125**, 377 (1978).
24. M. Schnitzer and E. H. Hansen, *Soil Sci.*, **109**, 333 (1970).
25. C. M. G. van den Berg and J. R. Kramer, *Anal. Chim. Acta*, **106**, 113 (1979).
26. M. S. Ardakani and F. J. Stevenson, *Soil Sci. Soc. Amer. J.*, **36**, 884 (1972).
27. R. F. C. Mantoura, A. Dickson and J. P. Riley, *Est. Coast. Mar. Sci.*, **6**, 387 (1978).
28. M. L. Crosser and H. E. Allen, *Soil Sci.*, **123**, 176 (1977).
29. P. MacCarthy, *J. Environ. Sci. Health*, **A12**, 43 (1977).
30. G. C. Smith, T. F. Rees, P. MacCarthy, and S. R. Daniel, *Soil Sci.*, **141**, 7 (1986).
31. P. MacCarthy and H. B. Mark, Jr., *Soil Sci. Soc. Amer. J.*, **40**, 267 (1976).
32. K. Matsuda and S. Ito, *Soil Sci. Plant Nutr.*, **16**, 1 (1970).
33. H. Zunino and J. P. Martin, *Soil Sci.*, **123**, 188 (1977).
34. J. R. Tuschall, Jr. and P. L. Brezonik, *Anal. Chim. Acta*, **149**, 47 (1983).
35. S. D. Young, B. W. Bache, and D. J. Linehan, *J. Soil Sci.*, **33**, 467 (1982).
36. J. A. Marinsky, S. Gupta, and P. Schindler, *J. Colloid Interface Sci.*, **89**, 401 (1982).
37. P. MacCarthy and G. C. Smith, "Metal Binding by Ligand Mixtures: A Quantitative Model," in D. D. Hemphill, Ed., *Trace Substances in Environmental Health*, University of Missouri, Columbia, 1978, pp. 472–480.
38. E. Tipping and M. A. Hurley, *Geochim. Cosmochim. Acta*, **56**, 3627 (1992).

18

CLAY–ORGANIC COMPLEXES AND FORMATION OF STABLE AGGREGATES

The interaction of organic substances with clay has a multitude of consequences that are reflected in the physical, chemical, and biological properties of the soil matrix. Jacks[1] probably overemphasized the significance of clay–organic complexes when he stated "the union of mineral and organic matter to form the organo–mineral complex (is) a synthesis as vital to the continuance of life as, and less understood than, photosynthesis." Nevertheless, his assessment serves to direct attention to the importance of clay–humus complexes in soil.

The point has been made elsewhere in the text that clays tend to stabilize organic matter, and that, other environmental factors being equal, a high correlation exists between the organic matter and clay contents of many soils. Some types of organic substances serve to bridge soil particles together, thereby forming stable aggregates. Pesticides of various types are adsorbed by clay (as well as organic matter) and rendered inactive. Clays are known to catalyze reactions of various adsorbed organic molecules. The activities of enzymes in soil may be modified through adsorption to clay.

By necessity, the scope of this chapter is limited to a cursory examination of clay–organic complexes. Several excellent reviews can be consulted for supplementary information.[2-6] The subject of soil aggregation is discussed in considerable detail by Emerson et al.,[7] Harris et al.,[8] and Oades.[9]

NATURE OF CLAY COLLOIDS

Clay minerals of major importance in soil can be grouped into four main categories: montmorillonites, illites, kaolinites, and the vermiculites. Some distinguishing characteristics of these common clay types are given in Table 18.1.

Table 18.1 Cation Exchange Capacities and Specific Surface Areas of Four Common Clay Minerals

Clay Type	Specific Surface Area	Cation Exchange Capacity
	meters^2/g	cmole$_c$/kg
Vermiculite	600–800	100–150
Montmorillonite	600–800	80–150
Illite	65–100	10–40
Kaolinite	7–30	3–15

Montmorillonite is an expanding lattice clay that provides internal as well as external surfaces for adsorption. It is a three-layer clay that consists of an Al oxyhydroxide layer sandwiched between two Si–oxide layers. A given clay crystal consists of several sheets of these three-layer molecules. The layers or sheets allow other substances (e.g., water) to penetrate between them thereby causing swelling and shrinking. This is the basis for the term "expanding lattice" clay. As with other 2:1 clays, negative charges on internal and external surfaces result from isomorphous substitution of Al for Si in the tetrahedral layer and of a divalent cation (e.g., Mg) for Al in the octahedral layer. These negative charges are satisfied by exchange cations. Differences in the cation exchange capacities for the clays shown in Table 18.1 are due partly to variations in the extent of ionic substitutions in the lattice.

Other charges on clays develop at crystal edges due to broken bonds, both negative and positive. The negative charges result from exposed OH groups, which dissociate with changes in pH and are termed "pH-dependent" charges, as opposed to fixed charges resulting from isomorphous substitution.

Illitic clays are also of the three-layer type. They have a higher negative charge per unit lattice than the montmorillonites and K^+ ions are located between the neighboring tetrahedral sheets, thereby holding them together so tightly that they do not shrink, swell, or permit entry of organic molecules into interlamellar spaces. Thus, the term "nonexpanding" 2:1 clays. These clays have lower cation exchange capacities and specific surface areas than the expanding types.

Kaolinite is a two-layer clay that consists of alternate layers of Si oxides and Al hydroxyoxides. In contrast to montmorillonite and other three-layer clays, the cationic and anionic exchange properties are believed to originate mainly from unsatisfied valences at the edges of particles. One surface layer of kaolinite consists of OH groups in octahedral position with Al, which affords a special opportunity for adsorption of certain organic molecules.

The source of negative charges on the broken edges of kaolinite, as well as other clays, is believed to arise from dissociation of a proton (H^+) from an exposed OH group. This is possible because oxygen atoms at the edges are in contact with one rather than two Si or Al atoms. The hydrogens of tetrahedral OHs (those associated with Si) are presumed to be more likely to dissociate than those of octahedral OH (those associated with Al). As one might expect, dissociation is strongly pH dependent.

Discrete positive sites are also possible at clay edges, particularly at low pHs, through protonation of OH groups (clay]–OH + H^+ → clay]–OH_2^+).

In the natural state, the surfaces of clay minerals are hydrated. This surface water is believed to be less dense (more ordered) than normal water and is said to have an "ice-like" structure. Adsorbed ions are to some extent associated with water molecules at the surface and are themselves hydrated. As will be shown later, hydrated and coordinated water molecules are believed to play an important role in adsorption reactions.

ADSORPTION OF DEFINED ORGANIC COMPOUNDS BY CLAY

Adsorption Isotherms

The adsorption of a solute from aqueous solution by a solid can be described by four basic types of adsorption isotherms (Fig. 18.1).[10] The L-type (for Langmuir) is the most common and describes the case where the solid has a high affinity for the solute. The S-type occurs when the solid has a high affinity for the solvent (e.g., water competes strongly with solute for adsorption sites). The C-type (constant partition) occurs when new sites become available as the solute is adsorbed. The H-type characteristically occurs when the solute has an unusually high affinity for the solid, such as would be the case for chemisorption of positively charged organic species by the negatively charged clay.

Adsorption isotherms for some select pesticides on soil clays are shown in Fig. 18.2.[11] The H-type isotherm for paraquat, a positively charged herbicide, is characteristic of an ion-exchange reaction. The L-type isotherm shown for prometone suggests monolayer adsorption and little, if any, competition with the solvent for adsorption sites on the clay. In contrast, the isotherm for Dasanit® indicates strong competition of water for adsorption sites. The isotherm for Lindane belongs to a subdivision of the C-type, the linear relationship being indicative of partition of the herbicide between solvent and clay surface.

Mathematical descriptions of adsorption isotherms (L-type) are conveniently carried out by application of the Freundlich and Langmuir equations.

The Freundlich adsorption equation is

$$x/m = KC^{1/n} \tag{1}$$

Fig. 18.1 Classification of adsorption isotherms according to Giles[10]

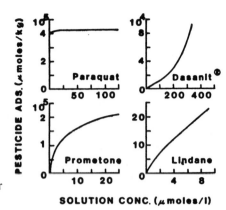

Fig. 18.2 Adsorption isotherms for four select herbicides. Adapted from Green[11]

or

$$\log (x/m) = \log K + 1/n \log C \qquad [2]$$

where x/m is the quantity of solute adsorbed per unit weight of adsorbent, C is the equilibrium concentration of the adsorbing compounds, and K and n are constants.

In practical terms, a straight line is obtained when the data are plotted as $\log(x/m)$ versus $\log C$. The intercept is equal to $\log K$ and the slope to $1/n$.

The constant K provides an indication of the extent of adsorption and has sometimes been used in correlation studies for determining the relative importance of the various soil properties on adsorption. A modified form of Eq. [1], with n taken as unity, has been used to express binding data for pesticide adsorption by soils (see Chapter 19).

Theoretically speaking, the Freundlich equation represents a situation in which the quantity of solute adsorbed increases indefinitely with increasing concentration. This will occur when multi-layers are formed. Also, heterogenous sorption sites are implied.

The Langmuir equation, which can be derived from the so-called Adair's equation for adsorption of a small molecule by a macromolecule (see Chapter 16), is as follows:

$$x/m = \frac{KbC}{1 + KC} \qquad [3]$$

or

$$C/x/m = \frac{1}{Kb} + \frac{C}{b} \qquad [4]$$

where x/m and C are the units defined above, K is a constant related to the bonding energy, and b is the adsorption maximum or total amount of solute capable of being adsorbed.

In this case, a straight line is obtained when $C/x/m$ is plotted against the equilibrium concentration, C. The Langmuir adsorption equation assumes monolayer adsorption on a uniform surface with no interactions between adsorbed molecules.

Adsorption Mechanisms

Several mechanisms are involved in the adsorption of organic compounds by clay minerals, the main ones being: 1) physical adsorption, or van der Waals forces, 2) electrostatic attraction or chemical adsorption, 3) H-bonding, and 4) coordination complexes. The latter two are considered as separate mechanisms although strictly speaking they fall within the other two categories. Two or more mechanisms may operate simultaneously, depending on the properties of the organic species, nature of the exchangeable cation on the clay, surface acidity, and moisture content of the system.

Most adsorption studies have been carried out with relatively simple clay-organic systems. However, it is reasonable to assume that the mechanisms involved will also apply to the native soil organic matter.

Physical Bonding (van der Waals Forces) Physical or van der Waals forces operate between all molecules, but are rather weak. Essentially, these forces result from fluctuations in the electric charge density of individual atoms. An electrically positive fluctuation in one atom tends to produce an electrically negative fluctuation in a neighboring atom and a net attractive force results. Attractive forces due to these fluctuations are presumed to exist between every pair of atoms or molecules.

A polar molecule has a permanent net asymmetry in the distribution of negative charges and molecules of opposite sign can line up through what is called a dipole–dipole interaction. The dipole moment of one molecule can produce a dipole in its neighbor, thereby leading to what is referred to as "induced" dipole interactions.

The following shows the attraction of two molecules due to a lineup of positive and negative charges.

$$\underline{\overline{\begin{array}{c} +-+-+- \\ -+-+-+ \end{array}}}$$

Adsorption due to physical forces can be of considerable importance in the adsorption of neutral polar and nonpolar molecules, particularly those that are high in molecular weight. This can be accounted for by the additive nature of these interactions. For many nonpolar organic compounds, van der Waals interactions may dominate the adsorption process.

Electrostatic Bonding Electrostatic bonding by soils and soil constituents (e.g., clay and organic matter) occurs through the process of cation exchange or protonation. The former occurs when positively charged organic cations replace inorganic cations on the exchange complex, as follows:

$$\boxed{\text{Clay}} - M^+ + R-NH_3^+ \rightleftharpoons \boxed{\text{Clay}} - \overset{+}{N}H_3R + M^+$$

The cationic property of a weakly basic organic molecule is strongly pH dependent. Thus, adsorption by this mechanism is related to both the basic character of the organic molecule and the pH of the system. Adsorption is also influenced by the properties of the organic cation (e.g., chain length) and the type of cation on the exchange complex.

As will be shown later, organic compounds that are weakly basic may be adsorbed at particle surfaces through protonation, a process in which an organic molecule assumes a positive charge by accepting an H^+ ion.

H-Bonding This is a linkage between two electronegative atoms through bonding with a single H^+ ion. The H^+ ion is a bare nucleus with a charge of $+1$ and has a strong tendency to share electrons with those atoms that contain an unshared electron pair, such as oxygen. The bond is weaker than ordinary ionic or covalent bonds but stronger than van der Waals forces of attraction.

Typical ways in which H-bonding may occur at clay surfaces are as follows:

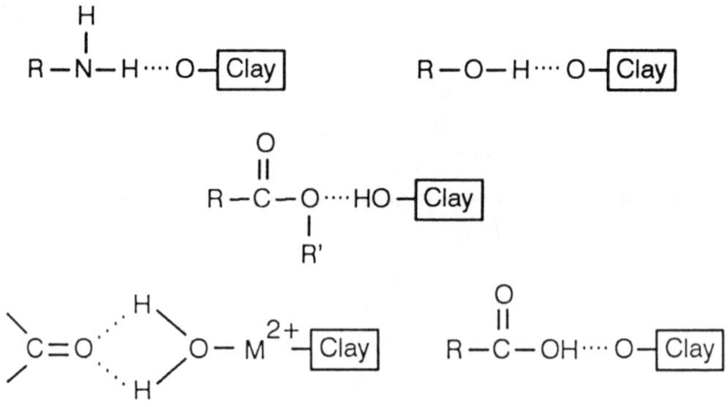

$$R-\underset{|}{\overset{H}{N}}-H \cdots O-\boxed{\text{Clay}} \qquad R-O-H \cdots O-\boxed{\text{Clay}}$$

$$R-\underset{|}{\overset{\overset{O}{\parallel}}{C}}-O \cdots HO-\boxed{\text{Clay}}$$
$$R'$$

$$\overset{\diagdown}{\underset{\diagup}{C}}=O \cdots \overset{H}{\underset{H}{\cdots}} \cdots \overset{\diagdown}{\underset{\diagup}{O}}-M^{2+}-\boxed{\text{Clay}} \qquad R-\overset{\overset{O}{\parallel}}{C}-OH \cdots O-\boxed{\text{Clay}}$$

A prime site for adsorption is surface oxygens or OHs; other possibilities include protons of adsorbed water molecules.

Coordination Bonding by this means is similar to that described in Chapter 16 for the complexation of polyvalent cations by organic matter. Essentially, the metal ion forms a bridge between the organic molecule and the soil constituent to which it is attached. For example, coordination bonding of a *s*-triazine herbicide may occur as follows:

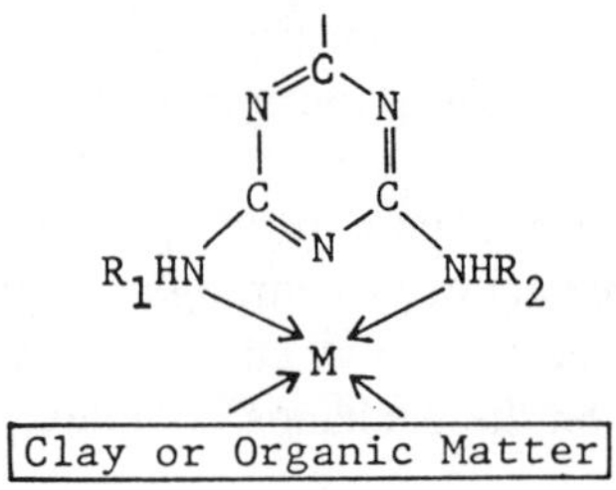

$$R_1HN-\underset{\underset{\boxed{\text{Clay or Organic Matter}}}{M}}{\text{triazine ring}}-NHR_2$$

A closely related bonding mechanism is where a polyvalent cation acts as a salt bridge between the soil component and the COOH group of an organic molecule.

$$\text{Clay or organic matter} - M - OOCR$$

Adsorption of Specific Classes of Organic Compounds

Extensive studies have been carried out on the adsorption of organic molecules by clay minerals and the following is an attempt to summarize the overall findings. Much of the early work came under the category of scientific inquiry; in recent years, the research has assumed increasing practical importance due to the introduction of synthetic organics into the environment, such as herbicides, insecticides, and fungicides. Specific reviews on pesticide interactions in soil are cited in Chapter 19.

Bailey and White[12] suggested four structural factors that determine the chemical character of a pesticide molecule and thus influence its adsorption on soil colloids:

1. Nature of functional group(s), such as alcoholic ($-OH$), ketonic ($>C=O$), carboxylic acid ($-COOH$), or amine ($-NH_2$)
2. Nature of substituting groups, which tend to alter the behavior of functional groups
3. Position of substituting groups with respect to the functional group that may enhance or hinder intramolecular bonding
4. Presence and magnitude of unsaturation in the molecule, which affects the lyophylic–lyophobic balance

As will be apparent later, these factors are in turn responsible for certain chemical properties that are important in adsorption to clays, notably acidity or basicity as reflected by the pK_a of the compound. Green[11] pointed out that emphasis on charge characteristics is warranted for the following reasons:

1. Clay surfaces are active in adsorption principally because of surface charge, the charge being predominantly negative in crystalline aluminosilicates, but positive in some clays, as described previously.
2. Adsorption reactions involving identifiable charge satisfaction are more easily characterized than adsorption due to the more obscure van der Waals forces (hence the large number of papers reporting detailed research on the nature of adsorption or organic cations and weak bases on montmorillonite).
3. The solvent of interest in pesticide–clay interactions is water, which is itself highly polar, so that the charge, polarity, or polarizability of a pesticide is pertinent not only to its interaction with the clay, but also controls its interaction with water; these properties largely determine a

compound's solubility in water and also the extent to which a pesticide can compete with water for adsorption sites on the clay.

Postulated adsorption mechanisms for some organic herbicides by montmorillonite are shown in Table 18.2.[13] Some notes regarding adsorption of specific classes of organic compounds are given below.

Organic Acids Negatively charged organic species are normally repelled by the negatively charged clay with little or no adsorption. However, the anionic character of organic acids is pH dependent and some adsorption is possible through H-bonding and van der Waals forces when the pH falls below the pK_a of the acidic group (molecule in undissociated form). As noted in Chapter 15, the pK_a of soil humic substances is strongly influenced by salt concentration; thus, the amount retained through H-bonding under natural soil conditions will be affected by the ionic composition of the soil solution.

Binding of organic anions by clay is possible when a polyvalent cation is present on the exchange complex, in which case the cation neutralizes the negative charge on the clay as well as the acidic group of the organic molecules to form a salt bridge (clay—M—OOCR). It is probably in this manner and through coordination that humic and fulvic acids are retained in the natural soil (see subsequent section). Acidic polysaccharides are undoubtedly bound in a similar manner.[14,15]

Huang et al.[16] found that the sorption capacity of soils for phenolic acids was significantly decreased by removal of sesquioxides. They concluded that the high sorption capacity of noncrystalline hydroxy Al and Fe complexes for phenolic acids was due to interaction of negatively charged carboxylate COO^- and phenolate O^- groups with positively charged $Al-OH_2^{0.5+}$ and $Fe-OH_2^{0.5+}$ sites. Acid sugars containing the COOH group (e.g., uronic acids) would be expected to be bound in the same way and to be favored by a low pH and the presence of Al.[17]

Organic Cations, Amines, and Amino Acids Positively charged organic molecules can be tightly bound to clay colloids through chemical adsorption. Retention of low-molecular-weight organic cations (less than about eight carbons) has been shown to be approximately equal to the CEC of the clay; for larger cations, adsorption is limited only by the surface area. Adsorption is accompanied by a corresponding release of the exchangeable cation. The adsorbed organic cation cannot normally be removed with water but may be replaced by leaching with a salt solution.

A factor of considerable importance in governing the adsorption of weak organic bases is whether or not the molecule can be protonated. The reaction for protonation of an organic amine is

$$RNH_2 + H_2O \rightleftharpoons RNH_3^+ + OH^-$$

Table 18.2 Postulated Adsorption Mechanisms for Some Organic Herbicides on Montmorillonite (Adapted from Bailey et al.[13])

Adsorbate family	R-N-H··· ··O-Clay	$\geq$C-H··· ··O-Clay	>C=O··· ··M^{z+}-Clay	$\underset{\text{R}}{\overset{\text{O}}{\overset{\|}{\text{R-C-O}}}}$··· ··HO-Clay	>C=O···$\overset{H}{\underset{H}{\overset{O}{\diagup\hspace{-6pt}\diagdown}}}$···O=C< ··$M^{z+}$-Clay	B + (H^+-Clay) → (HB^+-Clay)
s-Triazine	A_b	$A_a(?), A_b$	—	—	—	A_a
Substituted Ureas	A_b	—	A_b	—	$A_a(?), A_b$	A_a
Phenylcar-bamates	A_b	—	A_b, N_{Na}	A_b	$A_a(?), N_{Na}$	A_a
Anilide	A_b	A	A_b, N_{Na}	—	$A_a(?), N_{Na}$	A_a, N_{Na}
Phenylalkanoic acids	—	$A_a(?), A_b$	A_a, N_{Na}	—	$A_a(?), N_{Na}$	—

A = Adsorption mechanism applicable in both acidic and neutral systems
A_a = Mechanism applicable when acidity is such that pH$\leq$pK + 2
A_b = Mechanism applicable when acidity is such that pH>pK + 2
— = Appropriate group absent
N = No adsorption in either acidic or neutral systems
N_{Na} = No adsorption in the Na-montmorillonite system (pH 6.8)

for which the equilibrium constant is

$$K_b = \frac{[RNH_3^+][OH^-]}{[RNH_2]}$$

where H_2O is not shown because of the large amounts usually present. The equilibrium constant is often expressed in terms of a pK_b or $-\log K_b$.

For comparative purposes, the equilibrium constant is often given in terms of the appropriate acid dissociation constant, or more precisely, the pK_a as follows:

$$pK_a = 14 - pK_b \text{ at } 25°C$$

The pK_a value provides a convenient way of comparing the strength of acids or bases. The stronger the acid, the lower its pK_a; the stronger the base, the higher its pK_a.

Weakly basic compounds will occur in cationic forms only at pH values near or below their respective pK_a values (molecule in protonated form as RNH_3^+). The approximate relationship between pH, pK_a, and percentage of the molecule in the cationic form is as follows:

Relationship	Percentage of Molecules as	
	RNH_2	RNH_3^+
pH = pK_a	50	50
pH = pK_a + 1	90	10
pH = pK_a - 1	10	90

Within the pH range of most soils (5.0–8.0) many pesticides of the weakly basic type will occur in cationic form through gain of H^+ and will therefore be adsorbed by an ion-exchange mechanism.

For an amino acid, the relationship between the different forms are as follows:

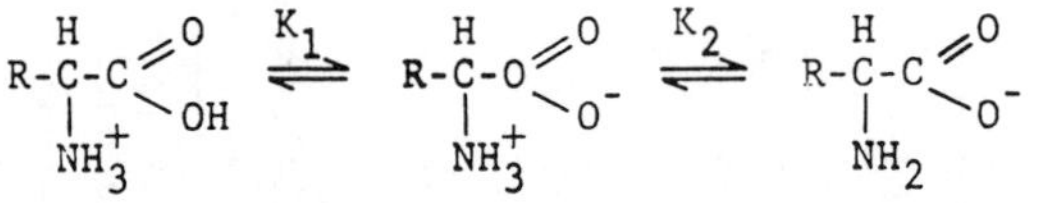

$$\underset{\text{Cation} \atop pH<pK_a}{\text{R-C-C}\begin{smallmatrix}H \\ | \\ NH_3^+\end{smallmatrix}\begin{smallmatrix}\nearrow O \\ \searrow OH\end{smallmatrix}} \overset{K_1}{\rightleftharpoons} \underset{\text{Zwitterion} \atop pH=pK_a}{\text{R-C-C}\begin{smallmatrix}H \\ | \\ NH_3^+\end{smallmatrix}\begin{smallmatrix}\nearrow O \\ \searrow O^-\end{smallmatrix}} \overset{K_2}{\rightleftharpoons} \underset{\text{Anion} \atop pH>pK_a}{\text{R-C-C}\begin{smallmatrix}H \\ | \\ NH_2\end{smallmatrix}\begin{smallmatrix}\nearrow O \\ \searrow O^-\end{smallmatrix}}$$

Some weak organic bases can be adsorbed by clay through direct protonation (molecule accepts an H^+ ion and assumes a positive charge). Sources of the proton include exchangeable H^+ on the exchange complex (reaction 1), hydration water of metal cations associated with clay mineral surfaces (reaction 2), and proton transfer from a protonated species already present (reaction 3).

$$\text{Clay}-H^+ + RNH_2 \rightarrow \text{Clay}-NH_3^+R \qquad [1]$$

$$\text{Clay}-M(H_2O)_x^{+n} + B \rightleftharpoons \text{Clay}-MOH(H_2O)_{x-1}^{n-1} + BH^+ \qquad [2]$$

$$\text{Clay}-AH^+ + B \rightleftharpoons \text{Clay}-H^+B + A \qquad [3]$$

The degree to which reaction 2 takes place is strongly influenced by the acidic properties of the hydrated metal ion.

As far as amino acids are concerned, adsorption occurs through cation exchange below the isoelectric point of the amino acid

$$\begin{pmatrix} R-CH-COOH \\ | \\ NH_3^+ \end{pmatrix}$$

As was the case with weak organic bases, amino acids may become cationic by protonation at clay surfaces.

Nonionic Polar Molecules A wide variety of uncharged organic molecules of polar character have been shown to be adsorbed by clay minerals. The main mechanism is by H-bonding, although nonpolar parts of the molecule may be further bound by van der Waals forces. Low-molecular-weight compounds are generally adsorbed only in very small amounts but uncharged polymers can be adsorbed in large amounts and strongly retained. The strong binding of high-molecular-weight polymers is undoubtedly due to the larger number of bonding points with the surface.

NATURALLY OCCURRING CLAY-ORGANIC COMPLEXES

Extensive studies have shown that much of the humified material in soil is firmly bound to colloidal clay. However, it is uncertain as to what proportion of the clay surface in any given soil is coated by organic substances—this will depend on organic matter content and the kind and amount of clay. In soils containing exceptional amounts of organic matter, such as many prairie grassland soils, practically all of the clay may be coated with a thin layer of organic matter.

Organic substances can be retained by clay minerals in two different ways:

1. By attachment to clay mineral surfaces, such as through cation and anion exchange, bridging by polyvalent cations (clay–metal–humus), H-bonding, van der Waals forces, and in other ways as described earlier
2. By penetration into the interlayer spaces of expanding-type clay minerals

With regard to item 2, adsorption of organic substances on interlamellar surfaces of montmorillonite causes a shift in the C-axis spacing as determined by X-ray diffraction. Whereas many defined organic compounds have been demonstrated in laboratory studies to be adsorbed in interlamellar surfaces (e.g., proteins), it has not been unequivocally established that such complexes occur naturally in soils and sediments. Presumably they do, because treatment of soils with acid solutions containing HF, which destroys hydrated silicate minerals, results in the solubilization of considerable quantities of organic matter, particularly from illuvial B horizons (see Chapter 2). Further evidence for interlamellar adsorption comes from the observation that recovery of amino acids from clay-rich subsoils by acid hydrolysis is incomplete unless preceded by pretreatment with HF (see Chapter 3).

Proportion of Soil Organic Matter Bound to Clay

Differential extraction procedures have been used in attempts to indicate the state of combinations of organic constituents in soil. Tyurin's scheme, as reported by Kononova,[18] is as follows:

Description	Types of Linkages
Humic substances soluble in dilute alkalies without prior removal of exchangeable Ca.	Free forms of polymeric complexes of humic and fulvic acids.
Humic substances which are only soluble in dilute alkalies after removal of exchangeable Ca with dilute mineral acids.	Polymeric complexes of humic and fulvic acids in linkage with Ca.
Humic substances soluble in dilute alkali only after alternate acid and base treatments.	Complexes of humic and fulvic acids linked with relatively stable hydrated sesquioxides.

Tyurin's work suggested that there were well-defined regularities in the way humic substances were combined in soils. In grassland soils, which are rich in exchangeable cations, the humic substances were believed to be linked mostly with Ca; free compounds were virtually absent. On the other hand, in forest soils the humic substances were predominately in the form of polymeric complexes readily solubilized by direct extraction with dilute alkali. The insoluble humic material in peat probably exists in polymeric forms held together by H-bonding, and so on.

In more modern times, the proportion of the soil C combined with clay has been estimated using the following approaches:

1. Removal of unbound organic matter by flotation in a liquid of density intermediate between the free material and the clay–organic complex. Solutions of density between 1.8 and 2.0 have been used, such as a benzene–bromoform mixture. The assumption is made that organic mat-

ter not removed by flotation occurs in intimate association with clay minerals

2. A combination of item 1 and further physical separations of organic matter into the different sand, silt, and clay fractions. This work is described in Chapter 2, for which references can be found. In most studies, much of the organic matter has been found to reside in the fine-clay fraction (<0.1 μm)

Results obtained by several workers for the proportion of the soil C contained in clay–organic complexes are given in Table 18.3. From 52 to 98 percent of the C in the soils examined was associated with clay. For some soils, the organic matter separated by flotation may have consisted largely of undecayed or partially modified plant remains.

Interactions Involving Humic and Fulvic Acids

Greenland[2] concluded that sufficient information was now available to enable the adsorption behavior of simple organic compounds at clay–mineral surfaces to be predicted with confidence. Whereas this statement may apply to many of the biochemical compounds that occur naturally in soils, it has limited appli-

Table 18.3 Proportion of Soil Organic C Contained in the Clay–Organic Complex[a]

Soil[b]	Method of Separation	Total C in Soil (%)	C in Clay-Organic Complex (% of total soil C)
Rendzina	Sedimentation in benzene-bromoform, s.g. 1.75	—	66.5
Podzol	Sedimentation in Toulet solution, s.g. 1.8	1.6	89.6
Chernozem		4.4	85.2
Silt under old pasture	Sedimentation in ethanol-bromoform, s.g. 2.0	2.34	77.5
Rendzina	"Flotation sieving"	5.8	54.3
Brown Earth		3.2	68.1
Red-Brown earth	Ultrasonic dispersion and sedimentation in bromoform pet. spirit, s.g. 2.0	2.23	71.5
Rendzina		5.8	68.4
Lateritic Red earth		1.7	97.8
Solodized Solonetz		1.04	76.4
Solonized Brown soil		0.58	51.6

[a] From Greenland.[3]

cation to the negatively charged, rather high-molecular-weight humic polymers, of which little is known. An understanding of the interaction between humic substances and clay is further hindered by the probability that hydrous oxides at clay surfaces may be involved.[2]

The possible types of bonds between humic substances and clays are summarized in Table 18.4[19]; most have been discussed in previous sections. Humic and fulvic acids contain a variety of reactive functional groups that are capable of combining with clay minerals, and, as shown by the schematic diagram of Fig. 18.3, the sorption of humic constituents by clay still provides an active organic surface for exchange with cations of the soil solution and for sorption of pesticides.[20]

Interactions With Clay Minerals Having Mica-Type Lattices (Montmorillonite, Illite, and Vermiculite) Since organic anions are normally repelled from negatively charged clay surfaces, adsorption of humic and fulvic acids by clay minerals such as montmorillonite occurs only when polyvalent cations are present on the exchange complex. Unlike Na^+ and K^+, polyvalent cations are able to maintain neutrality at the surface by neutralizing both the charge on the clay and the acidic functional group of the organic matter (e.g., COO^-). The importance of polyvalent cations in adsorbing humic and fulvic acids has often been emphasized.[21,22] Isotherms for the adsorption of humic acid by montmorillonite saturated with different cations are shown in Fig. 18.4.

The main polyvalent cations responsible for the binding of humic and fulvic

Table 18.4 Possible Types of Bonds Between Humic Substances and Minerals. Adapted from Francois[19]

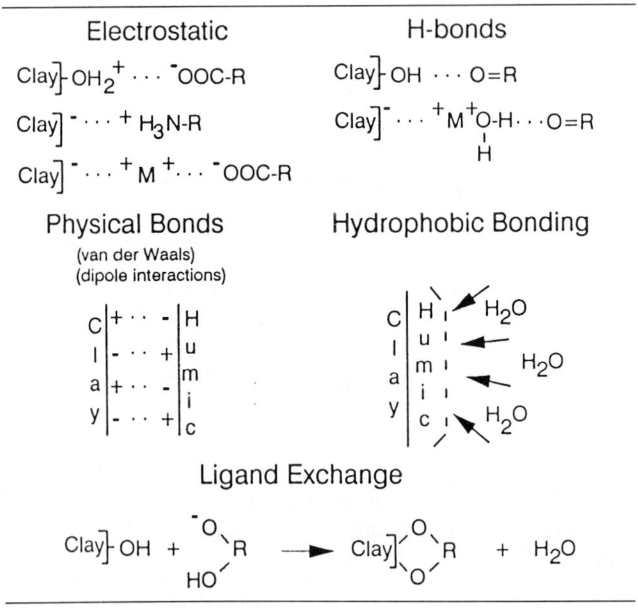

Electrostatic	H-bonds
$Clay\!\!]\!\!-OH_2^+ \cdots {^-}OOC\text{-}R$	$Clay\!\!]\!\!-OH \cdots O=R$
$Clay]^- \cdots {^+}H_3N\text{-}R$	$Clay]^- \cdots {^+}M{^+}O\text{-}H \cdots O=R$
$Clay]^- \cdots {^+}M^+ \cdots {^-}OOC\text{-}R$	

Physical Bonds	Hydrophobic Bonding
(van der Waals) (dipole interactions)	

Ligand Exchange

$Clay\!\!]\!\!-OH + {^-O\diagdown}R{\diagup}^{OH} \longrightarrow Clay]{\diagdown}^{O}_{O}\!R + H_2O$

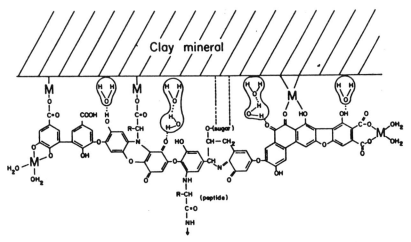

Fig. 18.3 Schematic diagram of clay–humate complex in soil. From Stevenson and Ardakani[20]

acids to soil clays are Ca^{2+}, Fe^{3+}, and Al^{3+}. The divalent Ca^{2+} ion does not form strong coordination complexes with organic molecules and would be effective only to the extent that a bridge linkage could be formed. As suggested by Greenland,[2] organic matter bound in this manner should be rather easily displaced by a monovalent cation, which may account for the small quantities of humic materials which can be displaced when Ca-saturated soils are leached with an NH_4^+-salt. In contrast, Fe^{3+} and Al^{3+} form strong coordination com-

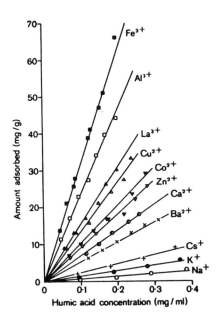

Fig. 18.4 Isotherms for the adsorption of humic acid by montmorillonite saturated with different cations at pH 7 and 25°C. From Theng and Scharpenseel[22]

plexes with humic substances, in which case displacement of the bound metal is difficult and may require extraction with a strong chelating agent.

Thus, two major types of interactions may be involved in the adsorption of organic polyanions by mica-type clay minerals. In the first type, the polyvalent cation acts as a bridge between two charged sites. In this case, hydration water of the exchange cation is not displaced but forms an H-bond with the organic anion (see item "electrostatic interactions" in Table 18.4). In the second type, the organic anion becomes coordinated to the cation, with displacement of a water molecule from the hydration shell (see "ligand exchange" in Table 18.4). For a long-chain organic molecule, several points of attachment to the clay particle are possible.

Other bonding forces may also operate between organic anions and the clay surface, including H-bonding between polar groups of the organic molecule and adsorbed water molecules or oxygens of the silicate surface. The strength of an individual bond is small but they are additive; thus, total adsorption energy can be appreciable. Rigorous drying, such as by desiccation at the soil surface or consumption of available moisture by plant roots, will tend to increase the bonding between humic material and clay be eliminating hydration water and bringing the humic matter in closer contact to the clay.

Another mechanism for adsorption of humic substances to mica-type clay minerals is by association with hydrous oxides at the surface. Greenland[2] suggested that, for many soils, hydrous oxides are equal in importance to mica-type surfaces in the sorption of humic substances. The mechanisms involved are similar to those described in the next section.

An important mechanism for retention of proteins and charged organic cations by expandable layer silicates is through adsorption on interlamellar spaces (see reviews mentioned earlier). Considerable controversy exists as to whether humic and fulvic acids are bound in this way in the natural soil; in laboratory studies, both positive[23,24] and negative[25] results have been obtained.

Interactions With Hydrous Oxide When clay minerals are coated with layers of hydrous oxides their surface reactions are dominated by these oxides rather than the clay. Opportunities exist for strong bonding through "coordination" (ligand exchange), as well as by "anion" exchange. Retention by anion exchange is possible because positive sites can exist on iron and Al oxides due to protonation of exposed OH groups under slightly acidic conditions, i.e., oxide]$-$OH + H$^+$ $\rightarrow$ oxide]$-$OH$_2^+$.

Organic anions bound by electrostatic forces would be expected to be readily removed by increasing the soil pH, or by leaching with NaCl of NH$_4$Cl. The fact that very little native soil organic matter is solubilized by these treatments (see Chapter 2) suggest that most of the organic matter is retained by supplementary mechanisms (e.g., ligand exchange).

Humic materials may interact with crystalline oxides in a similar manner. Allophanic material in soils developed in volcanic ash are strong adsorbents of humic substances, which accounts for the exceptionally high levels of organic matter in soils of high allophane content. Parfitt et al.[26] found that the sorption of fulvic acid on oxide surfaces was accompanied by displacement of OH groups by COO^- ions, indicating ligand exchange.

ROLE OF ORGANIC MATTER IN FORMING STABLE AGGREGATES IN SOIL

The term "soil structure" refers to the arrangement of soil particles into secondary particles or aggregates. Stable soil aggregates influence plant growth through their effects on: 1) aeration; 2) water penetration and retention; 3) mechanical impedance to roots; and 4) emergence of shoots. When the physical condition of the soil is such that water cannot enter or drain readily, germinating seeds fail to obtain the oxygen they require for respiration and shoots cannot break through the surface crust. Yields may be reduced accordingly, even though adequate nutrients are available in the soil.

Organic matter is considered to be of immense importance in forming good aggregates in a wide range of soil types, particularly those representative of the Mollisols, Alfisols, Ultisols, and Inceptisols. Organic matter is somewhat less important in the Oxisols, where hydrous oxides may play a predominant role. These relationships are illustrated in Fig. 18.5. A factor of some importance where organic matter plays a major role is the cohesion of clay and/or clay–humic complexes into larger particles, such as by sharing of intercrystaline ionic forces between clay particles or by bridging through the action of polyvalent cations and humus.

Aggregates do not directly influence plant growth but they alter the physical and chemical environments in which plant roots grow through their effects on porosity, aeration, moisture retention, and so on. Soils that are poorly aggregated have pores too small to permit the necessary movement of water and air. When well aggregated, even fine-textured soils permit adequate exchange of gases with the atmosphere.

The polysaccharide component of the soil has received the greatest attention in producing stable aggregates[7-9,27-31] but other organic substances (i.e., humic and fulvic acids) bonded to clay through association with Fe or Al may also be involved. For a study of the comparative effect of polysaccharides and humic substances on soil aggregation, the reader is referred to the paper by Swift.[29]

Haynes and Swift[30] found that, for a group of soils with variable cropping histories, aggregate stability was closely correlated with hot water-extractable carbohydrates, and a similar result was obtained by Kinsbursky et al.[31] for sewage-sludge amended soils. Foster,[27] using a staining technique specific for carbohydrates containing polyuronides, found that soil carbohydrates not only exist as coatings on clay platelets but that they occur in crevices of submicron sizes within mineral aggregates.

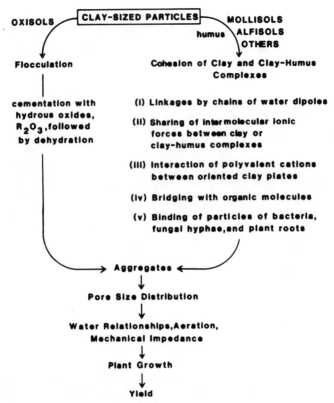

Fig. 18.5 Relationships between organic matter and the formation of soil aggregates. Adapted from an unpublished diagram by A.E. Erickson (personal communication)

Specific Action of Organic Substances

Organic constituents may influence soil aggregation in at least three different ways, as follows:

1. Organic substances serve as binding agents for the cohesion of clay particles, such as through H-bonding and coordination with polyvalent cations as described above. Flocculation of clay is considered to be an obligate prerequisite for aggregation through coprecipitation or flocculation with clay colloids. A wide variety of organic compounds may be involved, including humic and fulvic acids, which are probably linked to clay as a clay–metal–humus complex (see Fig. 18.3).

   Clay particles themselves may cohere and entrap or bridge between larger sand and soil grains. However, in the absence of organic coatings, clay is easily dispersed through the slaking action of water. Many investigators believe that the direct union of organic matter with clay is of considerable importance in the formation of aggregates of high stability.

2. Gelatinous organic materials surround soil particles and hold them together through a cementing or encapsulation action. Whereas a variety of compounds may be responsible, several lines of work indicate that a major role is played by the polysaccharides (see previous section). Living bacteria and other microorganisms have been shown in laboratory studies to bind soil particles together, and this has been attributed to the polysaccharides they secrete.

   For the most part, polysaccharides are readily attacked by microorganisms; thus, their effect on aggregation would be expected to be short lived. This explains why fresh organic matter in the form of plant or animal residues needs to be returned periodically to the soil in order to provide energy for resynthesis through the activities of microorganisms. *The action of polysaccharides and other gelatinous substances has been likened to the formation of "string of beads" when the polymers are charged and to the spreading of "coats of paint" when they are uncharged.*[3]

   Results of physical fractionations (Chapter 2), in combination with analyses of the various size fractions by ^{13}C–NMR spectroscopy (Chapter 11) have shown that the coarse clay (0.4–2 μm) generally contains the highest amount of polysaccharides. A review of physical fractionation, as related to the role of organic matter in the formation of stable aggregates, has been given by Baldock et al.[32]

3. Soil particles are held together through physical entanglement by fungal hyphae and microscopic plant roots. They mycelia of fungi are often extended throughout the soil and particles are entrapped and tied together. This is sometimes apparent with the naked eye; under magnification, small clumps of soil particles can often be seen clinging to the mycelia. Aggregation through this mechanism is ephemeral and dependent on maintaining a high content of fungal hyphae. The fibrous root systems of the grasses not only ramify and open up the soil but they encompass individual crumbs in a net-like web to form clusters resistant to the slaking action of water. *The action of fungal hyphae and roots in forming stable aggregates can be likened to binding by string or twine.*[3]

In concluding this section, it should be mentioned that one concept of aggregate formation is that clay is the principal binding agent and that organic materials do not act primarily to hold soil particles together but that their chief role is to modify the forces whereby clay particles themselves are attracted to one another. According to this view, it is the cohesive force between clay particles rather than the cementing action of organic substances that is the binding force in aggregation. This theory would appear to be inadequate for explaining aggregate formation is most agricultural soils.

Whereas relatively little is known as to how stable aggregates are formed, their synthesis is influenced by: 1) the kind and amount of organic matter in the soil, especially gums and mucilages, 2) the presence or absence of fungal hyphae and microscopic plant roots, 3) wetting and drying, 4) freezing and thawing, 5) nature of the cation on the exchange site, and 6) the burrowing action of soil animals, especially earthworms.

Theories of Soil Aggregation

At the outset it should be pointed out that diverse organic and inorganic constituents participate in the binding of soil particles into water stable aggregates and that the relative importance of each will depend on the environmental conditions under which aggregates are formed in any given soil type.

Observations from Dispersion of Soil Particles by Ultrasonic Vibration
Information concerning aggregation has come from studies using sonic vibration for dispersing soil particles.[33,34] The concepts that have evolved from these studies have been summarized by Bremner[33] as follows:

1. In soils well supplied with exchangeable cations, the fine-sand and silt-size microaggregates are mostly <250 μm in diameter and consist largely of clay and organic colloids linked together through polyvalent cations. These microaggregates can be represented as

$$[(C-P-OM)_x]_y$$

where C indicated clay, P the polyvalent metal ion (Ca^{2+}, Mg^{2+}, Fe^{3+}, etc.), OM the humified organic matter, and $C-P-OM$ the clay-size particles (<2 μm); x and y are finite whole numbers dictated by the size of the primary clay particle.

2. Stable microaggregates are formed by a mechanism that is a reversal of what occurs when soil particles are dispersed by water shaking. The reversible processes of dispersion (D) and aggregation (A) can be represented as follows:

$$[(C-P-OM)_x]_y \underset{A}{\overset{D}{\rightleftharpoons}} y(C-P-OM)_x \underset{A}{\overset{D}{\rightleftharpoons}} xy(C-P-OM)$$

The size, shape, and stability of the microaggregates are determined to a large extent by such factors as the absolute and relative amounts of clay and humified organic matter.

Clay Domain Theory of Emerson[35] According to this theory of aggregate formation, crumbs are formed from units of colloidal clay, or domains, and coarser particles of silt and sand (quartz) cemented together by humus. A domain was defined as "a group of clay crystals having suitable exchangeable cations which are oriented and sufficiently close together for the group to behave in water as a single unit."

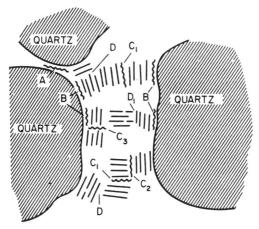

Fig. 18.6 Possible arrangements of organic matter, clay domains, and quartz to form a soil crumb. A: quartz–organic matter–quartz. B: quartz–organic matter–clay domain. C: clay domain–organic matter–clay domain (C_1 = face–face, C_2 = edge–face, C_3 = edge–edge). D: clay domain–clay domain, edge–face. From Emerson,[35] reproduced by permission of Oxford University Press, Oxford

Emerson's[35] model of a soil crumb is shown in Fig. 18.6. Four possible types of bonds are: A, quartz-organic matter-quartz; B, quartz–organic matter domain; C, domain–organic matter-domain (organic matter positioned between the faces of two clay domains, between two edges, and between an edge and a face); and D, domain–domain, edge-face.

Whereas the theory may explain many of the properties of soil crumbs, it fails to take into account the extensive occurrence of clay–organic complexes. In soils that are well supplied with organic matter, and which are often well-aggregated, most of the clay will be coated with organic matter. The "clay" domains of Emerson may in reality exist partly as "clay–humus" domains.

SOIL WETTABILITY

The condition of water-repellency on nonwettability due to the coating of sand grains with organic substances has been observed for citrus groves, burned-over areas of forest soils, and turf. In most cases, repellency has been associated with coarse-textured sandy soils.

The nonwettable condition of sandy soils has often been attributed to fats and waxes but evidence in support of this hypothesis is circumstantial. The study of Jamison[36] on the nonwettability of some Florida sandy soils would seem to eliminate lipid constituents as causative agents, because extraction of the problem soils with various organic solvents failed to improve wettability. On the other hand, Wander[37] concluded that the repellent nature of these soils was associated with the application of limestone or Mg-containing fertilizers.

He postulated that the added Ca or Mg formed insoluble soaps with soil fatty acids and these were responsible for the water-repellent character of the soils. The soaps, once formed, were only slightly soluble in ether and nonpolar solvents, which may explain why Jamison was unable to improve soil wettability by ether extraction. Soils in other areas of the world have been observed to have wetting characteristics similar to those of Florida.

In certain watersheds, wildfires have been shown to cause extreme water repellency in the underlying soil.[38-40] This has been attributed to vaporization or production of hydrophobic substances at the soil surface, which then migrate into the soil proper and condense in the cooler soil areas.

Miller and Wilkinson[41] concluded that fulvic acids synthesized by fungi were responsible for localized water-repellent areas on sand golf greens.

SUMMARY

A wide array of organic compounds, including amino acids, peptides, proteins, saccharides, and pesticides are absorbed to varying degrees by clays. Adsorption is determined to a large extent by the properties of the organic molecule, pH, kind and amount of clay, and nature of the exchange cation on the clay surface. Organic substances are protected against attack by microorganisms through adsorption by clays.

Organic substances of microbial and plant origin cement soil particles together to form stable aggregates, thereby influencing the physical properties of the soil. The coating of sand grains by organic compounds can lead to the condition of water-repellency.

REFERENCES

1. G. V. Jacks, *Soil Fert.*, **26,** 147 (1963).
2. D. J. Greenland, *Soil Sci.*, **111,** 34 (1971).
3. D. J. Greenland, *Soil Fert.*, **28,** 415, 521 (1965).
4. M. H. B. Hayes and F. L. Himes, "Nature and Properties of Humus–Mineral Complexes," in P. M. Huang and M. Schnitzer, Eds., *Interactions of Soil Minerals with Natural Organics and Microbes*, Special Publication 17, Soil Science Society of America, Madison, 1986, pp. 104–158.
5. B. K. G. Theng, *The Chemistry of Clay–Organic Reactions*, Hilger, London, 1974.
6. B. K. G. Theng, *Formation and Properties of Clay–Polymer Complexes*, Hilger, London, 1979.
7. W. W. Emerson, R. C. Foster, and J. M. Oades, "Organo–Mineral Complexes in Relation to Soil Aggregation and Structure," in P. M. Huang and M. Schnitzer, Eds., *Interactions of Soil Minerals with Natural Organics and Microbes*, Special Publication 17, Soil Science Society of America, Madison, 1986, pp. 521–548.
8. R. F. Harris, G. Chesters, and O. N. Allen, *Adv. Agron.*, **18,** 107 (1966).

9   J. M. Oades, *Plant Soil,* **76,** 319 (1986).
10 C. H. Giles, T. H. MacEwan, S. N. Nakiva, and D. Smith, *J. Chem. Soc.,* **1960,** 3973 (1960).
11 R. E. Green, "Pesticide-Clay-Water Interactions," in W. D. Guenzi, Ed., *Pesticides in Soil and Water,* American Society of Agronomy, Madison, 1975, pp. 3-36.
12 G. W. Bailey and J. L. White, *Residue Rev.,* **32,** 29 (1970).
13 G. W. Bailey, J. L. White, and T. Rothberg, *Soil Sci. Soc. Amer. Proc.,* **32,** 222 (1968).
14 C. E. Clapp and W. W. Emerson, *Soil Sci.,* **114,** 210 (1972).
15 R. L. Parfitt and D. J. Greenland, *Soil Sci. Soc. Amer. Proc.,* **34,** 862 (1970).
16 P. M. Huang, T. S. C. Wang, M. K. Wang, M. H. Wu and N. W. Hsu, *Soil Sci.,* **123,** 213 (1977).
17 R. L. Parfitt, *Soil Sci.,* **113,** 417 (1972).
18 M. Kononova, *Soil Organic Matter,* Pergamon Press, New York, 1966.
19 R. Francois, *Rev. Aquatic Sci.,* **3,** 41 (1990).
20 F. J. Stevenson and M. S. Ardakani, "Organic Matter Reactions Involving Micronutrients in Soils," in J. J. Mortvedt, P. M. Giordano, and W. L. Lindsay, Eds., *Micronutrients in Agriculture,* American Society of Agronomy, Madison, 1972, pp. 79-114.
21 L. T. Evans and E. W. Russell, *J. Soil Sci.,* **10,** 119 (1959).
22 B. K. G. Theng and H. W. Scharpenseel, "The Adsorption of ^{14}C Labelled Humic Acid by Montmorillonite," *Proc. Intern. Clay Mineral Conf.,* Mexico City, 1975, pp. 643-653.
23 M. Schnitzer and H. Kodama, *Soil Sci. Soc. Amer. Proc.,* **31,** 632 (1967).
24 F. M. Martinez and L. J. P. Rodriguez, *Z. Pflanz. Ernähr. Bodenk.,* **124,** 52 (1969).
25 B. K. G. Theng, *Geoderma,* **15,** 243 (1976).
26 R. L. Parfitt, A. R. Fraser, and V. C. Farmer, *J. Soil Sci.,* **28,** 289 (1977).
27 R. C. Foster, *Science,* **214,** 665 (1981).
28 M. V. Cheshire, C. P. Sparling, and C. M. Mundie, *Plant and Soil,* **76,** 339 (1984).
29 R. S. Swift, "Effects of Humic Substances and Polysaccharides on Soil Aggregation," in M. S. Wilson, Ed., *Advances in Soil Organic Matter Research; The Impact on Agriculture and the Environment,* Redwood Press Ltd., Wiltshire, England, 1991, pp. 153-162.
30 R. J. Haynes and R. S. Swift, *J. Soil Sci.,* **41,** 73 (1990).
31 R. S. Kinsbursky, D. Levanon, and B. Yaron, *Soil Sci. Soc. Amer. J.,* **53,** 1086 (1989).
32 J. A. Baldock, G. J. Currie, and J. M. Oades, "Organic Matter as Seen by Solid State ^{13}C NMR and Pyrolysis Tandem Mass Spectrometry," in M. S. Wilson, Ed., *Advances in Soil Organic Matter Research: The Impact on Agriculture and the Environment,* Redwood Press Ltd., Wiltshire, England, 1991, pp. 45-60.
33 J. M. Bremner, *Pontifciae Academiae Scient. Scripta Varia,* **32,** 143 (1968).
34 A. P. Edwards and J. M. Bremner, *J. Soil Sci.,* **18,** 47, 64 (1967).

35 W. W. Emerson, *J. Soil Sci.*, **10,** 235 (1959).
36 V. C. Jamison, *Soil Sci. Soc. Amer. Proc.*, **10,** 25 (1945).
37 I. W. Wander, *Science,* **110,** 229 (1949).
38 L. F. DeBano, L. D. Mann, and D. A. Hamilton, *Soil Sci. Soc. Amer. Proc.*, **34,** 130 (1970).
39 S. M. Savage, *Soil Sci. Soc. Amer. Proc.*, **38,** 652 (1974).
40 S. M. Savage, J. Osborn, J. Letey, and C. Heaton, *Soil Sci. Soc. Amer. Proc.*, **36,** 674 (1972).
41 R. H. Miller and J. F. Wilkinson, *Soil Sci. Soc. Amer. J.*, **41,** 1203 (1977).

19

ORGANIC MATTER REACTIONS INVOLVING PESTICIDES IN SOIL

Adsorption by organic matter is a key factor in the behavior of many pesticides in soil, including bioactivity, persistence, biodegradability, leachability, and volatility. It has been well established, for example, that the rate at which an adsorbable herbicide must be applied to the soil in order to achieve adequate weed control can vary as much as 20-fold, depending upon the nature of the soil and the amount of organic matter it contains. Soils that are black in color (e.g., Mollisols) have higher organic matter contents than those that are light in color (e.g., Alfisols), and pesticide application rates must often be adjusted upward on the darker soils in order to achieve the desired result. The dominant role of organic matter in adsorbing pesticides has been emphasized in several reviews on the subject.[1-6]

In this chapter, emphasis will be given to organic matter reactions involving herbicides in soil. However, the material covered will apply equally as well to pesticides in general.

Chemical formulas for herbicides mentioned in the text are given in Table 19.1. Type structures or examples of some of the more common herbicides are as follows:

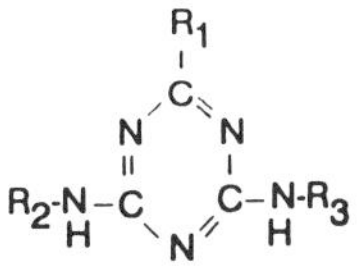

s-Triazines 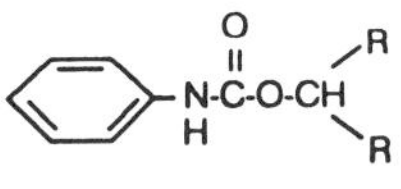   Phenylcarbamates 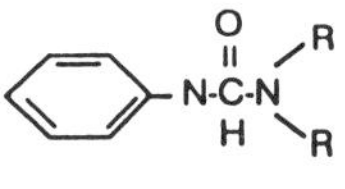   Substituted ureas

Amides — NH-C-R (O) on phenyl

Quaternary ammonium (diquat)

Phenoxyacetic acid — O-CH$_2$-COOH on phenyl

Three major factors determine the extent to which herbicides are adsorbed by soil organic matter, including: 1) physical–chemical characteristics of the adsorbents (chiefly humic colloids), 2) nature of the pesticides, and 3) properties of the soil system, such as clay mineral composition, pH, kinds and amounts of exchangeable cations, moisture, and temperature. Adsorption under any given set of soil conditions specified under 3) could better be evaluated if adequate information was available about the first two. The chemical and physical properties of the pesticide are nearly always known; accordingly, the main

Table 19.1 Chemical Formulas of Some Typical Pesticides

Common Name	Chemical Formula
s-Triazines	
Atrazine	2-chloro-4-ethylamino-6-isopropylamino-*s*-triazine
Simazine	2-chloro-4,6-bis(ethylamino)-*s*-triazine
Atratone	2-methoxy-4-ethylamino-6-isopropylamino-*s*-triazine
Ametryne	2-methylthio-4-ethylamino-6-isopropylamino-*s*-triazine
Prometon	2-methoxy-4,6-bis(isopropylamino)-*s*-triazine
Prometryn	2-methylthio-4,6-bis(isopropylamino)-*s*-triazine
Propazine	2-chloro-4,6-bis(isopropylamino)-*s*-triazine
Substituted Ureas	
Diuron	3-(3,4-dichlorophenyl)-1,1-dimethylurea
Monuron	3-(*p*-chlorophenyl)-1,1-dimethylurea
Fenuron	3-phenyl-1,1-dimethylurea
Linuron	3-(3,4-dichlorophenyl)-1-methoxy-1-methylurea
Neburon	1-butyl-3-(3,4-dichloropheny)-1-methylurea
Phenylcarbamate	
CIPC	isopropyl-*m*-chlorocarbanilate
Bipyridylium	
Quaternary Salts	
Diquat	6,7-dihydrodipyrido(1,2-a : 2′,1′-*c*)pyrazidinium salt
Paraquat	1,1′-dimethyl-4,4′dipyridinium salt
Others	
2,4-D	2,4-dichlorophenoxyacetic acid
Diphenamid	N,N-dimethyl-2,2-diphenylacetamide
Amitrole	3-amino-1,2,4-triazol
Lindane	1,2,3,4,5,6-hexachlorocyclohexane
DDT	1,1,1-trichloro-2,2-bis(*p*-chlorophenyl)ethane

limitation in understanding the role and function of organic matter in herbicide reactions is item 1).

The organic fraction of the soil also has the potential for promoting the nonbiological degradation of many herbicides,[7] as well as for forming strong chemical linkages with residues arising from their partial degradation by microorganisms.[4,5] These aspects of pesticide–soil organic matter interactions deserve further study because such processes play a key role in detoxification and protection of the environment. Chemical binding of pesticide-derived residues would increase their persistence in soil but probably in forms not harmful to the environment.

HUMUS CHEMISTRY IN RELATION TO PESTICIDE BEHAVIOR

Several aspects of organic matter chemistry require further elaboration regarding the fate of pesticides in soil, including 1) organic matter–clay interactions, 2) quantitative differences in the organic matter of soils, and 3) potential chemical reactions between pesticides and organic substances in soil. These items will be discussed briefly in the sections that follow.

Organic Matter Versus Clay as Adsorbent

Organic matter and clay are the soil components most often implicated in pesticide adsorption. However, individual effects are not as easily ascertained as might be supposed, for the reason that, in most soils, organic matter is intimately bound to the clay, probably as a clay–metal–organic complex. Thus, two major types of adsorbing surfaces are normally available to the pesticide, namely, clay–humus and clay alone. Accordingly, clay and organic matter function more as a unit than as separate entities and the relative contribution of organic and inorganic surfaces to adsorption will depend upon the extent to which the clay is coated with organic substances. As noted in Chapter 18 (e.g., see Fig. 18.3), interaction of organic matter with clay still provides an organic surface for adsorption.

Data published by Walker and Crawford[8] for adsorption of several *s*-triazines by 36 soils having widely variable organic matter contents (Fig. 19.1) suggest that, up to an organic matter content of about 8 percent, both organic and mineral surfaces are involved in adsorption; at higher organic matter contents, adsorption occurs mostly on organic surfaces. It should be noted, however, that the amount of organic matter required to coat the clay will vary from one soil to another and will depend on both kind and amount of clay in the soil. For soils having similar clay and organic matter contents, the contribution of organic matter will be highest when the predominant clay mineral is kaolinite and lowest when montmorillonite is the main clay mineral. Bailey and White[9] demonstrated that the adsorption capacity of clays for herbicides followed the order montmorillonite > illite > kaolinite.

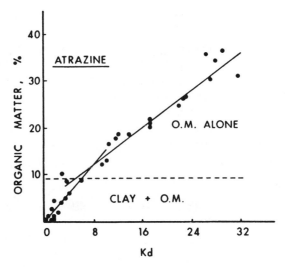

Fig. 19.1 Relation between organic matter content and atrazine adsorption by soil. Adapted from data published by Walker and Crawford[8]

Laboratory studies have, in general, corroborated field observations indicating that organic matter plays a major role in the performance of soil-applied pesticides. This work has generally involved multiple correlation analysis for pesticide adsorption by a series of soils with widely different properties, including organic matter content, clay content, clay mineral type, pH, and cation-exchange capacity (CEC). In a typical study, a given quantity of soil is added to a pesticide solution of known concentration, the mixture is allowed to equilibrate, and the concentration of the pesticide in the solution phase is estimated. The amount of pesticide adsorbed is subsequently calculated from the decrease in concentration and is usually expressed by such units as micromoles adsorbed per kilogram of soil (x/m). By repeating the measurements at several pesticide concentrations, an adsorption isotherm can be obtained by plotting the quantity adsorbed (x/m) versus the equilibrium concentration (C). In most instances, a straight line is obtained when the data are plotted as log (x/m) versus log C, according to the Freundlich adsorption equation,

$$x/m = KC^{1/n}$$

or

$$\log (x/m) = \log K + \frac{1}{n} \log C$$

where K and n are constants. The constant K provides a measure of the extent of adsorption and has been used in correlation studies aimed at determining the relative importance of the various soil parameters on adsorption.

A modified form of the Freundlich equation has often been used to express adsorption data. When n is near unity, the constant K (referred to as a distribution coefficient, K_d), is given by

$$K_d = \frac{\text{pesticide adsorbed } (\mu\text{mol/kg of soil})}{\text{pesticide in solution } (\mu\text{mol/L})}$$

Table 19.2 gives typical correlations between adsorption of some common herbicides and the soil variables of organic matter content, clay content, CEC, and pH. It can be seen that, in most cases, the correlation coefficient relating adsorption to organic matter content is considerably higher than for the other soil parameters, including clay content.

For nonionic organics of low water solubility (e.g., polychlorinated and

Table 19.2 Organic Matter, Clay, and Other Soil Properties Correlated with Adsorption Parameters[a]

		Correlation Coefficient[b]			
Compound	Number of Soils	Organic Matter	Clay	CEC	pH
s-Triazines					
Ametryne	34	0.41*	0.14	0.19	−0.37*
Atrazine	25	0.82**	0.65**	0.63**	−0.28
Propazine	25	0.74**	0.71**	0.69**	−0.41*
Prometon	25	0.26	0.60**	0.55**	−0.42*
Prometryn	25	0.40*	0.68**	0.63**	−0.49
Simazine	25	0.83**	0.77**	0.79**	−0.39
Simazine	65	0.72**	0.12	0.52**	0.04
Simazine	32	0.62**	0.27	0.54**	−0.35
Simazine	18	0.82**	0.48**	0.84**	−0.40
Substituted ureas					
Diuron	34	0.73**	0.37*	0.58**	0.10
Diuron	32	0.89**	0.28	0.56**	−0.03
Linuron	11	0.90**	0.06	0.57*	−0.14
Neburon	7	0.76*	−0.37	0.19	0.14
Picloram	6	0.90*	0.55	0.65	—
Phenylcarbamates					
CIPC	32	0.85**	0.16	0.38*	0.48*
Other					
Diphenamid	11	0.91**	0.16	0.60*	0.11

[a] From Stevenson[4] as compiled from literature data.
[b] *Significant at $p = 0.05$.
  **Significant at $p = 0.01$.

polycyclic hydrocarbons), the relative importance of clay and organic matter as adsorbents may depend on the moisture content of the soil. Chiou[10,11] pointed out that, in the presence of water, adsorption of nonionic organics by clay minerals is suppressed due to strong polar interaction of water with the clay, and sorption by the soil occurs primarily by solute interactions with organic matter. In dry soils, on the other hand, clays function as strong adsorbents and thereby can make a greater contribution to bonding than the organic matter.

Qualitative Differences in Soil Organic Matter

The fact that soils differ greatly in their organic matter contents is well known but it is not generally appreciated that major qualitative differences also exist, both with respect to the known classes of organic compounds (lipids, carbohydrates, proteins) but with the so-called humic substances (humic acid, fulvic acid, etc.). For example, the percentage of the organic matter as fats, waxes, and resins ranges from as little as 2 percent in some soils to over 20 percent in others, with the higher value being typical of forest humus layers and acid peats (see Chapter 7). The percentage of the organic matter as "protein" may vary from 15 to 45 percent; carbohydrate content from 5 to 25 percent.

Humified organic matter may comprise three-fourths of the total organic matter in some soils but less than one-third in others. The humic material of grassland soils is dominated by humic acids; that in forest soils is relatively rich in fulvic acids. The so-called brown humic acids are characteristic in the humic acids of peats and forest soils; gray humic acids are typical of the humic acids in grassland soils (see Chapter 2).

Differences in organic matter composition may have implications with respect to correlation studies of herbicide retention with organic matter content. The results of Hayes et al.[12] suggest that fulvic acids may be less effective in adsorbing s-triazine herbicides than humic acids. Dunigan and McIntosh[13] found that the ether- and alcohol-extractable components of soil organic matter (fats, waxes, and resins) had a negligible capacity to adsorb atrazine; a hot water-extractable component (presumably a polysaccharide) had a small adsorption capacity. Experiments with compounds representative of natural soil organic matter showed that polysaccharide-type constituents had rather low affinities for atrazine, a protein had an intermediate affinity, and humic acids and lignins had high affinities. Walker and Crawford,[8] in an experiment in which various decomposable organic materials were incubated with soils low in organic matter, found that both the type of material being decomposed and its stage of decomposition were important factors in the adsorption of s-triazines.

Abnormally high retention of herbicides has been observed in burned-over fields and those containing wind-blown carbon particles.[14] In general, activated charcoal tends to absorb pesticides, although the amount adsorbed varies greatly with the different compounds.[15]

Special Role of Fulvic Acids

Because of their low molecular weights and high acidities, fulvic acids are more soluble than humic acids, and they may have special functions with regard to herbicide transformations. First, they may act as transporting agents for certain pesticides in soils and natural waters. As noted later, soluble humic substances can act as carriers of xenobiotics in aquatic environments. According to Ballard,[16] the downward movement of the insecticide DDT in the organic layers of forest soils is due to water-soluble, humic-like substances.

Second, fulvic acids, by virtue of their high functional group content, may catalyze the chemical decomposition of certain herbicides. The suggestion has been made, for example, that these constituents might catalyze the hydroxylation of the chloro-s-triazines (see next section).

Potential Chemical Reactions Involving Pesticides and Organic Substances

There seems little doubt but that the organic fraction of the soil has the potential for promoting the nonbiological degradation of many pesticides. Organic compounds containing nucleophilic reactive groups of the type believed to occur in humic and fulvic acids (e.g., COOH, phenolic-, enolic, heterocyclic-, and aliphatic-OH, semiquinones, and others) are known to produce chemical changes in a wide variety of pesticides.[17] Of additional interest is that humic substances are rather strong reducing agents and have the capability of bringing about a variety of reductions and associated reactions.

Results of electron spin resonance (ESR) spectroscopy have provided evidence for the role of stable free radicals in humic and fulvic acids in chemical transformations of pesticides, such as charge–transfer (electron donor–acceptor) processes and cross-coupling reactions. This work has been reviewed by Senesi.[18] For the s-triazines, ESR results have been explained by assuming that electron deficient quinone-like structures in humic substances induce a single-electron transfer from the electron-rich nitrogen atoms of the herbicide, giving rise to the formation of radical cation and anion species in a charge–transfer complex, according to the following reaction[18]:

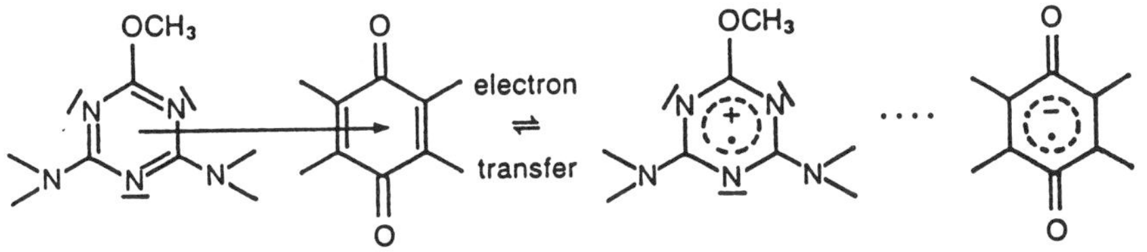

(s-triazine; electron–donor) (humic quinone; electron–acceptor) (radical cation and anion; charge-transfer complex)

Piccolo et al.,[19] also using ESR, found that adsorption of atrazine by humic acids increased with an increase in molecular size and was favored by a high degree of aromaticity.

Basic amino acids and similar compounds are able to catalyze the dehydrochlorination of DDT and lindane.[20] Certain chlorophyll degradation products (reduced prophyrins) can convert DDT to DDD.[21] Substances in soil organic matter which contain OH and NH_2 groups are potentially capable of being alkylated by the action of chlorinated aliphatic acids, as shown below.

$$R-NH_2 + Cl-CH_2-(CH_2)_n-COOH \rightarrow R-NH-(CH_2)_{n+1}-COOH$$
$$+ HCl$$

$$R-OH + Cl-CH_2-(CH_2)_n-COOH \rightarrow R-O-(CH_2)_{n+1}-COOH$$
$$+ HCl$$

Specific examples of nonbiological transformations brought about by the organic fraction of the soil includes hydroxylation of the chloro-s-triazines[7,22,23] and decomposition of 3-aminotriazol. With regard to the former, Armstrong and Conrad[7] concluded that hydrolysis of atrazine resulted from the sequence of events shown in Fig. 19.2. Adsorption was postulated to take place between a ring nitrogen (N) atom and a protonated COOH group. Hydrogen bonding of the ring N was believed to cause the withdrawal of electrons from the

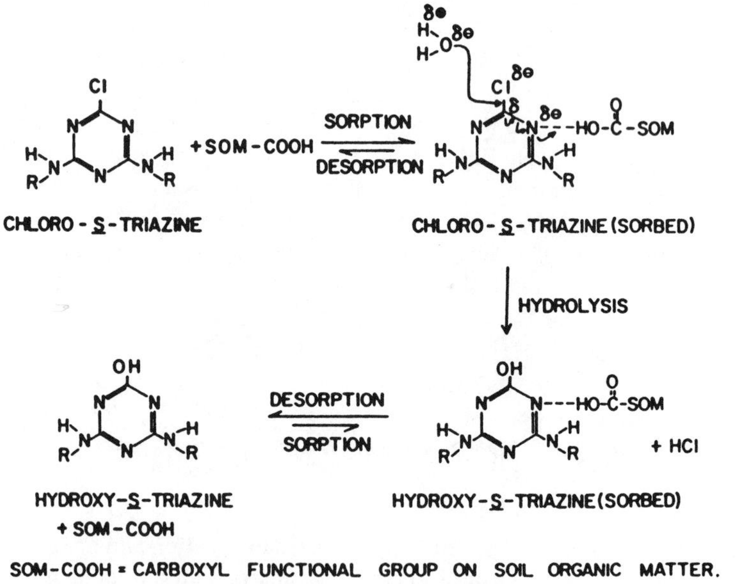

Fig. 19.2 Proposed model for sorption-catalyzed hydrolysis of the chloro-s-triazines by soil organic matter. From Armstrong and Conrad,[7] reproduced by permission of the Soil Science Society of America, Madison, Wisconsin

electron deficient carbon (C) atom bonded to the Cl, thereby enabling water to replace the Cl atom.

Chemical Binding of Pesticides and Their Decomposition Products

Substantial evidence exists to indicate that pesticide-derived residues can form stable chemical linkages with components of soil organic matter and that such binding greatly increases persistence of the pesticide residues. Two main mechanisms can be envisioned: 1) direct chemical attachment of the residues to reactive sites on colloidal organic surfaces, and 2) incorporation into the structures of newly formed humic and fulvic acids during the humification process. The reactions involved are similar to those described in Chapter 8 on the biochemistry of humus formation, where it was shown that many weakly basic compounds, including amino acids, pyrrols, amides, amines, and imines, have the ability to combine chemically with an array of carbonyl-containing substances, including reducing sugars, reductones, aldehydes, and ketones. Many of the common pesticides fall into one of these categories. Those pesticides which are basic in character, such as the s-triazines, have the potential for forming a chemical linkage with carbonyl constituents of soil organic matter; those containing the C=O group (e.g., the phenylcarbamates and substituted ureas) are theoretically capable of reacting with amino constituents. Condensation and conjugate reactions of pesticides with metabolic products have been postulated to constitute a form of pesticide transformations by microorganisms and higher plants.

Another factor to consider is that the partial degradation of many pesticides by microorganisms leads to the formation of chemically reactive intermediates that can combine with amino- or carbonyl-containing compounds, as illustrated in Fig. 19.3. Thus, loss of the side chain from the phenoxyalkanoic acids by enzymatic action leads to the formation of phenolic constituents which can either be oxidized further via the enzymatic route or undergo condensation

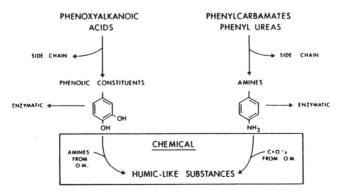

Fig. 19.3 Postulated chemical reactions between intermediate products of herbicide decomposition and constituents of soil organic matter. From Stevenson[4]

(probably as quinones) with amino compounds to form "humic-like" substances. On the other hand, amines (or chloroamines) produced by biological decomposition of such herbicides as the acylanilides, phenylcarbamates, and phenylureas may react with carbonyl constituents of soil organic matter. Entry into the C cycle by this mechanism may constitute a form of natural detoxification. By virtue of their high chemical reactivities, decomposition products of some pesticides can become part of the pool of precursor molecules for humus synthesis, and, in so doing, lose their identity.

In early studies, Bartha and his coworkers[24-26] concluded that the bulk of the chloroanilines liberated by partial degradation of phenylamide herbicides (acylanilides, phenylcarbamates, and phenylureas) becomes immobilized in soil by chemical bonding to organic matter. The chemically-bound residues could not be recovered by extraction with organic solvents or inorganic salts; partial release was achieved by base and acid hydrolysis.[26] The hydrolyzable portion of the humus-bound chloraniline was believed to be attached as the anil or anilinoquinone and the nonhydrolyzable part as heterocyclic ring structures or in ether linkages. Soil-bound chloroaniline residues are not easily attacked by microorganisms.[24,27] The reader is referred to the review of Bollag[28] for a discussion of the cross-coupling of xenobiotics with humic substances.

ADSORPTION MECHANISMS

Bonding mechanisms for the retention of pesticides by organic substances in soil include ion exchange, protonation, H-bonding, van der Waals forces, and coordination through an attached metal ion (ligand exchange). In addition, nonpolar molecules may be bound to hydrophobic surfaces through "hydrophobic bonding."

A brief description of bonding mechanisms is given in Chapter 18, those aspects that relate specifically to pesticide–organic matter interactions are discussed below.

Ion Exchange and Protonation

Like the secondary silicate minerals (e.g., montmorillonite, illite, kaolinite), soil organic colloids are negatively charged, although positive spots may be present under some conditions through free amino groups. Adsorption of pesticides by this mechanisms is largely restricted to those types which exist as cations (e.g., as $R\mathrm{NH}_3^+$), or which can become positively charged through protonation. The reaction for the protonated compound is

$$[\mathrm{OM}]-\mathrm{COO}^- + R-\mathrm{NH}_3^+ \rightarrow [\mathrm{OM}]-\mathrm{COO}^- \cdots {}^+\mathrm{NH}_3-R$$

Diquat and paraquat, being divalent, have the potential for reacting with more than one negatively charged site on soil humic colloids, such as through two COO^- ions (illustrated below for diquat).

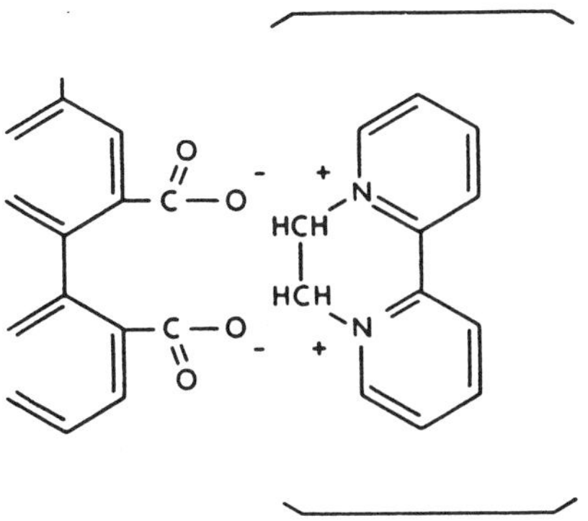

Diquat

Other possible combining sites on organic matter include a COO^- ion plus a phenolate ion combination and a COO^- (or phenolate ion) plus a free radical site.

Based on results of infrared studies, Khan[29] suggested that bipyridylium herbicides form charge-transfer complexes with humic substances, a result that could not be confirmed by Burns et al.[30] The studies of Khan showed that paraquat was adsorbed in greater amounts by humic and fulvic acids, and by an organo–clay complex, than was diquat.

Factors that influence the availability of exchange sites for adsorption include the presence of competing metal cations and pH. Soil pH has a direct bearing on the relative importance of organic matter and clay in retaining organic cations. Unlike clay, organic colloids have a strongly pH-dependent charge. Therefore, the contribution of organic matter to the CEC, and subsequently retention, will be higher in neutral and slightly alkaline soils than in acidic ones. For each unit change in pH, the change in CEC for organic matter is several fold greater than for clay.

In a typical temperate zone soil, the contribution of organic matter to the CEC may range from about 50 percent at pH 7 to only 15 percent at pH 5. A complicating factor in evaluating the relative importance of organic matter and clay in adsorbing diquat and paraquat is that these organic cations are known to be held within the interlayer surfaces of expanding clay minerals. Clay may prove to be of greater importance than organic matter in adsorbing diquat and paraquat, except perhaps in soils rich in organic matter.

Less basic compounds, such as the s-triazines, may become cationic through protonation. Whether or not protonation occurs will depend upon: 1) the nature of the herbicide in question as reflected by its pK_a, and 2) the proton-supplying power of the humic colloids. Reactions leading to adsorption, as postulated by Weber et al.,[31] are shown by the following equations:

$$T + H_2O \rightleftharpoons HT^+ + OH^- \qquad [1]$$

$$R-COOH + H_2O \rightleftharpoons R-COO^- + {}^+H_3O \qquad [2]$$

$$R-COO^- + HT^+ \rightleftharpoons R-COO-HT \quad [3]$$

$$R-COOH + T \rightleftharpoons R-COO-HT \quad [4]$$

where R is the organic colloids, T the s-triazine molecule, T^+ the protonated molecule, and H_3O^+ the hydronium ion.

Equation [1] represents pH-dependent adsorption through protonation in the soil solution while [2] represents ionization of the colloid COOH group. Ionic adsorption of the cationic s-triazine molecule, formed by reaction [1], is shown by equation [3]. Adsorption through direct protonation on the surface of the organic colloid is shown by reaction [4].

Soil pH has a profound effect on adsorption of not only the s-triazines but other weakly basic herbicides by organic matter, a typical example being Amitrole. The soil reaction governs not only the ionization of acidic groups on humic colloids but the relative quantity of the herbicide that occurs in cationic form, in accordance with reaction [1]. A more complete explanation of the effect of pH on protonation of weakly basic compounds and their adsorption by soil colloids is given in Chapter 18.

The pK_a of acidic groups in humic acids (notably COOH) is of the order of 4.8 to 5.2 at an ionic strength of 0.1 but higher at lower salt concentrations (see Chapter 15). Thus, it would appear that ion exchange would not be an important mechanism for adsorption of atrazine and simazine, which have pK_as of 1.68 and 1.65, respectively. It should be pointed out however, that the pH at the surface of soil organic colloids may be as much as two pH units lower than that of the liquid environment. The adsorption capacities of soil organic matter preparations for the s-triazines has been found to follow the order expected on the basis of their pK_a values.[31]

Ion exchange is but one of several mechanisms for adsorption of the s-triazines to organic colloids. Other possibilities include H-bonding between the secondary amino group of the s-triazine molecule and OH or C=O groups of the organic matter (a). Retention of the protonated s-triazine can also occur at a free radical site (b).

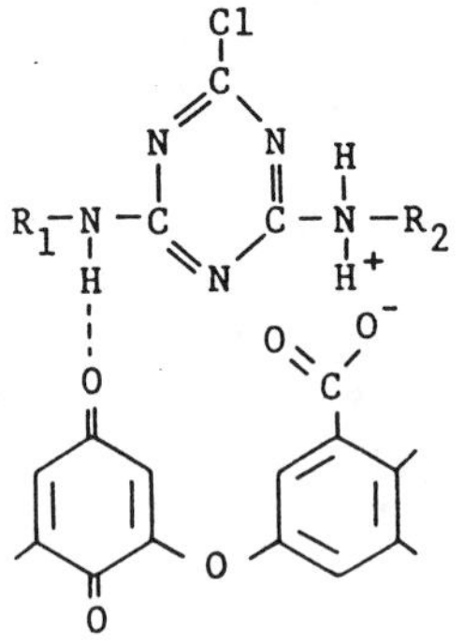

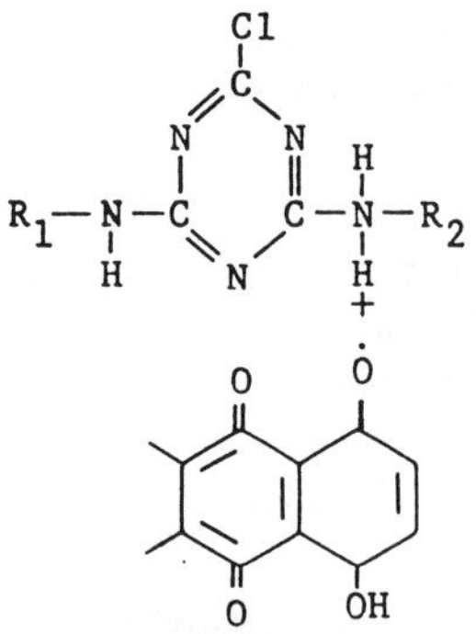

On the basis of an infrared study of some s-triazine-humic acid complexes, Sullivan and Felbeck[32] concluded that one secondary amino group was bound to either a C=O or quinone group of the humic acid whereas the other secondary amino group became protonated and was bound through ion exchange to a COO⁻ ion (see above diagram).

For anionic pesticides, such as the phenoxyalkanoic acids, repulsion by the predominantly negatively charged surface or organic colloids may occur. Positive adsorption of anionic herbicides at pH values below their pK_a values can be attributed to adsorption of the unionized form of the herbicide to organic surfaces, such as by H-bonding between the COOH group and C=O or NH groups of organic matter.

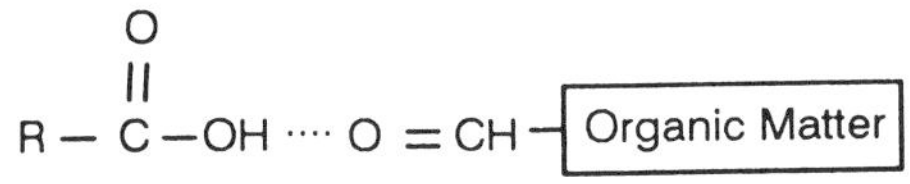

$$R-\overset{O}{\underset{\|}{C}}-OH \cdots O=CH-\boxed{\text{Organic Matter}}$$

H-Bonding, van der Waals Forces, and Coordination

Adsorption mechanism for retention of nonionic polar herbicides such as the phenylcarbamates and substituted ureas, are illustrated in Fig. 19.4. The great

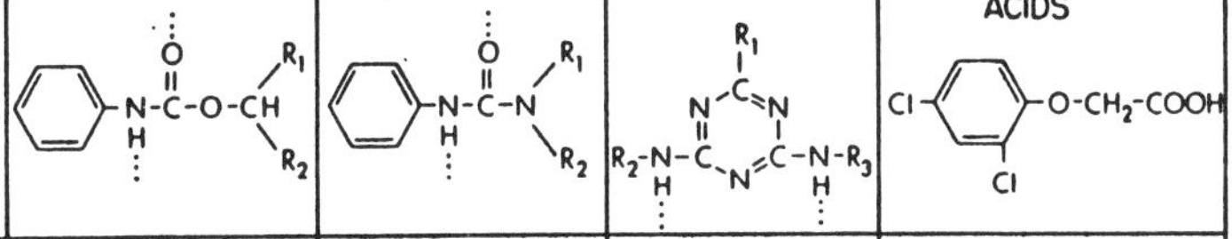

	PHENYLCARBAMATES	SUBST. UREAS	s-TRIAZINES	PHENOXYALKANOIC ACIDS
VAN DER WAALS	+	+	+	+
H-BONDING				
⟩NH⋯O=⟨ HA	+	+	+	−
−OH⋯O=	−	−	+ (R_1 = OH)	−
−C(=O)−O⋯HO− / HN=HA	+	−	−	−
⟩C=O⋯HN−H	+	+	−	+ (pH < pK_a)
LIGAND EXCHANGE				
⟩C=O⋯M^{z+}−HA	+	+	−	−
SALT LINKAGE				
−C(=O)−O−M−O−C(=O)− HA	−	−	−	+ (pH > 7.0)

Fig. 19.4 Typical bonding mechanisms for adsorption of some of the common herbicide types by soil organic matter. From Stevenson[4]

importance of H-bonding is suggested, with multiple sites being available on both herbicide and organic matter surface. Other adsorption mechanisms include van der Waals forces (physical adsorption), ligand exchange ($-M^{z+} \cdots O=C$), and, for herbicides containing an ionizable COOH group, a salt linkage through a divalent cation on the organic exchange site. For chlorinated phenoxyalkanoic acids, such as 2,4-D, H-bonding will be limited to acid conditions where COOH groups are unionized.

Considerable variation can be expected in the adsorption capacity of organic matter for nonionic polar herbicides, depending upon stearic effects and the number and kinds of electronegative atoms in the molecule.

Pi Bonding

Pi- (π) bonding results from the overlap of bonding orbitals perpendicular to the aromatic ring. These bonds are believed to be involved in the binding of alkenes, alkyenes, and aromatic compounds to soil organic matter.

Hydrophobic Bonding

Hydrophobic sorption has been proposed as a mechanism for retention of nonpolar organic compounds by soil organic matter.[10,11,33-38] Hydrophobic bonding increases as the solute becomes more and more nonpolar, or as water solubility decreases. Active surfaces for hydrophobic bonding include the fats, waxes, and resins, as well as aliphatic side chains on humic and fulvic acids (see Chapter 12 for typical structures).

There are two schools of thought as to the nature of the sorption interaction. One holds that retention of the solute occurs through "adsorption" onto hydrophobic surfaces of the organic matter and results from a "squeezing-out" of the molecule from solution and its accumulation on at the solid interphase where competition with the solvent is minimum.[8,34,36,38] The second scheme is that bonding occurs through "partition," in which the case the sorbed material penetrates into the interior of the organic phase.[10,11,33,35,37] Partition is distinguished from adsorption by retention of the sorbed material "within internal spaces" (voids) of organic matter components (e.g., humic substances); in adsorption, the sorbate occupies only the "surface" of the adsorbate.

In the model proposed by Wershaw,[37] humic substances are depicted as existing as membrane-like aggregates that are held together by weak bonding mechanisms, such as Pi bonding, H-bonding, and hydrophobic interactions. The membrane-like structure contains hydrophobic interiors to which hydrophobic compounds will partition and hydrophilic exteriors to which polar compounds will interact.

Arguments can be advanced both for[10,11,35,37] and against[36,38] the partition hypothesis. Arguments given by Chiou[10,11] in support of partition are as follows:

1. Sorption of nonpolar organics is correlated more strongly with the organic matter content of the soil than any other factor
2. Water competes favorably with neutral organics for adsorption on mineral surfaces
3. Sorption of nonpolar organics does not show competition with other similar organic compounds, as is typical of adsorption on surfaces (e.g., activated charcoal)
4. Highly hydrophobic organic molecules sorb to dissolved humic and fulvic acids, akin to a "partition-like" interaction mechanically similar to the action of a surfactant micelle, where sorbates are partitioned into a microscopic organic phase

Some of the above arguments are not unique to the partition model but apply equally as well to the physical adsorption model.[36,38] In practice, sorption of nonpolar organics may occur through both partition and hydrophobic adsorption, and, in some cases, in other ways as discussed earlier. Nonpolar compounds added to soils often demonstrate nonequilibrium behavior, such as long-term uptake and "tailing" upon elution from soil columns (see review of Pignatello[35]). Furthermore, the residues of long-time contaminated soils often occur in a slow-desorbing recalcitrant state. These results are consistent with the partition hypothesis.

In contrast to the above, Murphy et al.[39] found that sorption isotherms for some nonpolar organics on humic-coated minerals were nonlinear, implying that retention was due to adsorption onto rather than their partition into the surface organic phase. In other work, Steinberg et al.[40] concluded that the long-term persistence of 1,2-dibromoethane in soils (up to 19 yr.) was due to entrapment in intraparticle micropores of the soil, as evidenced by enhanced release upon pulverization of the soil. As Weber et al.[41] accurately pointed out, sorption processes in environmental systems comprise multiple reaction phenomena involving many different reaction mechanisms.

When retention of nonpolar organic compounds by soil is expressed as a function of organic C content, a constant K_{oc} is obtained which is characteristic property of the compound being adsorbed.

$$K_{oc} = \frac{K_d}{\% \text{ organic C in soil}}$$

where K_{oc} is the Freundlich K_d constant divided by percent C in the soil.

A direct relationship has been found to exist between K_{oc} and the compound's octanol–water partitions coefficient, K_{ow}. This allows for the estimation of K_{oc} from a knowledge of K_{ow}. It should be noted that K_{oc} can also be estimated from the compound's water solubility (i.e., the lower the solubility the higher the K_{oc}).

The utility of the above relationships is that the adsorption properties of a wide range of nonionic compounds by soil can be predicted without extensive screening. All that is required is a knowledge of the organic C content of the soil and the octanol–water partition coefficients (or water solubilities) for the compounds being tested. Caution must be exercised in using this approach, as exceptions can be expected and specific compounds may deviate from accepted rules. For example, some nonionic pesticides, such as the s-triazines and phenylureas, contain polar functional groups and can be bound through H-bonding (discussed above).

The different roles of organic matter and clay minerals in the sorption of nonpolar organics provide a basis for assessing the activity of soil-incorporated pesticides.[10,11] Under dry conditions, adsorption to clay minerals (in addition to retention by partitioning into organic matter) can lower the activity of the soil-applied pesticide. Upon wetting, the chemical activity may rise as a result of displacement by water of the pesticide adsorbed by the clay. Losses through vaporization would be expected to be similarly affected.

The presence of humic substances in water, even in trace amounts, can significantly enhance the solubilities and thereby environmental distribution of

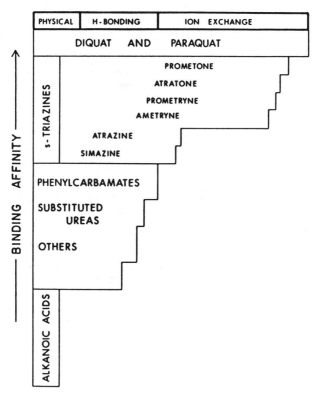

Fig. 19.5 Relative affinities of herbicides for soil organic matter surfaces. From Stevenson[4]

nonpolar xenobiotics.[33,42] The occurrence of these compounds in leachate waters and in the pore water of sediments has been attributed to partitioning onto organic surfaces.

RELATIVE AFFINITIES OF PESTICIDES FOR SOIL ORGANIC MATTER

The deliberations of the previous section serve to emphasize that the various pesticides differ greatly in their relative affinities for soil organic colloids. The approximate order for some common herbicides are given in Fig. 19.5. Thus, the cationic herbicides (diquat and paraquat) would be expected to be the most strongly bound, followed by those weakly basic types capable of being protonated under moderately acidic conditions. For the s-triazines, differences in adsorbability can be accounted for by variations in pK_a, with the more basic compounds (high pK_a) being absorbed the strongest. Herbicides in the next order of adsorption are those having very low pK_a values but which contain one or more polar groups suitable for H-bonding. Anionic pesticides may or may not be adsorbed, depending upon soil pH.

SUMMARY

Organic matter plays a major role in the adsorption of pesticides in soil. Adsorption depends upon the physical and chemical properties of the pesticide, nature of the organic matter, and properties of the soil system (organic matter content, kind and amount of clay, pH, type of exchange cations). The organic fraction of the soil also has the potential for promoting the nonbiological degradation of certain pesticides.

Information as to how pesticides react with soil organic matter may provide a rational basis for their effective use, thereby reducing undesirable side effects due to carryover, contamination of the environment, and, in the case of herbicides, phytotoxicity to subsequent crops.

REFERENCES

1 H. H. Cheng, Ed., *Pesticides in the Soil Environment: Processes, Impacts, and Modeling*, Soil Science Society of America, Madison, 1989.
2 G. W. Bailey and J. L. White, *J. Agr. Food Chem.*, **12,** 324 (1964).
3 B. L. Sawhney and K. Brown, Eds., *Reactions and Movement of Organic Chemicals in Soils*, Special Publication 22, Soil Science Society of America, Madison, 1989.
4 F. J. Stevenson, *J. Environ. Qual.*, **1,** 333 (1972).
5 S. B. Weed and J. B. Weber, "Pesticide–Organic Matter Interactions," in W. D.

Guenzi, Ed., *Pesticides in Soil and Water*, American Society of Agronomy, Madison, 1974, pp. 39–66.

6 N. Senesi and Y. Chen, "Investigations of Toxic Organic Substances with Humic Substances," in Z. Gerstl, Y. Chen, V. Mingelgrin, and B. Yaron, Eds., *Toxic Organic Chemicals in Porous Media*, Springer-Verlag, New York, 1989, pp. 37–90.

7 D. E. Armstrong and J. G. Conrad, "Nonbiological Degradation of Pesticides," in W. D. Guenzi, *Pesticides in Soil and Water*, American Society of Agronomy, Madison, 1974, pp. 123–131.

8 A. Walker and D. V. Crawford, "The Role of Organic Matter in Adsorption of the Triazine Herbicides by Soil," in *Isotopes and Radiation in Soil Organic Matter Studies*, International Atomic Energy Agency, Vienna, 1968, pp. 91–108.

9 G. W. Bailey and J. W. White, *Residue Rev.*, **32,** 29 (1970).

10 C. T. Chiou, "Roles of Organic Matter, Minerals, and Moisture in Sorption of Nonionic Compounds and Pesticides by Soil," in P. MacCarthy, C. E. Clapp, R. L. Malcolm, and P. R. Bloom, Eds., *Humic Substances in Soil and Crop Sciences: Selected Readings*, American Society of Agronomy, Madison, 1990, pp. 111–160.

11 C. T. Chiou, "Theoretical Considerations of the Partition Uptake of Nonionic Organic Compounds by Soil Organic Matter," in B. L. Sawhney and K. Brown, Eds., *Reactions and Movement of Organic Chemicals in Soils*, Special Publication 22, Soil Science Society of America, Madison, 1989, pp. 1–29.

12 M. H. B. Hayes, M. Stacey, and J. M. Thompson, "Adsorption of s-Triazine Herbicides by Soil Organic–Matter Preparations," in *Isotopes and Radiation in Soil Organic Matter Studies*, International Atomic Energy Agency, Vienna, 1968, pp. 75–90.

13 E. P. Dunigan and T. H. McIntosh, *Weed Sci.*, **19,** 279 (1971).

14 H. W. Hilton and Q. H. Yuen, *J. Agric. Food Chem.*, **11,** 230 (1963).

15 D. L. Coffey and G. F. Warren, *Weed Sci.*, **17,** 16 (1969).

16 T. M. Ballard, *Soil Sci. Soc. Amer. Proc.*, **35,** 145 (1971).

17 D. G. Crosby, "The Nonbiological Degradation of Pesticides in Soils," in *Proc. Int. Symp. on Pesticides in Soil*, Michigan State University, East Lansing, 1970, pp. 86–94.

18 N. Senesi, *Adv. Soil Sci.*, **14,** 77 (1990).

19 A. Piccolo, G. Celano, and C. De Simone, *Sci. Total Environ.*, **117/118,** 403 (1992).

20 K. A. Lord, *J. Chem. Soc. (London)*, **1948,** 1657 (1948).

21 R. P. Miskus, D. P. Blair, and J. E. Casida, *J. Agr. Food Chem.*, **13,** 481 (1965).

22 D. E. Armstrong and G. Chesters, *Environ. Sci. Tech.*, **2,** 683 (1968).

23 G.-C. Li and G. T. Felbeck, Jr, *Soil Sci.*, **114,** 201 (1972).

24 R. Bartha, *J. Agr. Food Chem.*, **19,** 385 (1971).

25 R. Bartha and D. Pramer, *Adv. Appl. Microbiol.*, **13,** 317 (1970).

26 T. S. Hsu and R. Bartha, *Soil Sci.*, **118,** 213 (1974).

27 H. Chiska and P. C. Kearney, *J. Agr. Food Chem.*, **18,** 854 (1970).

28 J.-M. Bollag, "Cross-Coupling of Humus Constituents and Xenobiotic Substances," in *Aquatic and Terrestrial Humic Materials*, R. F. Christman and E. T.

Gjessing, Eds., Ann Arbor Science Press, Ann Arbor, Michigan, 1983, pp. 127–141.

29. S. U. Khan, *Can. J. Soil Sci.*, **53,** 199 (1973).
30. E. G. Burns, M. H. B. Hayes, and M. Stacey, *Pesticide Sci.*, **4,** 201 (1973).
31. J. B. Weber, S. B. Weed, and T. M. Ward, *Weed Sci.*, **17,** 417 (1969).
32. J. D. Sullivan, Jr. and G. T. Felbeck, Jr., *Soil Sci.*, **106,** 42 (1968).
33. C. T. Chiou, R. L. Malcolm, T. I. Brinton, and D. E. Kile, *Environ. Sci. Technol.*, **20,** 502 (1986).
34. J. J. Hassett and W. L. Banwart, "The Sorption of Nonpolar Organics by Soils and Sediments," in B. L. Sawhney and K. Brown, Eds., *Reactions and Movement of Organic Chemicals in Soils*, Special Publication 22, Soil Science Society of America, Madison, 1989, pp. 31–44.
35. J. J. Pignatello, "Sorption Dynamics of Organic Compounds in Soils and Sediments," in B. L. Sawhney and K. Brown, Eds., *Reactions and Movement of Organic Chemicals in Soils*, Special Publication 22, Soil Science Society of America, Madison, 1989, pp. 45–80.
36. U. Mingelgrin and Z. Gerstl, *J. Environ. Qual.*, **12,** 1 (1983).
37. R. L. Wershaw, *J. Contaminant. Hydrol.*, **1,** 29 (1986).
38. W. G. MacIntyre and C. L. Smith, *Environ. Sci. Technol.*, **18,** 295 (1984).
39. E. M. Murphy, J. M. Zachara, and S. C. Smith, *Environ. Sci. Technol.*, **24,** 1507 (1990).
40. S. M. Steinberg, J. J. Pignatello, and B. L. Sawhney, *Environ. Sci. Technol.*, **21,** 1201 (1987).
41. W. J. Weber, Jr., P. M. McGinley, and L. E. Katz, *Environ. Sci. Technol.*, **26,** 1955 (1992).
42. I. Kögel-Knabner, P. Knabner, and H. Deschauer, "Dissolved Organic Matter as Carrier for Exogenous Organic Chemicals in Soils," in M. S. Wilson, Ed., *Advances in Soil Organic Matter Research: The Impact on Agriculture and the Environment*, Redwood Press Ltd., Wiltshire, England, 1991, pp. 121–128.

20

ROLE OF ORGANIC MATTER IN PEDOGENIC PROCESSES

Soil formation can be regarded as the resultant of the various soil-forming factors acting on the parent material over the course of time (see Chapter 1). Except for soils developed on former basin sediments or ancient sedimentary rocks (shales), the parent material is usually devoid of organic matter and consists of variable quantities of sand, silt, clay, and carbonates; in certain cases, aluminum oxides, gypsum ($CaSO_4 \cdot 2H_2O$), and more soluble salts may be present. High temperatures and adequate moisture promote mineral transformations but these effects are magnified by living organisms.

The biosphere and organic substances produced therein are directly or indirectly responsible for the following[1]:

1. Decay of organic matter leads to the formation of CO_2, which acts as a weathering agent due to its tendency to form carbonic acid with water ($H_2O + CO_2 \rightarrow H_2CO_3$).
2. Organic chelating agents produced in the biosphere act as weathering agents; furthermore, they bring about the mobilization and transport of metal ions. Organic matter also functions as a reducing agent.
3. The uptake of certain elements by plants and microorganisms induces a strong driving force for reactions leading to the release of these elements into the soil solution.
4. Microorganisms catalyze oxidation and reduction processes that otherwise proceed at extremely slow rates.

A whole series of events are involved in the mobilization and transport of Fe, Al, and the various trace elements in the pedosphere, including disintegration of parent material through the action of organic substances synthesized by microorganisms, eluviation of the metals to lower soil horizons during soil formation, uptake by plant roots and translocation into leaf tissue, incorporation into the raw humus layer of the soil through leaf fall, complexation during humification of organic remains in the top soil, and transport to natural waters as soluble organic matter–metal complexes.

The migration of organic constituents into the lower soil horizons begins at an early stage of soil development, often in association with clay or metal ions. Worm and root channels and ped surfaces become coated with dark-colored mixtures of humus and clay. Buol and Hole[2] observed that clay coatings on ped surfaces have considerably higher organic matter contents than material within the peds. In some soils, streaks or tongues result from the downward seepage of humus. Illuvial humus also appears as coatings on sand and silt particles. Localized accumulations of sesquioxides (Fe and Al) and humus are common. In many soils, a secondary maximum in humus coincides with an accumulation of clay.

In addition to leaching, organic matter can be transferred downward in soil through the action of soil animals. Earthworms, for example, can completely mix soil to depths of 2 ft. or more, transferring organic matter downward in the process. Burrowing animals move soil material low in organic matter from the deeper horizons to the surface and vice versa.

Gains of organic matter in the top soil exceed the decay rate for a time after soil development commences. Ultimately, steady-state conditions are attained in which losses through decay and transfer equal gains from return of plant and animal residues (see Chapter 1). Transfer of organic matter into the lower soil horizon continues for some time after equilibrium levels are reached in the surface layer; over long periods, the total quantity in the profile stabilizes and remains fairly constant even though additions may continue.

Emphasis will be given in this chapter to rock weathering and the translocation of sesquioxides but it should be noted that a variety of other processes are affected as well. For example, organic matter may be involved in the process known as gley formation, a condition that occurs under impeded drainage or a high water table and where the strong reducing conditions lead to a soil layer having a light gray color tinged with blue or green, due presumably to Fe^{2+}. One theory is that organic substances in the subsoil are activated by the higher pHs and, in the absence of oxygen, become oxidized at the expense of Fe^{3+}, with production of Fe^{2+}. Another theory is that organic compounds associated with particle surfaces serve as sources of energy for microorganisms, which in turn reduce Fe^{3+} to Fe^{2+}. Organic constituents may be involved in the process of clay movement in soils, through active dispersion of the clay by soluble organic constituents.

WEATHERING OF ROCKS AND MINERALS

Mineral weathering refers to the breakdown and alteration of primary and secondary minerals to more stable forms under the influence of climate and biological activity. Both physical disintegration and chemical decomposition processes are involved. Important chemical changes include oxidation, hydration, carbonization, and the solubilization and redeposition of breakdown products. The original mineral can be dissolved or transformed into new products. In some cases, new minerals are formed through "neoformation," as discussed later. Organic substances influence all of these processes in diverse ways.

Biological weathering comprises the physical and chemical transformations that are brought about by living organisms and their decomposition products. For biochemical weathering, attention has been focused on two types of compounds, carbonic acid, which is formed from the CO_2 released during decay of organic matter, and organic chelates of microbial and higher plant origin.

Mineral dissolution through chelation involves the removal of structural cations from silicate minerals through the formation of stable complexes with metal ions and is presumed to be far more effective than hydrolysis, and possibly dissolution of the minerals by carbonic acid. When complexed by organic ligands, metal ions are maintained in solution and can be transported into the lower soil horizons, and hence to lakes and streams in percolating waters.

An extensive literature has developed on the dissolution of rocks and minerals through the action of organic substances, as attested by recent reviews on the subject.[3-8] The initial stage is characterized by colonization of rock surfaces by algae, fungi, and lichens, all of which produce chelating agents effective in solubilizing di- and trivalent cations from silicate minerals. With time, organic chelates of plant origin, or synthesized by soil microorganisms, further contribute to the weathering process and serve as agents for the transport of Fe, Al, and other cations.

A unified concept of the role of organic substances in the translocation and biological availability of metal ions in soils has been advanced by Zunino and Martin.[3] The first stage was described as an attack on insoluble mineral matter by simple organic chelates (e.g., lichen acids, organic acids, phenols) excreted by pioneering microorganisms. These were termed type-I complexes. With time and through the action of microorganisms, the trace metals become sequestered by newly synthesized humic substances to form type-II complexes, which were believed to be considerably more stable than type I, thereby preventing loss of essential metal ions by percolation to groundwaters. With an increase in trace element content, a point is reached at which the chelating sites become saturated and complexes less stable are formed, the so-called type-III complexes. Simple organic compounds produced in present-day soils were believed to compete successfully for the metal ions of type III to form complexes of type I. The metals contained in type-I complexes become incorporated into the tissues of plants and microorganisms, which upon death and decay reenter the cycle.

A schematic diagram illustrating the role of organic matter in metal-ion translocation from parent rocks to biological systems is shown in Fig. 20.1.

By Simple Organic Chelates Produced by Microorganisms

Mineral weathering by organic acids and other chelating organic biochemicals occurs by two distinctive modes of action: 1) lowering of pH due to ionization

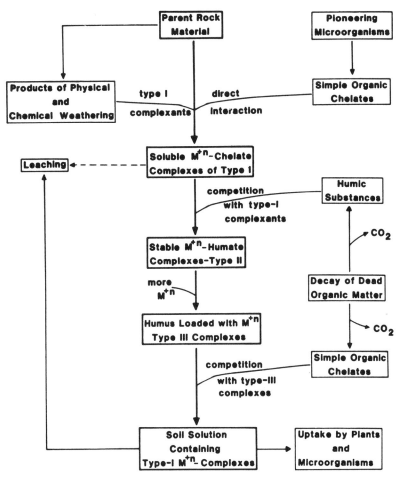

Fig. 20.1 Role of organic substances in metal-ion translocation. Adapted from a drawing by Zunino and Martin[3]

of COOH groups (COOH → COO$^-$ + H$^+$), and 2) formation of chelate complexes. Some compounds (i.e., formic and acetic acids) are effective only through the acidic (H$^+$) effect; others (i.e., citric acid) exert both acidic and complexation reactions (see Chapter 16).

Numerous organics produced by microorganisms can act as solubilizers of mineral matter in nature but organic acids appear to be the class of compounds mainly involved. This work, discussed in detail in Chapter 16, has shown that an unusually high proportion of the microorganisms associated with the "raw" soil of rock crevices and the interior of porous weathered stones produce copious amounts of organic acids, with decomposition of silicate minerals when tested under laboratory conditions. Fungi most active in dissolving silicates are those that produce citric or oxalic acids, both strong chelating agents.[5] A high proportion of the bacteria have been found to be gram-negative short rods capable of synthesizing 2-ketogluconic acid.[9,10] A summary of data obtained by Webley et al.[9] for the incidence of silicate-dissolving bacteria, actinomycetes, and fungi among the isolates from rock sequences and weathered stones is shown in Table 20.1.

The ability of lichens to dissolve mineral substances from rocks and absorb the nutrients within their tissues is well known. These organisms are widely distributed over the land masses of the world, including the Antarctic continent and desert regions of Australia and southwestern United States. It is likely that small amounts of lichen acids appear sporadically in these natural habitats. Their occurrence in most agricultural soils is doubtful.

In acid soils, mineral weathering may release large amounts of free Al^{3+} ions into the soil solution, creating toxic conditions for plant growth. However, when organic matter is present the free Al^{3+} is complexed, with amelioration of Al toxicities (see Chapter 2).

By Humic and Fulvic Acids

The possibility that humic and fulvic acids also participate in the chemical weathering of silicates was considered early in the history of soil science and evidence both for and against their involvement can be cited. Results of laboratory studies[11-13] have generally indicated that humic substances are equally as effective as simple organic acids in promoting the disintegration of silicate minerals. As can be seen from Table 20.2, data obtained by Baker[11] show that humic acids exhibited an activity of the same order as that of several known organic compounds. Accordingly, humic substances may affect metal ion mobility in environments where appreciable quantities of these substances are present in soluble forms. Due to their low molecular weights, fulvic acids may be particularly effective in dissolving silicate minerals.

Barman et al.[14] carried out kinetic studies on the solubilization of olivine, hornblend, tourmaline, biotite, and microcline by citric, oxalic, and salicylic acids (glycine also examined). An initial high rate of release of cations was attributed to two factors: 1) the dissolution of ultrafine particles, and 2) a diffusion controlled process involving dissolution of cations at or near the surface. The rate limiting step in the solubilization process, particularly for re-

Table 20.1 Incidence of Silicate-Dissolving Bacteria, Actinomycetes, and Fungi Among the Total Isolates From Rock Sequences and Weathered Stones[a]

	Total Isolates	Total Dissolving (%)			
		Ca Silicate	Wollastonite	Mg Silicate	Zn Silicate
Bacteria	265	83	57	65	Not tested
Actinomycetes	39	87	38	46	Not tested
Fungi	149	94	Not tested	76	96

[a]From Webley et al.[9]

Table 20.2 Action of Humic Acid and Various Organic Compounds in Solubilizing Metal Ions[a]

Sample[b]	Element Determined	Humic Acid	Salicylic Acid	Oxalic Acid	Pyrogallol	Alanine
		μg metal extracted in 1 hour				
Galena	Pb	200	130	95	35	<5
Sphalerite	Zn	30	30	20	8	20
Bornite	Cu	190	260	650	55	15
Chalcocite	Cu	3800	4450	9750	920	1530
Bismuthinite	Bi	550	180	4820	1640	55
Stibnite	Sb	45	<5	580	<5	<5
Pararammelsbergite	Ni	9800	10500	7620	2380	1730
Haematite	Fe	470	<3	80	20	20
Pyrolusite	Mn	1000	4200	15500	5150	520
Calcite	Ca	10500	11900	980	2040	1400
Copper	Cu	5700	5500	2620	1190	700
Lead	Pb	27400	41800	660	1470	240

[a] From Baker.[11]
[b] All extractants 0-1 percent w/v.

sistant minerals (e.g., microcline and tourmaline), was believed to be diffusion (dissociation) of the cation–organic complexes from the mineral surface, as illustrated below.

$$\text{Mineral} \begin{bmatrix} M^+ \\ M^+ \\ M^+ \end{bmatrix} + 3L^- \underset{}{\overset{\text{fast}}{\rightleftharpoons}} \text{Mineral} \begin{bmatrix} M^+L^- \\ M^+L^- \\ M^+L^- \end{bmatrix} \underset{}{\overset{\text{slow}}{\rightleftharpoons}} \text{Mineral} + 3ML$$

Results of a study showing the dissolution of Al and Si from microcline through the action of humic and fulvic acids as a function of time (pH 7) is shown in Fig. 20.2.

NEOGENESIS OF MINERALS

In addition to their effects on weathering, organic substances can retard or enhance the formation of secondary silicate minerals. This work, reviewed elsewhere,[4,5] has shown that a wide variety of organic acids (e.g., citric, maleic, tannic, 4-hydroxybenzoic) can hinder the precipitation of solid phase products of such elements as Al. The mechanism that has been postulated for this effect is that the ligand, through occupation of coordination sites of Al-hydroxides, impedes hydrolysis and polymerization, as illustrated for citrate below.[15]

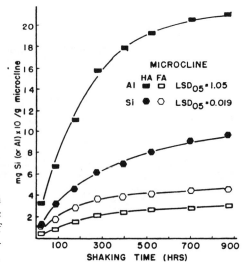

Fig. 20.2 Dissolution of Al and Si from microcline through the action of humic and fulvic acids (pH 7) as a function of time. From Tan[12], reproduced by permission of the Williams and Wilkins Co.

Kwong and Huang[16] found that the ability of several organic acids to perturb the precipitation of solid-phase Al followed the order: citrate > malate > tannate > aspartate > p-hydroxybenzoate.

Organic chelating agents also have the ability to distort the arrangement of the unit sheets normally found in crystalline Al-hydroxides, leading to the formation of short-range ordered precipitation products of Al. The action of citric acid in perturbing hydroxy–Al interlayering in montmorillonite was postulated by Kwong and Huang[15] to result from the formation of structures of the type depicted above. The formation of crystalline Al-hydroxides in soils and sediments can be hampered in environments (especially those that are acidic) where organic acids tend to accumulate. Organic acids can also influence the types of the Al-hydroxide polymorphs thus formed.

TRANSLOCATION OF MINERAL MATTER AND HORIZON DIFFERENTIATION

In humid and semihumid climates, the initial phase of soil formation leads to removal of exchangeable cations and soluble salts. The residue remaining behind consists of mixtures of silicate minerals and variable amount of silica and sesquioxides. Senstius[17] has concluded that subsequent weathering, mediated by decay products of organic matter, results in the formation of "climax" soils, the two extremes being the Laterite (now Oxisol) and the Podzol (now Spodosol). The overall reactions for these contrasting soil types are illustrated in Fig. 20.3. In the case of the Oxisol, the most important weathering agent is H_2CO_3 whereas with the Spodosol weathering results primarily from the action of organic chelates. The latter results in differential movement of metal ions according to their ability to form coordination complexes with organic ligands. Thus, Fe^{3+}, Al^{3+}, and other strongly chelated elements such as Cu^{2+} are eluted to a greater extent than Si and similarly weakly chelated ones.

Simonson[18] has outlined the role of organic matter in soil-forming processes. The downward movement of metal ions as soluble chelate complexes has been referred to by Swindale and Jackson[19] as "cheluviation" and typically occurs in Spodosols.

Spodosols have developed under climatic and biologic conditions that have resulted in the mobilization and transport of considerable quantities of Fe and Al into the subsoil. An organic-rich mineral layer of the soil (O and A horizons), consisting largely of decomposition products of forest litter, is underlain by a light-colored eluvial horizon (E), which has lost substantially more Fe and Al than Si. This horizon is, in turn, underlain by a dark-colored illuvial horizon, B, in which the major accumulation products are Fe, Al, and organic matter. Other soils with an E horizon also show evidence for transport of

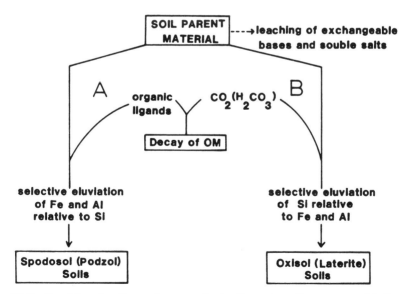

Fig. 20.3 Possible role of organic matter in the formation of two types of "climax" soils as postulated by Senstius.[17] Process A occurs under cool, moist conditions where incomplete oxidation of organic matter leads to the generation of chelates and the formation of Spodosols. Process B occurs in warm climates where decay of organic matter is complete and Oxisols are formed

sesquioxides in association with organic matter. A typical Spodosol profile as envisioned by Fink[20] is shown schematically in Fig. 20.4.

A preliminary step before appreciable translocation of Fe, Al, and organic matter can occur in the Spodosol is that a considerable portion of the exchangeable cations in the upper part of the solum must be displaced by H^+ and leached to lower horizons or out of the profile. This conditioning process is believed to take place rather rapidly on some parent materials, such as coarse-textured sediment low in exchangeable cations, but rather slowly with others, such as calcareous sediment.

Stobbe and Wright[21] reviewed the major processes involved in the formation

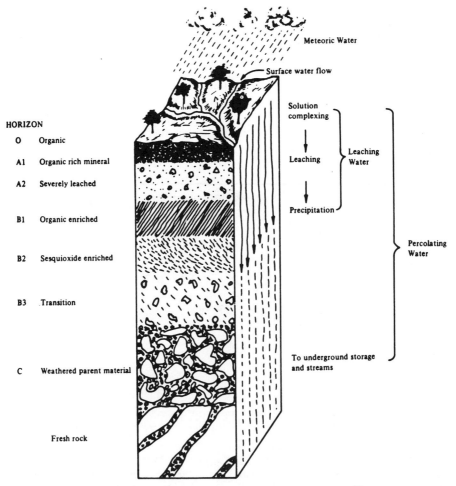

Fig. 20.4 A generalized landscape-Spodosol profile. From Finkl,[20] reproduced by permission from *The Encyclopedia of Soil Science, Part 1*, Dowden, Hutchinson, and Ross, Inc., Stroudsburg, Pennsylvania.

of Spodosols and reported that the prevailing concept at the time was that polyphenols, organic acids, and other complexing substances in percolating waters from the surface litter bring about the solution of sesquioxides, with formation of soluble metal–organic complexes. A second theory is that polymeric phenols (e.g., fulvic acids) are the primary agents. According to this concept, mobile organic colloids percolating downward through the soil profile form complexes with Fe and Al until a critical saturation level is attained, following which precipitation occurs. Partial decay of organic matter in the B horizon would further saturate the complex. Once started, accumulation would be self-perpetuating since the free oxides thus formed would cause further precipitation of the sesquioxide-humus complex. On periodic drying, the organic matter complex may harden, thereby restricting movement below the accumulation zone. De Coninck[22] concluded that, in the mobile state, metal-humate complexes are highly hydrated (i.e., they are hydrophilic). However, during transition to the solid state, hydration water is lost and the complexes become hydrophobic.

The concept that organic substances are solely responsible for the translocation of Fe and Al in Spodosols has not been universally accepted.[23-26] A recent theory is that, in some Spodosols, Fe and Al are transported as hydroxy Fe and Al silicate sols (i.e., Al as imogolite and allophane); soluble organic colloids migrating downward are then precipitated on the previously deposited imogolite and allophane.[23,24] Farmer et al.[25] attributed the formation of the B horizon in "Hydromorphic Humus Podzols" to coprecipitation arising from mixing of organic-rich surface waters with Al from groundwater. In other work, Taylor[26] attributed the formation of Spodosols to migration of soluble Si—Al and Fe(III)—Al hydroxy complexes, as opposed to Fe- and Al-organic complexes.

In all likelihood, several processes are involved in the formation of Spodosols, the relative importance of each being dependent on environmental conditions. Also, a host of compounds (e.g., biochemical compounds, humic substances) may be involved. Thus, the question is reduced to identifying the class or group of compounds that plays the major role.

Significance of Polyphenols

A popular theory at the present time is that polyphenols derived from the leaf litter are the main agents of podzolization. The studies of Bloomfield[27,28] show that aqueous extracts of tree litter from the forest floor of the Spodosol are able to dissolve ferric oxide and that this ability is due primarily to polyphenols in the extracts. According to Bloomfield, the solution of sesquioxides in the uppermost soil horizons results from the action of water-soluble, low-molecular-weight organic compounds (primarily polyphenols but to some extent organic acids and other substances) leached from the overlying surface litter. Following dissolution, Fe^{3+} is reduced to Fe^{2+} by the polyphenols, and it is in this form that the Fe is transported down the profile. Thus, two possible mechanisms are possible for the transport of Fe as Fe^{2+}: 1) complexation followed by reduction, and 2) reduction brought about under anaerobic conditions during saturation of the upper soil layers with water followed by chelation. These pathways are illustrated in Fig. 20.5.

Thus, the polyphenol theory adequately accounts for removal of Fe and Al from the A horizon. Precipitation in the B horizon is subsequently explained by either an assumed increase in pH or to partial mineralization of organic matter in the B horizon. In the latter case, continued microbial oxidation at some limiting depth of leaching or wetting would result in considerable accumulation of sesquioxides.

van Breeman and Brinkman[1] postulated the following reactions for precipitation of Fe oxides, where L refers to the complexing organic acid.

$$FeL^+ + 3H_2O \rightarrow Fe(OH)_3 + H^+ + H_2L$$

or

$$FeOHL° + 3H_2O \rightarrow Fe(OH)_3 + H_2L$$

Other environmental factors which may influence precipitation include mutual coagulation, changes in oxidation–reduction conditions (e.g., oxidation of Fe^{2+} to Fe^{3+}), and irreversible precipitation as a result of drying.

As Sesquioxide–Humus Sols

A theory that has received wide support by soil scientists is that the downward movement of sesquioxides in the Spodosol is due to the formation of negatively charged sols with humic substances, specifically, fulvic acid-type constituents formed in the overlying leaf litter through decay by microorganisms. The clas-

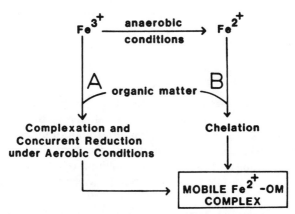

Fig. 20.5 Possible mechanisms for reduction of Fe^{3+} and movement in the ferrous state (Fe^{2+})

sical studies of Deb[29] show that humus can carry with it from 3 to 10 times its weight of iron oxide and that different fractions of humus vary in their ability to maintain oxides in solution. The humus-protected sol theory assumes that a fulvic acid solution percolating downward through the soil profile forms complexes with Fe and Al until a critical metal–fulvic acid ratio is reached, following which precipitation occurs. Environmental factors influencing precipitation would include those mentioned above. For example, partial decay of organic matter in the B horizon would further increase the metal–fulvic acid ratio. Once started, the accumulation process would be self-perpetuating since the free oxides thus formed would cause further precipitation of sesquioxide-humus soils. On periodic drying, the organic matter complex may harden, and this may further restrict movement below the accumulation zone. Considerable disagreement exists as to whether Fe in the complexes moves in the ferrous (Fe^{2+}) or ferric (Fe^{3+}) state.

The B horizon of the Spodosol serves as a rich source of fulvic acids but it is uncertain as to whether they originate in the forest litter or are formed *in situ*.

ORGANIC MATTER AND SOIL CLASSIFICATION

Criteria used in classifying soils nearly always include information regarding the physical and chemical characteristics of the diagnostic surface and subsurface horizons. These horizons are defined as layers within the soil that are roughly parallel to the earth's surface and that have properties unlike those of adjoining layers. Among the important criteria are organic matter content, color, structure, texture, consistency, and presence or absence of carbonates.

The importance of organic matter in the comprehensive system of soil classification, commonly referred to as the 7th Approximation, is due not only to the amounts present on a quantitative basis but to its effect on other soil properties, especially color. Organic matter is particularly important in assignments at the higher category levels of the classification system (order, suborder, great group, and subgroup), and that can be explained on the basis that the criteria for classification at these categories are based primarily on characteristics that can be observed in the field.

Pyrophosphate- or alkali-soluble organic matter and humus fraction ratios have been used as a means of discriminating between humus types on the forest floor and as a criteria for classification and recognition of Spodosol B horizons.[30-32] Extractable organic matter has also been used as a criteria for the classification of Histosols[33] and for identifying surface litter horizons of Alifsols. Limitations of alkali-soluble organic fractions as a criteria for soil classification has been discussed by Schuppli and McKeague.[34]

For a discussion of the formation/inheritance of pigmenting agents in soil (including organic matter), and of color science as applied to soil classification, the reader is referred to a recent publication entitled *Soil Color*.[35]

Organic Matter and the Diagnostic Surface Horizons

The diagnostic surface horizons include those referred to as epipedons, a term derived for the Greek epi (over) and pedon (soil). Organic matter is probably the most important single criterion for characterization and for establishing boundaries of the epipedon. This zone includes that part of the soil that has been darkened by organic matter (not to be confused with the A horizon). The illuvial B horizon, or part of it, is included with the epipedon when darkening by organic matter extends into or through this horizon.

Organic matter content determine whether any given soil horizon can be considered "organic" or not. If clay is absent, 20 percent or more organic matter must be present. When the mineral portion contains over 50 percent clay, at least 30 percent organic matter is needed. Intermediate contents of clay require proportional amounts of organic matter.

The 01 horizon is defined as the layer where original forms of most vegetative matter are visible to the naked eye. The 02 organic horizon is where extensive decomposition has occurred and few plant or animal parts can be recognized.

The epipedon where organic matter content is of greatest significance is in the identification of the histic epipedon, or peat layer. The horizon must have a thickness of 45 cm (30 cm if drained) and it must contain at least 20 percent organic matter if void of clay and 30 percent or more if the mineral portion contains 50 percent clay. Intermediate clay contents require proportional amounts of organic matter. Other restrictions are that the plow layer must contain more than 15 percent organic matter if clay is absent or 25 percent if the clay content of the mineral portion exceeds 50 percent. Once again, intermediate clay contents require proportional amounts of organic matter. If there is an overlying layer < 50 cm thick, any underlying organic horizon must contain sufficient organic matter to satisfy the requirements noted earlier. In this case, the histic epipedon is considered buried. Another requirement for the histic epipedon is that the layer must be saturated with water for more than 30 consecutive days during the year unless artificially drained.

The concept of mollic epipedon also centers on the properties that organic matter confers to the top layer of the solum. This epipedon is a thick, dark zone saturated with divalent cations, with a narrow C/N ratio, and with moderate to strong texture. All Mollisols, which includes most soils formally called Chermozen, Brunizem, and Chestnut, must have this layer. The role of organic matter is specific; that is, the epipedon must be at least one Munsell unit darker than the parent material and the zone must contain 1 percent organic matter throughout. Since a mollic epipedon is a mineral horizon, organic matter content must have a maximum as well as a minimum. The maximum corresponds to the lower limit for the histic epipedon.

With regard to organic matter, the requirements that the mollic epipedon must meet are as follows: 1) the C/N ratio must be 17 or less for virgin soils and 13 or less if the soil has been cultivated. The latter applies to soils cultivated 25 yr. or more. Normally, the C/N ratio decreases with depth. 2) The zone must contain at least 1 percent organic matter throughout. If the dark surface horizon is <7 in. thick in a virgin soil having a solum 18 inches or less, sufficient organic matter must be present to give an average of 1 percent organic matter or more to an Ap 7 in. thick.

Other examples where organic matter is used to establish the presence or absence of a particular type of epipedon can be found in any good book on soil classification.

Organic Matter and the Diagnostic Subsurface Horizons

Organic matter is often used as a criterion for the identification of horizons that form below the soil surface, a typical example being the spodic horizon. This horizon is one where amorphous materials composed of organic matter and Al, with or without Fe, have accumulated.

A spodic horizon is readily recognized in the field by its dark color and structure. The presence of organic complexes and a high cation-exchange capacity (CEC) is a further indication of a spodic horizon. Since most of the CEC is lost by ignition, the high value can be attributed to illuviated organic matter. The spodic horizon must also satisfy the relation (percentage extractable C + Fe + Al)/percentage clay > 0.15, which is used for confirmation.

PALEOHUMUS

Remnants of plant and animal life, as well as products produced from them through humification, have been observed in buried soils (paleosols) of all ages.[36] This organic matter, or paleohumus, is of interest in geology and pedology because of its importance as a stratigraphic marker and as a key to the environment of the geologic past. Thus, the occurrence of dark-colored humus zones, when used in conjunction with other pedological observations, has served as a basis for establishing the identification of buried soils, from which it has been possible to draw conclusions relative to climate, vegetative patterns, and the morphology of former land surfaces.

EFFECTS OF ACIDIC INPUTS (ACID RAIN) ON PEDOGENIC PROCESSES

Acid rain occurs over broad areas of northeastern United States, eastern Canada, and northern Europe. The most noticeable effect of acid rain has been a lowering of pH in numerous lakes in eastern North America and in Scandinavia. Accompanying the decrease in pH has been an increase in dissolved Al, which is toxic to aquatic organisms. As is the case with terrestrial soils, levels of the highly toxic Al^{3+} ion in solution can be reduced through complexation with organic substances, either from those contained in leachate waters or synthesized *in situ* by microorganisms (see Chapter 2). David and Driscoll[37] found that organic and F^- complexes were the dominant form of Al in the throughfall and leachates of the O, E, and B horizons of some Spodosol (Typic Haplothrod) soils in the Adirondack mountains of New York.

An additional concern of acid rain is the acidification of soils with subsequent solubilization and transport of Al to lakes and streams in drainage and groundwaters, thereby adversely affecting aquatic life.[37-39] The study of Vance and David[40] suggests that acidic inputs can effect both the amount and composition of organic matter in leachates of forest soils. The enrichment of Al in solutions draining the organic horizons of forest soils has been attributed to complexation by organic solutes.[39]

SUMMARY

Organic substances play an important role in weathering processes, the neogenesis of minerals, and in the transport of clay and metal ions in soil. The presence or absence of organic matter, as evidence by color, is an important criterion in the classification of soils.

REFERENCES

1 N. van Breeman and R. Brinkman, "Chemical Equilibria and Soil Formation," in G. H. Bolt and M. G. M. Bruggenwert, Eds., *Soil Chemistry*, Elsevier, New York, 1976, pp. 141–170.
2 S. W. Buol and F. D. Hole, *Soil Sci. Soc. Amer. Proc.*, **23,** 239 (1959).
3 H. Zunino and J. P. Martin, *Soil Sci.*, **123,** 65 (1977).
4 P. M. Huang and A. Violante, "Influence of Organic Acids on Crystallization and Surface Properties of Precipitation Products of Aluminum," in P. M. Huang and M. Schnitzer, Eds., *Interaction of Soil Minerals with Natural Organics and Microbes*, Special Publication 17, Soil Science Society of America, Madison, 1986, pp. 159–221.
5 M. Robert and J. Berthelin, "Role of Biological and Biochemical Factors in Soil Mineral Weathering," in P. M. Huang and M. Schnitzer, Eds., *Interaction of Soil Minerals with Natural Organics and Microbes*, Special Publication 17, Soil Science Society of America, Madison, 1986, pp. 453–495.
6 K. H. Tan, "Degradation of Soil Minerals by Organic Acids," in P. M. Huang and M. Schnitzer, Eds., *Interaction of Soil Minerals with Natural Organics and Microbes*, Special Publication 17, Soil Science Society of America, Madison, 1986, pp. 1–27.
7 K. H. Tan, *Clay Research,* **8,** 11 (1989).
8 F. J. Stevenson, "Organic Acids in Soil," in A. D. McLaren and G. H. Peterson, Eds., *Soil Biochemistry*, Marcel Dekker, New York, 1967, pp. 119–146.
9 D. M. Webley, M. E. K. Henderson, and I. F. Taylor, *J. Soil Sci.*, **14,** 102 (1963).

10 M. E. K. Henderson and R. B. Duff, *J. Soil Sci.*, **14,** 236 (1963).
11 W. E. Baker, *Geochim. Cosmochim. Acta,* **37,** 269 (1973).
12 K. H. Tan, *Soil Sci.,* **129,** 5 (1980).
13 M. Schnitzer and H. Kodama, *Geoderma,* **15,** 381 (1976).
14 A. K. Barman, C. Varadachari, and K. Ghosh, *Geoderma,* **53,** 45 (1992).
15 N. K. K. F. Kwong and P. M. Huang, *Soil Sci.,* **128,** 337 (1979).
16 N. K. K. F. Kwong and P. M. Huang, *Soil Sci. Soc. Amer. J.,* **43,** 1107 (1979).
17 W. W. Senstius, *Amer. Scientist,* **46,** 355 (1958).
18 R.W. Simonson, *Soil Sci. Soc. Amer. Proc.,* **23,** 152 (1959).
19 L. D. Swindale and M. L. Jackson, *Trans. 6th Intern. Congr. Soil Sci.,* E, 233 (1956).
20 C. W. Finkl, Jr., "Leaching," in R. W. Fairbridge and C. W. Finkl, Jr., Eds., *Encyclopedia of Soil Science, Part I*, Dowden, Hutchinson and Ross, Stroudsburg, 1970, pp. 260–265.
21 P. C. Stobbe and J. R. Wright, *Soil Sci. Soc. Amer. Proc.,* **23,** 161 (1959).
22 F. De Coninck, *Geoderma,* **24,** 101 (1980).
23 H. A. Anderson, W. L. Berrow, V. C. Farmer, A. Hepburn, J. D. Russell, and A. D. Walter, *J. Soil Sci.,* **33,** 125 (1982).
24 V. C. Farmer, J. D. Russell, and M. L. Berrow, *J. Soil Sci.,* **31,** 673 (1980).
25 V. C. Farmer, J. O. Skjemstad, and C. H. Thompson, *Nature,* **304,** 342 (1983).
26 R. M. Taylor, *Geoderma,* **42,** 65 (1988).
27 C. Bloomfield, *J. Sci. Food Agr.,* **8,** 389 (1957).
28 C. Bloomfield, "Organic Matter and Soil Dynamics," in E. G. Hallsworth and D. V. Crawford, Eds., *Experimental Pedology*, Butterworth, London, 1964, pp. 257–266.
29 B. C. Deb, *J. Soil Sci.,* **1,** 112 (1949).
30 L. J. Evans and B. H. Cameron, *Can. J. Soil Sci.,* **65,** 363 (1985).
31 L. E. Lowe, *Can. J. Soil Sci.,* **69,** 219 (1980).
32 J. A. McKeague, *Can. J. Soil Sci.,* **48,** 27 (1968).
33 N. O. Isirimah, D. R. Keeney, and G. B. Lee, *Soil Sci. Soc. Amer. Proc.,* **34,** 478 (1970).
34 P. A. Schuppli and J. A. McKeague, *Can. J. Soil Sci.,* **64,** 173 (1984).
35 J. M. Bigham and E. J. Ciolkosz, Eds., *Soil Color*, Special Publication 31, Soil Science Society of America, Madison, 1993.
36 F. J. Stevenson, *Soil Sci.,* **107,** 470 (1969).
37 M. David and C. T. Driscoll, *Geoderma,* **33,** 297 (1984).
38 E. M. Perdue, K. C. Beck, and J. H. Reuter, *Nature,* **260,** 418 (1976).
39 C. T. Driscoll, N. van Breemen, and J. Mulder, *Soil Sci. Soc. Amer. J.,* **49,** 437 (1985).
40 G. F. Vance and M. B. David, *Soil Sci. Soc. Amer. J.,* **53,** 1242 (1989).

INDEX

Acid rain, effect on pedogenic processes, 485–486
Absorption curves, *see* Spectroscopic approaches
Acid-base titration, *see* Potentiometric titration
Aluminum toxicities, 16, 382–383
Adsorption:
  mechanisms:
    coordination, 434
    H-bonding, 434, 465, 468
    hydrophobic bonding, 466–468
    ion-exchange and protonation, 436–439, 462–465
    Pi bonding, 466
    van der Waal's forces, 433, 465
  of defined organics by clay, 431–439
  of humic substances:
    by clay, 441–444
    by hydrous oxides, 444–445
  of pesticides:
    by clay, 431–432
    by organic matter, 453–469
Adsorption isotherms:
  classification of, 431–432
  Freundlich equation, 431–432, 456
  Langmuir equation, 432
Aggregation of soil:
  clay domain theory, 448–449
  effect of microorganisms, 447
  mechanisms, 446–450

  polysaccharides and, 142, 447
  role of organic matter, 17, 445–450
Aluminum:
  mobility in Spodosols, 479–483
  toxicity of, 16, 382–383
Alcoholic hydroxyl group:
  content in humic and fulvic acids, 226–229
  determination of, 222
Alkaline cupric oxide oxidation, 248–249
Alkaline nitrobenzene oxidation, 249
Alkaline permanganate oxidation, 237, 249–252
Alkanes, *see* Hydrocarbons
Amines:
  adsorption by clay, 436–438
  nitrosamines from, 109–110
  in soil, 87–88, 110
  in wet sediments, 12
Amino acids:
  adsorption by clay, 436–438
  as chelating agents, 380–381, 391
  determination of, 60, 71, 73–74
  distribution patterns in soil, 75–77
  extraction, 71
  free forms, 78–79
  of humic and fulvic acids, 77
  in humus synthesis, 204–205
  identification of, 73–74
  nitrite reactions with, 107
  nonprotein, 74
  quinone complexes of, 69–70

Amino acids (*Continued*)
  role in pesticide degradation, 459–460
  in root exudates, 79
  state in soil, 79–81
  stereochemistry of, 78
Amino sugars:
  amounts in soil, 81–82
  determination of, 60, 82–83
  glucosamine/galactosamine ratio, 83–84
  humus formation from, 205
  isolation of, 83–84
Ammonia, in soil hydrolysates, 59–66
Ammonia fixation:
  agricultural significance of, 102–103
  availability of fixed NH_3 to plants, 105–106
  mechanism of fixation, 103–105
Ammonium, *see* Fixed ammonium
Analytical pyrolysis, 256, 277–280
Anhydrides, 220–221
Apocrenic acid, 25, 30, 31
Aromaticity, of humic substances, 255, 269–272
Auxins, *see* Growth promoting substances

Baryta adsorption method, 215
Benzene carboxylic acids, in chemical degradation products, 250–252
Biochemical compounds:
  as chelating agents, 388–394, 474–476
  synthesis in soil, 76, 77–78, 145
Biochemistry of humus formation:
  lignin-protein theory, 28, 189–192, 197
  the Maillard reaction, 27–28, 206–208
  polyphenol theory, 188–190, 197–205
Biological stability:
  of humus, 6, 11–14
  of organic nitrogen, 63, 90–91, 298
Biological weathering, *see also* Soil formation
  by humic substances, 476–477
  by organic acids, 474–476
Biomass, *see* Microbial biomass
Bonding:
  of metal ions by humic substances, 394–400
  of organic compounds by clay, 431–439
  of pesticides by organic matter, 453–469
Borate complexes, 383
Brown humic acid, 29–30, 41, 44

Calcium acetate method, 217–218
Carbohydrates, *See also* Polysaccharides
  amino sugars, 81–84
  chromatographic separations, 159–163
  classification of, 142–144
  complex polysaccharides, 144, 146–153
  content in soils, 141, 162–163
  free sugars, 145–146
  hexose sugars, 141, 156–158
  of humic and fulvic acids, 48–51, 145
  in humus synthesis, 27–28, 206–208
  methylated sugars, 152
  pentose sugars, 141, 158
  significance in soil, 142
  state in soil, 144–145
  total reducing sugars, 154–155
  uronic acids, 141, 158–159, 163
Carbon, *see* Carbon transformations: Organic carbon
^{14}C-dating, 13–14
^{13}C-NMR spectroscopy:
  of humic substances, 2, 43, 265–272
  methods for, 263–264
  of organic matter *in situ*, 272–274
  theory of, 259–262
Carbon-bonded sulfur, 130–132
C/N ratio, 7, 100–102
C/N/P/S ratio, 113–114
C/P ratio: 113–114
C/S ratio, 115
Carbon transformations, 11–12
Carbonyl group:
  content in humic and fulvic acids, 226–229
  determination of, 222–224
Carboxyl group:
  content in humic and fulvic acids, 224–229
  determination of, 217–219
Carcinogens, 12, 88
Carotenoids, 12, 182–183
Cation-exchange capacity:
  contribution of organic matter, 297, 367–371
  relation to pesticide adsorption, 463–465
Chelation, *see also* Metal-organic matter complexes: Stability constants of metal complexes with humic and fulvic acids
  by biochemical compounds, 388–394, 474–476
  definition of, 380
  effect on phosphate solubilization, 126–127, 382
  by lichen acids, 393
  role in pedogenic processes, 478–483
  significance in soil, 381–383
Cheluviation, 479
Chemical structures:
  amino acids, 72
  carbohydrates, 142–144
  herbicides, 453–454
  humic and fulvic acids, 208–209, 285–301
  lignin, 195–196

INDEX **491**

Chitin, 81
Chlorophyll derivatives, see Porphyrins
Clay domain theory of soil aggregation, 448–449
Clay-fixed ammonium, see Fixed ammonium
Clay minerals:
  adsorption:
    of humic substances by, 441–444
    of organic compounds by, 431–439
    of pesticides by, 431–432
  properties of, 429–431
Clay-organic matter complexes, 298–299, 439–444
Coagulation, of humic substances, 372–373
Commercial humates, 18
Conditional stability constants, of metal-humate complexes, 409–410
Conductometric titration of humic and fulvic acids, 364
Crenic acid, 25, 30, 31
Cryoscopy, see Freezing point depression

Degradation products:
  by hydrolysis, 238–240
  by oxidation:
    alkaline cupric oxide, 248–249
    alkaline nitrobenzene, 249
    alkaline permanganate, 237, 249–252
    hypochlorite, 237, 253
    nitric acid, 237, 253
    peracetic acid, 253
  by pyrolysis, 255–256, 277–280
  by reduction:
    sodium amalgam, 237, 240–245
    zinc dust distillation and fusion, 237, 245–248
Diagenetic transformations, of humic substances, 231–232
Differential pulse polarography, 397
Dimethylarsine, 12
Dimethylselenide, 12
Dissolved organic carbon (DOC):
  fractionation of, 51–53
  isolation of fulvic acids, 51–52

E_4/E_6 ratio, 48, 214, 307
Electrometric titration:
  conductometric, 364
  high-frequency, 360
  potentiometric:
    in aqueous solution, 352–363
    of metal-organic matter complexes, 421–426
    in nonaqueous solvents, 365–366
  thermometric, 366

Electron microscopy, 339–341
Electron spin resonance (ESR) spectroscopy:
  of humic substances, 317–321
  of metal-humate complexes, 396
  nature of stable free radicals, 321–322
  of pesticide-organic matter complexes, 459
Electrophoresis, of humic substances, 373–374
Elemental content, of humic and fulvic acids, 212–214
Enzymes:
  occurrence in soils, 6
  role in humus formation, 198, 203–204
Equivalent weight, of humic substances, 364
Extraction:
  of lipids, 170–171
  of nitrogen compounds, 59–61, 71, 83
  of polysaccharides, 146–150
  of soil organic matter:
    with alkali, 34–37
    historical aspects of, 24–34
    with mild reagents, 37–40

Fats and waxes, see Lipids
Fatty acids, see Organic acids
Fluorescence spectroscopy:
  of humic and fulvic acids, 322
  of metal-humate complexes, 396
Fixed ammonium:
  amounts in soil, 96–100
  influence on C/N ratio, 100–102
Fractionation methods:
  dissolved organic carbon (DOC), 51–53
  humus, 41–45
  lipids, 171–172
  organic N, 59–66
  organic P, 125–126
  organic S, 130–133
  polysaccharides, 150–152
  trace elements, 384–386
Free amino acids, 78–79
Free amino groups, of humic acids, 225–226
Free radicals, in humic substances, 321
Freundlich adsorption equation, 431–432, 456
Frictional coefficient, 331–333
Fulvic acids:
  acidic properties of, 43–44, 227–228, 351–352
  adsorption of pesticides by, 459–460
  analytical pyrolysis of, 277–280
  biochemistry of formation, 188–210
  ^{13}C-NMR spectroscopy, 274–276
  complexation of metals, 394–400
  decomposition of silicate minerals by, 476–477

INDEX

Fulvic acids (*Continued*)
  definition of, 25, 30, 33, 49–51
  degradation products of, 236–254
  diagenetic transformations, 231–233
  electrometric titrations, 352–367
  electron spin resonance (ESR) spectroscopy, 318–321
  elemental content, 212–214
  equivalent weight, 364
  functional group content, 226–231
  gel filtration of, 47
  generic forms, 49–51, 266–269
  ^{1}H-NMR spectroscopy, 274–276
  infrared spectra of, 309–313
  molecular weight, 328, 338
  nitrogen of, 66, 68–69
  oxidation-reduction potential, 374–375
  purification, 49–51
  reductive properties, 401
  role in pedogenic processes, 482–483
  role in pesticide transformations, 459–462
  sorption by clay, 441–444
  sorption by hydrous oxides, 444
  type structures of, 286–287, 292–294
Functional groups, of humic and fulvic acids, 212–231
Fungal melanins, 43, 202
Fungi:
  decomposition of lignin by, 199–200
  synthesis of humic substances by, 201–203

Gaussian distribution function, 362–363, 419
Gel filtration:
  of humic and fulvic acids, 47, 343–347
  of polysaccharides, 150–151
Generic fulvic acids, 49–51, 266–269
Glucosamine/galactosamine ratio, 83–84
Glycerides, 180–181
Glycerophosphatides, 132
Gray humic acid, 29–30, 41, 44
Green humic acid, 43
Growth promoting substances, 18, 174

Herbicides, *see* Pesticides
Hexosamines, *see* Amino sugars
Hexose sugars, 141, 156–158
High frequency titration, of humic substances, 360
Histosols:
  forms of nitrogen in, 66–67
  lipid content of, 167–168
Humic acid/fulvic acid ratio, 45
Humic acids:
  acidic properties, 43–44, 227–278, 351–352

  adsorption by clays, 441–444
  adsorption by hydrous oxides, 444
  adsorption of pesticides by, 459–460
  analytical pyrolysis of, 277–280
  biochemistry of formation, 188–210
  ^{13}C-NMR spectroscopy of, 274–276
  coagulation of electrolytes, 372–373
  complexation of metal ions by, 394–400
  decomposition of silicate minerals by, 476–477
  definition of, 25, 30, 33
  degradation products of, 236–254
  diagenetic transformations of, 231–233
  electrometric titration of, 352–367
  electron spin resonance (ESR) spectroscopy of, 318–321
  elemental content, 212–214
  environmental significance of, 19–20
  equivalent weight, 364
  fractionation of, 48–49
  functional group content, 226–231
  gel filtration of, 47
  ^{1}H-NMR spectroscopy of, 274–276
  infrared spectra of, 309–313
  molecular weight, 330–338
  morphological features, 339–441
  nitrogen of, 66, 68–70
  purification, 45–49
  reduction properties, 401
  role in pesticide transformations, 459–462
  role in pedogenic processes, 482–483
  sorption by clay, 441–444
  sorption by hydrous oxides, 444
  type structures of, 285–291
  titration curves of, 355–358
Humic substances, *see also* Fulvic acids; Humic acids; Humin
  acidic properties, 43–44, 227–278, 351–352
  adsorption of pesticides by, 459–460
  biochemistry of formation, 188–210
  colloidal properties of, 325–347
  definition of, 32–33
  diagenetic transformations, 231–233
  environmental significance of, 19, 54
  extraction and fractionation, 34–35
  metal complexes of, *see* Metal-organic matter complexes,
  morphological features, 339–341
  in stream water, 51–54
  as a system of polymers, 44–45
Humin:
  composition, 43–44
  definition of, 30–31, 33, 41
  fractionation of, 43–44

INDEX **493**

Humus, *see* Biochemistry of humus formation: Organic matter
Hydrocarbons, 178–179
Hydrogen bonding, 434, 465, 468
Hydrophobic bonding, 466–468
Hydroxamate siderophores, 389–391
Hydroxyquinone, 213, 221, 321, *see also* Quinones
Hymatomelanic acid, 26, 31, 33
Hypochlorite oxidation, 237, 253

Infrared (IR) spectroscopy:
  of humic and fulvic acids:
    assignment of absorption bands, 308–309, 313
    classification of spectra, 311–313
    of methylated and acetylated derivatives, 313–317
  of metal-humate complexes, 395–396
  of pesticide-humate complexes, 465
Inositol hexaphosphates, *see* Inositol phosphates
Inositol phosphates, 119–122
Ion-exchange equilibrium method, 408–409

Job's method of determining stability constants, 409

K_d values, 457
K_{oc} values, 467–468
Ketones, 184

Langmuir adsorption equation, 432
Lichens, role in rock weathering, 474
Light fraction, of organic matter, 2–3, 54–55
Lignin:
  ammonia fixation by, 102
  decomposition of, 11–12, 199–201
  formation of humic substances from, 28, 188–205
  nitrite reactions with, 107–108
  structure of, 195–196
Lignin-protein theory, 28, 189–192, 197
Lignites, *see* Commercial humates
Lipids:
  adsorption of pesticides by, 466
  amounts in soil, 29, 166–168
  association with humic acids, 48
  association with humin, 43–44
  composition:
    carotenoids, 12, 182–183
    glycerides, 180–181
    hydrocarbons, 178–179
    ketones, 184
    organic acids, 176–178
    phospholipids, 123, 180–181
    phthalates, 184–185
    polycyclic hydrocarbons, 179–180
    porphyrins, 86, 183–184
    steroids and terpenoids, 181–182
    waxes, 175–176
  extraction and fractionation of, 170–172
  role in pesticide adsorption, 466
  significance in soil, 171–175
  sources of, 168–170
  supercritical gas extraction of, 48, 170
Litter, 2

Maillard reaction, 27–28, 206–208
Mean-residence-time (MRT), of modern humus, 13–14
Metal-organic matter complexes, *see also* Chelation: Trace elements
  characterization by:
    different pulse polarography, 397
    electron spin resonance spectroscopy (ESR), 396
    fluorescence spectroscopy, 396
    infrared spectroscopy (IR), 395–396
    molecular weight distribution, 397–398
    Mössbauer spectroscopy, 401
    potentiometric titration, 421–426
    thermogravimetric (TG) analysis of, 397
    ultraviolet spectroscopy, 396–397
  classification of, 378–379
  in formation of Spodosols, 478–483
  fractionation of, 384–386
  of humic and fulvic acids, 299, 394–400
    binding capacities, 399
    solubility characteristics, 398–399
    stability constants of, 405–407
  significance in soil, 381–383
Methane, formation in soil, 12
Methylmercury, 12
Methyl nitrite, 107
Microbial biomass:
  carbon of, 3–5
  nitrogen of, 89
  phosphorus of, 124–125
  sulfur of, 134–135
Micronutrients, *see* Trace elements
Molecular structure, *see* Chemical structures
Molecular weights:
  determination of:
    freezing point depression, 336–338
    gel filtration, 343–347
    light scattering, 338–339
    osmometry, 337–338
    small angle X-ray scattering, 342–343

Molecular weights:
  determination of *(Continued)*
    ultracentrifugation, 329–333
    viscosimetric measurements, 333–336
  of fulvic acids, 328, 338
  of humic acids, 330–338
Morphological features, of humic substances, 339–341
Møssbauer spectroscopy, 304, 401
Muramic acid, 83

Nitrite reactions:
  with amino acids, 107
  formation of nitrosamines, 109–110
  with humic substances, 107–109
Nitrogen, *see also* Organic nitrogen
  ammonia reactions with organic matter, 102–105
  C/N ratio, 7, 100–102
  clay-fixed ammonium, 96–100
  of humic and fulvic acids, 213
  isotopic abundance, 90
  nitrite reactions, 106–110
Nitrosamines, 12, 88, 109–110
Nitrous acid, *see* Nitrite
Nuclear magnetic resonance spectroscopy (NMR):
  ^{13}C-NMR:
    of humic substances, 265–272
    methods for, 263–264
    of organic matter *in situ*, 272–274
    theory of, 259–262
  ^{1}H-NMR, 274–276
  ^{15}N-NMR, 90, 276–277
  ^{32}P-NMR, 124, 277
Nucleic acids, 85–86, 122

Organic acids:
  adsorption by clay, 436
  amounts in soil, 176–178
  as chelating agents, 389–391, 474–476
  role in neogenesis of minerals, 477–478
  in wet sediments, 12
Organic matter:
  ammonia reactions with, 96–106
  associations in soil, 34, 474–476
  content in soil, 7–10
  contribution to cation-exchange capacity, 367–371
  correlation with pesticide retention, 453–469
  ^{14}C-dating of soils, 13–14
  ^{13}C-NMR spectroscopy of, 272–274
  decomposition processes, 11–12, 188–191
  effect on soil aggregation, 445–449
  extraction methods, 34–40
  fractionation schemes, 41–45
  function in soil, 14–17
  mean-residence-time (MRT) of, 13–14
  nitrite reactions with, 106–109
  physical fractionation of, 54–55
  pools of, 1–6
  in relation to soil classification, 483–485
  in relation to sustainable agriculture, 19
  role in pedogenic processes, 476–477
  trace element interactions, *see* Organic matter-metal complexes
  transformations in wet sediments, 11–12
Organic matter-metal complexes, *see* Metal-organic matter complexes
Organic nitrogen:
  amines, 87–88, 110
  amino acids, 71–80
  amino sugars, 81–84
  biological stability of, 63, 90–91, 298
  of the biomass, 89
  characterization by ^{15}N-NMR spectroscopy, 90, 276
  chlorophyll derivatives, 86, 183–184
  distribution patterns:
    effect of cultivation, 63
    in Histosols, 66
    in humic and fulvic acids, 66–70
    in mineral soils, 61–63
  glycerophosphatides, 86–87
  isotopes of, 90
  nitrosamines, 12, 88, 109–110
  nucleic acids, 85–86
  peptides and proteins, 70, 80
  stabilization of, 70, 90–91, 298
  unidentified forms (HUN fraction), 60–61
Organic phosphorus:
  C/P ratio, 113–114
  chemical fractionation of, 125–126
  determination of, 117–118
  forms in soil:
    inositol phosphates, 119–122
    metabolic phosphates, 123
    nucleic acids, 122
    phospholipids, 123, 180–181
  ^{31}P-NMR spectroscopy of, 124, 277
  of humic substances, 124
  of the soil biomass, 124–125
Organic soils, *see* Histosols
Organic sulfur:
  C/S ratio, 115
  determination of, 128–129
  forms in soil:
    amino acids, 132–133
    carbon-bonded, 130–132

ester sulfates, 130–132
  sulfolipids, 133
  fractionation of, 130–133
  of humic and fulvic acids, 133–134
  physical fractionation of, 134
  of the soil biomass, 134–135
  soluble forms, 135–136
  volatile forms, 136
Oxygen-containing functional groups, *see* Functional groups
Oxidation-reduction potential, of humic acids, 374–375

Paleohumus, 485
Pentose sugars, 141, 157–158
Peracetic oxidation, 253
Pesticides:
  adsorption:
    by clay, 431–433
    by organic matter, 18, 300, 453–469
  chemical reactions, 459–462
  hydrophobic bonding, 466–468
  K_d values, 457
  K_{oc} values, 467–468
  mechanisms, 462–468
  residues in soil, 461–462
Phenolic compounds, *see also* Polyphenols
  in chemical degradation products of humic substances, 237–244
  as chelating agents, 391–392
  from pesticides, 461–462
Phenolic hydroxyl:
  content in humic and fulvic acids, 226–228
  determination of, 221–222
Phospholipids:
  content in soils, 180–181
  nitrogen of, 86–87
  phosphorus of, 123
Phosphorus, *see* Organic phosphorus
Phthalates, 184–185
Physical fractionation, of organic matter, 54–55
Phytin, 119
Phytotoxicity, *see also* Toxicity
  of nitrite, 104
  of organic substances, 18, 173–174
Podzolization:
  role of humic substances in, 482–483
  role of polyphenols in, 481–482
Polycyclic hydrocarbons:
  in chemical degradation products of humic substances, 247–248
  content in soils, 179–180

Polyphenols:
  as chelating agents, 391–393
  in formation of humic substances, 188–190, 197–205
  from lignin degradation, 197–201
  role in pedogenesis, 481–482
  synthesis of microorganisms, 202–203
Polysaccharides:
  association with humic and fulvic acids, 48–49
  composition, 152–153
  extraction from soil, 146–150
  gel filtration of, 150–152
  molecular weight, 153
  properties, 152–153
  significance in soil, 142
Pools:
  of organic matter, 1–6
  of micronutrient cations, 384–386
Porphyrins, 86, 179–180
Potentiometric titration, *see* Electrometric titration
Proteins:
  adsorption by clays, 444
  of soil organic matter, 70, 80
Purine and pyrimidine bases, *see* Nucleic acids
Protonation, 462–464
Pseudopolysaccharides, 49, 153
Pyrolysis, *see* Analytical pyrolysis: Thermogravimetric analysis

Quinones:
  amino acid reactions with, 69, 204–205
  ammonia reactions with, 104–105
  determination of, 224–225
  free radicals of, 321
  role in humus formation, 197–205
  role in pesticide adsorption, 461–462
  as structural components of humic substances, 229–231

Radiocarbon dating, see ^{14}C-dating
Ratios:
  C/N, 7, 100–102
  C/N/P/S, 113–114
  C/P, 113–114
  C/S, 115
  E_4/E_6, 48, 214, 307
  glucosamine/galactosamine, 82–83
  H/C, 214
  humic acid/fulvic acid, 45
  O/C, 214

Reducing sugars, of soil organic matter, 153–154
Reduction:
 of metals by humic substances, 401
 of mercury, 20
Reductive cleavage:
 of flavonoids, 141–142
 of humic substances:
  by sodium amalgam, 237, 240–245
  by zinc-dust distillation and fusion, 237, 245–248
Rhizosphere:
 chelation in, 383
 humus synthesis in, 9

Scatchard plot, 414–417
Small angle X-ray scattering, 342–343
Sodium amalgam reduction, see Reductive cleavage
Soil aggregation: see Aggregation of soil
Soil biomass, see Microbial biomass
Soil carbohydrates, see Carbohydrates
Soil enzymes, 6
Soil formation:
 neogenesis of minerals, 477–478
 translocation of mineral matter, 478–473
 weathering of rocks and minerals, 473–477
Soil lipids, see Lipids
Soil organic matter, see Organic Matter
Soil structure, see Aggregation of Soil
Soil wettability, 175, 449–470
Sorption, see Adsorption
Spectroscopic approaches:
 electron spin resonance (ESR), 318–321, 396, 459
 fluorescence, 322, 396
 infrared (IR), 395–409
 nuclear magnetic resonance (NMR), 259–277
 ultraviolet (UV) and visible:
  E_4/E_6 ratio, 48, 214, 307
  quantitative analysis by, 305–306
Stability constants of metal-humate complexes:
 base titration method, 421–426
 continuous distribution models, 417–420
 determination of binding parameters, 406–408
 ion-exchange equilibrium method, 408–409
 Job's method, 409
 Scatchard plot approach, 414–417
Stable free radicals, see Free radicals
Steroids, 181–182
Structure, see Chemical structure

Sugar-amine condensation, see Maillard reaction
Sulfolipids, 133
Sulfur, see Organic sulfur

Terpenoids, 181–182
Thermogravimetric (TG) analysis, see also Analytical pyrolysis
 aromaticity estimates from, 255
 of humic substances, 255–256
 of metal-humate complexes, 397
Thermometric titration, 366
Titrations curves, see Electrometric titration
Total acidity, of humic substances, 215–217, 226–231, 365
Total hydroxyl:
 content in humic and fulvic acids, 226–229
 determination of, 219–221
Toxicity, see also Phytotoxicity
 of aluminum in acid soils, 16, 382–383
 of nitrites, 106
 nitrosamines, 109–110
Trace elements, see also Metal-organic matter complexes
 chelation by biochemical compounds, 388–394, 474–476
 complexation by humic and fulvic acids, 299, 394–400
 pools of, 384–386
 speciation in the soil solution, 386–388

Ultracentrifugation, 329–333
Ultraviolet (UV)/visible spectroscopy:
 color-producing groups, 304–305
 E_4/E_6 ratio, 48, 214, 307
 of humic and fulvic acids, 305
 of metal-humate complexes, 396–397
 quantitative analysis by, 305–306
Uronic acids, 141, 158–159, 163

van der Waal's forces, 433, 465–466
van Slyke reaction,
Viscosity, 333–336
Vitamins, 88–89

Waxes, 175–176
Weathering, see Biological weathering: Soil formation

X-ray diffraction, 339–341

Xenobiotics, 20

Zinc-dust distillation and fusion, 237, 245–248

ester sulfates, 130-132
    sulfolipids, 133
    fractionation of, 130-133
    of humic and fulvic acids, 133-134
    physical fractionation of, 134
    of the soil biomass, 134-135
    soluble forms, 135-136
    volatile forms, 136
Oxygen-containing functional groups, see Functional groups
Oxidation-reduction potential, of humic acids, 374-375

Paleohumus, 485
Pentose sugars, 141, 157-158
Peracetic oxidation, 253
Pesticides:
    adsorption:
        by clay, 431-433
        by organic matter, 18, 300, 453-469
    chemical reactions, 459-462
    hydrophobic bonding, 466-468
    K_d values, 457
    K_{oc} values, 467-468
    mechanisms, 462-468
    residues in soil, 461-462
Phenolic compounds, see also Polyphenols
    in chemical degradation products of humic substances, 237-244
    as chelating agents, 391-392
    from pesticides, 461-462
Phenolic hydroxyl:
    content in humic and fulvic acids, 226-228
    determination of, 221-222
Phospholipids:
    content in soils, 180-181
    nitrogen of, 86-87
    phosphorus of, 123
Phosphorus, see Organic phosphorus
Phthalates, 184-185
Physical fractionation, of organic matter, 54-55
Phytin, 119
Phytotoxicity, see also Toxicity
    of nitrite, 104
    of organic substances, 18, 173-174
Podzolization:
    role of humic substances in, 482-483
    role of polyphenols in, 481-482
Polycyclic hydrocarbons:
    in chemical degradation products of humic substances, 247-248
    content in soils, 179-180

Polyphenols:
    as chelating agents, 391-393
    in formation of humic substances, 188-190, 197-205
    from lignin degradation, 197-201
    role in pedogenesis, 481-482
    synthesis of microorganisms, 202-203
Polysaccharides:
    association with humic and fulvic acids, 48-49
    composition, 152-153
    extraction from soil, 146-150
    gel filtration of, 150-152
    molecular weight, 153
    properties, 152-153
    significance in soil, 142
Pools:
    of organic matter, 1-6
    of micronutrient cations, 384-386
Porphyrins, 86, 179-180
Potentiometric titration, see Electrometric titration
Proteins:
    adsorption by clays, 444
    of soil organic matter, 70, 80
Purine and pyrimidine bases, see Nucleic acids
Protonation, 462-464
Pseudopolysaccharides, 49, 153
Pyrolysis, see Analytical pyrolysis; Thermogravimetric analysis

Quinones:
    amino acid reactions with, 69, 204-205
    ammonia reactions with, 104-105
    determination of, 224-225
    free radicals of, 321
    role in humus formation, 197-205
    role in pesticide adsorption, 461-462
    as structural components of humic substances, 229-231

Radiocarbon dating, see ^{14}C-dating
Ratios:
    C/N, 7, 100-102
    C/N/P/S, 113-114
    C/P, 113-114
    C/S, 115
    E_4/E_6, 48, 214, 307
    glucosamine/galactosamine, 82-83
    H/C, 214
    humic acid/fulvic acid, 45
    O/C, 214